GEOMETRIC AND ALGEBRAIC STRUCTURES
IN DIFFERENTIAL EQUATIONS

GEOMETRIC AND ALGEBRAIC STRUCTURES IN DIFFERENTIAL EQUATIONS

Edited by

P. H. M. KERSTEN

Department of Applied Mathematics,
University of Twente,
Enschede, The Netherlands

and

I. S. KRASIL'SHCHIK

Moscow Institute of Municipal Economy
and Civil Engineering,
Moscow, Russia

Reprinted from
Acta Applicandae Mathematicae
Vol. 41, Nos. 1–3, December 1995

KLUWER ACADEMIC PUBLISHERS

DORDRECHT / BOSTON / LONDON

Library of Congress Cataloging-in-Publication Data

A C.I.P. Catalogue record for this book is available from the Library of Congress.

ISBN 978-94-010-6565-8 e-ISBN-13: 978-94-009-0179-7
DOI: 10.1007/978-94-009-0179-7

Published by Kluwer Academic Publishers,
P.O. Box 17, 3300 AA Dordrecht, The Netherlands.

Kluwer Academic Publishers incorporates
the publishing programmes of
D. Reidel, Martinus Nijhoff, Dr W. Junk and MTP Press.

Sold and distributed in the U.S.A. and Canada
by Kluwer Academic Publishers,
101 Philip Drive, Norwell, MA 02061, U.S.A.

In all other countries, sold and distributed
by Kluwer Academic Publishers Group,
P.O. Box 322, 3300 AH Dordrecht, The Netherlands.

Printed on acid-free paper

Table of Contents

Introduction

The idea of organizing a workshop on geometrical and algebraic structures of differential equations, the Proceedings of which are now in front of you, arose in summer 1992, when one of the editors (P.K.) was visiting the other editor (J.K.) in Moscow. The basic idea was to bring closer and intensify activities of mathematicians and mathematical physicists doing research in this field. The decision was taken that it would be a small workshop for specialists and the choice of the main part of participants was made for this reason by the organizing committee. We are glad that so many of them accepted the invitation to participate.

Beside those who contributed to the Proceedings, we want to express our gratitude Professors Dikii, Vinogradov, Prastaro, and Quispel, who delivered very interesting lectures but, for different reasons, were not able to submit corresponding texts.

We also want to thank the University of Twente and the Faculty of Applied Mathematics who provided us with all the necessary facilities so that the workshop could take place. It is our special duty to mention those who supported the workshop financially:

- Hewlett Packard Nederland B.V.
- Samenwerkingsverband FOM/STW Mathematische Fysica
- Vertrouwenscommissie Wiskundig Genootschap
- Koniklijke Nederlandse Akademie van Wetenschappen
- Stichting Universiteitsfonds Twente
- Faculty of Applied Mathematics, University of Twente
- Vakgroepfonds ADAM, University of Twente
- Mathematisch Research Institute

We are sure that without their help, it would have been impossible to organize this Workshop.

A word of thanks should be devoted to Professor M. Hazewinkel for his warm cooperation in publishing these Proceedings.

Enschede – Moscow,
August 1995

PAUL KERSTEN
JOSEPH KRASIL'SHCHIK

Acta Applicandae Mathematicae **41:** 3–19, 1995.

The Cohomology of Invariant Variational Bicomplexes

IAN M. ANDERSON
Department of Mathematics, Utah State University, Logan, UT 84322-3900, U.S.A.

and

JUHA POHJANPELTO
Department of Mathematics, Oregon State University, Corvallis, OR 97331-4605, U.S.A.

(Received: 30 November 1993)

Abstract. Let $\pi\colon E \to M$ be a fiber bundle and let Γ be an infinitesimal Lie transformation group acting on E. We announce various new results concerning the cohomology of the Γ invariant variational bicomplex $(\Omega_\Gamma^{*,*}(J^\infty(E)), \mathrm{d}_H, \mathrm{d}_V)$ and the associated Γ invariant Euler–Lagrange complex. As one application of our general theory, we completely solve the local invariant inverse problem of the calculus of variations for finite-dimensional infinitesimal Lie transformation groups.

Mathematics Subject Classifications (1991): 49N45, 58E30, 58G35.

Key words: inverse problem of calculus of variations, group actions, Lie algebra cohomology.

1. Introduction

The inverse problem of the calculus of variations is the problem of characterizing those systems of ordinary or partial differential equations which are the Euler–Lagrange equations for some variational problem. When a given system of equations possesses a Lie group G of symmetries, it is natural to ask if there is a Lagrangian for these equations with the same group G of symmetries. This is the invariant inverse problem of the calculus of variations.

To define this problem more precisely, let x^i, $i = 1, \ldots, n$, and u^α, $\alpha = 1, \ldots, m$, denote the independent and dependent variables. To a system of differential equations

$$F_\beta\Big(x^i, u^\alpha, \frac{\partial u^\alpha}{\partial x^i}, \frac{\partial u^\alpha}{\partial x^i \partial x^j}, \ldots\Big) = 0, \quad \beta = 1, \ldots, m, \tag{1.1}$$

we associate the $n+1$ form

$$\Delta = F_\beta \, \mathrm{d}u^\beta \wedge \mathrm{d}x^1 \wedge \mathrm{d}x^2 \wedge \cdots \wedge \mathrm{d}x^n. \tag{1.2}$$

The form Δ is called the *source form* for Equation (1.1). We call Δ a Euler–Lagrange form if there is a Lagrangian n form

$$\lambda = L\Big(x^i, u^\alpha, \frac{\partial u^\alpha}{\partial x^i}, \frac{\partial u^\alpha}{\partial x^i \partial x^j}, \dots\Big) dx^1 \wedge dx^2 \wedge \cdots \wedge dx^n$$

such that

$$\Delta = E(\lambda) = \frac{\delta L}{\delta u^\alpha}\, du^\alpha \wedge dx^1 \wedge dx^2 \wedge \cdots \wedge dx^n,$$

where $\delta L/\delta u^\alpha$ denotes the components of the classical Euler–Lagrange operator

$$\frac{\delta L}{\delta u^\alpha} = \frac{\partial L}{\partial u^\alpha} - D_i \frac{\partial L}{\partial u_i^\alpha} + \cdots .$$

A local Lie transformation group G acting on the total space E of dependent and independent variables is called a *distinguished* group of point symmetries for Equation (1.1) if the source form (1.2) is invariant under the prolonged action of G. If Δ is G invariant and Δ is derivable from some Lagrangian form λ, then we wish to know if Δ is derivable from a G invariant Lagrangian.

Examples of G invariant Euler–Lagrange forms which do not possess a G invariant Lagrangians abound. Examples from classical mechanics, where G is finite-dimensional, are given in [13, 14]. For the infinite-dimensional case, consider the potential Kadomtsev–Petviashvili (PKP) source form

$$\Delta_{\mathrm{PKP}} = \Big[u_{tx} + \tfrac{3}{2}\, u_x u_{xx} + u_{xxxx} + \tfrac{3}{4} s^2 u_{yy}\Big]\, du \wedge dt \wedge dx \wedge dy. \tag{1.3}$$

The PKP equation has an infinite-dimensional symmetry algebra Γ_{PKP} depending on five arbitrary functions of the time variable t [16]. It is a remarkable fact that all of these symmetries are distinguished symmetries for Δ_{PKP}. But there is no Γ_{PKP} invariant Lagrangian for (1.3). Also, with $n = 3$, the source form associated with the Cotton tensor (or the York curvature tensor)

$$\Delta = (\nabla_k R^i_h)\, dg_{ij} \wedge dx^j \wedge dx^h \wedge dx^k,$$

where R^i_j is the Ricci tensor of a metric g_{ij}, is invariant under the full pseudogroup of local diffeomorphisms (i.e. it is a natural source form). It is the Euler–Lagrange form of the Chern–Simons Lagrangian but it is not the Euler–Lagrange form of any natural Lagrangian [1]. Witten [24] begins his discussion of global aspects of current algebras with some invariant variational source forms which do not admit invariant Lagrangians. Of course, all these different Lagrangians are invariant, modulo divergences, but it is our contention that it is important to understand precisely when strictly invariant Lagrangians exist.

While the literature dealing with other aspects of the inverse problem is quite extensive, relatively little is known in general about the invariant inverse problem. Particular results can be found in [11–15, 19, 22].

In this paper, we announce a general approach, based upon properties of the variational bicomplex, for the solution of the local invariant inverse problem of the calculus of variations. Our methods yield a complete solution to this problem for finite-dimensional connected Lie transformation groups and, moreover, seem to be widely applicable to infinite-dimensional symmetry groups as well. In particular, our work clarifies the role of the Lie algebra cohomology of the symmetry group G in obstructing the local existence of G invariant Lagrangians. We also indicate other applications of our work, well beyond those of the invariant inverse problem of the calculus of variations. Full details of our results will appear elsewhere.

2. Invariant Variational Bicomplexes

In this section, we establish notation and give a brief description of the basic properties of the variational bicomplex. For more details we refer to [2] or [20].

Let $\pi\colon E \to M$ be an $n+m$-dimensional fiber bundle over an n-dimensional base manifold M, and let $\pi^\infty_M\colon J^\infty(E) \to M$ be the infinite jet bundle of local sections of $\pi\colon E \to M$. An adapted coordinate system $(x^i, u^\alpha) \to (x^i)$ of $\pi\colon E \to M$ induces coordinates $(x^i, u^\alpha, u^\alpha_i, u^\alpha_{ij}, \ldots)$ on $J^\infty(E)$.

The infinite jet bundle $\pi^\infty_M\colon J^\infty(E) \to M$ admits a natural flat connection $\mathcal{H}$ which is locally spanned by the total derivative vector fields

$$D_i = \frac{\partial}{\partial x^i} + u^\alpha_i \frac{\partial}{\partial u^\alpha} + u^\alpha_{ij} \frac{\partial}{\partial u^\alpha_j} + \cdots, \qquad i = 1, 2, \ldots, n.$$

Consequently, we can express the tangent bundle $T(J^\infty(E))$ as a direct sum

$$T(J^\infty(E)) = \mathcal{H} \oplus \mathcal{V}$$

of the bundle $\mathcal{H}$ of total vectors and the bundle $\mathcal{V}$ of π^∞_M vertical vectors. We let

$$\mathrm{Tot}\colon \mathcal{X}(J^\infty(E)) \longrightarrow \Gamma(J^\infty(E), \mathcal{H})$$

and

$$\mathrm{Vert}\colon \mathcal{X}(J^\infty(E)) \longrightarrow \Gamma(J^\infty(E), \mathcal{V})$$

denote the projections mapping smooth vector fields on $J^\infty(E)$ to their horizontal and vertical components.

The connection $\mathcal{H}$ also gives rise to a direct sum decomposition of the de Rham complex $\Omega^*(J^\infty(E))$ of $J^\infty(E)$

$$\Omega^*(J^\infty(E)) = \bigoplus_{r,s \geqslant 0} \Omega^{r,s}(J^\infty(E)),$$

into subspaces $\Omega^{r,s}(J^\infty(E))$ of horizontal degree r and vertical degree s. A form $\omega \in \Omega^k(J^\infty(E))$ belongs to $\Omega^{r,s}(J^\infty(E))$, where $k = r + s$, if

$$\omega(X_1, X_2, \ldots, X_k) = 0,$$

whenever either more than r of the vectors $X_1, X_2, \ldots, X_k$ are horizontal, or more that s of the vectors $X_1, X_2, \ldots, X_k$ are π^∞_M vertical.

We write $\mathrm{d} = \mathrm{d}_H + \mathrm{d}_V$ for the induced splitting of the exterior derivative into horizontal and vertical components.

Let

$$I\colon\ \Omega^{n,s}(J^\infty(E)) \longrightarrow \Omega^{n,s}(J^\infty(E)), \quad s \geqslant 1,$$

be the interior Euler operator given in local coordinates by

$$I(\omega) = \frac{1}{s}\,\mathrm{d}u^\alpha \wedge \left[\left(\frac{\partial}{\partial u^\alpha} \lrcorner\, \omega\right) - D_i\left(\frac{\partial}{\partial u^\alpha_i} \lrcorner\, \omega\right) + \cdots\right].$$

The spaces $\mathcal{F}^s(J^\infty(E))$ of functional s forms are, by definition, the images of $\Omega^{n,s}(J^\infty(E))$ under the mapping I. It is not difficult to see that $I \circ \mathrm{d}_H = 0$ and $I^2 = I$. The vertical derivative d_V induces a differential $\delta_V\colon \mathcal{F}^s(J^\infty(E)) \to \mathcal{F}^{s+1}(J^\infty(E))$ by $\delta_V = I \circ \mathrm{d}_V$.

The *augmented variational bicomplex* on $J^\infty(E)$ is then the double complex

$$
\begin{array}{ccccccccccccc}
 & & & & \uparrow{\scriptstyle \mathrm{d}_V} & & & & & & \uparrow{\scriptstyle \mathrm{d}_V} & & \uparrow{\scriptstyle \delta_V} & & \\
 & & 0 & \longrightarrow & \Omega^{0,3} & & & & \cdots & & \Omega^{n,3} & \xrightarrow{I} & \mathcal{F}^3 & \longrightarrow & 0 \\
 & & & & \uparrow{\scriptstyle \mathrm{d}_V} & & & & & & \uparrow{\scriptstyle \mathrm{d}_V} & & \uparrow{\scriptstyle \delta_V} & & \\
 & & 0 & \longrightarrow & \Omega^{0,2} & \xrightarrow{\mathrm{d}_H} & \Omega^{1,2} & \xrightarrow{\mathrm{d}_H} & \cdots\,\Omega^{n-1,2} & \xrightarrow{\mathrm{d}_H} & \Omega^{n,2} & \xrightarrow{I} & \mathcal{F}^2 & \longrightarrow & 0 \\
 & & & & \uparrow{\scriptstyle \mathrm{d}_V} & & \uparrow{\scriptstyle \mathrm{d}_V} & & \uparrow{\scriptstyle \mathrm{d}_V} & & \uparrow{\scriptstyle \mathrm{d}_V} & & \uparrow{\scriptstyle \delta_V} & & \\
 & & 0 & \longrightarrow & \Omega^{0,1} & \xrightarrow{\mathrm{d}_H} & \Omega^{1,1} & \xrightarrow{\mathrm{d}_H} & \cdots\,\Omega^{n-1,1} & \xrightarrow{\mathrm{d}_H} & \Omega^{n,1} & \xrightarrow{I} & \mathcal{F}^1 & \longrightarrow & 0 \\
 & & & & \uparrow{\scriptstyle \mathrm{d}_V} & & \uparrow{\scriptstyle \mathrm{d}_V} & & \uparrow{\scriptstyle \mathrm{d}_V} & & \uparrow{\scriptstyle \mathrm{d}_V} & & & & \\
0 & \longrightarrow & \mathbf{R} & \longrightarrow & \Omega^{0,0} & \xrightarrow{\mathrm{d}_H} & \Omega^{1,0} & \xrightarrow{\mathrm{d}_H} & \cdots\,\Omega^{n-1,0} & \xrightarrow{\mathrm{d}_H} & \Omega^{n,0} & & & &
\end{array}
$$

The *Euler–Lagrange complex* $\mathcal{E}^*(J^\infty(E))$ is, in turn, the edge complex of the augmented variational bicomplex on $J^\infty(E)$, namely,

$$
\begin{aligned}
0 &\longrightarrow R \longrightarrow \Omega^{0,0} \xrightarrow{\mathrm{d}_H} \Omega^{1,0} \xrightarrow{\mathrm{d}_H} \cdots \\
&\xrightarrow{\mathrm{d}_H} \Omega^{n-1,0} \xrightarrow{\mathrm{d}_H} \Omega^{n,0} \xrightarrow{\mathrm{E}} \mathcal{F}^1 \xrightarrow{\delta_V} \mathcal{F}^2 \xrightarrow{\delta_V} \cdots .
\end{aligned}
$$

Several important concepts in the calculus of variations can be interpreted in terms of the spaces and differentials occurring in the the Euler–Lagrange complex. For example, type $(n, 0)$ forms can be regarded as Lagrangians for variational problems, and source forms, the elements of $\mathcal{F}^1$, as systems of differential equations in n independent and m dependent variables. The mapping E: $\Omega^{n,0} \to \mathcal{F}^1$ agrees with the ordinary Euler–Lagrange operator, and the differential δ_V: $\mathcal{F}^1 \to \mathcal{F}^2$ is just the Helmholtz operator occurring in the inverse problem of calculus of variations.

The local cohomology of the augmented variational bicomplex and the Euler–Lagrange complex are well-known [3, 20, 21]:

THEOREM 2.1. *Let E be the trivial bundle $\mathbf{R}^{n+m} \to \mathbf{R}^n$. Then the rows and the columns of the augmented variational bicomplex are exact. The Euler–Lagrange complex is also exact.*

Invariant variational bicomplexes are constructed by introducing group actions on E, prolonging these actions to $J^\infty(E)$, and then restricting the variational bicomplex to the forms which are invariant under the prolonged actions. Since most of the groups which are of interest to us are connected Lie groups, we will only consider infinitesimal transformation groups. By an infinitesimal transformation group Γ on E, we mean a Lie subalgebra $\Gamma \subset \mathcal{X}(E)$ of the Lie algebra $\mathcal{X}(E)$ of all smooth vector fields on E. If X is any vector field on E, then we let $\operatorname{pr} X$ denote the prolongation of X to $J^\infty(E)$. The vector field $\operatorname{pr} X$ is the unique vector field on $J^\infty(E)$ which projects to X on E and which preserves the contact ideal on $J^\infty(E)$, that is, for any $\omega \in \Omega^{0,1}(J^\infty(E))$, $\mathcal{L}_{\operatorname{pr} X}\omega \in \Omega^{0,1}(J^\infty(E))$. See Olver [16] for the explicit prolongation formula. We write $\operatorname{pr}\Gamma$ for the infinite prolongation of Γ to $J^\infty(E)$,

$$\operatorname{pr}\Gamma = \{\operatorname{pr} X \mid X \in \Gamma\}.$$

The infinitesimal transformation group Γ is said to act projectably on E, if the one parameter transformation group generated by any $X \in \Gamma$ consists of fiber preserving transformations on E. In this case Lie differentiation with respect to $\operatorname{pr} X \in \operatorname{pr}\Gamma$ preserves the connection $\mathcal{H}$ and, consequently, commutes with the projection operators Tot and Vert, as well as with the horizontal and vertical differentials d_H, d_V, and the interior Euler operator I.

Let $\mathcal{U}$ be any open set in $J^\infty(E)$ and suppose that Γ acts projectably on E. Let $\Omega^{r,s}_{\operatorname{pr}\Gamma}(\mathcal{U})$ consist of all $\operatorname{pr}\Gamma$ invariant type (r, s) forms on $\mathcal{U}$, that is,

$$\Omega^{r,s}_{\operatorname{pr}\Gamma}(\mathcal{U}) = \{\omega \in \Omega^{r,s}(\mathcal{U}) \mid \mathcal{L}_{\operatorname{pr} X}\omega = 0 \text{ for all } X \in \Gamma\}.$$

We define the spaces $\mathcal{F}^s_{\operatorname{pr}\Gamma}(\mathcal{U})$ of Γ invariant functional forms in a similar fashion. The spaces $\Omega^{r,s}_{\operatorname{pr}\Gamma}(\mathcal{U})$ and $\mathcal{F}^s_{\operatorname{pr}\Gamma}(\mathcal{U})$ together with the differentials d_H, d_V, and δ_V form the *augmented* Γ *invariant variational bicomplex on* $\mathcal{U}$. The edge complex is now the Γ *invariant Euler–Lagrange complex*. Note, in particular, that the cohomology classes of the Γ invariant Euler–Lagrange complex $\mathcal{E}^*_{\mathrm{pr}\,\Gamma}(\mathcal{U})$ in degree $n+1$ are precisely the obstructions to constructing $\mathrm{pr}\,\Gamma$ invariant Lagrangians for $\mathrm{pr}\,\Gamma$ invariant source forms that satisfy the Helmholtz conditions.

EXAMPLE 2.2. Let E: $\mathbf{R}^n \times \mathbf{R}^m \to \mathbf{R}^n$ and let $\mathcal{U}^{n,m}_k$ be the open subset of $J^\infty(E)$ consisting of jets of sections $\mathbf{s} = \mathbf{s}(x^i)$ such that the vectors

$$\frac{\partial \mathbf{s}}{\partial x^{i_1}}, \frac{\partial^2 \mathbf{s}}{\partial x^{i_1} \partial x^{i_2}}, \ldots, \frac{\partial^k \mathbf{s}}{\partial x^{i_1} \partial x^{i_2} \ldots \partial x^{i_k}}$$

are pointwise linearly independent in $\mathbf{R}^m$. Gromov [8] calls such maps $\mathbf{s}$ *free maps to order* k. For example, with $n = 1$ and $k = m - 1$, $\mathbf{s}$ is a smooth curve in $\mathbf{R}^m$ whose derivatives determine a well-defined Frenet frame. When $k = 1$, $n = 2$ and $m = 3$, $\mathbf{s}$ determines a regular surface patch in $\mathbf{R}^3$, while when $k = 1$ and $n = m$, $\mathbf{s}$: $\mathbf{R}^n \to \mathbf{R}^n$ is a local diffeomorphism. We let $\Gamma_{\mathrm{diff}} = \mathcal{X}(\mathbf{R}^n)$ be the Lie algebra of all vector fields on the base $\mathbf{R}^n$ and we let $\Gamma_{\mathrm{eucl}} = \mathrm{e}(m)$ be the Lie algebra of the Euclidean group acting on the fiber $\mathbf{R}^m$. Let

$$\Gamma = \Gamma_{\mathrm{diff}} \oplus \Gamma_{\mathrm{eucl}}.$$

The variational bicomplex $\Omega^{*,*}_{\mathrm{pr}\,\Gamma}(\mathcal{U}^{1,m}_{m-1})$ is called the *natural variational bicomplex for regular curves*, $\Omega^{*,*}_{\mathrm{pr}\,\Gamma}(\mathcal{U}^{2,3}_1)$ is called the *natural variational bicomplex for regular surface patches in* $\mathbf{R}^3$ and $\Omega^{*,*}_{\mathrm{pr}\,\Gamma_{\mathrm{diff}}}(\mathcal{U}^{n,n}_1)$ is called the *natural variational bicomplex for local diffeomorphisms of* $\mathbf{R}^n$. The Lagrangian

$$\lambda_{\mathrm{ROT}} = \frac{1}{2\pi}\kappa\, ds = \frac{1}{2\pi}\frac{\ddot{u}^1 \dot{u}^2 - \ddot{u}^2 \dot{u}^1}{(\dot{u}^1)^2 + (\dot{u}^2)^2}\, dx$$

defines a nontrivial cohomology class in the Euler–Lagrange complex $\mathcal{E}^*_{\mathrm{pr}\,\Gamma}(\mathcal{U}^{1,2}_1)$. The integral of λ_{ROT} around a regular closed curve in the plane defines the rotation index of the curve. The intergrand in the Gauss–Bonnet formula is the Lagrangian

$$\lambda_{\mathrm{GB}} = \frac{1}{2\pi} K\, dA = \frac{1}{2\pi} K \sqrt{EG - F^2}\, dx^1 \wedge dx^2,$$

where $ds^2 = E\,(dx^1)^2 + 2F\, dx^1\, dx^2 + G\,(dx^2)^2$ is the first fundamental form of the surface and K its Gaussian curvature, defines a nontrivial cohomology class in the Euler–Lagrange complex $\mathcal{E}^*_{\mathrm{pr}\,\Gamma}(\mathcal{U}^{2,3}_1)$. The functional 2 form

$$\omega = \frac{u_{xx}}{u_x}\, du \wedge du_x \wedge dx - \frac{1}{u_x}\, du \wedge du_{xx} \wedge dx$$

is a cohomology class in the natural Euler–Lagrange complex $\mathcal{E}^*_{\mathrm{pr}\,\Gamma_{\mathrm{diff}}}(\mathcal{U}^{1,1}_1)$. As we shall see, the cohomology of the natural Euler–Lagrange complex for local diffeomorphisms of $\mathbf{R}^n$ is isomorphic to the Gelfand–Fuks cohomology of formal vector fields on $\mathbf{R}^n$. □

EXAMPLE 2.3. The symmetry algebra Γ_{PKP} of the PKP source form (1.3) is the infinite-dimensional Lie algebra spanned by five vector fields $X(f)$, $Y(g)$, $Z(h)$, $W(k)$ and $U(l)$, where f, g, h, k, l are arbitrary functions of the independent variable t ([6]). The vector field $X(f)$ is defined by

$$X(f) = f\frac{\partial}{\partial t} + \frac{2y}{3}f'\frac{\partial}{\partial y} + \left(\frac{x}{3}f' - \frac{2y^2s^2}{9}f''\right)\frac{\partial}{\partial x} - \\ -\left(\frac{u}{3}f' - \frac{x^2}{9}f'' + \frac{4xy^2s^2}{27}f''' - \frac{4y^4}{243}f''''\right)\frac{\partial}{\partial u},$$

where $s^2 = \pm 1$, with similar expressions, although not quite as complicated, for Y, Z, W and U. As we mentioned in the introduction, the PKP source form defines a nontrivial cohomology class in the Γ_{PKP} invariant Euler–Lagrange complex. □

EXAMPLE 2.4. Let $M = \mathbf{R}^n$ and let $E = T^*(M)$ with coordinates $(x^i, u_i) \to (x^i)$. Let $\Gamma = \Gamma_{\text{ga}} \oplus \mathbf{po}\,(1, n-1)$, where Γ_{ga} is the Abelian Lie algebra of vertical vector fields $Y = \phi(x)_{,i}(\partial/\partial u_i)$ and $\mathbf{po}(1, n-1)$ is the Lie algebra of the Poincaré group acting on E by pullback from the base. We call $\Omega^{*,*}_{\text{pr}\,\Gamma}(J^\infty(E))$ the *variational bicomplex for vector fields theories on* $\mathbf{R}^n$. Let

$$F = \mathrm{d}_H(u_i \mathrm{d}x^i) = u_{i,j}\,\mathrm{d}x^j \wedge \mathrm{d}x^i = F_{ij}\,\mathrm{d}x^i \wedge \mathrm{d}x^j.$$

For $n = 4$, the type $(2,0)$ and type $(4,0)$ forms $\alpha = F$ and $\lambda = F \wedge F$ are cohomology representatives in the invariant Euler–Lagrange complex $\mathcal{E}^*_{\text{pr}\,\Gamma}(J^\infty(E))$. When $n = 3$, the form α and the source form

$$\Delta = F_{ij}\,\mathrm{d}u_h \wedge \mathrm{d}x^i \wedge \mathrm{d}x^j \wedge \mathrm{d}x^h$$

are cohomology representatives in $\mathcal{E}^*_{\text{pr}\,\Gamma}(J^\infty(E))$. The Lagrangian form λ is a simple example of a topological Lagrangian and the source form Δ is the Chern–Simons mass term for the three-dimensional Abelian gauge theory [10]. This example is, of course, a special case of Yang–Mills fields on principal fiber bundles. □

The individual spaces $\Omega^{r,s}_{\text{pr}\,\Gamma}(\mathcal{U})$ in a given invariant variational bicomplex can often be explicitly described in terms of the differential invariants or, more generally, differential covariants, of Γ. In this regard, it is also quite important to be able to construct invariant or at least covariant bases for the spaces $\Omega^{1,0}_{\text{pr}\,\Gamma}(\mathcal{U})$ and $\Omega^{0,1}_{\text{pr}\,\Gamma}(\mathcal{U})$. We intent to deal with this topic in greater detail elsewhere.

Finally, it is possible to extend our considerations to nonprojectable infinitesimal Lie transformation groups, at least so far as to construct the corresponding invariant bicomplexes and Euler–Lagrange complexes. The extent to which the results which we announce in Sections 3 and 4 remain valid remains to be explored.

3. Exactness of the Interior Rows of the Γ Invariant Variational Bicomplex

Let Γ be a projectable infinitesimal Lie transformation group acting on the total space E. By introducing a certain class of invariant connections, the local d_H homotopy formula used in the proof of Theorem 2.1 can be modified to construct a Γ invariant d_H homotopy formula. This proves that the interior rows of the Γ invariant augmented variational bicomplex are exact.

DEFINITION 3.1. Let $\mathcal{U}$ be an open set in $J^\infty(E)$. A horizontal connection on the bundle $\mathcal{H}(\mathcal{U})$ of total vector fields is a $\mathbf{R}$-bilinear map which assigns to a pair of total vector fields X and Y on $\mathcal{U}$, a total vector $\nabla_X Y$ satisfying

(1) $\nabla_{fX} Y = f\nabla_X Y$; and

(2) $\nabla_X(fY) = X(f)\,Y + f\nabla_X Y$,

where f is any real-valued smooth function on $\mathcal{U}$.

If $\mathcal{R}_a = A^i_a D_j$, $a = 1, 2, \ldots, n$ is a basis for $\mathcal{H}$ on $\mathcal{U}$, then the components of ∇ with respect to this basis are given by $\nabla_{\mathcal{R}_a}\mathcal{R}_b = \Gamma^c_{ab}\mathcal{R}_c$. The connection coefficients Γ^c_{ab} are smooth functions on $\mathcal{U}$. We say ∇ is *torsion-free* if the torsion tensor $T(X, Y) = \nabla_X Y - \nabla_Y X - [X, Y]$ vanishes. The connection ∇ is Γ invariant if, for every $Z \in \Gamma$ and all total vector fields X and Y,

$$\mathcal{L}_{\mathrm{pr}\,Z}(\nabla_X Y) = \nabla_{(\mathcal{L}_{\mathrm{pr}\,Z} X)} Y + \nabla_X(\mathcal{L}_{\mathrm{pr}\,Z} Y).$$

Invariant, torsion-free horizontal connections on $\mathcal{H}$ can be constructed for a wide variety of infinitesimal transformation groups acting projectably on E. In fact, suppose that the prolongation $\mathrm{pr}\,\Gamma$ of Γ admits n functionally independent differential invariants $I^1, I^2 \ldots, I^n$ in some open set $\mathcal{U} \subset J^\infty(E)$. Then the set

$$\{\phi^1 = d_H I^1, \phi^2 = d_H I^2, \ldots, \phi^n = d_H I^n\} \tag{3.1}$$

forms a Γ invariant basis for the horizontal forms on $\mathcal{U}$. Let

$$\{\mathcal{R}_1, \mathcal{R}_2, \ldots, \mathcal{R}_n\} \tag{3.2}$$

be a basis for the distribution of total vector fields on $\mathcal{U}$ dual to $\{\phi^1, \phi^2, \ldots, \phi^n\}$. These total vector fields are Ovsiannikov's operators of invariant differentiation. See [17] and [18]. As the forms ϕ^a are d_H-closed and $\mathrm{pr}\,\Gamma$ invariant, the vector fields $\mathcal{R}^a$ commute among themselves and with the elements of $\mathrm{pr}\,\Gamma$, that is,

$$[\mathcal{R}_a, \mathcal{R}_b] = 0 \quad \text{and} \quad [\mathrm{pr}\,X, \mathcal{R}_a] = 0$$

for all a, b, and $X \in \Gamma$. With respect to the basis $\mathcal{R}_a$ for $\mathcal{H}$, we let ∇ be the unique horizontal connection on horizontal vector fields satisfying

$$\nabla_{\mathcal{R}_a}\mathcal{R}_b = 0 \tag{3.3}$$

for all a, b. This connection is evidently torsion-free and Γ invariant.

When the dimension of the base manifold M is one, it immediately follows from Theorem 2.1 that the interior rows of the Γ invariant variational bicomplex are locally exact. When E: $\mathbf{R}^n \times \mathbf{R}^m \to \mathbf{R}^n$ and Γ is a projectable subgroup of the general linear group on E, then the horizontal homotopy operators used to prove Theorem 2.1 are already $\operatorname{pr}\Gamma$ invariant. In the case when the dimension of the base manifold M is greater than one or the group action is more complicated than that of the general linear group, we have the following theorem.

THEOREM 3.2. *Let Γ be an infinitesimal transformation group acting projectably on the total space E and let $\mathcal{U}$ be an open set in $J^\infty(E)$. Suppose that*

(1) *∇ is a Γ invariant torsion-free horizontal connection on $\mathcal{U}$; and*
(2) *$\{\mathcal{R}_a\}$ is a basis for the distribution $\mathcal{H}(\mathcal{U})$ of horizontal vector fields on $\mathcal{U}$.*

Then the interior rows of the augmented $\operatorname{pr}\Gamma$ *invariant variational bicomplex*

$$0 \longrightarrow \Omega^{0,s}_{\operatorname{pr}\Gamma}(\mathcal{U}) \xrightarrow{d_H} \Omega^{1,s}_{\operatorname{pr}\Gamma}(\mathcal{U}) \cdots \Omega^{n-1,s}_{\operatorname{pr}\Gamma}(\mathcal{U}) \xrightarrow{d_H} \Omega^{n,s}_{\operatorname{pr}\Gamma}(\mathcal{U}) \xrightarrow{I} \mathcal{F}^{s}_{\operatorname{pr}\Gamma}(\mathcal{U}) \longrightarrow 0,$$

where $s \geqslant 1$, are exact.

To prove this theorem, we use the invariant horizontal connection ∇ to modify the construction of the local horizontal homotopy operators for the variational bicomplex. A special case of this construction already appears in [3]. Let us simply note here that the connection ∇ can be used to define a horizontal connection $\widehat{\nabla}$ on the full tangent bundle of $J^\infty(E)$, that is, an $\mathbf{R}$-bilinear map which assigns to each total vector field X on $\mathcal{U}$ and each arbitrary vector field Y on $\mathcal{U}$ a vector field $\widehat{\nabla}_X Y$ on $\mathcal{U}$ such that properties (1) and (2) of Definition 3.1 hold. This connection is defined by

$$\widehat{\nabla}_X Y = \nabla_X \operatorname{Tot} Y + \operatorname{Vert}[X, \operatorname{Vert} Y].$$

It can be intrinsically characterized by the property that $\widehat{\nabla}_X \operatorname{pr} Y = 0$ for all generalized vertical vector fields (or, evolutionary vector fields) Y on E. Let $\omega \in \Omega^{r,s}(J^\infty)$. Then $\widehat{\nabla}_X\omega$ is defined in the usual way and it is easy to prove that $\widehat{\nabla}_X\omega$ is a form of type (r, s); $\widehat{\nabla}_X\omega$ is Γ invariant whenever ω and ∇ are Γ invariant; and the horizontal differential of ω is given by

$$\begin{aligned} &d_H(\omega)(X_1, X_2, \ldots, X_{k+1}) \\ &\quad = \sum_{i=1}^{r+1}(-1)^{i+1}(\widehat{\nabla}_{\operatorname{Tot} X_i}\omega)(X_1, \ldots, \widehat{X}_i, \ldots, X_{r+1}), \end{aligned}$$

where $k = r+s$ and $X_1, X_2, \ldots, X_k$ are arbitrary vector fields on $J^\infty(E)$. These properties of $\widehat{\nabla}$ are the starting point for our proof of Theorem 3.2.

Standard elementary homological algebra arguments now prove the following corollary.

COROLLARY 3.3. *Under the hypothesis* (1) *and* (2) *of Theorem* 3.2, *the cohomology of the* Γ *invariant Euler–Lagrange complex is isomorphic to the cohomology of the* Γ *invariant de Rham complex on* $\mathcal{U}$:

$$H^p(\mathcal{E}^*_{\mathrm{pr}\,\Gamma}(\mathcal{U})) \cong H^p(\Omega^*_{\mathrm{pr}\,\Gamma}(\mathcal{U})).$$

Now assume that Γ is finite-dimensional. Then the Frobenius theorem combined with a simple dimension count shows that any regular point $\sigma \in J^\infty(E)$ of the prolonged group action has a neighborhood $\mathcal{U}$ in which there exist n functionally independent invariants of $\mathrm{pr}\,\Gamma$. Combined with Theorem 3.2, this suffices to prove the following result.

THEOREM 3.4. *Let* Γ *be a finite-dimensional infinitesimal Lie transformation group acting projectably on E. Then any regular point* $\sigma \in J^\infty(E)$ *of the group action has a neighborhood* $\mathcal{U}$ *such that the interior rows of the* Γ *invariant variational bicomplex* $\Omega^{*,*}_{\mathrm{pr}\,\Gamma}(\mathcal{U})$ *are exact and therefore* $H^p(\mathcal{E}^*_{\mathrm{pr}\,\Gamma}(\mathcal{U})) \cong H^p(\Omega^*_{\mathrm{pr}\,\Gamma}(\mathcal{U}))$.

Theorem 3.2 is also widely applicable beyond the finite-dimensional case. For example, for the natural variational bicomplex for surfaces, one can use the Christoffel symbols defined in terms of the first fundamental form for the horizontal connection ∇. For the natural variational bicomplex for local diffeomorphisms one has the covariant differentiation operators $\mathcal{R}_a = v^i_a D_i$, where (v^i_a) is the matrix inverse to u^a_i. For the invariant variational bicomplex defined by the PKP symmetry algebra, we are able to construct invariant differentiation operators

$$\begin{aligned} \mathcal{R}_x &= u_{xxx}^{-1/4} D_x, \qquad \mathcal{R}_y = u_{xxx}^{-1/2} D_y - u_{xxy} u_{xxx}^{-3/2} D_x, \\ \mathcal{R}_z &= u_{xxx}^{-3/4} D_t + \tfrac{2}{3} s^2 u_{xxy} u_{xxx}^{-7/4} D_y + \left(\tfrac{3}{2} u_x u_{xxx}^{-3/4} - \tfrac{3}{4} s^2 u_{xxy}^2 u_{xxx}^{-11/4}\right) D_x. \end{aligned} \tag{3.4}$$

For the variational bicomplex for vector field theories, the original horizontal homotopy operator is already Γ invariant and we simply apply Theorem 3.2 with $\mathcal{R}_a = D_a$. In fact we are, as yet, unable to construct examples of projectable infinitesimal groups for which the interior rows of the associated invariant variational bicomplex fail to be locally exact.

4. Cohomology of Invariant Variational Bicomplexes for Finite-Dimensional Infinitesimal Transformation Groups

Let $\mathbf{g}$ be a Lie algebra (finite or infinite-dimensional), and let $\Lambda^r(\mathbf{g})$ stand for the vector space of r multilinear alternating functionals on $\mathbf{g}$. As usual, we identify $\Lambda^0(\mathbf{g})$ with $\mathbf{R}$. We recall that the mapping

$$\mathrm{d}\colon\ \Lambda^r(\mathbf{g}) \longrightarrow \Lambda^{r+1}(\mathbf{g})$$

given by

$$\begin{aligned} &\mathrm{d}\alpha(X_0, X_1, \ldots, X_r) \\ &\quad = \sum_{i \leqslant j} (-1)^{i+j} \alpha([X_i, X_j], X_0, \ldots, \widehat{X_i}, \ldots, \widehat{X_j}, \ldots, X_r), \end{aligned} \tag{4.1}$$

where $\alpha \in \Lambda^r(g)$, makes $\Lambda^*(\mathbf{g})$ into a differential complex. The cohomology of the Lie algebra $\mathbf{g}$, $H^*(\mathbf{g})$, is, by definition, the cohomology of the complex $(\Lambda^*(\mathbf{g}), \mathrm{d})$.

Our next result is based upon the following general considerations. Let $\mathbf{g}$ be a Lie algebra, and suppose that Γ is an infinitesimal $\mathbf{g}$ transformation group acting on a manifold M. Write τ for the homomorphism $\tau\colon \mathbf{g} \to \Gamma$. A point $p \in M$ is called a regular point of the infinitesimal group action if the restriction map $\tau_p\colon \mathbf{g} \to T_p(M)$, defined by $\tau_p(X) = (\tau X)(p)$, is one-to-one. Choose a point $p \in M$, and let $\mathcal{U} \subset M$ be a neighborhood of p. We define a mapping

$$\rho_{\mathcal{U}}\colon \Omega^*_\Gamma(\mathcal{U}) \longrightarrow \Lambda^*(\mathbf{g})$$

by the assignment

$$\rho_{\mathcal{U}}(\omega)(\xi_1, \xi_2, \ldots, \xi_r) = (-1)^r \omega(\tau(\xi_1), \tau(\xi_2), \ldots, \tau(\xi_r))(p), \tag{4.2}$$

where $\omega \in \Omega^r_\Gamma(\mathcal{U})$ and $\xi_1, \xi_2, \ldots, \xi_r \in \mathbf{g}$. It is not difficult to prove that $\rho_{\mathcal{U}}$ is a cochain map.

THEOREM 4.1. *Let* $\mathbf{g}$ *be a finite-dimensional Lie algebra, and let* Γ *be an infinitesimal* $\mathbf{g}$ *transformation group acting on a manifold M. Suppose that* $p \in M$ *is a regular point of the group action. Then there is a neighborhood* $\mathcal{U}$ *of p such that the induced mapping in cohomology,* $\rho^*_{\mathcal{U}}\colon H^*(\Omega^*_\Gamma(\mathcal{U})) \to H^*(\mathbf{g})$, *is an isomorphism.*

Sketch of Proof. The proof of Theorem 4.1 proceeds in three steps. The first step consists of finding a local invariant coframe $\{\Theta^1, \ldots, \Theta^l, \mathrm{d}I^1, \ldots, \mathrm{d}I^{n-l}\}$, where $l = \dim(\mathbf{g})$ and $n = \dim(M)$, in a neighborhood $\mathcal{U}$ of the point p, where $I^1, \ldots, I^{n-l}$ are functionally independent local differential invariants of Γ. The invariant one forms $\Theta^1, \ldots, \Theta^l$ complementary to $\mathrm{d}I^1, \ldots, \mathrm{d}I^{n-l}$ can be chosen so that the subalgebra $\mathcal{S}^* \subset \Omega^*_\Gamma(\mathcal{U})$ over $\mathbf{R}$ they generate is closed under exterior differentiation, that is, $(\mathcal{S}^*, \mathrm{d})$ is a differential complex. In fact, the structure constants for Θ^i are just the negative of those for $\mathbf{g}$. In the second step of the proof it is shown, by explicitly constructing an invariant homotopy operator, that the inclusion map $i\colon \mathcal{S}^* \to \Omega^*_\Gamma(\mathcal{U})$ induces an isomorphism in the cohomology. The final step of the proof consists of showing that the restriction $\rho_{\mathcal{U},\mathcal{S}} : \mathcal{S}^* \to \Lambda^*(\mathbf{g})$ is an isomorphism. We emphasize the fact that this proof is constructive and readily lends itself to studying particular examples. □

We apply Theorem 4.1 to the problem at hand. Suppose that Γ is an infinitesimal $\mathbf{g}$ transformation group acting on the total space E of the fiber bundle

$\pi\colon E \to M$. Suppose that $\sigma \in J^\infty(E)$ is a regular point of the prolonged group action $\operatorname{pr}\Gamma$. Since $\mathbf{g}$, by assumption, is finite-dimensional we can find a finite k such that the point $\sigma^k = \pi_k^\infty(\sigma) \in J^k(E)$ is a regular point of the kth prolongation $\operatorname{pr}^k\Gamma$ of Γ.

COROLLARY 4.2. *Suppose that* $\mathbf{g}$ *is a finite-dimensional Lie algebra, and let* Γ *be an infinitesimal* $\mathbf{g}$ *transformation group acting on the total space* E *of the fiber bundle* $\pi\colon E \to M$. *Suppose that* $\sigma \in J^\infty(E)$ *is a regular point of the prolonged group action* $\operatorname{pr}\Gamma$. *Then there is a neighborhood* $\mathcal{U} \subset J^\infty(E)$ *of* σ *such that the* Γ *invariant cohomology* $H^*(\Omega^*_{\operatorname{pr}\Gamma}(\mathcal{U}))$ *is isomorphic with the Lie algebra cohomology of* $\mathbf{g}$.

We also note that by the reasoning above any cohomology class in $H^*(\Omega^*_{\operatorname{pr}\Gamma}(\mathcal{U}))$ can be represented by a form in $\Omega^*_{\operatorname{pr}^k\Gamma}(\mathcal{U}^k)$, where k is the smallest integer such that the kth prolongation of Γ is regular at σ^k.

Consider now the case when Γ acts projectably on E. Then by Theorems 3.4 and 4.2, we have the following.

THEOREM 4.3. *Suppose that* $\mathbf{g}$ *is a finite-dimensional Lie algebra, and let* Γ *be an infinitesimal* $\mathbf{g}$ *transformation group acting projectably on the total space* E. *Suppose that* $\sigma \in J^\infty(E)$ *is a regular point of the prolonged group action* $\operatorname{pr}\Gamma$. *Then there is a neighborhood* $\mathcal{U} \subset J^\infty(E)$ *of* σ *such that the cohomology of the* Γ *invariant Euler–Lagrange complex* $\mathcal{E}^*_{\operatorname{pr}\Gamma}(\mathcal{U})$ *is isomorphic with the Lie algebra cohomology of* $\mathbf{g}$,

$$H^p(\mathcal{E}^*_{\operatorname{pr}\Gamma}(\mathcal{U})) \cong H^p(\mathbf{g}).$$

In particular, we have characterized the obstructions to the solution of the local Γ invariant inverse problem to the calculus of variations.

COROLLARY 4.4. *Let* $\mathcal{U}$ *be as in Theorem* 4.3 *and suppose that* $H^{n+1}(\mathbf{g}) = 0$. *Then every* Γ *invariant source form on* $\mathcal{U}$ *which is the Euler–Lagrange form of some Lagrangian is the Euler–Lagrange form of a* Γ *invariant Lagrangian.*

EXAMPLE 4.5. Let F be a m-dimensional configuration space, let $E\colon \mathbf{R}\times F \to \mathbf{R}$. A source form Δ on $J^\infty(E)$ defines a system of ordinary differential equations on F. Let $\mathbf{g}$ be finite-dimensional, semisimple Lie algebra and Γ a projectable infinitesimal $\mathbf{g}$ transformation group on E. By the classical Whitehead lemma ([9]), $H^2(\mathbf{g}) = 0$ and so every Γ invariant, locally variational, source form on $J^\infty(E)$ is locally the Euler–Lagrange form of a Γ invariant Lagrangian.

It is important to emphasize the local nature of our results. For example, let Γ be the one-dimensional infinitesimal group generated by $X = \partial/\partial x$ and suppose that Δ is a locally variational, Γ invariant source form. By Theorem 4.3, Γ invariant Lagrangians always exist locally but, according to a theorem of

Tulczyjew [22], there may be obstructions in $H^1(F) \oplus H^2(F)$ to the existence of global Γ invariant Lagrangians. Indeed, let $F = \mathbf{R}^2 - \{0\}$ and consider the source form

$$\Delta = \left[\ddot{u} + \frac{v}{u^2+v^2}\right] \mathrm{d}u \wedge \mathrm{d}x + \left[\ddot{v} - \frac{u}{u^2+v^2}\right] \mathrm{d}v \wedge \mathrm{d}x.$$

This source form is the Euler–Lagrange form for the global x dependent Lagrangian

$$\lambda_1 = \left[-\frac{1}{2}(\dot{u}^2+\dot{v}^2) + x\frac{v\dot{u}-u\dot{v}}{u^2+v^2}\right] \mathrm{d}x$$

and the local Γ invariant Lagrangian

$$\lambda_2 = \left[-\frac{1}{2}(\dot{u}^2+\dot{v}^2) - \arctan\left(\frac{u}{v}\right)\right] \mathrm{d}x$$

but there is no global Γ invariant Lagrangian for Δ. □

EXAMPLE 4.6. Consider E: $\mathbf{R}^2 \times \mathbf{R} \to \mathbf{R}^2$, and let Γ be the infinitesimal Euclidean group $\mathbf{e}(2)$ acting on the base space $\mathbf{R}^2$. With respect to the coordinates (x, y, u, u_x, u_y) of the first jet bundle $J^1(E)$, the generators for the first prolongation of $\mathbf{e}(2)$ are

$$\mathrm{pr}^1 t_x = \frac{\partial}{\partial x}, \qquad \mathrm{pr}^1 t_y = \frac{\partial}{\partial y},$$

$$\mathrm{pr}^1 r = y\frac{\partial}{\partial x} - x\frac{\partial}{\partial y} + u_y\frac{\partial}{\partial u_x} - u_x\frac{\partial}{\partial u_y}.$$

A point $\sigma \in J^1(E)$ is a regular point of the prolonged group action if at σ, $u_x^2 + u_y^2 \neq 0$. The prolonged group action admits the two functionally independent invariants $I^1 = u$ and $I^2 = \sqrt{u_x^2 + u_y^2}$. An invariant coframe is given by

$$\mathrm{d}I^1 = \mathrm{d}u, \qquad \mathrm{d}I^2 = \frac{u_x \mathrm{d}u_x + u_y \mathrm{d}u_y}{I^2}, \qquad \Theta^1 = \frac{u_x \mathrm{d}u_y - u_y \mathrm{d}u_x}{(I^2)^2},$$

$$\Theta^2 = \frac{u_x \mathrm{d}x + u_y \mathrm{d}y}{I^2}, \qquad \Theta^3 = \frac{-u_y \mathrm{d}x + u_x \mathrm{d}y}{I^2}.$$

Let $p \in J^1(E)$ be the point with coordinates $x = y = u = u_y = 0$, $u_x = 1$. Then under ρ, defined as in (4.2), this invariant coframe is mapped to

$$\rho(\mathrm{d}I^1) = 0, \qquad \rho(\mathrm{d}I^2) = 0, \qquad \rho(\Theta^1) = \delta^r,$$

$$\rho(\Theta^2) = -\delta^x, \qquad \rho(\Theta^3) = -\delta^y,$$

where $\{\delta^x, \delta^y, \delta^r\}$ is the basis for $\Lambda^1(\mathbf{e}(2))$ dual to $\{t_x, t_y, r\}$.

The cohomology of the Lie algebra $\mathbf{e}(2)$ is easily calculated from the definitions. In degree 1, the cohomology is generated by δ^r, in degree 2, by $\delta^x \wedge \delta^y$, and in degree 3, by $\delta^r \wedge \delta^x \wedge \delta^y$. According to Corollary 4.2, these cohomology classes give rise to nontrivial cohomology classes in the $\mathbf{e}(2)$ invariant de Rham complex on $J^1(E)$. Explicitly, the class of δ^r pulls back to the class generated by Θ^1, the class of $\delta^x \wedge \delta^y$ to the class generated by $\Theta^2 \wedge \Theta^3 = \mathrm{d}x \wedge \mathrm{d}y$, and the class of $\delta^r \wedge \delta^x \wedge \delta^y$ to the class of

$$\frac{u_y \mathrm{d}u_x - u_x \mathrm{d}u_y}{u_x^2 + u_y^2} \wedge \mathrm{d}x \wedge \mathrm{d}y.$$

From the isomorphism in Corollary 3.3, we conclude that the cohomology of the $\mathbf{e}(2)$ invariant Euler–Lagrange complex is generated by

$$\alpha = \frac{(u_x u_{xy} - u_y u_{xx})\,\mathrm{d}x + (u_x u_{yy} - u_y u_{xy})\,\mathrm{d}y}{u_x^2 + u_y^2}$$

in degree 1, by $\mathrm{d}x \wedge \mathrm{d}y$ in degree 2, and by

$$\Delta_0 = \frac{(u_x^2 - u_y^2)u_{xy} + u_x u_y (u_{yy} - u_{xx})}{(u_x^2 + u_y^2)^2}\, \mathrm{d}u \wedge \mathrm{d}x \wedge \mathrm{d}y$$

in degree 3. In particular, if Δ is an $\mathbf{e}(2)$ invariant source form which satisfies the Helmholtz conditions, then there is an $\mathbf{e}(2)$ invariant Lagrangian λ and a constant c such that $\Delta = c\Delta_0 + E(\lambda)$. □

5. Cohomology of Invariant Variational Bicomplexes for Infinite-Dimensional Infinitesimal Transformation Groups

It is an outstanding problem as to what extend the results of Section 4 can be generalized to the case of the infinite-dimensional infinitesimal transformation groups. As a first step in this direction, we tentatively introduce the continuous cohomology of vector fields with support at a point as the replacement for the ordinary Lie algebra cohomology in Theorem 4.3.

Let M be a manifold and fix a point $p \in M$. Let $x = (x^1, x^2, \dots, x^n)$ be a coordinate system at p. We define the maps $\delta^i_J\colon \mathcal{X}(E) \to \mathbf{R}$, $i = 1, 2, \dots, n$, $|J| \geqslant 0$, by

$$\delta^i_J(X) = (-1)^{|J|} \frac{\partial^{|J|} X^i}{\partial x^J}(p),$$

where X^i is the ith component of the vector field X and $J = (j_1 j_2 \dots j_k)$ is a multi-index of length $k = |J|$. The algebra $\Lambda^*_{c,p}(\mathcal{X}(M))$ of continuous alternating forms on $\mathcal{X}(M)$ with support at p is the exterior algebra generated by $\{\delta^i_I\}$. The differential d: $\Lambda^r_{c,p}(\mathcal{X}(M)) \to \Lambda^{r+1}_{c,p}(\mathcal{X}(M))$ is given by formula (4.1), where now $X_0, X_1, \dots, X_r \in \mathcal{X}(M)$ and $\alpha \in \Lambda^r_{c,p}(\mathcal{X}(M))$. The continuous cohomology with support at a point $H^*_{c,p}(\mathcal{X}(M))$ of $\mathcal{X}(M)$ is, by definition, the cohomology of the complex $(\Lambda^*_{c,p}(\mathcal{X}(M)), d)$.

Next let $\Gamma \subset \mathcal{X}(M)$ be a finite or infinite-dimensional infinitesimal transformation group acting on M. The complex $\Lambda^*_{c,p}(\Gamma)$ is obtained by restricting forms in $\Lambda^*_{c,p}(\mathcal{X}(M))$ to Γ. We call the cohomology $H^*_{c,p}(\Gamma)$ of the complex $\Lambda^*_{c,p}(\Gamma)$ the continuous cohomology of Γ with support at p. When $\mathbf{g}$ is a finite-dimensional Lie algebra and Γ is an infinitesimal $\mathbf{g}$ transformation group acting on M, then it is a simple matter to show that for any regular point p the usual cohomology of $\mathbf{g}$ and the continuous cohomology of Γ with support at p coincide.

Now suppose that Γ acts on the total space of some fiber bundle $\pi\colon E \to M$. Fix a point $\sigma \in J^\infty(E)$ and let $\sigma_0 = \pi^\infty_E(\sigma)$. We define a mapping ϱ from the $\operatorname{pr}\Gamma$ invariant de Rham complex on an open set $\mathcal{U} \subset J^\infty(E)$ into the complex $\Lambda^*_{c,\sigma_0}(\Gamma)$ by

$$\varrho(\omega)(X_1, X_2, \dots, X_r) = (-1)^r \omega(\operatorname{pr} X_1, \operatorname{pr} X_2, \dots, \operatorname{pr} X_r)(\sigma),$$

where $\omega \in \Omega^r_{\operatorname{pr}\Gamma}(\mathcal{U})$ and $X_1, X_2, \dots, X_r \in \Gamma$. One can check that ϱ is a cochain mapping. In many examples, we can show that ϱ defines an isomorphism in cohomology although we are unable to prove this in general. However, when ϱ induces an isomorphism in cohomology we can conclude $H^p(\mathcal{E}^*_{\operatorname{pr}\Gamma}(\mathcal{U})) \cong H^p(\Lambda^*_{c,p}(\mathbf{g}))$.

EXAMPLE 5.1. For the natural variational bicomplexes for plane and space curves it is easy to check that ϱ gives an isomorphism in cohomology and to compute the cohomology $H^*_{c,p}(\Gamma)$ directly. The following table summarizes our results.

	$p=1$	2	3	4	5	6	7	8	9	$p \geqslant 10$
$H^p(\mathcal{E}^*_{\operatorname{pr}\Gamma}(\mathcal{U}^{1,2}_1))$	1	1	2	1	1	1	0	0	0	0
$H^p(\mathcal{E}^*_{\operatorname{pr}\Gamma}(\mathcal{U}^{1,3}_2))$	0	0	3	0	0	3	0	0	1	0

The case $p = 1$ was first calculated by Cheung [5]. The Lagrangian λ_{ROT} generates the one cohomology class in $H^1(\mathcal{E}^*_{\operatorname{pr}\Gamma}(\mathcal{U}^{1,2}_1))$.

For the natural variational bicomplex $\Omega^{*,*}_{\Gamma_{\mathrm{diff}}}(\mathcal{U}^{n,n}_1)$ for local diffeomorphism, we are able to prove that ϱ gives an isomorphism in cohomology so that, combined with Theorem 4.4, we have

$$H^p(\mathcal{E}^*_{\operatorname{pr}\Gamma}(\mathcal{U}^{n,n}_1)) \cong H^p(\Lambda_{c,p}(\mathcal{X}(\mathbf{R}^n)).$$

This latter cohomology is, by definition, the Gelfand–Fuks cohomology of formal vector fields on $\mathbf{R}^n$. This is computed in [4, 7]. Alternatively, we can calculate the Gelfand–Fuks cohomology from the d_V cohomology of the natural variational bicomplex $\Omega^{*,*}_{\Gamma_{\mathrm{diff}}}(\mathcal{U}^{n,n}_1)$. □

We are currently applying these ideas to the natural variational bicomplexes for regular surfaces patches in $\mathbf{R}^3$ and $\mathbf{R}^4$.

EXAMPLE 5.2. Let E: $\mathbf{R}^3 \times \mathbf{R} \to \mathbf{R}^3$, and let Γ_{PKP} be the symmetry algebra of the potential Kadomtsev-Petviashvili equation. Let $\mathcal{U} \subset J^\infty(E)$ be the open set

$$\mathcal{U} = \{(t, x, y, u, u_t, u_x, u_y, u_{tt}, \ldots) \mid u_{xxx} \neq 0\}.$$

Using the invariant differentiation operators (3.4), we can explicitly construct a basis for $\Omega^*_{\mathrm{pr}\,\Gamma_{\mathrm{PKP}}}(\mathcal{U})$ and thereby prove that ϱ is surjective. We have yet to prove that ϱ induces an isomorphism in cohomology. This is tantamount to proving that the complex constructed from the kernel of ϱ is acyclic. Nevertheless, after some long calculations it is found that the continuous cohomology of Γ_{PKP} with support at σ is three-dimensional in degree 4. All these cohomology classes pull back to classes in $H^4(\Omega^*_{\mathrm{pr}\,\Gamma_{\mathrm{PKP}}}(\mathcal{U}))$, which, in turn, give rise to three cohomology classes in degree 4 in the $\mathrm{pr}\,\Gamma_{\mathrm{PKP}}$ invariant Euler–Lagrange complex. Remarkably enough, two of the cohomology classes in $\mathcal{E}^4_{\mathrm{pr}\,\Gamma_{\mathrm{PKP}}}(\mathcal{U})$ are represented by the source forms

$$\begin{aligned} \Delta_1 &= \left(u_{tx} + \tfrac{3}{2}u_x u_{xx} + \tfrac{3}{4}s^2 u_{yy}\right) \mathrm{d}u \wedge \mathrm{d}t \wedge \mathrm{d}x \wedge \mathrm{d}y, \\ \Delta_2 &= u_{xxxx}\, \mathrm{d}u \wedge \mathrm{d}t \wedge \mathrm{d}x \wedge \mathrm{d}y, \end{aligned}$$

both of which appear in the PKP source form. □

EXAMPLE 5.3. For the variational bicomplex for vector field theories on $\mathbf{R}^n$, we again find that ϱ induces an isomorphism in cohomology and, in this case, the continuous cohomology ring with support at a point is generated by a single two form arising from the curvature two form F. □

Acknowledgement

Part of this work was completed during the second author's visit at Utah State University. He would like to extend his thanks to the faculty of the mathematics department at USU for a productive and enjoyable year. This work is supported, in part, by grant DMS-91000674 from the National Science Foundation.

References

1. Anderson , I. M.: Natural variational principles on Riemannian manifolds, *Ann. Math.* **120** (1984), 329–370.
2. Anderson, I. M.: Introduction to the variational bicomplex, in M. Gotay, J. Marsden and V. Moncrief (eds), *Mathematical Aspects of Classical Field Theory*, Contemporary Mathematics 132, Amer. Math Soc., Providence, 1992, pp. 51–73.
3. Anderson, I. M.: *The Variational Bicomplex*, Academic Press, Boston (to appear).

4. Bott, R.: Notes on Gel'fand–Fuks cohomology and characteristic classes (notes by M. Mostow and J. Perchik), Eleventh Holiday Symposium, New Mexico State Univ., Las Cruces, 1973.
5. Cheung, W. S.: Higher order conservation laws and a higher order Noether's theorem, *Adv. Appl. Math.* **8** (1987), 446–485.
6. David, D., Kamran, N., Levi, D., and Winternitz, P.: Symmetry reduction for the Kadomtsev–Petviashvili equation using a loop algebra, *J. Math. Phys.* **27** (1986), 1225–1237.
7. Fuks, D. B.: *Cohomology of Infinite Dimensional Lie Algebras*, Consultants Bureau, New York, 1986.
8. Gromov, M.: *Partial Differential Relations*, Springer-Verlag, Berlin, 1986.
9. Hilton, P J. and Stammbach, U.: *A Course in Homological Algebra*, Springer-Verlag, New York, 1971.
10. Jackiw, R.: Topological investigations of quantized gauge theories, in A. S. Wrightman and P. A. Anderson (eds), *Current Algebras and Anomalies*, Princeton Series in Physics, Princeton University Press, Princeton, 1985, pp. 210–360.
11. Lévy-Leblond, J. M.: Group-theoretical foundations of classical mechanics: the Lagrangian gauge problem, *Comm. Math. Phys.* **12** (1969), 64–79.
12. López, M. C., Noriega, R. J., and Schifini, C. G.: The equivariant inverse problem and the uniqueness of the Yang–Mills equations, *J. Math. Phys.* **30** (1989), 2382–2387.
13. Marmo, G., Saletan, E. J., and Simoni, A.: On obtaining strictly invariant Lagrangians from gauge invariant Lagrangians, *Nuovo Cim.* **96B** (1986), 159– 163.
14. Marmo, G. and Morandi, G.: Inverse problem with symmetries and the appearance of cohomologies in classical Lagrangian dynamics, *Rep. Math Phys.* **3** (1989), 389–410.
15. Marmo, G., Morandi, G., and Rubano, C.: Symmetries in the Lagrangian and Hamiltonian formalism: the equivariant inverse problem, in B. Gruber and F. Iachelle (eds), *Symmetries in Science* 111, Plenum, New York, 1989, pp. 243–309.
16. Olver, P. J.: *Applications of Lie Groups to Differential Equations*, Springer, New York, 1986.
17. Olver, P. J.: Differential invariants, *Acta Appl. Math.* **41** (1995), 271–284 (this issue).
18. Ovsiannikov, L. V.: *Group Analysis of Differential Equations*, Academic Press, New York, 1982.
19. Noriega, R. J. and Schifini, C. G.: The equivariant inverse problem and the Maxwell equations, *J. Math. Phys.* **28** (1987), 815–817.
20. Tsujishita, T.: On variation bicomplexes associated to differential equations, *Osaka J. Math.* **19** (1982), 311–363.
21. Tulczyjew,W. M.: The Euler–Lagrange resolution, *Lecture Notes in Math.* 836, Springer-Verlag, New York, 1980, pp. 22–48.
22. Tulczyjew, W. M.: Cohomology of the Lagrange complex, *Ann. Scuola. Norm. Sup. Pisa* **14** (1987), 217–227.
23. Whiston, G. S.: On the gauge variance of action functions under transformations on space-time, *Internat. J. Theoret. Phys.* **5** (1972), 391–401.
24. Witten, E.: Global aspects of current algebra, *Nuclear Phys. B* **223** (1983), 422–432.

Acta Applicandae Mathematicae **41:** 21–43, 1995.

The Use of Factors to Discover Potential Systems or Linearizations

GEORGE BLUMAN and PATRICK DORAN-WU
Department of Mathematics, University of British Columbia, Vancouver, B.C., Canada, V6T IZ2

(Received: 28 February 1994)

Abstract. Factors of a given system of PDEs are solutions of an adjoint system of PDEs related to the system's Fréchet derivative. In this paper, we introduce the notion of potential conservation laws, arising from specific types of factors, which lead to useful potential systems. Point symmetries of a potential system could yield nonlocal symmetries of the given system and its linearization by a noninvertible mapping.

We also introduce the notion of linearizing factors to determine necessary conditions for the existence of a linearization of a given system of PDEs.

Mathematics Subject Classifications (1991): 35A30, 58G35, 35K55, 22E65, 58B25.

Key words: nonlocal symmetries, potential symmetries, linearization.

1. Introduction

Consider a system of N partial differential equations (PDEs) $R\{u\}$ given by

$$G^\sigma\Big(x, u, \underset{1}{u}, \underset{2}{u}, \ldots, \underset{k}{u}\Big) = 0, \quad \sigma = 1, 2, \ldots, N, \tag{1.1}$$

with independent variables $x = (x_1, x_2, \ldots, x_n)$ and dependent variables $u = (u^1, u^2, \ldots, u^m)$; $\underset{j}{u}$ denotes the set of coordinates corresponding to all jth-order partial derivatives of u with respect to x (a coordinate in $\underset{j}{u}$ is denoted by

$$u^\gamma_{i_1 i_2 \ldots i_j} = \frac{\partial^j u^\gamma}{\partial x_{i_1} \partial x_{i_2} \cdots \partial x_{i_j}}$$

with $\gamma = 1, 2, \ldots, m$; $i_j = 1, 2, \ldots, n$; $j = 1, 2, \ldots, k$).

DEFINITION 1.1. *A symmetry* of a given system of PDEs $R\{u\}$ is a transformation mapping any solution of $R\{u\}$ into another solution.

This definition is strictly topological and, in particular, coordinate-free. Consequently, $R\{u\}$ should admit a wide range of continuous symmetries: its family of solutions is expected to be invariant under a wide range of continuous deformations. For the rest of this paper we consider continuous symmetries characterized by infinitesimal generators whose forms allow such symmetries to be discovered and utilized algorithmically.

DEFINITION 1.2. A (*Lie*) *point symmetry* admitted by $R\{u\}$ is characterized by an infinitesimal generator of the form

$$\mathbf{X} = \sum_{\mu=1}^{m} \eta^{\mu}\left(x, u, \underset{1}{u}\right) \frac{\partial}{\partial u^{\mu}} \tag{1.2}$$

with η linear in the coordinates of $\underset{1}{u}$:

$$\eta^{\mu} = \alpha^{\mu}(x, u) - \sum_{i=1}^{n} \xi_i(x, u) u_i^{\mu}. \tag{1.3}$$

An infinitesimal generator (1.2) corresponds to a one-parameter Lie group of point transformations

$$\begin{aligned} x_i^* &= x_i + \varepsilon \xi_i(x, u) + \mathrm{O}(\varepsilon^2), \quad i = 1, 2, \ldots, n, \\ u^{\mu^*} &= u^{\mu} + \varepsilon \alpha^{\mu}(x, u) + \mathrm{O}(\varepsilon^2), \quad \mu = 1, 2, \ldots, m. \end{aligned} \tag{1.4}$$

Under the action of (1.2), a solution $u = \theta(x)$ of $R\{u\}$ is mapped into the one-parameter family of solutions

$$u = \Phi(x; \varepsilon) = \mathrm{e}^{\varepsilon \mathrm{U}} u\big|_{u=\theta(x)}, \tag{1.5}$$

where U is the prolongation operator given by

$$\mathrm{U} = \mathbf{X} + (D_i \eta^{\mu}) \frac{\partial}{\partial u_i^{\mu}} + \cdots + (D_{i_1} D_{i_2} \cdots D_{i_j} \eta^{\mu}) \frac{\partial}{\partial u^{\mu}_{i_1 i_2 \cdots i_j}} + \cdots$$

in terms of total differential operators

$$D_i = \frac{\partial}{\partial x_i} + u_i^{\gamma} \frac{\partial}{\partial u^{\gamma}} + \cdots + u^{\gamma}_{i i_1 i_2 \cdots i_l} \frac{\partial}{\partial u^{\gamma}_{i_1 i_2 \cdots i_l}} + \cdots, \quad i = 1, 2, \ldots, n.$$

(Summation over a repeated index is assumed throughout this paper.)

Lie [1–7] gave an algorithm to find the infinitesimal generators of a given system $R\{u\}$: If $u^* = u + \varepsilon \eta(x, u, \underset{1}{u}) + \mathrm{O}(\varepsilon^2)$, then

$$\begin{aligned} &G^{\sigma}\left(x, u^*, \underset{1}{u}^*, \underset{2}{u}^*, \ldots, \underset{k}{u}^*\right) \\ &\quad = G^{\sigma}\left(x, u, \underset{1}{u}, \underset{2}{u}, \ldots, \underset{k}{u}\right) + \varepsilon \sum_{\rho=1}^{m} \mathcal{L}_{\rho}^{\sigma}[u] \eta^{\rho} + \mathrm{O}(\varepsilon^2), \end{aligned}$$

where $\mathcal{L}[u]$ is the Fréchet derivative of $R\{u\}$. One can prove the following theorem:

THEOREM 1.3. **X** *is admitted by* $R\{u\}$, *including its differential consequences, if and only if the equations*

$$\sum_{\rho=1}^{m} \mathcal{L}_{\rho}^{\sigma}[u]\eta^{\rho} = 0, \quad \sigma = 1, 2, \ldots, N, \tag{1.6}$$

are satisfied for any solution $u = \theta(x)$ *of* $R\{u\}$.

The determining Equations (1.6) form an overdetermined linear system of PDEs with $n + m$ unknowns $\alpha^1, \alpha^2, \ldots, \alpha^m$; $\xi_1, \xi_2, \ldots, \xi_n$. There exist various symbolic manipulation programs [8–14] which perform one or more of the following functions automatically and/or interactively: set up determining equations, find the dimension (if finite) of their solution space, and solve them explicitly.

For a given system $R\{u\}$, point symmetries can yield various applications including the discovery of new solutions from known solutions (Equation (1.5)), the construction of specific invariant solutions [1–7], and the generation of conservation laws through Noether's theorem [4–7]. In addition, one can determine algorithmically whether or not $R\{u\}$ can be linearized by an *invertible* point transformation and construct an explicit linearization when one exists [15, 16, 6].

DEFINITION 1.4. A *local symmetry* admitted by $R\{u\}$ is characterized by an infinitesimal generator of the form

$$\mathbf{X} = \sum_{\mu=1}^{m} \eta^{\mu}\left(x, u, \underset{1}{u}, \underset{2}{u}, \ldots, \underset{p}{u}\right)\frac{\partial}{\partial u^{\mu}}. \tag{1.7}$$

DEFINITION 1.5. A local symmetry of the form (1.7) is a *contact symmetry* when $m = p = 1$; a *Lie–Bäcklund* (*higher, higher order, generalized*) *symmetry* [5, 6], when it is not a point or contact symmetry.

The algorithm for determining local symmetries of a given $R\{u\}$ involves solving the corresponding determining Equations (1.6).

DEFINITION 1.6. A *nonlocal symmetry* of $R\{u\}$ is a continuous symmetry admitted by $R\{u\}$ which is not characterized by an infinitesimal generator of local type (1.7).

In order to have algorithms to compute or utilize nonlocal symmetries most effectively, one should be able to characterize them in terms of infinitesimal generators of point symmetries in some coordinate frame. There exists an algorithm to find a class of such nonlocal symmetries (*potential symmetries*) provided system $R\{u\}$ contains at least one PDE expressed as a conservation law. This allows one to introduce potential variables v and a related potential system $S\{u, v\}$ [6,

17, 18]. Consequently, one can extend the known applications and calculations of point symmetries to potential symmetries. Moreover, one can discover algorithmically linearizations by *noninvertible* mappings [6, 19].

In this paper, we develop the notions of potential conservation laws which are useful for discovering potential systems, and linearizing factors which identify linearizable systems. We clarify and extend results presented in recent papers [20–22]. A complete potential symmetry analysis is given for the nonlinear diffusion equation.

2. Potential Symmetries

Suppose one PDE of $R\{u\}$, without loss of generality $G^N = 0$, is a conservation law

$$\sum_{i=1}^{n} D_i f^i\Big(x, u, \underset{1}{u}, \underset{2}{u}, \ldots, \underset{k-1}{u}\Big) = 0.$$

Then $R\{u\}$ is the system given by

$$G^\sigma\Big(x, u, \underset{1}{u}, \underset{2}{u}, \ldots, \underset{k}{u}\Big) = 0, \quad \sigma = 1, 2, \ldots, N-1, \tag{2.1}$$

$$\sum_{i=1}^{n} D_i f^i\Big(x, u, \underset{1}{u}, \underset{2}{u}, \ldots, \underset{k-1}{u}\Big) = 0. \tag{2.2}$$

If $n = 2$, let $x_1 = x$, $x_2 = t$. Through (2.2), one can introduce auxiliary potential variables v and form potential system $S\{u, v\}$ given by

$$f^1 = \frac{\partial v}{\partial t}, \qquad f^2 = -\frac{\partial v}{\partial x}. \tag{2.3}$$

If $n \geqslant 3$, then through (2.2), one can introduce n auxiliary potential variables $v = (v^1, v^2, \ldots, v^n)$ and up to $\frac{1}{2}n(n-1)(n-2)$ nontrivial constants $\{\alpha_{ijk}\}$ and form a gauge-dependent auxiliary system (potential system) $S\{u, v\}$ of $N+n-1$ PDEs:

$$\begin{aligned} f^i\Big(x, u, \underset{1}{u}, \underset{2}{u}, \ldots, \underset{k-1}{u}\Big) &= \sum_{j,k=1}^{n} \in_{ijk} \alpha_{ijk} \frac{\partial v^j}{\partial x_k}, \quad i = 1, 2, \ldots, n, \\ G^\sigma\Big(x, u, \underset{1}{u}, \underset{2}{u}, \ldots, \underset{k}{u}\Big) &= 0, \quad \sigma = 1, 2, \ldots, N-1, \end{aligned} \tag{2.4}$$

where $\in_{ijk}$ is the permutation symbol and $\{\alpha_{ijk}\}$ satisfies conditions

$$\alpha_{ijk} = \alpha_{kji}, \tag{2.5}$$

$$\sum_{j,k=1}^{n} |\in_{ijk} \alpha_{ijk}| \neq 0, \quad i = 1, 2, \ldots, n. \tag{2.6}$$

Note that conditions (2.5), (2.6) effectively mean that for particular choices of gauge, one need only have $n-1$ potential variables.

More generally, one can introduce $\frac{1}{2}n(n-1)$ potential variables $v = (\Psi^{12}, \Psi^{13}, \ldots, \Psi^{1n}, \Psi^{23}, \ldots, \Psi^{2n}, \ldots, \Psi^{n-1,n})$, where Ψ^{ij} $(i<j)$ are components of an antisymmetric tensor, such that

$$f^i\Big(x, u, \underset{1}{u}, \underset{2}{u}, \ldots, \underset{k-1}{u}\Big) = \sum_{i<j}(-1)^j \frac{\partial \Psi^{ij}}{\partial x_j} + \sum_{j<i}(-1)^{i-1}\frac{\partial \Psi^{ji}}{\partial x_j}, \quad i,j = 1,2,\ldots,n, \tag{2.7}$$

and form a corresponding potential system of $N+n-1$ PDEs with $m+\frac{1}{2}n(n-1)$ dependent variables $u = (u^1, u^2, \ldots, u^m)$, Ψ^{ij} $(i<j)$. Since (2.7) is underdetermined, one can impose suitable constraints (a choice of gauge) on the potentials Ψ^{ij} to make system (2.7) a determined system [18].

If $(u(x), v(x))$ solves $S\{u,v\}$, then $u(x)$ solves $R\{u\}$; if $u(x)$ solves $R\{u\}$, then through integrability conditions (2.3), (2.4), or (2.7) there exists some (nonunique) $v(x)$ such that $(u(x), v(x))$ solves $S\{u,v\}$. Since $v(x)$ is not unique, it follows that an invertible point or contact transformation in (x,u)-space could yield a noninvertible nonlocal transformation in (x,u,v)-space and, vice-versa, an invertible point transformation in (x,u,v)-space could yield a noninvertible nonlocal transformation in (x,u)-space. Consequently, the study of potential system $S\{u,v\}$ through qualitative or quantitative methods which are not coordinate-dependent, may yield new results for $R\{u\}$ and vice-versa. In particular, a symmetry of $S\{u,v\}$ $(R\{u\})$ defines a symmetry of $R\{u\}$ $(S\{u,v\})$; a point symmetry of $S\{u,v\}$ $(R\{u\})$ could induce a nonlocal symmetry of $R\{u\}$ $(S\{u,v\})$.

DEFINITION 2.1. A *potential symmetry* of $R\{u\}$, related to potential system $S\{u,v\}$, is a point symmetry of $S\{u,v\}$ which does not project onto a point symmetry of $R\{u\}$.

The proof of the following theorem follows immediately:

THEOREM 2.2. *A potential symmetry of $R\{u\}$ is a nonlocal symmetry of $R\{u\}$. In particular, suppose*

$$\begin{aligned}\mathbf{X}^S &= \left[\alpha^\mu(x,u,v) - \xi_i^S(x,u,v)u_i^\mu\right]\frac{\partial}{\partial u^\mu} + \\ &\quad + \left[\beta^\nu(x,u,v) - \xi_i^S(x,u,v)v_i^\nu\right]\frac{\partial}{\partial v^\nu}\end{aligned}$$

is a point symmetry of $S\{u,v\}$. Then $\mathbf{X}^S$ induces a potential symmetry of $R\{u\}$ if and only if at least one component of (α, ξ^S) depend essentially on v; otherwise $\mathbf{X}^S$ projects onto a point symmetry of $R\{u\}$, namely

$$\mathbf{X} = \left[\alpha^\mu(x,u) - \xi_i^S(x,u)u_i^\mu\right]\frac{\partial}{\partial u^\mu}.$$

Conversely,

$$\mathbf{X}^R = [\alpha^\mu(x,u) - \xi_i^R(x,u)u_i^\mu]\frac{\partial}{\partial u^\mu}$$

yields a nonlocal symmetry of $S\{u,v\}$ *if and only if*

$$\mathbf{X}^S = \mathbf{X}^R + \zeta^\nu\Big(x, u, \underset{1}{u}, \underset{1}{v}\Big)\frac{\partial}{\partial v^\nu}$$

is not a point symmetry of $S\{u,v\}$ *for any choice of* ζ.

Now let $v^{(1)} = v$; $S^{(1)} = S\{u, v^{(1)}\}$. Suppose a PDE of $S^{(1)}$ is a conservation law. Then one can introduce further potential variables (and gauge constants and/or constraints) $v^{(2)}$ and form potential system $S^{(2)} = S^{(2)}\{u, v^{(1)}, v^{(2)}\}$ of $N + 2(n-1)$ PDEs with $n^{(2)}$ dependent variables, $N + 2(n-1) \leqslant n^{(2)} \leqslant m + n(n-1)$. Point symmetries of $S^{(2)}$ could yield additional nonlocal symmetries of $R\{u\}$. Continuing this process with other conservation laws one could obtain potential variables $v^{(1)}, v^{(2)}, \ldots, v^{(J)}$ and corresponding potential systems $S^{(1)}, S^{(2)}, \ldots, S^{(J)}\{u, v^{(1)}, v^{(2)}, \ldots, v^{(J)}\}$. Potential system $S^{(J)}$ would involve $N + J(n-1)$ PDEs with $n^{(J)}$ dependent variables, $N + J(n-1) \leqslant n^{(J)} \leqslant m + \frac{1}{2}Jn(n-1)$. At any step $J \geqslant 2$, a point symmetry of $S^{(J)}$ could yield a potential symmetry of $S^{(J-1)}$ which is either a point symmetry or nonlocal symmetry of $R\{u\}$. If a potential symmetry of $S^{(J-1)}$ yields a point symmetry of $R\{u\}$, then a 'lost' point symmetry of $R\{u\}$ is 'recovered'. (A point symmetry of $R\{u\}$ is said to be 'lost' in $S^{(K)}$ if it does not induce a point symmetry of $S^{(K)}$.)

3. Conservation Laws, Potential Conservation Laws, and Potential Systems

Up to now, in order to obtain potential systems, we assumed that at least one PDE of a given system is a conservation law. The question of how to construct conservation laws yielding useful potential systems naturally arises. After defining the adjoint of an operator, we state some known theorems concerning the discovery of conservation laws.

DEFINITION 3.1. The *adjoint* of the differential operator $\mathcal{L}[u]$ is the differential operator $\mathcal{L}^*[u]$ which satisfies

$$\int_\Omega V^\sigma \mathcal{L}_\rho^\sigma[u] W^\rho \,\mathrm{d}x = \int_\Omega W^\rho \mathcal{L}_\sigma^{*\rho}[u] V^\sigma \,\mathrm{d}x \tag{3.1}$$

on any domain $\Omega \subset \mathbf{R}^n$, for every pair of k times differentiable functions

$$V(x) = (V^1(x), V^2(x), \ldots, V^N(x)),$$
$$W(x) = (W^1(x), W^2(x), \ldots, W^m(x))$$

with compact support in Ω.

In particular, $\mathcal{L}^*[u]$ is the adjoint of $\mathcal{L}[u]$ if $V^\sigma \mathcal{L}^\sigma_\rho[u] W^\rho - W^\rho \mathcal{L}^{*\rho}_\sigma[u] V^\sigma$ is a divergence expression.

THEOREM 3.2. *Suppose there exists a set of factors (multipliers, characteristics) $\{\lambda^\sigma(x, u, \underset{1}{u}, \underset{2}{u}, \ldots, \underset{p}{u})\}$ where the components of $u(x)$ are arbitrary p-times differentiable functions such that*

$$\sum_{\sigma=1}^{N} \lambda^\sigma G^\sigma = \sum_{i=1}^{n} D_i f^i \tag{3.2}$$

holds for some $\{f^i(x, u, \underset{1}{u}, \underset{2}{u}, \ldots, \underset{q}{u})\}$. Then

$$\sum_{\sigma=1}^{N} \mathcal{L}^{*\rho}_\sigma[u] \lambda^\sigma = 0, \quad \rho = 1, 2, \ldots, m, \tag{3.3}$$

must hold for any solution of $R\{u\}$ and its differential consequences, where $\mathcal{L}^*[u]$ is the adjoint of the Fréchet derivative $\mathcal{L}[u]$ of $R\{u\}$ [5, 23].

THEOREM 3.3. *If the Fréchet derivative of $R\{u\}$ is selfadjoint, i.e. $\mathcal{L}[u] = \mathcal{L}^*[u]$, then system $R\{u\}$ is the set of Euler–Lagrange equations for some variational principle with Lagrangian L* [5, 23].

THEOREM 3.4 (Noether's Theorem [5, 6, 23–25]). *If $L(x, u, \underset{1}{u}, \underset{2}{u}, \ldots, \underset{l}{u})$ is a Lagrangian for $R\{u\}$ then $\{\lambda^\sigma\}$ yields a set of factors for a conservation law of $R\{u\}$ if both*

(1) $$\mathbf{X} = \sum_{\mu=1}^{m} \lambda^\mu\Big(x, u, \underset{1}{u}, \underset{2}{u}, \ldots, \underset{p}{u}\Big) \frac{\partial}{\partial u^\mu}$$

is a local symmetry of $R\{u\}$;

(2) $$\sum_{\sigma=1}^{m} \mathcal{M}_\sigma[u] \lambda^\sigma = \sum_{i=1}^{n} D_i A^i$$

for some $\{A^i(x, u, \underset{1}{u}, \underset{2}{u}, \ldots, \underset{q}{u})\}$, where $\mathcal{M}[u]$ is the Fréchet derivative of Lagrangian L.

If (1) and (2) hold then the resulting conservation law is $\sum_{i=1}^{n} D_i[W^i[u, \lambda] - A^i] = 0$, where

$$\begin{aligned} W^i[u, \lambda] &= \lambda^\gamma \left[\frac{\partial L}{\partial u^\gamma_i} + \cdots + (-1)^{l-1} D_{i_1} \cdots D_{i_{l-1}} \frac{\partial L}{\partial u_{ii_1 \cdots i_{l-1}}} \right] + \\ &\quad + (D_{i_1} \lambda^\gamma) \left[\frac{\partial L}{\partial u^\gamma_{i_1 i}} + \cdots + (-1)^{l-2} D_{i_2} \cdots D_{i_{l-1}} \frac{\partial L}{\partial u^\gamma_{i_1 i i_2 \cdots i_{l-1}}} \right] + \\ &\quad + \cdots + (D_{i_1} \cdots D_{i_{l-1}} \lambda^\gamma) \frac{\partial L}{\partial u^\gamma_{i_1 \cdots i_{l-1} i}}. \end{aligned}$$

Note that if $\mathcal{L}[u]$ is not selfadjoint, then only Theorem 3.2 holds. Unlike Theorem 3.4, it yields no explicit formula for a conservation law of $R\{u\}$.

In principle any conservation law of $R\{u\}$ leads to an associated potential system: Suppose a set of factors $\{\lambda^\sigma(x, u, \underset{1}{u}, \underset{2}{u}, \ldots, \underset{p}{u})\}$ exists with $\lambda^M \neq 0$ so that (3.2) holds.

Consider the new system $\widehat{R}_M\{u\}$ given by

$$G^\sigma = 0, \quad \sigma = 1, 2, \ldots, M-1, M+1, \ldots, N,$$
$$\sum_{i=1}^{n} D_i f^i = 0. \tag{3.4}$$

It follows that each solution of $R\{u\}$ is a solution of $\widehat{R}_M\{u\}$. On the other hand, each solution of $\widehat{R}_M\{u\}$ is a solution of $R\{u\}$ or factor system $\widetilde{R}_M\{u\}$ given by

$$G^\sigma = 0, \quad \sigma = 1, 2, \ldots, M-1, M+1, \ldots, N,$$
$$\lambda^M\left(x, u, \underset{1}{u}, \underset{2}{u}, \ldots, \underset{p}{u}\right) = 0. \tag{3.5}$$

Suppose there are solutions of $\widetilde{R}_M\{u\}$ which are not solutions $R\{u\}$. (This can only happen when $\lambda^M = 0$ has a solution $u(x)$.) Since the solution set of $\widehat{R}_M\{u\}$ is the union of the solution sets of $\widetilde{R}_M\{u\}$ and $R\{u\}$, one would expect $\widehat{R}_M\{u\}$ to lose symmetries of $R\{u\}$. This leads to the consideration of only certain types of factors in order to discover useful potential systems:

DEFINITION 3.5. A *potential factor* is a factor which does not vanish for any $u(x)$, i.e.

$$\lambda^M\left(x, u, \underset{1}{u}, \underset{2}{u}, \ldots, \underset{p}{u}\right) = 0 \tag{3.6}$$

has no solutions $u(x)$.

DEFINITION 3.6. A *potential conservation law* of $R\{u\}$ is a conservation law of $R\{u\}$ arising from a set of factors with at least one potential factor.

DEFINITION 3.7. Let $\widehat{R}_M\{u\}$ be a system (3.4) associated with a potential conservation law of $R\{u\}$ with potential factor λ^M. A corresponding potential system (see (2.3), (2.4), or (2.7)) $\widehat{S}_M\{u, v\}$ is a *useful potential system.*

If a potential system arising from $R\{u\}$ is not a useful potential system, then one would expect it to yield no potential symmetries of $R\{u\}$. For example, let $R\{u\}$ be the linear wave equation

$$u_{xx} - x^{-4} u_{tt} = 0. \tag{3.7}$$

The factor $\lambda = 2u_t$ yields

$$2u_t(u_{xx} - x^{-4}u_{tt}) = D_x(2u_x u_t) - D_t((u_x)^2 + x^{-4}(u_t)^2)$$

and, hence, one obtains conservation law and system $\widehat{R}^1\{u\}$ given by

$$D_x(2u_x u_t) - D_t((u_x)^2 + x^{-4}(u_t)^2) = 0. \tag{3.8}$$

Correspondingly, one has the factor system $\widetilde{R}^1\{u\}$ given by

$$u_t = 0 \tag{3.9}$$

and potential system $\widehat{S}^1\{u, v\}$ given by

$$v_t = 2u_x u_t, \qquad v_x = x^{-4}(u_t)^2 + (u_x)^2. \tag{3.10}$$

On the other hand, factor $\lambda = 1$ yields conservation law

$$D_x(u_x) - D_t(x^{-4}u_t) = 0 \tag{3.11}$$

with corresponding factor system $\widetilde{R}^2\{u\}$ given by the equation

$$1 = 0 \tag{3.12}$$

and potential system $\widehat{S}^2\{u, v\}$ given by

$$v_t = u_x, \qquad v_x = x^{-4}u_t. \tag{3.13}$$

Conservation law (3.8) is not a potential conservation law; in particular $u_t = 0$ has solutions (thus $\lambda = u_t$ is not a potential factor) and almost all solutions of $u_t = 0$ do not solve Equation (3.7). On the other hand, conservation law (3.11) is a potential conservation law ($\widehat{R}^2\{u\} \equiv R\{u\}$; there are no solutions $u(x)$ of Equation (3.12)).

It is interesting to compare the point symmetries of $R\{u\}$, $\widehat{S}^1\{u, v\}$, and $\widehat{S}^2\{u, v\}$ given by (3.7), (3.10), and (3.13), respectively:

(1) $R\{u\}$ admits an infinite-parameter group which leads to its mapping to the wave equation $u_{xt} = 0$ [17, 6].

(2) $\widehat{S}^1\{u, v\}$ admits

$$\mathbf{X}_1^{\widehat{S}^1} = u_t\frac{\partial}{\partial u} + v_t\frac{\partial}{\partial v},$$

$$\mathbf{X}_2^{\widehat{S}^1} = [u + 2(tu_t - xu_x)]\frac{\partial}{\partial u} + 2(tv_t - xv_x)\frac{\partial}{\partial v},$$

$$\mathbf{X}_3^{\widehat{S}^1} = (xu - x^2u_x)\frac{\partial}{\partial u} + (u^2 - x^2v_x)\frac{\partial}{\partial v},$$

$$\mathbf{X}_4^{\widehat{S}^1} = u\frac{\partial}{\partial u} + 2v\frac{\partial}{\partial v},$$

$$\mathbf{X}_5^{\widehat{S}^1} = x\frac{\partial}{\partial u} + 2u\frac{\partial}{\partial v},$$

$$\mathbf{X}_6^{\widehat{S}^1} = \frac{\partial}{\partial u},$$

$$\mathbf{X}_7^{\widehat{S}^1} = \frac{\partial}{\partial v}. \tag{3.14}$$

($\mathbf{X}_4^{\widehat{S}^1}, \mathbf{X}_5^{\widehat{S}^1}, \mathbf{X}_6^{\widehat{S}^1}$, and $\mathbf{X}_7^{\widehat{S}^1}$ are trivial since independent variables are invariant.)

Note that each infinitesimal generator of $\widehat{S}^1\{u,v\}$ is admitted by $\widetilde{R}\{u\}$ ($u_t = 0$) as well as $R\{u\}$. From the form of (3.14) we see that $\widehat{S}^1\{u,v\}$ yields no potential symmetries of $R\{u\}$, as to be expected.

(3) $\widehat{S}^2\{u,v\}$ admits a four-parameter group [17, 6] given by infinitesimal generators

$$\mathbf{X}_1^{\widehat{S}^2} = u_t\frac{\partial}{\partial u} + v_t\frac{\partial}{\partial v}, \qquad \mathbf{X}_2^{\widehat{S}^2} = (tu_t - xu_x)\frac{\partial}{\partial u} + (tv_t - xv_x - 2v)\frac{\partial}{\partial v},$$

$$\mathbf{X}_3^{\widehat{S}^2} = [3tu - xv + (t^2 + x^{-2})u_t - 2xtu_x]\frac{\partial}{\partial u} + \\ + [(t^2 + x^{-2})v_t - 2xtv_x - (tv + x^{-1}u)]\frac{\partial}{\partial v},$$

$$\mathbf{X}_4^{\widehat{S}^2} = u\frac{\partial}{\partial u} + v\frac{\partial}{\partial v}.$$

Infinitesimal generator $\mathbf{X}_3^{\widehat{S}^2}$ yields a potential symmetry of $R\{u\}$.

Factor $\lambda = 2u_t$ yields a conservation law and corresponding potential system $\widehat{S}\{u,v\}$ where $R\{u\}$ is the wave equation

$$u_{xx} - \frac{1}{c^2(x)}u_{tt} = 0.$$

For any wave speed $c(x)$ one can show that this potential system has no potential symmetries [26] and that the point symmetries of $\widehat{S}\{u,v\}$ project onto point symmetries of both $R\{u\}$ and $u_t = 0$. This is to be expected, since $\lambda = 2u_t$ is not a potential factor of $R\{u\}$.

For physical equations one can have factors depending on u yielding potential conservation laws. For example, consider the equation of one-dimensional planar gas dynamics $R\{u\}$ given by ($u = (v, p, \rho)$):

$$G^1 = \rho_t + v\rho_x + \rho v_x = 0, \tag{3.15a}$$

$$G^2 = \rho(v_t + vv_x) + p_x = 0, \tag{3.15b}$$

$$G^3 = \rho(p_t + vp_x) + B(p,\rho)v_x = 0, \tag{3.15c}$$

where $B(p,\rho)$ satisfies some constitutive relation. Let $\lambda = (\lambda^1, \lambda^2, \lambda^3) = (v, 1, 0)$. Then the resulting conservation law is a potential conservation law since $\lambda^2 = 1$ is a potential factor. It yields $\widehat{R}_2\{u\}$ given by the system of PDEs (3.15a, c) and

$$vG^1 + G^2 = D_t(\rho v) + D_x(p + \rho v^2) = 0. \tag{3.16}$$

4. Complete Potential Symmetry Analysis of the Nonlinear Diffusion Equation

As a prototypical example we consider the nonlinear diffusion equation $R\{u\}$ given by

$$u_t = \left(K(u)u_x\right)_x. \tag{4.1}$$

4.1. Potential systems of $R\{u\}$

First we determine all potential systems of (4.1) arising from potential conservation laws for any diffusivity $K(u)$. The Fréchet derivative of (4.1) is given by

$$\begin{aligned} \mathcal{L}[u] &= K(u)D_x^2 + 2K'(u)u_x D_x - D_t + \left[K''(u)u_x^2 + K'(u)u_{xx}\right] \\ &= D_x^2 \cdot K(u) - D_t. \end{aligned} \tag{4.2}$$

The adjoint of (4.2) is $\mathcal{L}^*[u] = K(u)D_x^2 + D_t \neq \mathcal{L}[u]$. From Theorem 3.2, if a factor $\lambda(x,t,u,u_x,u_t)$ yields a conservation law of (4.1), then

$$\mathcal{L}^*[u]\lambda = K(u)D_x^2\lambda + D_t\lambda = 0 \tag{4.3}$$

must hold for any solution of (4.1), including its differential consequences. One can show that for *any* $K(u)$ the only solutions of (4.3) are $\lambda = c_1 x + c_2$ for arbitrary constants c_1, c_2.

Hence there are at most two potential systems arising from potential factors $\lambda = 1$, $\lambda = x$; factor $\lambda = 1$ obviously yields a potential conservation law from the form of (4.1); factor $\lambda = x$ yields the potential conservation law $x[u_t - (K(u)u_x)_x] = D_t[xu] - D_x[x(L(u))_x - L(u)] = 0$, where $K(u) = L'(u)$. Consequently, we obtain useful potential systems

$$S^1\{u,v\}: \begin{cases} v_x = u, \\ v_t = \left(L(u)\right)_x, \end{cases} \tag{4.4}$$

and

$$S^2\{u,V\}: \begin{cases} V_x = xu, \\ V_t = x\left(L(u)\right)_x - L(u). \end{cases} \tag{4.5}$$

4.1.1. *Potential Systems of* $S^1\{u, v\}$

The Fréchet derivative of $S^1\{u, v\}$ is the operator

$$\mathcal{L}[u, v] = \begin{bmatrix} 1 & -D_x \\ K'(u)u_x + K(u)D_x & -D_t \end{bmatrix}$$

with adjoint given by

$$\mathcal{L}^*[u, v] = \begin{bmatrix} 1 & -K(u)D_x \\ D_x & D_t \end{bmatrix}.$$

Then two cases arise out when solving

$$\mathcal{L}^*[u, v] \begin{bmatrix} \lambda^1(x, t, u, v) \\ \lambda^2(x, t, u, v) \end{bmatrix} = 0 \tag{4.6}$$

where $(u(x, t),\ v(x, t))$ is any solution of $S^1\{u, v\}$:

(1) $K(u)$ *arbitrary*: Here one can show that the only solution of (4.6) is $(\lambda^1, \lambda^2) = (0, 1)$, leading to potential system

$$T^1\{u, v, w\}: \quad \begin{cases} v_x = u, \\ w_x = v, \\ w_t = L(u). \end{cases} \tag{4.7}$$

The Fréchet derivative of (4.7) is

$$\mathcal{L}[u, v, w] = \begin{bmatrix} 1 & -D_x & 0 \\ 0 & 1 & -D_x \\ K(u) & 0 & -D_t \end{bmatrix}$$

with adjoint

$$\mathcal{L}^*[u, v, w] = \begin{bmatrix} 1 & 0 & K(u) \\ D_x & 1 & 0 \\ 0 & D_x & D_t \end{bmatrix}.$$

Then one can show that the system

$$\mathcal{L}^*[u, v, w] \begin{bmatrix} \lambda^1(x, t, u, v, w) \\ \lambda^2(x, t, u, v, w) \\ \lambda^2(x, t, u, v, w) \end{bmatrix} = 0, \tag{4.8}$$

where $(u(x, t), v(x, t), w(x, t))$ is any solution of (4.7), only has the trivial solution $(\lambda^1, \lambda^2, \lambda^3) = (0, 0, 0)$.

(2) $K(u) = u^{-2}$: Here $(\lambda^1, \lambda^2) = (u^{-1}F^1(v,t), F^2(v,t))$, where $(F^1(v,t), F^2(v,t))$ are arbitrary functions satisfying the linear system

$$F^1 = \frac{\partial F^2}{\partial v}, \qquad \frac{\partial F^1}{\partial v} = -\frac{\partial F^2}{\partial t}. \tag{4.9}$$

In Section 5, we show how these factors indicate the linearization of the system

$$v_x = u, \qquad v_t = u^{-2}u_x. \tag{4.10}$$

4.1.2. *Potential Systems of* $S^2\{u, V\}$

The Fréchet derivative of $S^2\{u, V\}$ is the operator

$$\mathcal{L}[u,V] = \begin{bmatrix} x & -D_x \\ x(K'(u)u_x + K(u)D_x) - K(u) & -D_t \end{bmatrix}$$

with adjoint given by

$$\mathcal{L}^*[u,V] = \begin{bmatrix} x & -K(u)(2 + xD_x) \\ D_x & D_t \end{bmatrix}.$$

Then two cases arise when seeking factors $(\lambda^1(x,t,u,V), \lambda^2(x,t,u,V))$ satisfying the corresponding adjoint Equations (3.3):

(1) $K(u)$ *arbitrary*: Here the only solution of (3.3) is $(\lambda^1, \lambda^2) = (0, x^{-2})$. These factors yield the potential conservation law ($x^{-2} \neq 0$ if $x \in \mathbf{R}$)

$$x^{-2}\left[V_t - x\left(L(u)\right)_x + L(u)\right] = D_t[x^{-2}V] - D_x\left[x^{-1}L(u)\right] = 0,$$

leading to potential system

$$T^2\{u,V,W\}\colon \begin{cases} W_x = x^{-2}V, \\ W_t = x^{-1}L(u), \\ V_x = xu. \end{cases}$$

It is unnecessary to seek factors for $T^2\{u,V,W\}$, since one can show that $T^1\{u,v,w\}$ and $T^2\{u,V,W\}$ are equivalent through the mapping

$$v = x^{-1}V + W, \qquad w = xW. \tag{4.11}$$

However, $S^1\{u,v\}$ and $S^2\{u,V\}$ are not invertibly equivalent since, as will be seen in Section 4.2, for any $K(u)$ these systems admit point symmetry Lie algebras of different dimension.

(2) $K(u) = u^{-2}$: Here

$$(\lambda^1, \lambda^2) = \left(\frac{1}{xu}, \frac{V}{x^2}\right)$$

is the only other solution of Equations (3.3). Obviously λ^1 is a potential factor. These factors yield the potential conservation law

$$\frac{1}{xu}(V_x - xu) + \frac{V}{x^2}(V_t - u^{-1} - xu^{-2}u_x) = D_x\left[\frac{V}{xu} - x\right] + D_t\left[\frac{V^2}{2x^2}\right] = 0,$$

which, in turn, yields systems

$$\widehat{S}_1^2\colon\ \{u, V\}\colon\quad \begin{cases} V_t = u^{-1} + xu^{-2}u_x, \\ D_x\left(\frac{V}{xu} - x\right) + D_t\left(\frac{V^2}{2x^2}\right) = 0, \end{cases} \tag{4.12}$$

$$\widehat{S}_2^2\colon\ \{u, V\}\colon\quad \begin{cases} V_x = xu, \\ D_x\left(\frac{V}{xu} - x\right) + D_t\left(\frac{V^2}{2x^2}\right) = 0. \end{cases} \tag{4.13}$$

Since λ^2 is not a potential factor, only $\widehat{S}_1^2$ leads to a useful potential system, given by

$$\overline{T}_1^2\{u, V, \mathcal{W}\}\colon\quad \begin{cases} V_t = u^{-1} + xu^{-2}u_x, \\ \mathcal{W}_t = 2\left(x - \frac{V}{xu}\right), \\ \mathcal{W}_x = \frac{V^2}{x^2}. \end{cases} \tag{4.14}$$

The Fréchet derivative of $\overline{T}_1^2\{u, V, \mathcal{W}\}$ is given by

$$\mathcal{L}[u, V, \mathcal{W}] = \begin{bmatrix} xu^{-2}D_x - (u^{-2} + 2xu^{-3}u_x) & -D_t & 0 \\ \frac{2}{xu^2}V & -2(xu)^{-1} & -D_t \\ 0 & 2x^{-2}V & -D_x \end{bmatrix}$$

with adjoint

$$\mathcal{L}^*[u, V, \mathcal{W}] = \begin{bmatrix} -u^{-2}(2 + xD_x) & \frac{2}{xu^2}V & 0 \\ D_t & -2(xu)^{-1} & 2x^{-2}V \\ 0 & D_t & D_x \end{bmatrix}.$$

It turns out that the only solution of

$$\mathcal{L}^*[u, V, \mathcal{W}] \begin{bmatrix} \lambda^1(x, t, u, V, \mathcal{W}) \\ \lambda^2(x, t, u, V, \mathcal{W}) \\ \lambda^3(x, t, u, V, \mathcal{W}) \end{bmatrix} = 0,$$

when $(u(x, t), V(x, t), \mathcal{W}(x, t))$ solves (4.14), is $(\lambda^1, \lambda^2, \lambda^3) = (x^{-2}, 0, 0)$. The resulting potential conservation law yields system

$$\overline{U}_1^2\{u, V, \mathcal{W}, \mathcal{Z}\}\colon\quad \begin{cases} \mathcal{W}_x = x^{-2}V^2, \\ \mathcal{W}_t = 2\left(x - \frac{V}{xu}\right), \\ \mathcal{Z}_x = x^{-2}V, \\ \mathcal{Z}_t = -\frac{1}{xu}. \end{cases}$$

One can show that this potential system admits only trivial factors of the form $\lambda^\sigma(x, t, u, V, \mathcal{W}, \mathcal{Z})$.

The factors and potential systems arising for the nonlinear diffusion equation $R\{u\}$ are summarized by the following diagrams:

Case I: $K(u)$ arbitrary

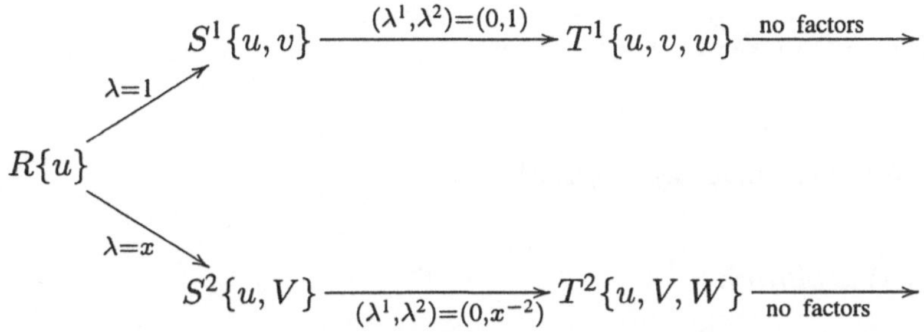

Case II: $K(u) = u^{-2}$

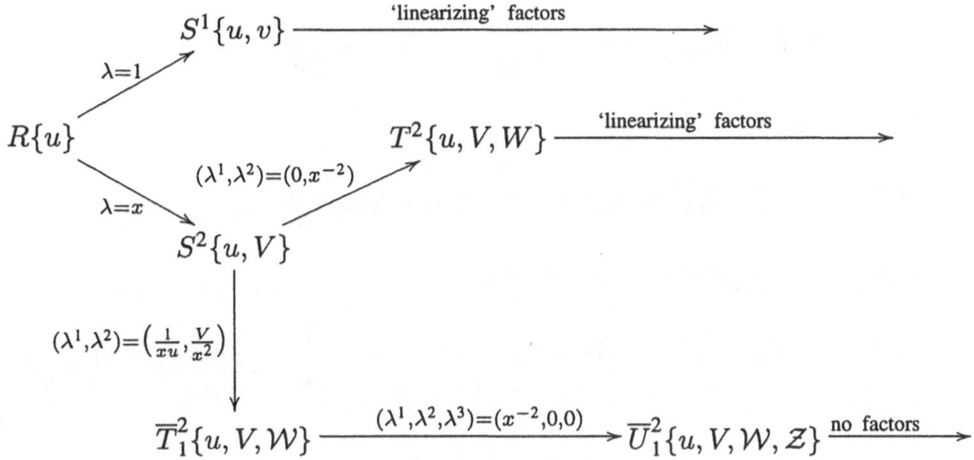

4.2. SYMMETRY CLASSIFICATION OF $R\{u\}$

The infinitesimal generators of symmetries of $R\{u\}$ arising from point symmetries of systems $R\{u\}$, $S^1\{u, v\}$, $T^1\{u, v, w\}$, $S^2\{u, V\}$, $\overline{T}_1^2\{u, V, \mathcal{W}\}$, and $\overline{U}_1^2\{u, V, \mathcal{W}, \mathcal{Z}\}$ depend on the form of diffusivity $K(u) \neq$ const, modulo scaling and translations in u.

4.2.1. *Point Symmetries of $R\{u\}$*

(1) *$K(u)$ arbitrary*:

$$\mathbf{X}_1^R = u_x \frac{\partial}{\partial u}, \qquad \mathbf{X}_2^R = u_t \frac{\partial}{\partial u}, \qquad \mathbf{X}_3^R = (x u_x + 2t u_t) \frac{\partial}{\partial u}.$$

(2) $K(u) = u^\lambda$:

$$\mathbf{X}_1^R,\ \mathbf{X}_2^R,\ \mathbf{X}_3^R,\ \mathbf{X}_4^R = (2u - \lambda x u_x)\frac{\partial}{\partial u}.$$

(3) $K(u) = u^{-4/3}$:

$$\mathbf{X}_1^R, \ldots, \mathbf{X}_4^R,\ \mathbf{X}_5^R = (3xu + x^2 u_x)\frac{\partial}{\partial u}.$$

4.2.2. *Point Symmetries of* $S^1\{u, v\}$

(1) $K(u)$ *arbitrary*:

$$\mathbf{X}_1^{S^1} = \mathbf{X}_1^R + v_x\frac{\partial}{\partial v}, \qquad \mathbf{X}_2^{S^1} = \mathbf{X}_2^R + v_t\frac{\partial}{\partial v},$$

$$\mathbf{X}_3^{S^1} = \mathbf{X}_3^R + (xv_x + 2tv_t - v)\frac{\partial}{\partial v}, \qquad \mathbf{X}_4^{S^1} = \frac{\partial}{\partial v}.$$

(2) $K(u) = u^\lambda$:

$$\mathbf{X}_1^{S^1}, \ldots, \mathbf{X}_4^{S^1},\ \mathbf{X}_5^{S^1} = \mathbf{X}_4^R + ((2+\lambda)v - \lambda x v_x)\frac{\partial}{\partial v}.$$

(3) $K(u) = \frac{1}{1+u^2}\mathrm{e}^{a \arctan u}$, $a = \mathrm{const}$:

$$\mathbf{X}_1^{S^1}, \ldots, \mathbf{X}_4^{S^1},\ \mathbf{X}_5^{S^1} = [(u^2+1) + vu_x + atu_t]\frac{\partial}{\partial u} + [x + vv_x + atv_t]\frac{\partial}{\partial v}.$$

(4) $K(u) = u^{-2}$:

$$\mathbf{X}_2^{S^1}, \ldots, \mathbf{X}_5^{S^1},\ \mathbf{X}_\infty^{S^1} = [F^1(v,t)u_x + u^2 F^2(v,t)]\frac{\partial}{\partial u} + F^1(v,t)v_x\frac{\partial}{\partial v},$$

where $(F^1(v,t),\ F^2(v,t))$ is an arbitrary solution of the linear system

$$\frac{\partial F^1}{\partial t} = \frac{\partial F^2}{\partial v}, \qquad \frac{\partial F^1}{\partial v} = F^2. \tag{4.15}$$

4.2.3. *Point Symmetries of* $T^1\{u, v, w\}$

(1) $K(u)$ *arbitrary*:

$$\mathbf{X}_1^{T^1} = \mathbf{X}_1^{S^1} + w_x\frac{\partial}{\partial w}, \qquad \mathbf{X}_2^{T^1} = \mathbf{X}_2^{S^1} + w_t\frac{\partial}{\partial w},$$

$$\mathbf{X}_3^{T^1} = \mathbf{X}_3^{S^1} + (xw_x + 2tw_t - 2w)\frac{\partial}{\partial w}, \qquad \mathbf{X}_4^{T^1} = \mathbf{X}_4^{S^1} + x\frac{\partial}{\partial w},$$

$$\mathbf{X}_5^{T^1} = \frac{\partial}{\partial w}.$$

(2) $K(u) = u^\lambda$:

$$\mathbf{X}_1^{T^1}, \ldots, \mathbf{X}_5^{T^1},\ \mathbf{X}_6^{T^1} = \mathbf{X}_5^{S^1} + [2(1+\lambda)w - \lambda x w_x]\frac{\partial}{\partial w}.$$

(3) $K(u) = \frac{1}{1+u^2}\,\mathrm{e}^{a \arctan u}$, $a = \text{const}$:

$$\mathbf{X}_1^{T^1}, \ldots, \mathbf{X}_5^{T^1},\ \mathbf{X}_6^{T^1} = \mathbf{X}_5^{S^1} + [\tfrac{1}{2}(x^2 - v^2) + vw_x + atw_t]\frac{\partial}{\partial w}.$$

(4) $K(u) = u^{-2}$:

$$\mathbf{X}_2^{T^1}, \ldots, \mathbf{X}_6^{T^1},\ \mathbf{X}_\infty^{T^1} = \mathbf{X}_\infty^{S^1} + [F^3(v,t) - vF^1(v,t) + F^1(v,t)w_x]\frac{\partial}{\partial w},$$

where $(F^1(v,t), F^2(v,t), F^3(v,t))$ is an arbitrary solution of the linear system

$$\frac{\partial F^3}{\partial v} = F^1, \qquad \frac{\partial F^3}{\partial t} = F^2, \qquad \frac{\partial F^1}{\partial v} = F^2. \tag{4.16}$$

(5) $K(u) = u^{-4/3}$:

$$\mathbf{X}_1^{T^1}, \ldots, \mathbf{X}_6^{T^1},\ \mathbf{X}_7^{T^1} = \mathbf{X}_5^{R} + (xv - w + x^2 v_x)\frac{\partial}{\partial v} + (x^2 w_x - xw)\frac{\partial}{\partial w},$$

(6) $K(u) = u^{-2/3}$:

$$\mathbf{X}_1^{T^1}, \ldots, \mathbf{X}_6^{T^1},\ \mathbf{X}_7^{T^1} = (3uv + wu_x)\frac{\partial}{\partial u} + (v^2 + wv_x)\frac{\partial}{\partial v} + ww_x\frac{\partial}{\partial w}.$$

4.2.4. *Point Symmetries of* $S^2\{u, V\}$

(1) $K(u)$ *arbitrary*:

$$\mathbf{X}_1^{S^2} = \mathbf{X}_2^{R} + V_t\frac{\partial}{\partial V}, \qquad \mathbf{X}_2^{S^2} = \mathbf{X}_3^{R} + (xV_x + 2tV_t - 2V)\frac{\partial}{\partial V},$$

$$\mathbf{X}_3^{S^2} = \frac{\partial}{\partial V}.$$

(2) $K(u) = u^\lambda$:

$$\mathbf{X}_1^{S^2},\ \mathbf{X}_2^{S^2},\ \mathbf{X}_3^{S^2},\ \mathbf{X}_4^{S^2} = \mathbf{X}_4^{R} + (2(\lambda+1)V - \lambda x V_x)\frac{\partial}{\partial V}.$$

(3) $K(u) = u^{-4/3}$:

$$\mathbf{X}_1^{S^2}, \ldots, \mathbf{X}_4^{S^2}, \ \mathbf{X}_5^{S^2} = \mathbf{X}_5^{R} + x^2 V_x \frac{\partial}{\partial V}.$$

4.2.5. *Point Symmetries of* $\overline{T}_1^2\{u, V, \mathcal{W}\}$

Here $K(u) = u^{-2}$ with admitted infinitesimal generators

$$\mathbf{X}_1^{\overline{T}_1^2} = \mathbf{X}_1^{S^2} + \mathcal{W}_t \frac{\partial}{\partial \mathcal{W}}, \qquad \mathbf{X}_2^{\overline{T}_1^2} = \mathbf{X}_2^{S^2} + (x\mathcal{W}_x + 2t\mathcal{W}_t - 3\mathcal{W})\frac{\partial}{\partial \mathcal{W}},$$

$$\mathbf{X}_3^{\overline{T}_1^2} = \mathbf{X}_4^{S^2} + 2(x\mathcal{W}_x - \mathcal{W})\frac{\partial}{\partial \mathcal{W}}, \qquad \mathbf{X}_4^{\overline{T}_1^2} = \frac{\partial}{\partial \mathcal{W}}.$$

4.2.6. *Point Symmetries of* $\overline{U}_1^2\{u, V, \mathcal{W}, \mathcal{Z}\}$

Here $K(u) = u^{-2}$ with admitted infinitesimal generators

$$\mathbf{X}_1^{\overline{U}_1^2} = \mathbf{X}_1^{\overline{T}_1^2} + \mathcal{Z}_t \frac{\partial}{\partial \mathcal{Z}}, \qquad \mathbf{X}_2^{\overline{U}_1^2} = \mathbf{X}_2^{\overline{T}_1^2} + (x\mathcal{Z}_x + 2t\mathcal{Z}_t - \mathcal{Z})\frac{\partial}{\partial \mathcal{Z}},$$

$$\mathbf{X}_3^{\overline{U}_1^2} = \mathbf{X}_3^{\overline{T}_1^2} + 2x\mathcal{Z}_x \frac{\partial}{\partial \mathcal{Z}}, \qquad \mathbf{X}_4^{\overline{U}_1^2} = \mathbf{X}_4^{\overline{T}_1^2}, \qquad \mathbf{X}_5^{\overline{U}_1^2} = \frac{\partial}{\partial \mathcal{Z}}.$$

We now analyze the above symmetries in view of the material presented in Sections 2 and 3:

When $K(u) = u^{-4/3}$, the point symmetry $\mathbf{X}_5^R$ is 'lost' in $S^1\{u, v\}$, since it induces no point symmetry of $S^1\{u, v\}$. In particular, $\mathbf{X}_5^R$ induces a nonlocal symmetry of $S^1\{u, v\}$ which is represented by the infinitesimal generator

$$\mathbf{X}_5^{S^1} = \mathbf{X}_5^{R} + (x^2 v_x + xv - D_x^{-1} v)\frac{\partial}{\partial v}.$$

On the other hand, $S^1\{u, v\}$ yields potential symmetries of $R\{u\}$ given by $\mathbf{X}_5^{S^1}$ when $K(u) = \frac{1}{1+u^2} e^{a \arctan u}$, and by $\mathbf{X}_\infty^{S^1}$ when $K(u) = u^{-2}$. The latter symmetry leads directly to the linearization of $R\{u\}$ by a noninvertible mapping [6, 15, 16].

For any $K(u)$, $T^1\{u, v, w\}$ 'covers' $R\{u\}$ and $S^1\{u, v\}$, since the point symmetries of $T^1\{u, v, w\}$ project onto all point symmetries of both $R\{u\}$ and $S^1\{u, v\}$. In particular, the point symmetry $\mathbf{X}_5^R$, 'lost' in $S^1\{u, v\}$, is 'recovered' as a point symmetry of $T^1\{u, v, w\}$. Moreover, if $K(u) = u^{-2/3}$, the point symmetry $\mathbf{X}_7^{T^1}$ yields a potential symmetry of $S^1\{u, v\}$ and a (new) nonlocal symmetry of $R\{u\}$.

For any $K(u)$, the point symmetry $\mathbf{X}_1^R$ is 'lost' in $S^2\{u, V\}$, since it induces no point symmetry of $S^2\{u, V\}$. One can show that $\mathbf{X}_1^R$ induces a nonlocal symmetry of $S^2\{u, V\}$ which is represented by the infinitesimal generator

$$\mathbf{X}^{S^2} = \mathbf{X}_1^R + (xu - D_x^{-1}u)\frac{\partial}{\partial V}.$$

All other point symmetries of $R\{u\}$ induce point symmetries of $S^2\{u, V\}$ and, in turn, $S^2\{u, V\}$ yields no potential symmetries of $R\{u\}$.

Since $T^2\{u, V, W\}$ is equivalent to $T^1\{u, v, w\}$, through mapping (4.11) it follows that each point symmetry of $T^1\{u, v, w\}$ accordingly maps into a point symmetry of $T^2\{u, V, W\}$. In particular, it is interesting to note that the point symmetry $\mathbf{X}_1^R$ 'lost' in $S^2\{u, V\}$ is 'recovered' as the point symmetry

$$\mathbf{X}^{T^2} = \mathbf{X}_1^R + (V_x - x^{-1}V - W)\frac{\partial}{\partial V} + (W_x + x^{-1}W)\frac{\partial}{\partial W}$$

of $T^2\{u, V, W\}$.

Finally, the potential systems $\overline{T}_1^2\{u, V, \mathcal{W}\}$, $\overline{U}_1^2\{u, V, \mathcal{W}, \mathcal{Z}\}$, which only arise for $K(u) = u^{-2}$, are disappointing since they do not 'recover' $\mathbf{X}_1^R$ as a point symmetry, their point symmetries yield no nonlocal symmetries of R{u}, and, unlike $T^2\{u, V, W\}$, do not lead directly to the linearization of $R\{u\}$.

5. Linearizing Factors

Suppose $R\{u\}$ is a linear system of PDEs given by

$$G^\sigma = \sum_{\rho=1}^{m} L_\rho^\sigma[x]u^\rho = 0, \quad \sigma = 1, 2, \ldots, N. \tag{5.1}$$

Let $L^*[x]$ be the adjoint of linear operator $L[x]$. From the definition of the adjoint, the proof of the following theorem is obvious:

THEOREM 5.1. *A set of factors* $\lambda(x) = (\lambda^1(x), \lambda^2(x), \ldots, \lambda^N(x))$ *yields a conservation law for* (5.1) *if and only if*

$$\sum_{\sigma=1}^{N} L_\sigma^{*\rho}[x]\lambda^\sigma(x) = 0, \quad \rho = 1, 2, \ldots, m.$$

Theorem 5.1 combined with the observation that conservation laws are invariant under contact transformations [27] leads one to consider u-dependent factors which yield conservation laws. In particular, if $R\{x, u\}$ is linearizable by an invertible contact transformation it is necessary that it admit u-dependent factors of the form

$$\lambda^\sigma\Big(x, u, \underset{1}{u}, \underset{2}{u}, \ldots, \underset{p}{u}\Big) = \mathbf{A}_\rho^\sigma\Big(x, u, \underset{1}{u}, \underset{2}{u}, \ldots, \underset{p}{u}\Big)F^\rho(X),$$

where

$$\mathbf{A}^{\sigma}_{\rho}\left(x, u, \underset{1}{u}, \underset{2}{u}, \ldots, \underset{p}{u}\right), \quad \sigma, \rho = 1, 2, \ldots, N,$$

are specific functions of the components of $(x, u, \underset{1}{u}, \underset{2}{u}, \ldots, \underset{p}{u})$ and $F(X) = (F^1(X), F^2(X), \ldots, F^N(X))$ are arbitrary functions satisfying a linear system

$$\sum_{\sigma=1}^{N} L^{*\rho}_{\sigma}[X]F^{\sigma} = 0, \quad \rho = 1, 2, \ldots, m; \tag{5.2}$$

$X = (X_1(x, u), X_2(x, u), \ldots, X_n(x, u))$ yields independent variables for a resulting linear system (X can depend on components of $\underset{1}{u}$ in the scalar case) given by

$$\sum_{\rho=1}^{m} L^{\sigma}_{\rho}[X]U^{\rho} = 0, \quad \sigma = 1, 2, \ldots, N,$$

with dependent variables $U = (U^1, U^2, \ldots, U^m)$; $L[X]$ is the adjoint of linear operator $L^*[x]$. This yields necessary conditions for linearizing $R\{u\}$ [22] and leads to the following definition:

DEFINITION 5.2. Factors $\lambda^{\sigma} = \lambda^{\sigma}(x, u, \underset{1}{u}, \underset{2}{u}, \ldots, \underset{p}{u})$, $\sigma = 1, 2, \ldots, N$, are *linearizing factors* for $R\{u\}$ provided corresponding adjoint Equations (3.3) can be expressed in the form (5.2).

If a given system $R\{u\}$ admits linearizing factors, then it is unnecessary to determine a corresponding conservation law since such a conservation law does not help in finding an explicit linearization of $R\{u\}$. In particular one must still apply specific symmetry algorithms [6, 15, 16] to construct linearizations when they exist. We now consider four examples:

5.1. Nonlinear diffusion equation

The nonlinear diffusion system (4.4), for $K(u) = u^{-2}$, admits linearizing factors with arbitrary functions satisfying (4.9). The application of linearization algorithms [6, 15, 16] yields the mapping of (4.4) to the heat equation, which is the adjoint equation of (4.9).

5.2. Burgers' equation

Burgers' equation $u_{xx} - uu_x - u_t = 0$, written in conservation form ($\lambda = 2$) $D_x(2u_x - u^2) - D_t(2u) = 0$, yields potential system $S\{u, v\}$ given by

$$v_x = 2u, \qquad v_t = 2u_x - u^2. \tag{5.3}$$

One can show that (5.3) admits linearizing factors

$$(\lambda^1, \lambda^2) = \mathrm{e}^{-v/4}\Big(\tfrac{1}{2}\,uF^1(x,t) + F^2(x,t),\ F^1(x,t)\Big),$$

where $(F^1(x,t),\ F^2(x,t))$ is an arbitrary solution of the linear system

$$\frac{\partial F^1}{\partial x} = F^2, \qquad \frac{\partial F^1}{\partial t} = -\frac{\partial F^2}{\partial x}. \tag{5.4}$$

Again the application of linearization algorithms [6, 15, 16] yields the mapping of (5.3) to the heat equation, which is the adjoint equation of (5.4).

5.3. Nonlinear Telegraph Equation

The nonlinear telegraph system

$$v_t = u_x, \qquad v_x = u^{-2}u_t + 1 - u^{-1} \tag{5.5}$$

admits linearizing factors $(\lambda^1, \lambda^2) = (F^1(X,T),\ u^{-1}F^2(X,T))$, where $(X,V) = (x-v, t-\log u)$, and $(F^1(X,T),\ F^2(X,T))$ is an arbitrary solution of the linear system

$$\frac{\partial F^1}{\partial X} + \frac{\partial F^2}{\partial T} - F^2 = 0, \qquad \frac{\partial F^1}{\partial T} + \frac{\partial F^2}{\partial X} = 0. \tag{5.6}$$

The application of linearization algorithms [6, 15, 16] yields the mapping of (5.5) to a linear system, which is the adjoint of (5.6).

5.4. Nonlinear Diffusion Equation Revisited

For arbitrary $K(u)$, the nonlinear diffusion system (4.7) admits no factors of the form $\lambda^\sigma(x,t,u,v,w)$, $\sigma = 1,2,3$. However, when $K(u) = u^{-2}$, it does admit linearizing factors

$$(\lambda^1, \lambda^2, \lambda^3) = (u^{-1}F^1(v,t), u^{-2}u_xF^1(v,t) - F^2(v,t), uF^3(v,t)),$$

where $(F^1(v,t), F^2(v,t), F^3(v,t))$ is an arbitrary solution of the linear system

$$\frac{\partial F^1}{\partial v} = F^2, \qquad \frac{\partial F^1}{\partial t} = -\frac{\partial F^2}{\partial v}, \qquad F^3 = -F^1. \tag{5.7}$$

Again, application of linearization algorithm [6, 15, 16] leads to the mapping of the nonlinear diffusion system (4.7), when $K(u) = u^{-2}$, to the adjoint of system (5.7), namely the linear heat equation system.

6. Remarks

Other approaches to obtain nonlocal symmetries by 'covering systems' [28–30] or to obtain linearizations [31] appear to be restricted to PDEs with two independent variables. The approaches presented in this paper to obtain useful potential systems or linearizing factors clearly extend to systems of PDEs with three or more independent variables. A specific example of such a potential system yielding potential symmetries will be presented in a future paper.

References

1. Lie, S.: Über die Integration durch bestimmte Integrable von einer Klasse linearer partieller Differentialgleichungen, *Arch. Math.* **6** (1881), 328–368; also *Gesammelte Abhandlungen*, Vol. III, B. G. Teubner, Leipzig, 1922, pp. 492–523.
2. Ovsiannikov, L. V.: *Group Properties of Differential Equations*, Novosibirsk, 1962 (in Russian).
3. Bluman, G. W. and Cole, J. D.: *Similarity Methods for Differential Equations*, Appl. Math. Sci., 13, Springer-Verlag, New York, 1974.
4. Ovsiannikov, L. V.: *Group Analysis of Differential Equations*, Academic Press, New York, 1982.
5. Olver, P. J.: *Applications of Lie Groups to Differential Equations*, GTM, 107, Springer-Verlag, New York, 1986.
6. Bluman, G. W. and Kumei, S.: *Symmetries and Differential Equations*, Appl. Math. Sci., 81, Springer-Verlag, New York, 1989.
7. Stephani, H.: *Differential Equations: Their Solution Using Symmetries*, Cambridge University Press, Cambridge, 1989.
8. Schwarz, F.: Automatically determining symmetries of partial differential equations, *Computing* **34** (1985), 91–106.
9. Kersten, P. H. M.: *Infinitesimal Symmetries: A Computational Approach*, CWI Tract No. 34, Centrum voor Wiskunde en Informatica, Amsterdam, 1987.
10. Head, A. K.: LIE: A muMATH Program for the Calculation of the LIE Algebra of Differential Equations, CSIRO Division of Material Sciences, Clayton, Australia, 1990, 1992.
11. Reid, G. J.: Algorithms for reducing a system of PDEs to standard form, determining the dimension of its solution space and calculating Taylor series solution, *Europ. J. Appl. Math.* **2** (1991), 293–318.
12. Champagne, B., Hereman, W., and Winternitz, P.: The computer calculation of Lie point symmetries of large systems of differential equations, *Comput. Phys. Comm.* **66** (1991), 319–340.
13. Wolf, T. and Brand, A,: The computer algebra package CRACK for investigating PDEs, in J. F. Pommaret (ed.), *Proc. ERCIM Advanced Course on Partial Differential Equations and Group Theory*, Bonn, 1992.
14. Hereman, W.: Review of symbolic software for the computation of Lie symmetries of differential equations, *Euromath Bull.* **1**(2) (1994), 45–82.
15. Kumei, S. and Bluman, G. W.: When nonlinear differential equations are equivalent to linear differential equations, *SIAM J. Appl. Math.* **42** (1982), 1157–1173.
16. Bluman, G. W. and Kumei, S.: Symmetry-based algorithms to relate partial differential equations: I. Local symmetries, *Europ. J. Appl. Math.* **1** (1990), 189–216.
17. Bluman, G. W. and Kumei, S.: On invariance properties of the wave equation, *J. Math. Phys.* **28** (1987), 307–318.
18. Bluman, G.W., Kumei, S., and Reid, G. J.: New classes of symmetries for partial differential equations, *J. Math. Phys.* **29** (1988), 806–811, 2320.
19. Bluman, G. W. and Kumei, S.: Symmetry-based algorithms to relate partial differential equations: II. Linearization by nonlocal symmetries, *Europ. J. Appl. Math.* **1** (1990), 217–223.
20. Bluman, G.W.: Use and construction of potential symmetries, *J. Math. Comput. Modelling* **8**(10) (1993), 1–14.
21. Bluman, G. W.: Potential symmetries and linearization, in P. A. Clarkson (ed.), *Proc. NATO Advanced Research Workshop: Applications of Analytic and Geometric Methods to Nonlinear Differential Equations*, Kluwer Acad, Publ., 1993, pp. 363–373.

22. Bluman, G. W.: Potential symmetries and equivalent conservation laws, in N. H. Ibragimov *et al.* (eds), *Proc. Workshop Modern Group Analysis: Advanced Analytical and Computational Methods in Mathematical Physics*, Kluwer Acad. Publ., 1993, pp. 71–84.
23. Vinogradov, A. M.: Symmetries and conservation laws of partial differential equations: basic notions and results, *Acta Appl. Math.* **15** (1989), 3–21.
24. Noether, E.: Invariante Variationsprobleme, *Nachr. König. Gesell. Wissen. Göttingen, Math-Phys. Kl.*, (1918), 235–257.
25. Bessel-Hagen, E.: Über die Erhaltungssätze der Elektrodynamik, *Math. Ann.* **84** (1921), 258–276.
26. Ma, A.: Potential symmetries of the wave equation, MS Thesis, University of British Columbia, 1990.
27. Bluman, G. W.: Invariance of conserved forms under contact transformations, Preprint, 1992.
28. Krasil'shchik, I. S. and Vinogradov, A. M.: Nonlocal symmetries and the theory of coverings: an addendum to A. M. Vinogradov, Local symmetries and conservation laws, *Acta Appl. Math.* **2** (1984), 79–96.
29. Krasil'shchik, I. S. and Vinogradov, A. M.: Nonlocal trends in the geometry of differential equations: symmetries, conservation laws, and Bäcklund transformations, *Acta Appl. Math.* **15** (1989), 161–209.
30. Khor'kova, N. G.: Conservation laws and nonlocal symmetries, *Mat. Zametki* **41** (1988), 134–144.
31. Mikhailov, A. V., Shabat, A. B., and Sokolov, V. V.: The symmetry approach to classification of integrable equations, in V. E. Zakharov (ed.), *What Is Integrability*, Springer-Verlag, Berlin, 1991, pp. 115–184.

Acta Applicandae Mathematicae **41:** 45–56, 1995.

A Method for Computing Symmetries and Conservation Laws of Integro-Differential Equations

V. N. CHETVERIKOV
Department of Applied Mathematics, Moscow State Technical University, ul. 2-aja Baumanskaja 5, 107005 Moscow, Russia

and

A. G. KUDRYAVTSEV
Department of Theoretical Problems, Russian Academy of Sciences, ul. Vesnina 12, 121002 Moscow, Russia

(Received: 11 July 1994)

Abstract. A method for computing symmetries and conservation laws of integro-differential equations is proposed. It resides in reducing an integro-differential equation to a system of boundary differential equations and in computing symmetries and conservation laws of this system. A geometry of boundary differential equations is constructed like the differential case. Results of the computation for the Smoluchowski's coagulation equation are given.

Mathematics Subject Classifications (1991): 45K05, 58A20, 58G20, 58G35.

Key words: integro-differential equations, boundary differential equations, jet spaces, symmetries, conservation laws.

1. Introduction

For constructing a theory of symmetries and conservation laws of integro-differential equations (IDEs), it is natural to adopt the ideas from a similar theory for differential equations [1–6] and to use the analogy between these two kinds of equations. But there are several ways to do it. For instance, one can pass to an equivalent system of differential equations in moments with both infinite number of dependent variables and infinite number of equations [7–10]. Or one can solve integro-differential equations for symmetries introducing the concept of functionally independent variables in the integro-differential case and equating the coefficients of these variables to zero [11–14].

For this purpose, we use the concept of covering [6]. Each definite integral in the equation is replaced by the corresponding difference of values of the antiderivative on boundary sets. The integro-differential equation takes the boundary (or functional) differential form (see Section 2 below). We show (Section 3) that a geometric theory of boundary differential equations (BDEs) can be constructed in just the same way as the analogous theory of differential equations [3,5,6]. In particular, this allows to define and to compute higher symmetries of BDEs (and, hence, IDEs). Section 4 contains computational results for the first-order symmetries of the coagulation kinetic equation. In the case of conservation laws, we formulate only basic principles of the theory and give the results of a realization of these principles for the coagulation equation (Section 5).

A more extensive version of this work will be published in *Advances in Soviet Mathematics* [15].

2. Examples of Transformation Integro-Differential Equations into Boundary Differential Equations

The fundamental theorem calculus is the basic technique which we intend to use here.

EXAMPLE 1. Consider the one-dimensional nonlinear integral Gammershtein's equation of the second kind:

$$u(x) = \int_a^b K(x, s, u(s))\mathrm{d}s, \tag{2.1}$$

where $K(x, s, u)$ is a given function, $x \in [a, b]$, $s \in [a, b]$. Introduce a dependent variable v by

$$v_s'(x, s) = K(x, s, u(s)), \qquad v(x, a) = 0. \tag{2.2}$$

Then

$$u(x) = v(x, b), \tag{2.3}$$

and Equation (2.1) is equivalent to the system of Equations (2.2)–(2.3).

Note that the new variable v is nonlocal, because it depends on all values of a solution u on the interval $[a, b]$ (see (2.2)). For this reason, we shall say that the system (2.2)–(2.3) is a covering of Equation (2.1).

The system (2.2)–(2.3) has a boundary form. It involves two independent variables x, s, two dependent variables u, v, and the restrictions of the dependent variable v onto the boundary sets $\{s = a\}$ and $\{s = b\}$. The dependent variable u appears as $u(x)$ and as $u(s)$.

To transform the system further, we introduce the following maps of the manifold $M = [a, b] \times [a, b]$:

$$\begin{aligned} g_a&\colon\ M \longrightarrow M\colon\ (x, s) \longmapsto (x, a), \\ g_b&\colon\ M \longrightarrow M\colon\ (x, s) \longmapsto (x, b), \\ g_c&\colon\ M \longrightarrow M\colon\ (x, s) \longmapsto (s, x). \end{aligned}$$

In addition, let us think of $u(x)$ as of a function in x and s satisfying the equation $u'_s = 0$. Then, using the notation $v_1 = v'_x$, $v_2 = v'_s$, $u_c = g_c{}^*(u)$, etc., the system (2.2)–(2.3) can be written as

$$u_2 = 0, \qquad v_2 = K(x, s, u_c), \qquad v_a = 0, \qquad u = v_b, \tag{2.4}$$

where K is the same as in (2.1).

The symbols u, v, u_1, v_1, u_2, v_2, u_a, v_a, u_b, v_b, u_c, v_c, etc., can be considered as symbols of the coordinates of some generalized jet space (see below). The subscripts 1 and 2 are symbols of the derivatives with respect to x and s, respectively. The letters a, b or c indicate the action of the homomorphisms $g_a{}^*, g_b{}^*$, or $g_c{}^*$.

Like in the case of differential equations, one must learn to prolong any system of the form (2.4). In order to find new (prolonged) equations, we should use two sorts of acts: (a) differentiations and (b) actions of homomorphisms. So for system (2.4), new equations will involve new variables of the form

$$u_{cb} = g_b{}^*[g_c{}^*(u)] = (g_c \circ g_b)^*(u),$$

$$u_{2c} = g_c{}^*(u'_s), \qquad u_{c2} = [g_c{}^*(u)]'_s, \quad \text{etc.}$$

Note that $u_{2c} \neq u_{c2}$, since from the first equation of (2.4), we have $u_{2c} = 0$, but $u_{c2} = [u(s, x)]'_s = u_1(s, x) = u_{1c}$. Hence, it is necessary to consider the whole semigroup generated by the mappings of the form g_a, g_b, g_c and the identity map g_e of M (the identity element of the semigroup) and to take into account some natural principle to rearrange the subscripts in u and v.

For Equation (2.1), the semigroup consists of 14 members

$$g_e, \quad g_a, \quad g_b, \quad g_c, \quad g_a \circ g_c, \quad g_c \circ g_a, \quad g_b \circ g_c, \quad g_c \circ g_b, \quad g_a \circ g_c \circ g_a,$$
$$g_a \circ g_c \circ g_b, \quad g_b \circ g_c \circ g_a, \quad g_b \circ g_c \circ g_b, \quad g_c \circ g_a \circ g_c, \quad g_c \circ g_b \circ g_c$$

with the relations

$$g_a \circ g_b = g_a, \quad g_b \circ g_a = g_b, \quad g_a^2 = g_a, \quad g_b^2 = g_b, \quad g_c^2 = g_e,$$
$$g_a \circ g_c \circ g_b \circ g_c = g_a \circ g_c \circ g_b, \quad g_a \circ g_c \circ g_a \circ g_c = g_a \circ g_c \circ g_a,$$
$$g_b \circ g_c \circ g_b \circ g_c = g_b \circ g_c \circ g_b, \quad g_b \circ g_c \circ g_a \circ g_c = g_b \circ g_c \circ g_a,$$
$$g_c \circ g_a \circ g_c \circ g_a = g_a \circ g_c \circ g_a, \quad g_c \circ g_b \circ g_c \circ g_b = g_b \circ g_c \circ g_b,$$
$$g_c \circ g_a \circ g_c \circ g_b = g_b \circ g_c \circ g_a, \quad g_c \circ g_b \circ g_c \circ g_a = g_a \circ g_c \circ g_b.$$

Let G_1 denote this semigroup.

EXAMPLE 2. Consider the coagulation kinetic equation

$$\frac{\partial u(x,t)}{\partial t} = \frac{1}{2}\int_0^x K(x-z, z)u(x-z, t)u(z,t)\,\mathrm{d}z - $$
$$-u(x,t)\int_0^\infty K(x,z)u(z,t)\,\mathrm{d}z, \tag{2.5}$$

where $K(x, z)$ is a known function, such that $K(z, x) = K(x, z)$ for any $x \geqslant 0$ and $z \geqslant 0$.

Doing like in Example 1, we obtain a system of boundary differential equations with an infinite semigroup. That is why we choose another way and introduce nonlocal variables v, w by the equations

$$\begin{aligned} &v'_z(x, z, t) - v'_x(x, z, t) = K(x, z)u(x, t)u(z, t), \\ &v(z, x, t) = -v(x, z, t), \\ &w'_z(x, z, t) = K(x, z)u(z, t), \\ &w(x, 0, t) = 0. \end{aligned} \tag{2.6}$$

Using notations of Example 1 and $u_3 = u'_t$, we obtain the system

$$\begin{aligned} &v_2 - v_1 = Kuu_c, \qquad v_c = -v, \qquad w_2 = Ku_c, \qquad w_a = 0, \\ &u_2 = 0, \qquad u_3 = -v_a - w_b u, \qquad (xKu)_{cb} = 0. \end{aligned} \tag{2.7}$$

Here the sixth equation has been obtained by rewriting Equation (2.5) in the variables v and w. The last equation follows from the convergence of the second integral in (2.5).

To define the semigroup of the system, it is enough to show how its generators g_a, g_b, g_c act on functions of x, z, t. We define

$$\begin{aligned} g_a^*(f)(x, z, t) &= f(x, 0, t), \qquad g_c^*(f)(x, z, t) = f(z, x, t), \\ g_b^*(f)(x, z, t) &= \lim_{z \to \infty} f(x, z, t), \end{aligned}$$

and consider only the functions f for which the mentioned limit and the analogous limits for $g_c^*(f)$ and $g_c^*(g_b^*(f))$ exist. (Here we think of $g_b^*(f)$ as of a function in x, t.)

Note that the semigroup generated by g_a, g_b, g_c, $g_e = \mathrm{id}_M$ coincides with the one from Example 1.

Remark. The second example demonstrates that there are many BDEs for the same IDE, although solutions of each BDE are in one-to-one correspondence with solutions of the IDE. The form of BDE, the semigroup of boundary maps and even the group of classical symmetries of BDE depend on the choice of nonlocal variables. On the other side, the change of BDE is a nonlocal transformation. If equations $\mathcal{Y}$ and $\mathcal{Y}_1$ connect by a nonlocal transfomation, then a symmetry of $\mathcal{Y}_1$ is called a nonlocal symmetry of $\mathcal{Y}$. Therefore, the group of nonlocal symmetries is independent of the choice of nonlocal variables, and for this approach it is correct to seek a group of not classical but nonlocal symmetries.

3. Generalized Jet Spaces, Boundary Differential Equations, and Their Symmetries

In this section, we generalize certain concepts of the geometry of differential equations to the case of boundary differential equations. Our exposition follows the corresponding part of the book [3], and we turn our attention to the points distinguishing the two theories.

Let π: $E \to M$ be a smooth bundle, $\Gamma(\pi)$ be the totality of sections of the bundle π, G be a semigroup (or a monoid) of certain maps of the base manifold M, such that the identity map of M is a member of G: $\mathrm{id}_M = g_e \in G$. We say that two sections h_1: $M \to E$ and h_2: $M \to E$ are (k, G)-equivalent (in particular, for $k = \infty$) at a point $x \in M$, if for any map $g \in G$ submanifolds $(h_1 \circ g)(M)$ and $(h_2 \circ g)(M)$ are tangent to each other with the order $\geqslant k$ at the point $(h_1 \circ g)(x) = (h_2 \circ g)(x) \in E$. The set of all the sections which are (k, G)-equivalent to a section h at a point x is called the (k, G)-jet of h at the point x and is denoted by $[h]_x^{(k,G)}$. The set of all the (k, G)-jets of sections of a bundle π is a smooth manifold with singularities. We denote this manifold by $J^k(\pi; G)$ and call it the generalized jet space.

Remark. If the semigroup is trivial, i.e. $G = \{\mathrm{id}_M\}$, then $J^k(\pi; G)$ is the ordinary jet space $J^k\pi$.

Introduce coordinates $\{x_i, u^j_{\sigma g}\}$ on $J^\infty(\pi; G)$. Let $\{x_i\}$ be coordinates in a neighborhood U of a point $x \in M$, the module $\Gamma(\pi|_U)$ having a basis, $\{h^j\}$ be coordinates of a section $h \in \Gamma(\pi|_U)$ in this basis, $\sigma = (i_1, \dots, i_n)$ be a multi-index, $g \in G$. The formula

$$u^j_{\sigma g}([h]_x^{(\infty,G)}) = g^*\left(\frac{\partial^{|\sigma|} h^j}{\partial x_1^{i_1} \dots \partial x_n^{i_n}}\right)(x)$$

for various j, σ, g defines functions on $J^\infty(\pi; G)$, which together with $\{x_i\}$ give the coordinate system on $J^\infty(\pi; G)$.

Like in the differential case [3], the manifolds M and $J^k(\pi; G)$, $k = 0, 1, \dots, \infty$, are mutually connected by natural maps

$$\pi_{k,s}\colon J^k(\pi; G) \longrightarrow J^s(\pi; G), \quad k > s, \quad \text{and} \quad \pi_k\colon J^k(\pi; G) \longrightarrow M,$$

where

$$\pi_{k,s}([h]_x^{(k,G)}) = [h]_x^{(s,G)}, \qquad \pi_k([h]_x^{(k,G)}) = x.$$

Each section $h \in \Gamma(\pi)$ defines a section

$$j_k(h)\colon M \longrightarrow J^k(\pi; G)\colon j_k(h)(x) = [h]_x^{(k,G)}.$$

The Cartan distribution $\mathcal{C}$ on $J^k(\pi; G)$ is defined as a linear span of the subspaces

$$L(x_{k+1}) = T_{x_k}(j_k(h)(M)), \qquad \pi_{k+1,k}(x_{k+1}) = x_k.$$

A differential form, which vanishes on the Cartan distribution, is called a Cartan form.

By a system of boundary partial differential equations (or simply an equation) of the order $\leqslant k$ imposed on sections of π with a semigroup G, we mean a submanifold $\mathcal{Y} \subset J^k(\pi; G)$. A section $h \in \Gamma(\pi)$ is a solution of the equation $\mathcal{Y} \subset J^k(\pi; G)$, if $j_k(h)(M) \subset \mathcal{Y}$. The set of all jets $[h]_x^{(k+s,G)}$, such that for any map $g \in G$ the submanifold $(j_k(h) \circ g)(M)$ is tangent to $\mathcal{Y}$ at the point $[h]_x^{(k,G)}$ with the order $\geqslant s$, is denoted by $\mathcal{Y}^{(s)}$ and is called the sth prolongation of the equation $\mathcal{Y}$. We have $\pi_{k+l,k+s}(\mathcal{Y}^{(l)}) \subset \mathcal{Y}^{(s)}$ for $l \geqslant s \geqslant 0$ and define the infinite prolongation of the equation $\mathcal{Y}$ as the inverse limit of the chain of maps

$$\mathcal{Y}^{(0)} \underset{\pi_{k+1,k}}{\longleftarrow} \mathcal{Y}^{(1)} \longleftarrow \cdots \longleftarrow \mathcal{Y}^{(s-1)} \underset{\pi_{k+s,k+s-1}}{\longleftarrow} \mathcal{Y}^{(s)} \longleftarrow \cdots,$$

which is denoted by $\mathcal{Y}^\infty$.

The direct limit of the chain of the injections

$$C^\infty(M) \underset{\pi_0^*}{\longrightarrow} \mathcal{F}_0(\pi; G) \underset{\pi_{1,0}^*}{\longrightarrow} \mathcal{F}_1(\pi; G) \longrightarrow \cdots \longrightarrow \mathcal{F}_k(\pi; G) \underset{\pi_{k+1,k}^*}{\longrightarrow} \mathcal{F}_{k+1}(\pi; G) \longrightarrow \cdots,$$

where $\mathcal{F}_k(\pi; G) = C^\infty(J^k(\pi; G))$, is an $\mathbb{R}$-algebra and is denoted by $\mathcal{F}(\pi; G)$. By smooth functions on $J^\infty(\pi; G)$, we mean elements of $\mathcal{F}(\pi; G)$. Smooth functions on an equation $\mathcal{Y}$ are defined in the same way.

Boundary differential operators and their linearizations are defined by analogy with differential operators. Let $\pi\colon E_\pi \to M$, $\xi\colon E_\xi \to M$ be smooth bundles, $\mathcal{F}_k(\pi, \xi; G)$ be the $\mathbb{R}$-algebra of sections of the bundle $\pi_k{}^*(\xi)$. Every element $\varphi \in \mathcal{F}_k(\pi, \xi; G)$ can be identified with the nonlinear boundary differential operator Δ_φ of degree $\leqslant k$, which maps sections of the bundle π into sections of the bundle ξ, by means of the equality

$$\Delta_\varphi(h) = j_k(h)^*(\varphi),$$

where h is an arbitrary section of the bundle π. If $\varphi \in \mathcal{F}_k(\pi, \xi; G)$, then the linearization l_φ of the corresponding operator Δ_φ can be written in coordinates $\{x_i, u^j_{\sigma g}\}$ on $J^\infty(\pi; G)$ as

$$l_\varphi(\psi) = \sum_{\beta,j,\sigma,g} \frac{\partial \varphi^\beta}{\partial u^j_{\sigma g}} (g^* \circ D_\sigma)(\psi^j),$$

where

$$\psi \in \mathcal{F}_k(\pi, \xi; G), \quad g \in G,$$

$$\sigma = (i_1, \dots, i_n), \qquad D_\sigma = D_1^{i_1} \circ \cdots \circ D_n^{i_n},$$

D_i being the total drivative with respect to x_i.

Any map g from the semigroup G induces a monomorphism g^* of the algebra of functions on $\mathcal{Y}^\infty$. A derivation of the algebra of functions on $\mathcal{Y}^\infty$ which preserves the module of Cartan forms of $\mathcal{Y}^\infty$ and commutes with any monomorphism $g^*, g \in G$ is called a higher symmetry of the equation $\mathcal{Y}$. Any nontrivial symmetry is identified with a section of the bundle $\pi_k{}^*(\pi)$ for some k and π. If a system of equations $\mathcal{Y}$ has the form $\Delta(h) = 0$, where Δ is a boundary differential operator, then the equation for a symmetry $\varphi \in \Gamma(\pi_k{}^*(\pi))$ can be written as

$$l_\Delta^{\mathcal{Y}}(\varphi) = 0, \tag{3.1}$$

where $l_\Delta^{\mathcal{Y}}$ is the restriction of the operator l_Δ to the manifold $\mathcal{Y}^\infty$ (cf. [3, 5]).

4. Group Analysis of the Coagulation Kinetic Equation

The theory presented above will be employed in this and the next section. We shall find the symmetry group of Equation (2.5) and shall use the symmetries for reducing the equation.

The Smoluchowski equation (2.5) describes the time evolution of the size distribution of particles coagulating by two-body collisions. This equation was first applied to small, suspended particles which collide and coagulate by virtue of their Brownian motion and has subsequently been applied to interacting polymers and to other physical systems.

To compute symmetries of this equation we supersede it with the system (2.7) and use formula (3.1), where $\mathcal{Y}$ is the submanifold which is defined in the jet space $J^1(\pi; G_1)$ by Equations (2.7), π and G_1 are the bundle and the semigroup from Example 2 of Section 2. In this case, Equation (3.1) takes the form

$$\begin{aligned} &D_2V - D_1V = K(Uu_c + uU_c), \qquad V_c = -V, \\ &D_2W = KU_c, \qquad W_a = 0, \qquad D_2U = 0, \\ &D_3U = -V_a - W_bu - w_bU, \qquad (xKU)_{cb} = 0, \end{aligned} \tag{4.1}$$

where U, V, W, $U_c = g_c{}^*(U)$, $V_a = g_a{}^*(V)$, etc., are components of the symmetry $X = U\partial_u + V\partial_v + W\partial_w + U_c\partial_{u_c} + \cdots$, ∂_a is a derivation of $\mathcal{F}(\pi; G)$ with respect to a, D_i is the restriction of the corresponding total derivative onto $\mathcal{Y}^\infty$, and the maps g_a, g_b, g_c are described in Example 2.

The computation of the symmetries of the first order is performed in the case when U, V, W are functions of x, z, t, u, u_c, u_1, u_{1c}, v, v_a, v_1, v_{1a}, v_3, v_{3a}, w, w_b, w_1, w_{1b}, w_3, w_{3b}. In other words, we consider only variables which appear in Equations (2.7). We obtain

$$\begin{aligned} U &= \eta - \xi^1 u_1 - \xi^3 u_3, \\ V &= \eta^3 - \xi_c^1 Kuu_c - (\xi^1 + \xi_c^1)v_1 - \xi^3 v_3 = \eta^3 - \xi^1 v_1 - \xi_c^1 v_2 - \xi^3 v_3, \\ W &= \eta^4 - \xi_c^1 Ku_c - \xi^1 w_1 - \xi^3 w_3 = \eta^4 - \xi^1 w_1 - \xi_c^1 w_2 - \xi^3 w_3, \end{aligned} \tag{4.2}$$

where the functions ξ^1, ξ^3, η, η^3, η^4 are defined by the form of the function $K(x, z)$. For an arbitrary K we have

$$\begin{aligned} \xi^1 &= 0, \qquad \xi^3 = -C_1 t + C_2, \\ \eta &= C_1 u, \qquad \eta^3 = 2C_1 v, \qquad \eta^4 = C_1 w. \end{aligned} \tag{4.3}$$

For homogeneous function K, i.e. when $K(\lambda x, \lambda z) = \lambda^\sigma K(x, z)$, we have

$$\begin{aligned} \xi^1 &= Cx, \qquad \xi^3 = -[C_1 + C(1 + \sigma/2)]t + C_2, \\ \eta &= (C_1 - C\sigma/2)u, \ \eta^3 = (2C_1 + C)v, \\ \eta^4 &= [C_1 + C(1 + \sigma/2)]w \end{aligned} \tag{4.4}$$

with arbitrary constants C, C_1, C_2.

It is easy to see that the symmetries (4.2)–(4.4) are obtained by lifting the fields

$$\xi^1 \partial_x + \xi^3 \partial_t + \eta \partial_u$$

from the space of variables (x, t, u) onto the jet space $J^\infty(\pi, G_1)$.

The case of a homogeneous kernel K is of particular interest to specialists [16]. For such a kernel, the symmetry algebra is spanned by operators

$$\begin{aligned} &X_1 = \partial_t, \qquad X_2 = t\partial_t - u\partial_u, \\ &X_3 = x\partial_x - (1 + \sigma/2)t\partial_t - (1/2\sigma)u\partial_u. \end{aligned} \tag{4.5}$$

Like differential equations (see [1]), integro-differential equations can be reduced by means of symmetries. The symmetry algebra (4.5) have been used by authors to reduce Equation (2.5). The reduction of the integro-differential equation (2.5) and of the system (2.7), which is equivalent to (2.5), gives the same results.

The reduced equations for (2.5) are

$$\begin{aligned} &\frac{1}{2} \int_0^x K(x - z, z) F(x - z) F(z) \, dz - \\ &\quad - F(x) \int_0^\infty K(x, z) F(z) \, dz = 0, \end{aligned} \tag{4.6}$$

$$\begin{aligned} -F(x) &= \frac{1}{2} \int_0^x K(x - z, z) F(x - z) F(z) \, dz - \\ &\quad - F(x) \int_0^\infty K(x, z) F(z) \, dz, \end{aligned} \tag{4.7}$$

$$\begin{aligned} &\beta^{-1} \left[\xi F'(\xi) - \left(\alpha - \frac{\sigma}{2} \right) F(\xi) \right] \\ &\quad = \frac{1}{2} \int_0^\xi K(\xi - s, s) F(\xi - s) F(s) \, ds - \\ &\quad - F(\xi) \int_0^\infty K(\xi, s) F(s) \, ds, \end{aligned} \tag{4.8}$$

$$
\begin{aligned}
&-[\xi F'(\xi) + (1+\sigma)F(\xi)] \\
&\quad = \frac{1}{2}\int_0^{\xi} K(\xi - s, s)F(\xi - s)F(s)\,\mathrm{d}s - \\
&\quad -F(\xi)\int_0^{\infty} K(\xi, s)F(s)\,\mathrm{d}s.
\end{aligned} \tag{4.9}
$$

Corresponding fields from the algebra (4.5) and the forms of invariant solutions one can see in Table I.

TABLE I.

Equation	Field	Form of inv. solutions
(4.6)	∂_t	$F(x)$
(4.7)	$t\partial_t - u\partial_u$	$t^{-1}F(x)$
(4.8)	$x\partial_x - \beta t\partial_t + \left(\alpha - \frac{\sigma}{2}\right)u\partial_u$	$t^{-(\alpha-1)/\beta}F(\beta)$
(4.9)	$\partial_t + x\partial_x - (1+\sigma)u\partial_u$	$\mathrm{e}^{-(1+\sigma)t}F(\eta)$

where α is an arbitrary constant,

$$
\beta = \alpha + 1 + \frac{\sigma}{2}, \qquad \xi = xt^{1/\beta}, \qquad \eta = xe^{-t}.
$$

Remark. Equations (4.6) and (4.8), in the case $\sigma/2 - \alpha = 2$, were considered in [17] and [16, 18], respectively.

5. Conservation Laws of Boundary Differential Equations

In this section, we formulate ideas for constructing a theory of conservation laws of boundary differential equations and present a result which a realization of these ideas gives in the case of the coagulation kinetic equation (2.5).

We shall start from the fact that conservation laws are necessary to find integral conserved quantities. Namely, let $\mathcal{Y}$ be a boundary differential equation, the manifold M of independent variables of $\mathcal{Y}$ having the form $M = M_0 \times \mathbb{R}$, where the coordinate on $\mathbb{R}$ corresponds the time variable t. The integral $\int_{M_0} \omega(u)$ of a differential form $\omega(u)$ on M depending on solutions u of $\mathcal{Y}$ is called a conserved quantity if its value is independent of t for any u.

To obtain the equality

$$
\frac{\mathrm{d}}{\mathrm{d}t}\int_{M_0} \omega(u) = \int_{M_0} \frac{\mathrm{d}}{\mathrm{d}t}\omega(u) = 0
$$

for any u without computing the value of the integral we can use only general statements. In the boundary differential case, those statements are the generalized Stokes theorem and the equality

$$
\int_{M_0} f^*(\omega) = I_f \int_{M_0} \omega,
$$

where f is a diffeomorphism of the manifold M_0, the number I_f is equal to -1 if f changes the orientation of M_0 and $+1$ otherwise.

These statements are equivalent to the fact that pairs of forms $f^*(\omega_1) - I_f\omega_1$ and 0, $\sum_i g_i^*(\omega_2)$ and $d\omega_2$ have the same values of integrals over corresponding manifolds (M_0 or ∂M_0), where f is a diffeomorphism of M_0, $\{g_i\}$ are projections of the manifold M_0 onto its boundary ∂M_0, the sets $\{g_i(M_0)\}$ cover the whole ∂M_0 and intersect only at sets of lesser dimension.

Construct a bicomplex whose differentials map differential forms ω_1 and ω_2 into the corresponding pairs of forms (see above). This bicomplex is

$$\begin{array}{ccccccccc}
 & & 0 & & 0 & & 0 & & \\
 & & \uparrow & & \uparrow & & \uparrow & & \\
0 & \longrightarrow & \Lambda^n & \xrightarrow{\partial_1} & \Lambda^n & \xrightarrow{\partial_2} & \Lambda^n & \longrightarrow & \cdots \\
 & & \uparrow{\scriptstyle d} & & \uparrow{\scriptstyle -d} & & \uparrow{\scriptstyle d} & & \\
0 & \longrightarrow & \Lambda^{n-1} & \xrightarrow{\partial_1} & \Lambda^{n-1} & \xrightarrow{\partial_2} & \Lambda^{n-1} & \longrightarrow & \cdots \\
 & & \uparrow & & \uparrow & & \uparrow & & \\
 & & \cdots & & \cdots & & \cdots & &
\end{array}$$

where Λ^i is the module of differential forms on M depending on solutions of $\mathcal{Y}$, the vertical complexes are the de Rham complexes on M. The differentials ∂_i, $i = 1, \dots, n$, are the action on forms of some members of the semigroup ring $\mathbb{Z}(G)$, where G is the semigroup of the equation $\mathcal{Y}$. So

$$\partial_1(\omega) = f^*(\omega) - I_f\omega, \qquad \partial_2(\omega) = \sum_i g_i^*(\omega),$$

where f and $\{g_i\}$ are elements of G and are induced by a diffeomorphism of M_0 and projections M_0 onto ∂M_0 of the above-mentioned form. The cohomologies of the bicomplex in the term $\Lambda_1^n \oplus \Lambda_2^{n-1} \oplus \cdots$, where the subscript references to the number of a column, are called the conservation laws of the equation $\mathcal{Y}$.

If forms $\omega_1 \in \Lambda^n$, $\omega_2 \in \Lambda^{n-1}$, etc., define a conservation law and $\partial_2(\omega_2) = 0$, then the integral of ω_2 over M_0 gives a conserved quantity. Really, if $N = M_0 \times [t_0, t_1]$ and $N_1 = \partial M_0 \times [t_0, t_1]$, then

$$\int_{M_0} \omega_2 \Big|_{t=t_0} + \int_{N_1} \omega_2 - \int_{M_0} \omega \Big|_{t=t_1} = \int_N d\omega_2 = \int_N [f^*(\omega_1) - I_f(\omega_1)] = 0$$

and $\omega_2|_{N_1} = \partial_2(\omega_2)|_{N_1} = 0$.

For Example 2, we have

$$\partial_1 = g_e^* + g_c^*, \qquad \partial_2 = (g_b^* - g_a^*) \circ (g_e^* - g_c^*), \qquad \partial_3 = (g_b^* - g_a^*) \circ g_c^*.$$

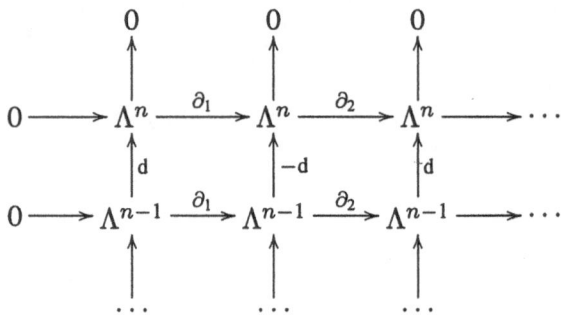

If $(zv)_b = 0$, θ is such a function of z that $\theta(\infty) - \theta(0) = 1$, and θ' is its derivative, then the forms

$$\begin{aligned}\omega_1 &= \tfrac{1}{2}xv_1 \mathrm{d}x \wedge \mathrm{d}z \wedge \mathrm{d}t,\\ \omega_2 &= xu\theta'\mathrm{d}z \wedge \mathrm{d}x + \left(xwu - x\theta(v_a + w_b u) - xv - \tfrac{1}{2}zv\right)\mathrm{d}t \wedge \mathrm{d}x -\\ &\quad -\tfrac{1}{2}\,xv\,\mathrm{d}t \wedge \mathrm{d}z,\end{aligned}$$

define the conservation law of mass

$$\int_0^\infty xu(t,x)\,\mathrm{d}x.$$

We have proved that there are no other conservation laws of Eequation (2.5) which are defined by differential forms on the first jets.

Acknowledgement

We would like to express our gratitude to A. M. Vinogradov for an offer to study these problems and for an advise to use coverings.

References

1. Ovsiannikov, L. V.: *Group Analysis of Differential Equations*, Academic Press, New York, 1982.
2. Ibragimov, N. H.: *Group Transformations in the Mathematical Physics*, Reidel, Dordrecht, 1985.
3. Krasil'shchik, I. S., Lychagin, V. V., and Vinogradov, A. M.: *Geometry of Jet Spaces and Nonlinear Partial Differential Equations*, Gordon and Breach, New York, 1986.
4. Olver, P.: *Applications of Lie Groups to Differential Equations*, Springer, New York, 1986.
5. Vinogradov, A. M.: Local symmetries and conservation laws, *Acta Appl. Math.* **2**(1) (1984), 21–78.
6. Krasil'shchik, I. S. and Vinogradov, A. M.: Nonlocal symmetries and the theory of coverings, *Acta Appl. Math.* **2**(1) (1984), 79–96.
7. Taranov, V. B.: On symmetry of one-dimensional high-freguency motions of collisionless plasma, *Zh. Tekh. Fiz.* **46**(6) (1976), 1271–1277 (in Russian).
8. Bounimovich, A. I. and Krasnoslobodsev, A. V.: Group invariant solutions of kinetic equations, *Izv. Akad. Nauk SSSR, Mekh. Zhidk. Gaza* **4** (1982), 135–140 (in Russian).
9. Bounimovich, A. I. and Krasnoslobodsev, A. V.: On some invariant transformations of kinetic equations, *Vestnik Moskov. Univ., Ser. 1, Mat. Mekh.* **4** (1983), 69–72 (in Russian).
10. Bobylev, A. V.: Exact solutions of the nonlinear Boltzmann equations and the theory of relaxation of the Maxwell gas, *Teoret. Mat. Fiz.* **60**(2) (1984), 280–310 (in Russian).
11. Meleshko, S. V.: Group properties of equations of motions of a viscoelastic medium, *Model. Mekh.* **2 (19)**(4) (1988), 114–126 (in Russian).
12. Grigoriev, Yu. N. and Meleshko, S. V.: Group theoretical analysis of kinetic Boltzmann equation and its models, *Arch. Mech.* **42**(6) (1990), 693–701.
13. Senashov, S. I.: Group classification of an equation of a viscoelastic rod, *Model. Mekh.* **4 (21)**(1) (1990), 69–72 (in Russian).
14. Kovalev, V. F., Krivenko, S. V., and Pustovalov, V. V.: Symmetry group for kinetic equations for collisionless plasma, *Pisma Zh. Eksper. Teoret. Fiz.* **55**(4) (1992), 256–259 (in Russian).
15. Chetverikov, V. N. and Kudryavtsev, A. G.: Modelling integro-differential equations and a method for computing their symmetries and conservation laws, *Adv. Sov. Math.* (to appear in 1995).

16. Drake, R. L.: A general mathematical survey of the coagulation, in G. M. Hidy and J. R. Brock (eds), *Topics in Current Aerosol Research, Part 2*, International Reviews in Aerosol Physics and Chemistry, Vol. 3, Pergamon, New York, 1972.
17. Vinokurov, L. I. and Kac, A. V.: Power solutions of the kinetic equation for the stationary coagulation of atmospheric aerosols, *Izv. Akad. Nauk SSSR, Ser. Fiz. Atmosfer. Okeana* **16**(6) (1980), 601 (in Russian).
18. Lushnikov, A. A.: Evolution of coagulating systems, *J. Colloid Interface Sci.* **45**(3) (1973), 549–556.

Acta Applicandae Mathematicae **41:** 57–98, 1995.

Multiparameter Quantum Groups and Multiparameter R-Matrices

MICHIEL HAZEWINKEL
CWI, PO Box 94079, 1090 GB Amsterdam, The Netherlands

(Received: 16 August 1994)

Abstract. There exists an $\binom{n}{2}+1$ parameter quantum group deformation of GL_n which has been constructed independently by several (groups of) authors. In this note, I give an explicit R-matrix for this multiparameter family. This gives additional information on the nature of this family and facilitates some calculations. This explicit R-matrix satisfies the Yang–Baxter equation. The centre of the paper is Section 3 which describes all solutions of the YBE under the restriction $r^{ab}_{cd}=0$ unless $\{a,b\}=\{c,d\}$. One kind of the most general constituents of these solutions precisely corresponds to the $\binom{n}{2}+1$ parameter quantum group mentioned above. I describe solutions which extend to an enhanced Yang–Baxter operator and, hence, define link invariants. The paper concludes with some preliminary results on these link invariants.

Mathematics Subject Classifications (1991): 16W30, 58F07, 57M25.

Key words: multiparameter quantum group, multiparameter R-matrix, Yang–Baxter equation, Hopf algebra, bialgebra, PBW-algebra, knot invariant, link invariant, confluent rewriting system, fundamental communication relations (FCR), diamond lemma, generalized quantum space.

0. Introduction and Statement of Main Results

This paper is concerned with multiparameter R-matrices and corresponding quantum groups and knot and link invariants. The starting point is an $\binom{n}{2}+1$ parameter deformation of the bialgebra of polynomials on the $n\times n$ matrices

$$K[t^1_1,t^1_2,t^1_3,\ldots,t^2_1,t^2_2,\ldots,t^n_1,\ldots,t^n_n]=K[t],\quad t^i_j\longmapsto t^i_k\otimes t^k_j,\quad \varepsilon(t^i_j)=\delta^i_j,$$

where δ^i_j is the Kronecker delta. Here K is an arbitrary ground field and the Einstein summation convention is in force, i.e. $t^i_k\otimes t^k_j$ stands for $\sum_{k=1}^n t^i_k\otimes t^k_j$. This $\binom{n}{2}+1$ parameter deformation has apparently been independently constructed in various ways by many (groups of) authors, published and unpublished, all more or less in the winter of 1990/1991. I know of several (including myself) and the construction is so natural that quite likely there are more, [1, 3, 5, 9, 11, 14–21]. (Not all these papers deal with the full family and [3], in fact, describes a quantum group which does not fit in this family at all.)

Perhaps the most natural point of view is to take two 'most general' n-dimensional quantum spaces

$$\mathbb{A}=K\langle X^1,\ldots,X^n\rangle/(X^iX^j=q^{ij}X^jX^i),$$

$$\mathbb{B} = K\langle Y_1, \ldots, Y_n \rangle / (Y_i Y_j = q_{ij} Y_j Y_i).$$

Here the $q^{ij} = (q^{ji})^{-1}$, $q^{ii} = 1$, $q_{ji} = (q_{ij})^{-1}$, $q_{ii} = 1$ are arbitrary parameters (viewed as elements of K or as (Laurent) variables). Now look for a maximal quotient $K\langle t \rangle / I$, of $K\langle t \rangle$, $t^i_j \mapsto t^i_k \otimes t^k_j$, to co-act on the left on $\mathbb{A}$ and on the right on $\mathbb{B}$ by the standard formulas

$$X^i \longmapsto t^i_k \otimes X^k, \qquad Y_j \longmapsto Y_k \otimes t^k_j.$$

For the resulting bialgebra $K\langle t \rangle / I$ to be nice, in the sense that the underlying algebra is PBW (Poincaré–Birkhoff–Witt), certain relations must hold between the q^{ij} and q_{kl}, viz. that after a possible permutation of the $1, \ldots, n$ (a renumbering of the variables), $q^{ij} q_{ij} = \rho \neq -1$, for all $i < j$. This material, which can also be found in [1] and other papers, is recalled in Sections 1 and 2 below.

The heart of the paper is Section 3. In it I consider the Yang–Baxter equation

$$R_{12} R_{13} R_{23} = R_{23} R_{13} R_{12}, \quad R = (r^{ab}_{cd}) \tag{0.1}$$

and describe all invertible solutions which satisfy the additional condition

$$r^{ab}_{cd} = 0 \text{ unless } \{a, b\} = \{c, d\}. \tag{0.2}$$

These solutions are constructed with blocks, consisting of several components, which are fitted together in certain not entirely trivial ways, cf. Theorem 3.35 for a precise discription. For instance, a block consisting of three components with only one element looks as follows:

$$\begin{array}{r|ccc|ccc|ccc|}
 & 11 & 12 & 13 & 21 & 22 & 23 & 31 & 32 & 33 \\
11 & \rho_1 & & & & & & & & \\
12 & & x_{21}^{-1} z & & y & & & & & \\
13 & & & x_{31}^{-1} z & & & & y & & \\
\hline
21 & & & & x_{21} & & & & & \\
22 & & & & & \rho_2 & & & & \\
23 & & & & & & x_{32}^{-1} z & & y & \\
\hline
31 & & & & & & & x_{31} & & \\
32 & & & & & & & & x_{32} & \\
33 & & & & & & & & & \rho_3
\end{array} \tag{0.3}$$

where the ρ_1, ρ_2, ρ_3 are all three solutions of $X^2 = yX + z$ (but not necessarily all three are equal). Given any $n^2 \times n^2$ matrix R, there is a natural bialgebra $K\langle t \rangle / I(R)$, $t^i_j \mapsto t^i_k \otimes t^k_j$. Here, $I(R)$ is the ideal generated by the fundamental commutation relations (FCR) of [6]

$$R T_1 T_2 = T_2 T_1 R, \tag{0.4}$$

where $T = (t^i_j)$, $T_1 = T \otimes I_n$, $T_2 = I_n \otimes T$.

Multiplying a solution of (0.1) with an invertible scalar, produces another solution and does not affect the relations defined by (0.4). Thus, the parameter z in (0.3) (or rather its $n^2 \times n^2$ generalization) can be normalized to 1 (by multiplying with $(\sqrt{z})^{-1}$). The two roots of $X^2 = yX + z$ are then q and $-q^{-1}$. If all the ρ_i are now equal to q, the invertible $n^2 \times n^2$ matrix like (0.3) precisely defines the $\binom{n}{2} + 1$-parameter deformation of Sections 1 and 2. This is the main result of Section 4. Having an explicit invertible R-matrix that satisfies the YBE (0.1), for this $\binom{n}{2}+1$-parameter quantum matrix algebra has a number of considerable advantages. For instance, it immediately follows that the rewriting rules (0.4) are confluent which greatly simplifies the proof that this $\binom{n}{2} + 1$ parameter quantum matrix algebra is a PBW algebra. It also helps with the matter of defining a quantum determinant and the definition of an antipode on the bialgebra obtained by making the quantum determinant invertible, thus obtaining an $\binom{n}{2} + 1$ parameter quantum group. This is not further explored here, but see [4, 12, 13, 6].

It also seems from (0.3) that $\binom{n}{2}$ parameters of the $\binom{n}{2}+1$, viz. the x_{ij}, $i > j$, are rather trivial and that there is only one real parameter viz. y (or q; $y = q - q^{-1}$ if $z = 1$). This does not mean that the general quantum matrix algebra ($z = 1$, x_{ij} arbitrary) and the classical one ($z = 1 = x_{ij}$) are isomorphic; they are not. All the same, the x_{ij} do seem less basic than q. I do not know how to make this intuition more precise except in the case of the link invariants defined by the enhanced Yang–Baxter operator that is associated to (0.3), cf. below.

Each block of a solution of (0.1) (assuming (0.2)) defines a scalar. If all those scalars are equal (and only then) the solution gives rise to an enhanced Yang–Baxter operator $(\tau R, \nu, \alpha, \beta)$ in the sense of [22] and, hence, gives rise to a link invariant. In this setting, the $\binom{n}{2}$ extra parameters x_{ij}, $i > j$, are indeed trivial. They do not show up in the link invariant in the sense that if the $n^2 \times n^2$ generalization of (0.3) (even with both q and $-q^{-1}$ occurring for the ρ_i; we are taking $z = 1$) is extended to an enhanced Yang–Baxter operator, which can always be done, than the resulting link invariant is the same as one obtained with all $x_{ij} = 1 = z$ (but possibly a different n). This 'triviality of the x_{ij}' result only applies to 'one type II block' solutions of (0.1). Even in the case of a two size 1 block solution of (0.1), nontrivially fitted together, a nontrivial link invariant appears. Though, of course, the two constituents themselves give nothing. (An $n = 1$ solution of (0.1) always defines a trivial link invariant.) Mixing and fitting together different blocks of both different and the same types seems to promise a rich collection of probably new link invariants. This matter remains to be explored.

As indicated above, the general solution of the Yang–Baxter equation under condition (0.2) consists of blocks which are fitted together in certain ways, each block consisting of several components. In an earlier preprint version of this paper, I mistakenly concluded that each component would be of size one or that a whole block would consist of just one component. This oversight was spotted and corrected by Dr Nico van de Hijligenberg. I am most grateful to him for this and for the considerable amount of work he did in checking the whole manuscript in his characteristic thorough way and the work he put in towards the necessary corrections. In essence, the correction means that in the 'S-formulation' (see Section 5) certain diagonal scalars (those with all four upper and lower indices equal) in the general solution according to the original preprint, can be replaced by scalar matrices (that same scalar multiplying an identity matrix).

1. Generalized Quantum Space $\mathbb{A}_q^n$

The coordinate ring is $K\langle X^1, X^2, \ldots, X^n\rangle/I_n$, where I_n is the ideal generated by the elements

$$X^a X^b - q^{ab} X^b X^a, \tag{1.1}$$

where $q^{ab} = (q^{ba})^{-1}$ and $q^{aa} = 1$ for all $a, b \in \{1, \ldots, n\}$. Thus, depending on one's point of view, $\mathbb{A}_q^n$ is a family of algebras parametrized by $\binom{n}{2}$ parameters or an algebra over $K[q^{ab}, (q^{ab})^{-1}; a > b]$, the ring of commutative Laurent polynomials in $\binom{n}{2}$ variables q^{ab}, $a > b$.

If $q^{ab} = 1$, for all a, b, one refinds the coordinate ring $K[X^1, X^2, \ldots, X^n]$. The algebra $\mathbb{A}_q^n$ is graded and it is a graded deformation of $\mathbb{A}_0^n = K[X^1, \ldots, X^n]$ in the sense that $\dim(\mathbb{A}_q^n)_m = \dim(\mathbb{A}_0^n)_m$ for all q where a lower m indicates the homogeneous part of degree m. Also $\mathbb{A}_q^n$ is a PBW algebra in the sense that the monomials

$$(X^1)^{i_1} \cdots (X^n)^{i_n}, \quad i_j \in \mathbb{N} \cup \{0\} \tag{1.2}$$

form a basis of $\mathbb{A}_q^n$. Indeed it is obvious from (1.1) that every element can be written as a sum of elements of the form (1.2); to prove the other half, it suffices by the diamond lemma, [2], to prove that all the 'overlaps'

$$X^a(X^b X^c), \quad (X^a X^b)X^c$$

are confluent, i.e. give the same results when using the rewriting rules (1.1). Now

$$\begin{aligned} X^a(X^b X^c) &= q^{bc}(X^a X^c)X^b = q^{bc}q^{ac}X^c(X^a X^b) = q^{bc}q^{ac}q^{ab}X^c X^b X^a, \\ (X^a X^b)X^c &= q^{ab}X^b(X^a X^c) = q^{ab}q^{ac}(X^b X^c)X^a = q^{ab}q^{ac}q^{bc}X^c X^b X^a. \end{aligned}$$

So this is indeed the case.

2. Generalized Matrix Quantum Algebras

Consider the left-coaction of

$$K\langle t\rangle = K\langle t_1^1, \ldots, t_n^1; \ldots; t_1^n, \ldots, t_n^n\rangle$$

on

$$K\langle X\rangle = K\langle X^1,\dots,X^n\rangle$$

given by the usual formula

$$X^i \longmapsto t^i_k \otimes X^k \tag{2.1}$$

(summation implied).

Now look at what relations are needed between the t's in order that this becomes a co-action of some quotient of $K\langle t\rangle$ on $\mathbb{A}^n_q$. This means that the relations $X^aX^b = q^{ab}X^bX^a$ must be preserved. The image of $X^aX^b - q^{ab}X^bX^a$ under (2.1) is

$$t^a_{r_1}t^b_{r_2} \otimes X^{r_1}X^{r_2} - q^{ab}t^b_{s_1}t^a_{s_2} \otimes X^{s_1}X^{s_2} \tag{2.2}$$

The coefficient of X^rX^r in (2.2) is

$$t^a_rt^b_r - q^{ab}t^b_rt^a_r \tag{2.3}$$

and the coefficient of X^rX^s, $r < s$ in (2.2) is

$$t^a_rt^b_s - q^{ab}t^b_rt^a_s + (q^{rs})^{-1}t^a_st^b_r - (q^{rs})^{-1}q^{ab}t^b_st^a_r. \tag{2.4}$$

Let us count the number of independent relations.

(i) For $a = b$ no relations arise from (2.3).

(ii) If $a \neq b$, then the relations (2.3) fall in groups of two

$$t^a_rt^b_r = q^{ab}t^b_rt^a_r, \qquad t^b_rt^a_r = q^{ba}t^a_rt^b_r, \tag{2.5}$$

which are equivalent because $q^{ba} = (q^{ab})^{-1}$. Thus, there are precisely

$$n\binom{n}{2} = \frac{1}{2}n^2(n-1)$$

relations resulting from (2.3). And these are independent.

(iii) If $a = b$ in (2.4), no relations result.

(iv) If $r = s$ in (2.4), the relations (2.4) are implied by (2.3).

(v) For $a \neq b$, $r \neq s$, the relations (2.4) fall into groups of four (or groups of two if one takes $r < s$), viz.

$$\begin{aligned}
t^a_rt^b_s - q^{ab}t^b_rt^a_s + (q^{rs})^{-1}t^a_st^b_r - q^{ab}(q^{rs})^{-1}t^b_st^a_r &= 0,\\
t^b_rt^a_s - q^{ba}t^a_rt^b_s + (q^{rs})^{-1}t^b_st^a_r - q^{ba}(q^{rs})^{-1}t^a_st^b_r &= 0,\\
t^a_st^b_r - q^{ab}t^b_st^a_r + (q^{sr})^{-1}t^a_rt^b_s - q^{ab}(q^{sr})^{-1}t^b_rt^a_s &= 0,\\
t^b_st^a_r - q^{ba}t^a_st^b_r + (q^{sr})^{-1}t^b_rt^a_s - q^{ba}(q^{sr})^{-1}t^a_rt^b_s &= 0.
\end{aligned} \tag{2.6}$$

These four relations are all the same, e.g., the second is obtained from the first by multiplication of the first by $-q^{ba}$ and the fourth results from the first by multiplication of the first by $(-q^{ba})(q^{sr})^{-1}$. These relations only involve the four products $t^a_r t^b_s$, $t^b_r t^a_s$, $t^a_s t^b_r$, $t^b_s t^a_r$ and they are the only relations in which these four (for given a, b, r, s) are involved. Thus, there are precisely

$$\frac{n^2(n-1)^2}{4}$$

independent relations of this type. In total we therefore have

$$\tfrac{1}{2}n^2(n-1) + \tfrac{1}{4}n^2(n-1)^2 = \tfrac{1}{4}n^2(n^2-1)$$

quadratic relations.

To make the dimension of the degree two part of $K\langle t\rangle/I$ equal to that of the degree two part of $K[t]$, we need

$$n^4 - \left(n^2 + \frac{n^2(n^2-1)}{2}\right) = \tfrac{1}{2}n^2(n^2-1)$$

relations, so that precisely half of them are missing. There are a variety of ways to add the missing relations. An extremely elegant one is to make $K\langle t\rangle/I$ also act on the right on the dual of the quantum space $\mathbb{A}^n_q$, [16]. This, however, does not result in the most general quantum matrix algebra. To obtain that, consider a second, a priori completely different, quantum space

$$\mathbb{B}^n_q = K\langle X_q, \ldots, X_n\rangle/(X_b X_a = q_{ba} X_a X_b,\ a, b \in \{1, \ldots, n\}) \tag{2.7}$$

on which a suitable quotient of $k\langle t\rangle$ is supposed to act on the right by

$$X_i \longmapsto X_j \otimes t^j_i, \tag{2.8}$$

where, of course, $q_{ba} = q_{ab}^{-1}$, $q_{aa} = 1$, $q_{ab} \neq 0$.

(NB, the q_{ba} are a second set of parameters, which have, a priori, nothing to do with the q^{ab}.) The requirement that the action (2.8) be compatible with the commutation relations $X_b X_a = q_{ba} X_a X_b$ of $\mathbb{B}^n_q$, gives necessary relations on the t^i_j which are completely analogous to those produced by having $k\langle t^i_j\rangle$ act on the left on $K\langle X^a\rangle$ as above. They are

$$t^r_a t^r_b = q_{ab} t^r_b t^r_a, \tag{2.9}$$

$$t^r_a t^s_b - q_{ab} t^r_b t^s_a + (q_{rs})^{-1} t^s_a t^r_b - q_{ab}(q_{rs})^{-1} t^s_b t^r_a = 0. \tag{2.10}$$

In case $q_{ab} = -(q^{ab})^{-1}$, relations (2.4) and (2.10) coincide. But generically they are independent.

2.11. LEMMA. *Let I_L in $K\langle t\rangle$ be the two sided-ideal generated by the elements* (2.3) *and* (2.4), *let I_R be the two-sided ideal generated by the relations* (2.9) *and* (2.10). *Both I_L and I_R are bialgebra ideals in $K\langle t\rangle$ and, hence, so is I, the two sided ideal generated by I_L and I_R together.*

The proof of this is contained in Appendix 1.

Remark. There is also a more elegant way to see that I_L and I_R are bialgebra ideals. Let $A = \mathbb{A}_q^n$. The dual space is $A^! = K\langle X_1, \ldots, X_n\rangle/J$, where J is generated by X_j^2, $X_iX_j = -q^{ij}X_jX_i$. It is now a simple mater to check that $A^! \bullet A$, as defined in [16], is precisely $K\langle t\rangle/I_L$. Now $A^! \bullet A$ is always a bialgebra ([16, Section 5]), for any quadratic algebra A. The results above now brings the additional bit of information that $A^! \bullet A$ is, in fact, the largest quotient of $K\langle t\rangle$ which co-acts on the left on $\mathbb{A}_q^n$.

Assume from now on that $q^{ab} + q_{ba}^{-1} \neq 0$ for all a, b. Then the relations (2.4) and (2.10) combine to give

$$\begin{aligned} t_s^b t_r^a = {} & (q^{sr} + q_{sr}^{-1})^{-1}(q^{ba}q^{sr} - q_{sr}^{-1}q_{ba}^{-1})t_s^a t_r^b + \\ & +(q^{sr} + q_{sr}^{-1})^{-1}(q^{ba} + q_{ba}^{-1})t_r^a t_s^b. \end{aligned} \tag{2.12}$$

Now order the t_b^a as follows. Choose an ordering on the set of indices $\{1, \ldots, n\}$ and define

$$t_b^a < t_d^c \iff \begin{cases} a < c, \\ \text{or } a = c \text{ and } b < d. \end{cases} \tag{2.13}$$

Then it follows from $t_r^a t_s^a = q_{rs} t_s^a t_r^a$ and (2.12) that every monomial in $K\langle t\rangle$ can be written modulo I in the form

$$t_{j_1}^{i_1} t_{j_2}^{i_2} \ldots t_{j_m}^{i_m} \quad t_{j_1}^{i_1} \leqslant t_{j_2}^{i_2} \leqslant \ldots \leqslant t_{j_m}^{i_m}. \tag{2.14}$$

2.15. DEFINITION. An algebra A over K is a PBW algebra if there are elements $x_1, \ldots, x_m$ in A such that the monomials

$$x_1^{r_1} x_2^{r_2} \ldots x_m^{r_m}, \quad r_i \in \mathbb{N} \cup \{0\}$$

form a basis of A over K.

It does not yet follow that $K\langle t\rangle/I$ is a PBW algebra. All we know so far is that (for any ordering of the indices $a, b, \ldots$) the monomials (2.14) generate the algebra and that the monomials of degree 2

$$t_{j_1}^{i_1} t_{j_2}^{i_2} \quad t_{j_1}^{i_1} \leqslant t_{j_2}^{i_2}$$

are independent (as they should be for a PBW algebra).

2.16. EXAMPLE OF A PBW ALGEBRA. Let $\mathfrak{g}$ be a Lie algebra over K and $U\mathfrak{g}$ its universal enveloping algebra. Let $x_1, \ldots, x_m$ be a basis over K for $g \subset U\mathfrak{g}$ (as a vector space). Then by the PBW-theorem (Poincaré–Birkhoff–Witt). The

$$x_1^{r_1} \ldots x_m^{r_m}, \quad r_i \in \mathbb{N} \cup \{0\}$$

are a basis for $U\mathfrak{g}$ over K. Thus, $U\mathfrak{g}$ is a PBW algebra. This is, of course, the result which suggested the phrase 'PBW-algebra'. If $\mathfrak{g}$ is Abelian, then $U\mathfrak{g} = S\mathfrak{g}$ the symmetric algebra of $\mathfrak{g}$ over K, viz.

$$S\mathfrak{g} = K[x_1, \dots, x_m]$$

2.17. THEOREM [1]. *Let K, q_{ab}, q^{ab}, t, I be as before, then $K\langle t\rangle/I$ is a PBW algebra with generators t^i_j, $i,j = 1,\dots,n$ if and only if $q^{ab} + q^{-1}_{ab} \neq 0$ for all a,b and there is a total ordering on the index set I (possibly different from $1 < 2 < \dots < n$) such that*

$$q^{ab}/q_{ba} = q^{cd}/q_{dc} = \rho \neq -1 \quad \text{for all} \quad a < b,\ c < d. \tag{2.18}$$

Thus, we get an $\binom{n}{2} + 1$ parameter family of PBW deformations of the polynomial algebra $K[t^1_1, \dots, t^n_n]$. Note that I is a graded ideal so that $M_q = K\langle t\rangle/I$ is also graded. Give the t^i_j degree 1, then

$$\begin{aligned} \dim(M_q)_r &= \#\left\{(r_1, \dots, r_m) \mid r_i \in \mathbb{N} \cup \{0\},\ \sum_{i=1}^{m} r_i = r\right\} \\ &= \dim K[t^1_1, \dots, t^n_n]_r, \end{aligned}$$

where $m = n^2$, and A_r denotes the homogeneous component of degree r of a graded algebra A.

The Hilbert–Poincaré series of a graded algebra A is by definition equal to

$$H_A(t) = \sum_{r=1}^{\infty} \dim(A_r) t^r. \tag{2.19}$$

Thus, the Hilbert–Poincaré series of every $K\langle t\rangle/I$ satisfying (2.18) is equal to that of the polynomial algebra $K[t]$ and the $M_q = K\langle t\rangle/I$ are a deformation of the graded algebra $K[t]$ in the sense of graded algebras.

2.20. *Proof of the necessity of* (2.18). By the remark just below 2.10, we already know that we must have $q^{ab} + q^{-1}_{ab} \neq 0$ to get the right amount of linear independent monomials of degree 2.

Take $s = a$, $r = b$ in (2.12) to get

$$t^b_a t^a_b = q^{ba} q_{ab} t^a_b t^b_a. \tag{2.21}$$

Now use (2.21) and (2.12) and $t^a_r t^b_r = q^{ab} t^b_r t^a_r$, $t^a_r t^a_s = q_{rs} t^a_s t^a_r$ to calculate $t^c_a t^b_a t^a_b$ in two ways for $a \neq b \neq c \neq a$

$$\begin{aligned} t^c_a(t^b_a t^a_b) &= q^{ba} q_{ab} (t^c_a t^a_b) t^b_a \\ &= q^{ba} q_{ab} (q^{ab} + q^{-1}_{ab})^{-1} (q^{ca} q^{ab} - q^{-1}_{ab} q^{-1}_{ca}) t^a_a (t^c_b t^b_a) + \\ &\quad + q^{ba} q_{ab} (q^{ab} + q^{-1}_{ab})^{-1} (q^{ca} + q^{-1}_{ca}) t^a_b (t^c_a t^b_a) \end{aligned}$$

$$
\begin{aligned}
&= q^{ba}q_{ab}(q^{ab}+q_{ab}^{-1})^{-1}(q^{ca}q^{ab}-q_{ab}^{-1}q_{ca}^{-1})(q^{ba}+q_{ba}^{-1})^{-1}\times\\
&\times(q^{cb}q^{ba}-q_{cb}^{-1}q_{ba}^{-1})t_a^a t_b^b t_a^c+\\
&+q^{ba}q_{ab}(q^{ab}+q_{ab}^{-1})^{-1}(q^{ca}q^{ab}-q_{ab}^{-1}q_{ca}^{-1})\times\\
&\times(q^{ba}+q_{ba}^{-1})^{-1}(q^{cb}+q_{cb}^{-1})t_a^a t_a^b t_b^c+\\
&+q^{ba}q_{ab}(q^{ab}+q_{ab}^{-1})^{-1}(q^{ca}+q_{ca}^{-1})q^{cb}t_b^a t_a^b t_a^c.
\end{aligned}
$$

On the other hand,

$$
\begin{aligned}
(t_a^c t_a^b)t_b^a &= q^{cb}t_a^b(t_a^c t_b^a)\\
&= q^{cb}(q^{ab}+q_{ab}^{-1})^{-1}(q^{ca}q^{ab}-q_{ca}^{-1}q_{ab}^{-1})(t_a^b t_a^a)t_b^c+\\
&\quad+q^{cb}(q^{ab}+q_{ab}^{-1})^{-1}(q^{ca}+q_{ca}^{-1})(t_a^b t_b^a)t_a^c\\
&= q^{cb}(q^{ab}+q_{ab}^{-1})^{-1}(q^{ca}q^{ab}-q_{ca}^{-1}q_{ab}^{-1})q^{ba}t_a^a t_a^b t_b^c+\\
&\quad+q^{cb}(q^{ab}+q_{ab}^{-1})^{-1}(q^{ca}+q_{ca}^{-1})q^{ba}q_{ab}t_b^a t_a^b t_a^c.
\end{aligned}
$$

It follows that the coefficient of $t_a^a t_b^b t_a^c$ must be zero, which gives

$$q^{ca}q^{ab}-q_{ab}^{-1}q_{ca}^{-1}=0 \quad\text{or}\quad q^{cb}q^{ba}-q_{cb}^{-1}q_{ba}^{-1}=0. \tag{2.22}$$

Let $\rho_{ab}=q^{ab}q_{ab}=q^{ab}q_{ba}^{-1}$. Then (2.22) says

$$\rho_{ab}=\rho_{ac} \quad\text{or}\quad \rho_{ab}=\rho_{cb} \tag{2.23}$$

(This holds for all triples $a\neq b\neq c\neq a$.) Choose a fixed i,j say $i=1$, $j=2$ and let $\rho=\rho_{ij}$. Then (2.23) implies

$$\rho_{ab}=\rho \quad\text{or}\quad \rho_{ab}=\rho^{-1}, \qquad \text{for all } a,b \tag{2.24}$$

(but (2.24) is strictly weaker than (2.23)).

If $\rho=\rho^{-1}$ (i.e. $\rho=\pm1$), then for all a,b, $\rho_{ab}=q^{ab}/q_{ba}=\rho$ and any ordering works. If $\rho\neq\rho^{-1}$ define

$$i>j \iff \rho_{ij}=\rho \tag{2.25}$$

Then $i>j$, $j>k\Rightarrow\rho_{ij}=\rho$ and $\rho_{jk}=\rho$, so that by (2.23) (with $a=i$, $b=k$, $c=j$) $\rho_{ik}=\rho$, i.e. $i>k$, proving that the order defined by (2.25) is transitive. For this order, we have

$$\frac{q^{ij}}{q_{ji}}=\rho_{ij}=\rho \quad \text{for } i>j.$$

This finishes the proof of the necessity of Theorem 2.17. The sufficiency can now be handled by the Diamond lemma [2], which says, in this case, that if all the overlaps $(t_b^a t_s^r)t_v^u - t_b^a(t_s^r t_v^u)$ are zero, then the monomials (2.14) are a basis. Though there is a good deal of symmetry which can be exploited, this still involves quite a number of cases and rather lengthy calculations for each case. We shall use a different approach, cf. Corollary 4.25.

3. A Rather General Candidate R-Matrix

Let $R = (r^{ab}_{cd})$ be an $n^2 \times n^2$ matrix over K. In this section, we examine a fairly general R-matrix whose form is inspired by the kind of commutation relations of Section 2 and study when it satisfies the Yang–Baxter equation

$$R_{12}R_{13}R_{23} = R_{23}R_{13}R_{12}. \tag{3.1}$$

Here, $R\colon\ V \otimes V \to V \otimes V$, where V has basis $e^1, \dots, e^n$, is given by

$$R(e^i \otimes e^j) = r^{ij}_{kl} e^k \otimes e^l,$$

$$R_{12} = R \otimes \mathrm{Id}, \qquad R_{23} = \mathrm{Id} \otimes R,$$

and

$$R_{13}(e^i \otimes e^j \otimes e^k) = r^{ik}_{mn} e^m \otimes e^j \otimes e^n.$$

In terms of the entries r^{ab}_{cd} of R, Equation (3.1) says

$$r^{ab}_{k_1k_2} r^{k_1c}_{uk_3} r^{k_2k_3}_{vw} = r^{bc}_{l_1l_2} r^{al_2}_{l_3w} r^{l_3l_1}_{uv}, \tag{3.2}$$

for all $a, b, c, u, v, w \in \{1, 2, \dots, n\}$.

Now consider a general R-matrix with the requirement that

$$r^{ab}_{cd} = 0 \quad \text{unless } \{a, b\} = \{c, d\}. \tag{3.3}$$

Thus, the only possibly nonzero entries are of the form r^{ab}_{ab}, r^{ab}_{ba}, r^{aa}_{aa} (and r^{ba}_{ba}, r^{ba}_{ab}), $a \neq b$.

This is more or less inspired by the commutation relations of Section 2 and, as we shall see in Section 4, it is possible to choose the r^{ab}_{cd} such that the commutation relations of Section 2 are reproduced. It is somewhat remarkable that the requirement that an R-matrix of type (3.3) satisfy YB is practically (but not quite) equivalent to the requirement that it gives the right number of relations in degree 2 and that then these are precisely the commutation relations of Section 2 above.

The following lemma drastically reduces the number of equations (3.2) that must be examined (from n^6 to $6n^3$).

3.4. LEMMA. *Let R be an $n^2 \times n^2$ matrix satisfying* (3.3). *Then both sides of* (3.2) *are zero unless* $\{a, b, c\} = \{u, v, w\}$.

Proof. If a term on the left-hand side of (3.2) is nonzero we must have $\{a, b\} = \{k_1, k_2\}$, $k_3 \in \{k_1, c\}$ so $\{k_1, k_2, k_3\} \subset \{a, b, c\}$. Further $u \in \{k_1, c\}$,

$v, w \in \{k_2, k_3\}$ so $\{u, v, w\} \subset \{k_1, c, k_2, k_3\} = \{k_1, k_2, k_3\} \subset \{a, b, c\}$. Similarly $\{k_2, k_3\} = \{v, w\}$, $k_1 \in \{u, k_3\}$ so $\{k_1, k_2, k_3\} \subset \{u, v, w\}$; $\{a, b\} = \{k_1, k_2\}$, $c \in \{u, k_3\}$ so $\{a, b, c\} \subset \{k_1, k_2, k_3, u\} \subset \{u, v, w\}$.

The argument that for a nonzero term on the right-hand side we must have $\{a, b, c\} = \{u, v, w\} = \{l_1, l_2, l_3\}$ is quite similar. Indeed $\{b, c\} = \{l_1, l_2\}$, $l_3 \in \{a, l_2\}$ so $\{l_1, l_2, l_3\} \subset \{a, b, c\}$; $\{u, v\} = \{l_1, l_3\}$, $w \in \{a, l_2\}$, so $\{u, v, w\} \subset \{l_1, l_2, l_3, a\} \subset \{a, b, c\}$; and $\{l_1, l_3\} = \{u, v\}$, $l_2 \in \{l_3, w\}$, so $\{l_1, l_2, l_3\} \subset \{u, v, w\}$; $\{b, c\} = \{l_1, l_2\}$, $a \in \{l_3, w\}$ so $\{a, b, c\} \subset \{l_1, l_2, l_3, w\} \subset \{u, v, w\}$.

3.5. LEMMA. *Let R be an $n^2 \times n^2$ matrix satisfying* (3.3). *Then*

$$\det(R) = \prod_{i=1}^{n} r_{ii}^{ii} \prod_{i<j} (r_{ij}^{ij} r_{ji}^{ji} - r_{ji}^{ij} r_{ij}^{ji}).$$

Proof. Immediate.

3.6. THE R-EQUATIONS

Many of the equations (3.2), assuming (3.3), are automatically satisfied. Take, for example, $a \neq b \neq c \neq a$, $u = a$, $v = b$, $w = c$. Then the nonzero left-hand terms must have $k_1 = a = u$, $k_3 = c$ and, hence, $k_2 = b$ so the LHS is equal to $r_{ab}^{ab} r_{ac}^{ac} r_{bc}^{bc}$. For the RHS, we must have $l_3 = a$, $l_2 = c$, hence $l_1 = b$ and so the RHS is $r_{bc}^{bc} r_{ac}^{ac} r_{ab}^{ab}$ and so this equation is automatically satisfied. As it turns out, there remain the following equations

$$r_{bc}^{bc}(r_{ba}^{ab} r_{ca}^{ac}) = r_{bc}^{bc}(r_{ba}^{ab} r_{cb}^{bc} + r_{ca}^{ac} r_{bc}^{cb})$$
$$(a \neq b \neq c \neq a, u = b, v = c, w = a), \tag{R1}$$

$$r_{ab}^{ab}(r_{ca}^{ac} r_{ab}^{ba} + r_{ba}^{ab} r_{cb}^{bc}) = r_{ab}^{ab}(r_{cb}^{bc} r_{ca}^{ac})$$
$$(a \neq b \neq c \neq a, u = c, v = a, w = b), \tag{R2}$$

$$r_{ab}^{ab} r_{ba}^{ba} r_{ca}^{ac} + r_{ba}^{ab} r_{ba}^{ab} r_{cb}^{bc} = r_{bc}^{bc} r_{cb}^{cb} r_{ca}^{ac} + r_{cb}^{bc} r_{cb}^{bc} r_{ba}^{ab}$$
$$(a \neq b \neq c \neq a, u = c, v = b, w = a), \tag{R3}$$

$$r_{ac}^{ac} r_{ca}^{ac} r_{ac}^{ca} = 0 \quad (a = b \neq c, u = a, v = c, w = a), \tag{R4}$$

$$r_{aa}^{aa} r_{aa}^{aa} r_{ca}^{ac} = r_{aa}^{aa} r_{ca}^{ac} r_{ca}^{ac} + r_{ac}^{ac} r_{ca}^{ac} r_{ca}^{ca}$$
$$(a = b \neq c, u = c, v = w = a), \tag{R5}$$

$$r_{aa}^{aa} r_{aa}^{aa} r_{ac}^{ca} = r_{aa}^{aa} r_{ac}^{ca} r_{ac}^{ca} + r_{ca}^{ca} r_{ac}^{ac} r_{ac}^{ca}$$
$$(a \neq b = c, u = b, v = b, w = a), \tag{R6}$$

$$r^{ab}_{ba}r^{ba}_{ab}r^{ab}_{ba} = r^{ab}_{ba}r^{ba}_{ab}r^{ba}_{ab} \quad (a = c \neq b, u = a, v = b, w = a) \tag{R7}$$

All the other cases either give nothing or give back one of these seven types of equations. For the complete detailed analysis, cf. Appendix 2.

3.7. A SOLUTION FAMILY

Take

$$r^{ij}_{ij} = x_{ij} \quad \text{for } i > j, \qquad r^{ij}_{ij} = x^{-1}_{ji}\lambda^u\lambda_d \quad \text{for } i < j,$$

$$r^{ii}_{ii} = \lambda^u, \qquad r^{ij}_{ji} = \lambda^u - \lambda_d \quad \text{if } i < j, \qquad r^{ij}_{ji} = 0 \quad \text{for } i > j.$$

It is a straightforward matter to check that these r's satisfy (R1)–(R7).

There are $\binom{n}{2}$ parameters x_{ij}, $i < j$ and two more parameters λ^u, λ_d. One of these can be eliminated by dividing all parameters by an arbitrary number.

Thus, we have here an $\binom{n}{2}+1$ parameter family and this is, in fact, the $\binom{n}{2}+1$ parameter family of Section 2 above. The connections are

$$q^{ab} = x^{-1}_{ab}\lambda^u, \qquad q_{ba} = x^{-1}_{ab}, \quad a > b. \tag{3.8}$$

3.9. 'PARTIAL ORDERING' $\{1,\dots,n\}$

We assume that R is invertible. Define for $a, b \in \{1,\dots,n\}$:

$$a \leqslant b \iff r^{ab}_{ba} \neq 0. \tag{3.10}$$

3.11. LEMMA. *The relation defined by* (3.10) *is a 'partial order'.*

Proof. We have to show transitivity. Let $r^{ab}_{ba} \neq 0 \neq r^{bc}_{cb}$, i.e. $a \leqslant b$, $b \leqslant c$ and we have to show $r^{ac}_{ca} \neq 0$ (which is $a \leqslant c$).

By (R7), there are four cases to be considered

$$r^{ab}_{ba} \neq 0, \qquad r^{ba}_{ab} = 0, \qquad r^{bc}_{cb} \neq 0, \qquad r^{cb}_{bc} = 0, \tag{3.11.1}$$

$$r^{ab}_{ba} \neq 0, \qquad r^{ba}_{ab} = 0, \qquad r^{bc}_{cb} = r^{cb}_{bc} \neq 0, \tag{3.11.2}$$

$$r^{ab}_{ba} = r^{ba}_{ab} \neq 0, \qquad r^{bc}_{cb} \neq 0, \qquad r^{cb}_{bc} = 0, \tag{3.11.3}$$

$$r^{ab}_{ba} = r^{ba}_{ab} \neq 0, \qquad r^{bc}_{cb} = r^{cb}_{bc} \neq 0. \tag{3.11.4}$$

In case (1) by the invertibility of R (cf. Lemma 3.5), also $r^{ab}_{ab} \neq 0 \neq r^{ba}_{ba}$. Hence, by (R2), $r^{ab}_{ba}r^{bc}_{cb} = r^{bc}_{cb}r^{ac}_{ca}$ and, hence, $r^{ac}_{ca} = r^{ab}_{ba} \neq 0$.

In case (2), also $r^{ab}_{ab} \neq 0 \neq r^{ba}_{ba}$ and using (R2) with a and b interchanged gives $r^{bc}_{cb}r^{ab}_{ba} = r^{ac}_{ca}r^{bc}_{cb}$ so that again $r^{ac}_{ca} = r^{ab}_{ba} \neq 0$.

In case (3), by the invertibility of R, $r^{bc}_{bc} \neq 0 \neq r^{cb}_{cb}$ and hence by (R1) $r^{ab}_{ba}r^{ac}_{ca} = r^{ab}_{ba}r^{bc}_{cb}$ and, hence, $r^{ac}_{ca} = r^{bc}_{cb} \neq 0$.

In case (4), suppose that $r_{ca}^{ac} = 0$. Then, by invertibility of R, $r_{ac}^{ac} \neq 0 \neq r_{ca}^{ca}$.

Now use (R3) with b and c interchanged to obtain

$$r_{ac}^{ac} r_{ca}^{ca} r_{ba}^{ab} + r_{ca}^{ac} r_{ca}^{ac} r_{bc}^{cb} = r_{cb}^{cb} r_{bc}^{bc} r_{ba}^{ab} + r_{bc}^{cb} r_{bc}^{cb} r_{ca}^{ac}.$$

By (R4), $r_{cb}^{cb} = r_{bc}^{bc} = 0$ (because $r_{cb}^{bc} r_{bc}^{cb} \neq 0$); hence this would give

$$r_{ac}^{ac} r_{ca}^{ca} r_{ba}^{ab} = 0, \quad \text{i.e. } r_{ba}^{ab} = 0,$$

a contradiction. Hence, $r_{ca}^{ac} \neq 0$, concluding the proof of the lemma. □

We note that the relation $\leqslant$ does not satisfy the antisymmetry, i.e. it does not satisfy: $a \leqslant b$ and $b \leqslant a$ implies $a = b$. For this reason, we wrote 'partial ordering', the consequences of this will be examined in more detail in Section 3.15.

3.12. Blocks

Still assuming that R is invertible, define two indices $a, b \in \{1, \ldots, n\}$ to be *connected* (notation $\sim$) if $a \leqslant b$ or $b \leqslant a$ in the ordering of (3.9) above.

3.13. LEMMA. *Connectedness is an equivalence relation.*

Remark. This is not immediately implied by Lemma 3.11. It adds information, e.g., to the case $a \leqslant b, a \leqslant c$, by stating that then b and c are connected.

Proof of Lemma 3.13. Suppose that $a \sim b$ and $b \sim c$, we prove that $a \sim c$. There are four cases to consider

$$r_{ba}^{ab} \neq 0,\ r_{cb}^{bc} \neq 0. \text{ Then } a \leqslant b, b \leqslant c, \text{ hence } a \leqslant c \text{ and } r_{ca}^{ac} \neq 0, \tag{3.13.1}$$

$$r_{ab}^{ba} \neq 0, r_{bc}^{cb} \neq 0. \text{ Then } b \leqslant a, c \leqslant b, \text{ hence } c \leqslant a \text{ and } r_{ac}^{ca} \neq 0. \tag{3.13.2}$$

The other two cases involve more work:

$$r_{ba}^{ab} \neq 0, \qquad r_{bc}^{cb} \neq 0. \tag{3.13.3}$$

As in the case of the proof of Lemma 3.11, there are (by (R7)) four possible subcases to consider.

$$r_{ba}^{ab} \neq 0, \qquad r_{ab}^{ba} = 0, \qquad r_{bc}^{cb} \neq 0, \qquad r_{cb}^{bc} = 0, \tag{3.13.3.1}$$

$$r_{ba}^{ab} \neq 0, \qquad r_{ab}^{ba} = 0, \qquad r_{bc}^{cb} = r_{cb}^{bc} \neq 0, \tag{3.13.3.2}$$

$$r_{ba}^{ab} = r_{ab}^{ba} \neq 0, \qquad r_{bc}^{cb} \neq 0, \qquad r_{cb}^{bc} = 0, \tag{3.13.3.3}$$

$$r_{ba}^{ab} = r_{ab}^{ba} \neq 0, \qquad r_{bc}^{cb} = r_{cb}^{bc} \neq 0. \tag{3.13.3.4}$$

In the last three subcases, Lemma 3.11 is immediately applicable. It remains to deal with (3.13.3.1). In this case, $r_{bc}^{bc} \neq 0$ by invertibility. □

Now use (R2) after the permutation $b \mapsto c \mapsto b$, $a \mapsto a$ to find

$$r_{ac}^{ac}(r_{ba}^{ab}r_{ac}^{ca} + r_{ca}^{ac}r_{bc}^{cb}) = r_{ac}^{ac}(r_{bc}^{cb}r_{ba}^{ab}). \tag{3.13.3.5}$$

Now if $r_{ca}^{ac} = r_{ac}^{ca} = 0$, $r_{ac}^{ac} \neq 0$ by invertibility. Hence, the RHS of (3.13.3.5) is not equal to zero so that also r_{ac}^{ca} or r_{ca}^{ac} must be nonzero, yielding a contradiction. By consequence, $a \sim c$.

The final case is

$$r_{ab}^{ba} \neq 0, \qquad r_{cb}^{bc} \neq 0, \tag{3.13.4}$$

Again there are four subcases

$$r_{ab}^{ba} \neq 0, \qquad r_{ba}^{ab} = 0, \qquad r_{cb}^{bc} \neq 0, \qquad r_{bc}^{cb} = 0, \tag{3.13.4.1}$$

$$r_{ab}^{ba} \neq 0, \qquad r_{ba}^{ab} = 0, \qquad r_{cb}^{bc} = r_{bc}^{cb} \neq 0, \tag{3.13.4.2}$$

$$r_{ab}^{ba} = r_{ba}^{ab} \neq 0, \qquad r_{cb}^{bc} \neq 0, \qquad r_{bc}^{cb} = 0, \tag{3.13.4.3}$$

$$r_{ab}^{ba} = r_{ba}^{ab} \neq 0, \qquad r_{cb}^{bc} = r_{bc}^{cb} \neq 0. \tag{3.13.4.4}$$

Again, Lemma 3.11 immediately takes care of (3.13.4.2)–(3.13.4.4) and only (3.13.4.1) remains. In this case if $r_{ca}^{ac} = r_{ac}^{ca} = 0$, $r_{ac}^{ac} \neq 0$, which by (R1) (with a and b interchanged) would imply $r_{ab}^{ba}r_{cb}^{bc} = 0$, contradicting (3.13.4.1). Hence, $r_{ca}^{ac} \neq 0$ or $r_{ac}^{ca} \neq 0$ and we are done.

3.14. DEFINITION. An equivalence class $B \subset \{1, \ldots, n\}$ under the equivalence relation of connectedness will be called a block.

3.15. STRUCTURE OF BLOCKS I

In this subsection and the next, the structure of blocks is examined. More precisely, if B is a block, the submatrix $R_B = (r_{cd}^{ab})_{a,b,c,d \in B}$ is determined. After that, we will examine how blocks can fit together.

A block is a totally ordered subset of $\{1, \ldots, n\}$. However, due to the lack of the antisymmetry property of the ordering relation $\leqslant$, it is possible that inside a block elements a and b exist that cannot be separated. By this, we mean that there may be elements a and b that satisfy the condition $a \leqslant b$ and $b \leqslant a$. In this case, we will say that a and b are *strongly connected* (notation $a \simeq b$).

3.16. DEFINITION. An equivalence class $C \subset B$ under the equivalence relation of strong connectedness will be called a component.

The first step in constructing the general R-matrix is determining the submatrix $R_C = (r_{cd}^{ab})_{a,b,c,d \in C}$, where C is a component of a block B.

3.17. PROPOSITION. *Let C be a component of a block B, then there is a $\lambda \neq 0$ such that for all $a, b \in C$ $(a \neq b)$:*

$$r_{aa}^{aa} = r_{bb}^{bb} = r_{ba}^{ab} = r_{ab}^{ba} = \lambda, \qquad r_{ab}^{ab} = r_{ba}^{ba} = 0. \tag{3.18}$$

Proof. By assumption $r_{ba}^{ab} \neq 0 \neq r_{ab}^{ba}$. Hence, $r_{ab}^{ab} = r_{ba}^{ba} = 0$ by (R4), and $\lambda = r_{ba}^{ab} = r_{ab}^{ba}$ by (R6). Putting this in (R5) gives

$$r_{aa}^{aa} r_{aa}^{aa} r_{ba}^{ab} = r_{aa}^{aa} r_{ba}^{ab} r_{ba}^{ab}. \tag{3.19}$$

By invertibility of R (cf. Lemma 3.5), $r_{aa}^{aa} \neq 0$. Hence, $r_{aa}^{aa} = r_{ba}^{ab} = \lambda$ and switching a, b, also $r_{bb}^{bb} = \lambda$. Hence, (3.18) holds for these particular $a, b \in C$. Now let $c \in C, a \neq c \neq b$. The same argument as given above can be applied with c substituted for b which proves the proposition.

3.20. STRUCTURE OF BLOCKS II

Let B be a block, it consists of several components $C_1, C_2, \ldots, C_p$. Since all elements of B are connected, we may assume that the components are numbered such that $C_1 < C_2 < \cdots < C_p$, i.e. $i < j$, $a \in C_i$ and $b \in C_j$ implies $a < b$. Here $a < b$ stands for $a \leqslant b$ and not $b \leqslant a$. The structure of the submatrices R_{C_j} follows from the preceding proposition, the next proposition describes the structure of the submatrix R_B.

3.21. PROPOSITION. *Let B be a block with components $C_1 < C_2 < \cdots < C_p$ and let λ_j be the scalar that corresponds to the submatrix R_{C_j} according to Proposition 3.17 (for all $1 \leqslant j \leqslant p$), then there are scalars $y \neq 0$ and $z \neq 0$ such that for all $i < j$, $a \in C_i$ and $b \in C_j$:*

$$r_{ab}^{ba} = 0, \qquad r_{ba}^{ab} = y \quad \textit{and} \quad r_{ab}^{ab} r_{ba}^{ba} = z. \tag{3.22}$$

Furthermore, the scalars λ_j satisfy the quadratic equation

$$(\lambda_j)^2 = y\lambda_j + z. \tag{3.23}$$

Proof. According to Proposition 3.17, we already know that for all $a, b \in C_j$

$$r_{aa}^{aa} = r_{bb}^{bb} = r_{ba}^{ab} = r_{ab}^{ba} = \lambda_j, \qquad r_{ab}^{ab} = r_{ba}^{ba} = 0.$$

We take elements a and b of C_i and c of C_j $(i < j)$, then $a \simeq b < c$ so

$$r_{ca}^{ac} \neq 0 \neq r_{cb}^{bc}, \qquad r_{ac}^{ca} = 0 = r_{bc}^{cb}. \tag{3.24}$$

It follows from (3.24) that

$$r_{bc}^{bc} \neq 0 \neq r_{cb}^{cb}, \qquad r_{ac}^{ac} \neq 0 \neq r_{ca}^{ca}. \tag{3.25}$$

Now use (R1) to see that

$$r_{ca}^{ac} = r_{cb}^{bc} = y_{i,j} \quad \text{(defining } y_{i,j}\text{)}. \tag{3.26}$$

Consider (R5)

$$(\lambda_i)^2 y_{i,j} = \lambda_i (y_{i,j})^2 + r_{ac}^{ac} r_{ca}^{ca} y_{i,j}, \tag{3.27}$$

and similarly with a and b interchanged to find

$$(\lambda_i)^2 y_{i,j} = \lambda_i (y_{i,j})^2 + r_{bc}^{bc} r_{cb}^{cb} y_{i,j} \tag{3.28}$$

which gives us the definition of $z_{i,j}$ as $z_{i,j} = r_{ac}^{ac} r_{ca}^{ca} = r_{bc}^{bc} r_{cb}^{cb}$. Take $a \in C_i$, $b \in C_j$ and $c \in C_k$ with $i < j < k$, then by using (R1) and (R2), it follows that

$$r_{ba}^{ab} = y_{i,j} = r_{cb}^{bc} = y_{j,k} = r_{ca}^{ac} = y_{i,k}. \tag{3.29}$$

By this y is well defined. Using this in (R3) gives

$$r_{ab}^{ab} r_{ba}^{ba} y + y^3 = z_{i,j} y + y^3 = r_{bc}^{bc} r_{cb}^{cb} y + y^3 = z_{j,k} y + y^3 \tag{3.30}$$

and, hence, as $y \neq 0$, $z_{i,j} = z_{j,k}$. Switching b and c in (R3) now gives $z_{i,k} = z_{j,k}$ and this establishes the first part of the proposition. The last part of Proposition 3.21 now follows directly from (R5) and (R6).

3.31. PROPOSITION. *Let $B_1, \ldots, B_m$ be the blocks of $\{1, \ldots, n\}$, then there are z_{st}, $s, t \in \{1, \ldots, m\}$, $z_{st} = z_{ts}$, such that*

$$r_{ab}^{ab} r_{ba}^{ba} = z_{st} \quad \textit{for all} \quad a \in B_s,\ b \in B_t\ (s \neq t). \tag{3.32}$$

Proof. Choose $c \in B_s, d \in B_t$ and set

$$z_{st} = r_{cd}^{cd} r_{dc}^{dc}. \tag{3.33}$$

If $\#B_s = \#B_t = 1$ there is nothing more to prove. If $\#B_s = 1$, $\#B_t > 1$, let $b \in B_t$, $b \neq d$. Then $r_{db}^{bd} \neq 0$ or $r_{bd}^{db} \neq 0$ and in both cases (R3) gives

$$r_{bc}^{bc} r_{cb}^{cb} = r_{cd}^{cd} r_{dc}^{dc}, \tag{3.34}$$

establishing the result in this case. The case $\#B_s > 1$, $\#B_t = 1$ goes the same. Finally, if $a \neq c$, $a \in B_s$, $b \neq d$, $b \in B_t$, then we get again $r_{ba}^{ab} = r_{bc}^{cb}$ and also because $r_{ca}^{ac} \neq 0$ or $r_{ac}^{ca} \neq 0$

$$r_{ab}^{ab} r_{ba}^{ba} = r_{bc}^{bc} r_{cb}^{cb},$$

which combined with (3.33) gives (3.32).

It will now turn out that the various properties which have been derived are, in fact, also sufficient to guarantee a solution of the YBE. This leads to the following description of all invariable solutions of the YBE under the restriction $r^{ab}_{cd} = 0$, unless $\{a, b\} = \{c, d\}$.

3.35. THEOREM. *Divide the set of indices* $\{1, \dots, n\}$ *into blocks and divide these blocks into components. Further choose numbers* $\in K$ *as follows*:

(i) *For each block* B_s *consisting of a single component* C *choose* $\lambda_s \in K$, $\lambda_s \neq 0$.
(ii) *For each block* B_s *with more than one component, choose* $y_s \in K$, $z_s \in K$, $z_s \neq 0$, $y_s \neq 0$ *and for each component* C^s_j *in* B_s *choose a* λ^s_j *satisfying* $(\lambda^s_j)^2 = \lambda^s_j y_s + z_s$.
(iii) *For each two blocks* B_s, B_t, $s \neq t$ *choose* $z_{st} \in K$, $z_{st} \neq 0$, $z_{st} = z_{ts}$.
(iv) *For each* $a, b \in B_s$ *with* $a > b$ *choose* $x_{ab} \in K$, $x_{ab} \neq 0$.
(v) *For each* $a \in B_s$ *and* $b \in B_t$ *with* $s > t$ *choose* $x_{ab} \in K$, $x_{ab} \neq 0$.

Now define the r^{ab}_{cd} *as follows*

(vi) *If* $a, b \in C^s_j \subset B_s$, $a \neq b$, $r^{aa}_{aa} = r^{bb}_{bb} = r^{ab}_{ba} = r^{ba}_{ab} = \lambda^s_j$, $r^{ab}_{ab} = r^{ba}_{ba} = 0$.
(vii) *If* $a, b \in B_s$, $a < b$, $r^{ab}_{ba} = y_s$, $r^{ba}_{ab} = 0$, $r^{ab}_{ab} = z_s x^{-1}_{ba}$, $r^{ba}_{ba} = x_{ba}$.
(viii) *If* $a \in B_s$, $b \in B_t$, $s < t$, $r^{ab}_{ab} = x_{ab}$, $r^{ba}_{ba} = z_{st} x^{-1}_{ab}$, $r^{ab}_{ba} = r^{ba}_{ab} = 0$.
(ix) $r^{ab}_{cd} = 0$ *unless* $\{a, b\} = \{c, d\}$.

Then the r^{ab}_{cd} *thus specified constitute a solution of the* YBE.

Moreover, up to a permutation of $\{1, \dots, n\}$ (*nonunique as a rule*) *every solution satisfying* (ix) *is thus obtained.*

Proof. After a permutation of indices, if necessary, the 'partial order' defined by $a \leqslant b \Leftrightarrow r^{ab}_{ba} \neq 0$ is compatible with the natural order of $\{1, \dots, n\}$. The statement that all solutions under the restriction (ix) are obtained by the recipe (i)–(viii) above is now the content of the lemmas and formulas (3.10)–(3.34). It remains to show that if $R = (r^{ab}_{cd})$ is constructed by this recipe, then it is indeed a solution. This is a fairly straightforward verification of (R1)–(R7).

The six equations (R1). If a, b, c do not all belong to the same block, at most one of the three pairs r^{ab}_{ba}, r^{ba}_{ab}; r^{bc}_{cb}, r^{cb}_{bc}; r^{ac}_{ca}, r^{ca}_{ac} can be nonzero. As each term in an (R1) equation involves a product of elements from different pairs, all terms in an (R1) equation are zero in this case. It remains to check the case that a, b, c all belong to the same block. If they all belong to the same component, then $r^{bc}_{bc} = 0$ and both sides are zero. If they belong to different components, then if $a < b < c$, $r^{cb}_{bc} = 0$ and $r^{ac}_{ca} = r^{bc}_{cb} = y_s$; if $a < c < b$, $r^{bc}_{cb} = 0$ and $r^{ab}_{ba} = y_s = r^{cb}_{bc}$; if $b < a < c$, $r^{ab}_{ba} = 0 = r^{cb}_{bc}$; if $b < c < a$, $r^{ac}_{ca} = 0 = r^{ab}_{ba}$; if $c < a < b$, $r^{ac}_{ca} = 0 = r^{bc}_{cb}$; if $c < b < a$, $r^{ab}_{ba} = 0 = r^{ac}_{ca}$; so (R1) holds in all six cases. If two of them are in the same component, then there also are six cases to be investigated: if $a \simeq b < c$ $r^{ac}_{ca} = r^{bc}_{cb} = y_s$; if $a \simeq c < b$ $r^{ab}_{ba} = r^{cb}_{bc} = y_s$; if

$b < a \simeq c \; r^{cb}_{bc} = r^{ab}_{ba} = 0$; if $c < a \simeq b \; r^{ac}_{ca} = r^{bc}_{cb} = 0$; if $b \simeq c < a$ or $a < b \simeq c$ then $r^{bc}_{bc} = 0$.

The six equations (R2). As in the case of (R1) if a, b, c, $a \neq b \neq c \neq a$, do not all belong to the same block, all terms are zero, and, also again, if a, b, c all belong to the same component, then $r^{ab}_{ab} = 0$. If two of them are in the same component, then if $a \simeq b$ (R2) is trivial since $r^{ab}_{ab} = 0$. If $a \simeq c < b$, $r^{ba}_{ab} = r^{bc}_{cb} = 0$; if $b \simeq c < a$, $r^{ab}_{ba} = r^{ac}_{ca} = 0$; if $a < b \simeq c$, $r^{ab}_{ba} = r^{ac}_{ca} = y_s$ and if $b < a \simeq c \; r^{ba}_{ab} = r^{bc}_{cb} = y_s$. It remains to deal with the case that a, b, c all belong to a block B_s and to different components. If $a < b < c$, $r^{ba}_{ab} = 0$ and $r^{ab}_{ba} = y_s = r^{ac}_{ca}$; if $a < c < b$, $r^{bc}_{cb} = 0 = r^{ba}_{ab}$; if $b < a < c$, $r^{ab}_{ba} = 0$, $r^{ba}_{ab} = y_s = r^{bc}_{cb}$; if $b < c < a$, $r^{ac}_{ca} = 0 = r^{ab}_{ba}$; if $c < a < b$, $r^{ac}_{ca} = 0 = r^{bc}_{cb}$; if $c < b < a$; $r^{ac}_{ca} = 0 = r^{ab}_{ba}$. Thus, (R2) holds in all cases.

The six equations (R3). If a, b, c do not belong to the same block, both the second term on the left and the second term on the right are equal to zero. Take $a \in B_s$, $b \in B_t$, $c \in B_u$, if $s \neq u$ then (R3) is trivial since $r^{ac}_{ca} = 0$ and if $t \neq s = u$, then $r^{ab}_{ab} r^{ba}_{ba} = r^{bc}_{bc} r^{cb}_{cb} = z_{st}$. What remains is the case $s = t = u$. If a, b and c belong to the same component C^s_j both sides are equal to $(\lambda^s_j)^3$ since $r^{ab}_{ab} = r^{bc}_{bc} = 0$. If two of them are in the same component, then again there are six cases to be considered: if $a \simeq b < c$, $y_s(\lambda^s_j)^2 = y_s z_s + \lambda^s_j (y_s)^2$; if $c < a \simeq b$, $r^{ac}_{ca} = r^{bc}_{cb} = 0$; if $c \simeq a < b$, $r^{bc}_{cb} = 0$ and $r^{ab}_{ab} r^{ba}_{ba} = r^{bc}_{bc} r^{cb}_{cb} = z_s$; if $b < c \simeq a$, $r^{ab}_{ba} = 0$ and $r^{ab}_{ab} r^{ba}_{ba} = r^{bc}_{bc} r^{cb}_{cb} = z_s$; if $a < b \simeq c$, $y_s z_s + \lambda^s_j (y_s)^2 = y_s (\lambda^s_j)^2$; if $b \simeq c < a$, $r^{ac}_{ca} = r^{ab}_{ba} = 0$. Finally, if a, b, c all belong to different components of a block B_s the first term on the left and the first term on the right are either equal to zero ($c < a$) or equal to $z_s y_s$ ($a < c$). The other terms are zero unless $a < b < c$ and then both are equal to $(y_s)^3$. By this (R3) holds in all cases.

The two equations (R4). If a and c are not in the same block $r^{ac}_{ca} = 0$. If they are in the same component of a block, $r^{ac}_{ac} = 0$; if they are in the same block but in different components $r^{ac}_{ca} r^{ca}_{ac} = 0$.

The two equations (R5) If a and c are not both in the same block $r^{ac}_{ca} = 0$ and all terms are zero. If a and c are in the same component of a block B_s, $r^{aa}_{aa} = \lambda_s = r^{ac}_{ca}$ and $r^{ac}_{ac} = r^{ca}_{ca} = 0$ so that (R5) holds. Finally, if a and c are in different components of B_s, all terms are zero unless $a > c$ and then $r^{aa}_{aa} = \lambda^s_j$; $r^{ac}_{ca} = y_s$, $r^{ac}_{ac} r^{ca}_{ca} = z_s$ by (viii) and (R5) holds because λ^s_j solves $X^2 = X y_s + z_s$.

The two equations (R6). Exactly the same argument as (R5).

The two equations (R7). $r^{ab}_{ba} r^{ba}_{ab} = 0$ unless a and b belong to the same component of a block B_s and then $r^{ab}_{ba} = r^{ba}_{ab} = \lambda^s_j$.

3.36. Some examples

In case of a solution consisting of only one block we speak of an *irreducible* solution, a solution consisting of several blocks is called *reducible*. There are two kinds of blocks which are rather special. The first is the one that consists of only one component and the second one is build from components that contain only one element, we shall denote these blocks by blocks of type I and type II, respectively.

	11	12	13	21	22	23	31	32	33
11	λ								
12		zx_{21}^{-1}		y					
13			zx_{31}^{-1}				y		
21				x_{21}					
22					λ				
23						zx_{32}^{-1}		y	
31							x_{31}		
32								x_{32}	
33									μ

$n = 3$; one block of type II
$(\lambda^2 = \lambda y + z;\ \mu^2 = \mu y + z;\ \lambda, \mu, x_{ij}, z \neq 0;\ p = 5)$

	11	12	13	21	22	23	31	32	33
11	λ								
12				λ					
13							λ		
21		λ							
22					λ				
23								λ	
31			λ						
32						λ			
33									λ

$n = 3$; one block of type I
$(\lambda \neq 0;\ p = 1)$

	11	12	13	14	21	22	23	24	31	32	33	34	41	42	43	44
11	λ_1															
12		$z_1x_{21}^{-1}$			y_1											
13			$z_{12}x_{31}^{-1}$													
14				$z_{12}x_{41}^{-1}$												
21					x_{21}											
22						λ_1										
23							$z_{12}x_{32}^{-1}$									
24								$z_{12}x_{42}^{-1}$								
31									x_{31}							
32										x_{32}						
33											λ_2					
34												$z_2x_{43}^{-1}$			y_2	
41													x_{41}			
42														x_{42}		
43															x_{43}	
44																μ_2

$n = 4$; two blocks of type II of size 2
($\lambda_1^2 = \lambda_1 y_1 + z_1$; $\lambda_2^2 = \lambda_2 y_2 + z_2$; $\mu_2^2 = \mu_2 y_2 + z_2$; x_{ij}, λ_i, μ_2, z_i, $z_{12} \neq 0$; $p = 11$)

	11	12	13	21	22	23	31	32	33
11	λ_1								
12		$z_{12}x_{21}^{-1}$							
13			$z_{13}x_{31}^{-1}$						
21				x_{21}					
22					λ_2				
23						$z_{23}x_{32}^{-1}$			
31							x_{31}		
32								x_{32}	
33									λ_3

$n = 3$, three blocks of type I of size 1
($p = 9$, all parameters $\neq 0$; if all blocks are of size 1, R is simply any invertible diagonal matrix)

	11	12	13	14	21	22	23	24	31	32	33	34	41	42	43	44
11	λ_1															
12					λ_1											
13			zx_{31}^{-1}						y							
14				zx_{41}^{-1}									y			
21		λ_1														
22						λ_1										
23							zx_{32}^{-1}			y						
24								zx_{42}^{-1}						y		
31									x_{31}							
32										x_{32}						
33											λ_2					
34												zx_{43}^{-1}			y	
41													x_{41}			
42														x_{42}		
43															x_{43}	
44																λ_3

$n = 4$, one block of size 4 with three components, two of size 1 and one of size 2 ($\lambda_i^2 = \lambda_i y + z$; λ_i, y, z, $x_{ij} \neq 0$; $p = 7$)

	11	12	13	14	21	22	23	24	31	32	33	34	41	42	43	44
11	λ_1															
12					λ_1											
13			zx_{31}^{-1}						y							
14				zx_{41}^{-1}									y			
21		λ_1														
22						λ_1										
23							zx_{32}^{-1}			y						
24								zx_{42}^{-1}						y		
31									x_{31}							
32										x_{32}						
33											λ_3					
34															λ_2	
41													x_{41}			
42														x_{42}		
43												λ_2				
44																λ_2

$n = 4$, one block with two components of size 2
($\lambda_i^2 = \lambda_i y + z; \lambda_i, z, y, x_{ij} \neq 0; p = 6$)

In the examples above, p is the number of parameters that are present in the R-matrix. An irreducible solution has $p = 1$ in case of type I and $p = \binom{n}{2} + 2$ in case of type II, where n is the size of the block. In the reducible cases, the number of parameters can increase drastically to a maximum of n^2; in that case there are n blocks of size 1 and R is simply any invertible diagonal matrix. This is, in a way, the most degenerate case.

3.37. Concluding comments for Section 3

Any solution of the YBE, in fact any $n^2 \times n^2$ matrix R, can be used to define a bialgebra by commutation relations $RT_2T_2 = T_2T_1R$, cf. below. The 'standard' quantum group of type A_{n-1} corresponds to the case of one block of type II of size n with $y = q - q^{-1}$, $r_{aa}^{aa} = \lambda = q$ for all a, $z = 1$, $x_{ab} = 1$ for all $a > b$.

As we shall see, the irreducible case of type II, with r_{aa}^{aa} for all a equal to the same solution λ of $X^2 = yX + z$ corresponds to the $\binom{n}{2} + 1$ multiparameter quantum group of Section 2. In this case, there are $p = \binom{n}{2} + 2$ parameters, but one is superfluous because multiplication by a scalar is irrelevant both for the YBE and for the commutation relations defined by an R.

The structure of the R-matrix for the $\binom{n}{2} + 1$ parameter quantum group is illuminating. There are $\binom{n}{2}$ 'diagonal parameters' and these define what in several ways seems to be a rather nonessential (though definitely not trivial in the technical sense) deformation of the matrix algebra. The phrase 'rather nonessential' here is intuitive and should be given precise meaning. One fact in this direction is that the extra $\binom{n}{2}$ parameters (the x_{ij}) do not appear to give any more sensitive Turaev-type knot invariants; they simple drop out of the defining trace formula, even though the relevant braid group representations are different.

The irreducible type II R-matrix with mixed r_{aa}^{aa}, meaning that some of the r_{aa}^{aa} are equal to one solution of $X^2 = (q - q^{-1})X + 1$ and some to the other one, give rise to bialgebras with nilpotents (so not quantum groups in the accepted sense of the word); they also give the same polynomial Turaev-type knot invariants (for a lower size R-matrix).

The known classical R-matrices of type B^1, C^1, D^1, A^2 do not arise as special cases of those of Theorem 3.35. These classical R-matrices do, however, satisfy a very similar condition to the one considered here. Let σ be the involution on $\{1, \dots, n\}$ given by $\sigma(i) = n + 1 - i$. Then these R matrices of type B^1, C^1, D^1, A^2 satisfy

$$r_{cd}^{ab} = 0 \quad \text{unless} \quad \{a, b\} = \{c, d\} \quad \text{or} \quad b = \sigma(a),\ d = \sigma(c). \tag{3.38}$$

It looks possible to extend the analysis of this section to the case of all solutions of the YBE satisfying (3.38).

It seems likely that the $\binom{n}{2}+1$ parameter quantum R-matrix is maximal though this remains to be proved. Possibly it will thus be possible to find the maximal families for type B^1, C^1, C^1, A^2 as well.

Work on all these matters is in progress.

4. The R-Matrix Bialgebras Defined by the Fairly General R-Matrix of Section 3

Let R again be any matrix satisfying

$$R^{ab}_{cd} = 0 \quad \text{unless} \quad \{a,b\} = \{c,d\}. \tag{4.1}$$

We investigate the commutation relations defined by

$$RT_1T_2 = T_2T_1R, \tag{4.2}$$

where

$$T = \begin{pmatrix} t_1^1 & \cdots & t_n^1 \\ \vdots & & \vdots \\ t_1^n & \cdots & t_n^n \end{pmatrix}, \quad T_1 = T \otimes I_n, \quad T_2 = I_n \otimes T.$$

Then the relations (4.2) written out become

$$r^{ab}_{i_1i_2} t^{i_1}_c t^{i_2}_d = r^{j_1j_2}_{cd} t^b_{j_2} t^a_{j_1}. \tag{4.3}$$

Let $I(R)$ be the two-sided ideal in $K\langle t\rangle$ generated by the relations (4.3). Then $I(R)$ is a bialgebra ideal, cf., e.g., [10].

4.4. THEOREM. *Let R be a solution of the YBE consisting of one type II block of size n such that, moreover, r^{aa}_{aa} = constant for all $a \in \{1,\dots,n\}$, then R defines a multiparameter quantum matrix algebra as described in Section 2 above.*

Proof. Recall that the quantum matrix algebra in question arises by taking the maximal quotient of $K\langle t_1^1,\dots,t_n^n\rangle$ that acts from the left on a quantum space $K\langle X^1,\dots,X^n\rangle$, $X^iX^j = q^{ij}X^jX^i$ by the usual matrix action and from the right on a quantum space $K\langle Y_1,\dots,Y_n\rangle$, $Y_kY_l = q_{kl}Y_lY_k$, where $q^{ii} = 1$, $q^{ij} = (q^{ji})^{-1}$, $q_{kk} = 1$, $q_{kl} = (q_{lk})^{-1}$ and the q^{ij} and q_{kl} are related by

$$q^{ij}q_{ij} = \rho \neq -1 \quad (i<j) \tag{4.5}$$

and the relations defining the quantum matrix algebra are

$$t^r_a t^r_b = q_{ab} t^r_b t^r_a, \tag{4.6}$$

$$t^r_a t^s_b - q_{ab} t^r_b t^s_a + (q_{rs})^{-1} t^s_a t^r_b - q_{ab}(q_{rs})^{-1} t^s_b t^r_a = 0, \tag{4.7}$$

$$t^r_a t^s_a = q^{rs} t^s_a t^r_a, \tag{4.8}$$

$$t^r_a t^s_b - q^{rs} t^s_a t^r_b + (q^{ab})^{-1} t^r_b t^s_a - (q^{rs})(q^{ab})^{-1} t^s_b t^r_a = 0. \tag{4.9}$$

Choose y, z, x_{ij}, $i < j$, as in Theorem 3.35. Let λ^u, $-\lambda_d$ be the two solutions of $X^2 = Xy + z$ and take

$$r^{aa}_{aa} = \lambda^u, \qquad r^{ab}_{ab} = x_{ab} \quad \text{for } a > b,$$
$$r^{ba}_{ba} = \lambda^u \lambda_d x^{-1}_{ab} \quad \text{for } a > b, \tag{4.10}$$
$$r^{ab}_{ba} = \lambda^u - \lambda_d \quad \text{for } a < b, \qquad r^{ab}_{ba} = 0 \quad \text{for } a > b,$$

as described by Theorem 3.35. (One can also take $r^{aa}_{aa} = -\lambda_d$ for all a; that gives an isomorphic matrix algebra.)

The nontrivial relations resulting from 4.3 are

$$a = b, \quad c = d, \quad r^{aa}_{aa} t^a_c t^a_c = r^{cc}_{cc} t^a_c t^a_c, \tag{4.11}$$
$$a = b, \quad c \neq d, \quad r^{aa}_{aa} t^a_c t^a_d = r^{cd}_{cd} t^a_d t^a_c + r^{dc}_{cd} t^a_c t^a_d, \tag{4.12}$$
$$a \neq b, \quad c = d, \quad r^{ab}_{ab} t^a_c t^b_c + r^{ab}_{ba} t^b_c t^a_c = r^{cc}_{cc} t^b_c t^a_c, \tag{4.13}$$
$$a \neq b, \quad c \neq d, \quad r^{ab}_{ab} t^a_c t^b_d + r^{ab}_{ba} t^b_c t^a_d = r^{cd}_{cd} t^b_d t^a_c + r^{dc}_{cd} t^b_c t^a_d. \tag{4.14}$$

Because $r^{aa}_{aa} = r^{cc}_{cc} = \lambda^u$, (4.11) holds. Now take

$$q^{ab} = x_{ab}(\lambda^u)^{-1}, \qquad q_{ba} = x_{ab}\lambda_d^{-1} \quad \text{for } a < b. \tag{4.15}$$

Notice that indeed $q^{ab}q_{ab} = x_{ab}(\lambda^u)^{-1}(x^{-1}_{ab}\lambda_d) = \lambda_d(\lambda^u)^{-1} = \rho =$ constant.

Substituting the values of (4.10) in (4.12), we obtain for $d < c$

$$\lambda^u t^a_c t^a_d = x_{cd} t^a_d t^a_c + (\lambda^u - \lambda_d) t^a_c t^a_d$$

so that indeed

$$t^a_c t^a_d = \lambda_d^{-1} x_{cd} t^a_d t^a_c = \lambda_d^{-1} x_{dc} t^a_d t^a_c = q_{cd} t^a_d t^a_c, \tag{4.16}$$

which is (4.6). And for $c < d$, we get

$$\lambda^u t^a_c t^a_d = \lambda^u \lambda_d x^{-1}_{cd} t^a_d t^a_c,$$

which gives

$$t^a_c t^a_d = \lambda_d x^{-1}_{cd} t^a_d t^a_c = q^{-1}_{dc} t^a_d t^a_c = q_{cd} t^a_c t^a_d,$$

which is the same as (4.16).

Now substitute the values of (4.10) in (4.13). There are again two cases to consider.

If $a < b$ we find

$$\lambda_d \lambda^u x^{-1}_{ab} t^a_c t^b_c + (\lambda^u - \lambda_d) t^b_c t^a_c = \lambda^u t^b_c t^a_c,$$

which gives (using 4.15)

$$t^a_c t^b_c = (\lambda^u)^{-1} x_{ab} t^b_c t^a_c = q^{ab} t^b_c t^a_c,$$

which is (4.8).

If $a > b$ we find

$$x_{ab}t^a_c t^b_c = \lambda^u t^b_c t^a_c$$

which gives

$$t^b_c t^a_c = (\lambda^u)^{-1} x_{ab} t^a_c t^b_c = q^{ba} t^a_c t^b_c.$$

Finally substitute the values of (4.10) in (4.14). Note that (4.14) really embodies four equations between the $t^a_c t^b_d$, $t^a_d t^b_c$, $t^b_c t^a_d$, $t^b_d t^a_c$; namely, the one written down and the three obtained by switching a and b, switching c and d, and switching both.

Taking $a < b$, $c < d$, we find

$$\lambda^u \lambda_d x_{ab}^{-1} t^a_c t^b_d + (\lambda^u - \lambda_d) t^b_c t^a_d = \lambda^u \lambda_d x_{cd}^{-1} t^b_d t^a_c. \tag{4.17}$$

Switching a and b in (4.14) and then substituting gives

$$x_{ab} t^b_c t^a_d = \lambda^u \lambda_d x_{cd}^{-1} t^a_d t^b_c, \tag{4.18}$$

$$\lambda^u \lambda_d x_{ab}^{-1} t^a_d t^b_c + (\lambda^u - \lambda_d) t^b_d t^a_c = x_{cd} t^b_c t^a_d + (\lambda^u - \lambda_d) t^b_d t^a_c. \tag{4.19}$$

Finally, switching both a, b and c, d and then substituting gives

$$x_{ab} t^b_d t^a_c = x_{cd} t^a_c t^b_d + (\lambda^u - \lambda_d) t^a_d t^b_c. \tag{4.20}$$

Observe that (4.18) and (4.19) are identical. It is easily checked that

$$x_{ab}(\lambda^u \lambda_d)^{-1}(4.17) + (x_{cd}^{-1})(4.20) - (\lambda_d^{-1} - (\lambda^u)^{-1})(4.18)$$

has equal left- and right-hand sides. Thus (4.17)–(4.20) are equivalent to (4.17)–(4.18).

Multiply (4.17) by $x_{ab}(\lambda^u \lambda_d)^{-1}$ to find

$$t^a_c t^b_d + x_{ab} \lambda_d^{-1} t^b_c t^a_d - x_{ab} \lambda_u^{-1} t^b_c t^a_d - x_{ab} x_{cd}^{-1} t^b_d t^a_c = 0 \tag{4.21}$$

and now use (4.18) to rewrite the third term to find

$$t^a_c t^b_d + x_{ab} \lambda_d^{-1} t^b_c t^a_d - \lambda_d x_{cd}^{-1} t^a_d t^b_c - x_{ab} x_{cd}^{-1} t^b_d t^a_c = 0. \tag{4.22}$$

Because $a < b$, $c < d$, we have by (4.15) that

$$q_{ab}^{-1} = q_{ba} = x_{ab} \lambda_d^{-1}, \qquad q_{cd} = (q_{dc})^{-1} = (x_{cd} \lambda_d^{-1})^{-1} = \lambda_d x_{cd}^{-1},$$

$$q_{ab}^{-1} q_{cd} = x_{ab} \lambda_d^{-1} \lambda_d x_{cd}^{-1} = x_{ab} x_{cd}^{-1},$$

so that (4.22) is identical with (4.7).

Now use (4.18) to rewrite the second term in (4.21). This gives

$$t^a_c t^b_d + \lambda^u x^{-1}_{cd} t^a_d t^b_c - x_{ab}\lambda^{-1}_u t^b_c t^a_d - x_{ab} x^{-1}_{cd} t^b_d t^a_c = 0. \tag{4.23}$$

Again, as $a < b$, $c < d$, we have by (4.15) that

$$\begin{aligned} q^{ab} &= (\lambda^u)^{-1} x_{ab}, \qquad (q^{cd})^{-1} = ((\lambda^u)^{-1} x_{cd})^{-1} = \lambda^u x^{-1}_{cd}, \\ q^{ab}(q^{cd})^{-1} &= (\lambda^u)^{-1} x_{ab} \lambda^u x^{-1}_{cd}, \end{aligned} \tag{4.24}$$

so that (4.23) is identical with (4.9).

This finishes the proof of the theorem. (Though not necessary, given what has been shown about the rank of the various groups of relations involved, it is in fact now not difficult to show that inversely the groups of relations (4.7)–(4.9) imply the group (4.14), i.e. (4.17)–(4.20).)

4.25. COROLLARY. *Let $M^{n\times n}_q$ be the multiparameter quantum matrix algebra of Section 2, i.e. $M^{n\times n}_q = K\langle t\rangle / I$ when I is the ideal of the relations (4.6)–(4.9). Then $M^{n\times n}_q$ is a PBW algebra with the same Hilbert–Poincaré series as $K[t^1_1, \dots, t^n_n]$.*

Proof. We already know that the dimension of the degree 2 part is exactly right viz. $n^2 + \binom{n^2}{2}$. The commutation relations are of the form

$$T_1 T_2 = R^{-1} T_2 T_1 R.$$

Now R satisfies the YBE, i.e.

$$R_{12} R_{13} R_{23} = R_{23} R_{13} R_{12}. \tag{4.26}$$

Now for the triple product $T_1 T_2 T_3$,

$$T_1 = T \otimes I \otimes I, \qquad T_2 = I \otimes T \otimes I, \qquad T_3 = I \otimes I \otimes T,$$

we have that

$$\begin{aligned} T_1(T_2T_3) &= T_1 R^{-1}_{23} T_3 T_2 R_{23} = R^{-1}_{23}(T_1T_3)T_2R_{23} = R^{-1}_{23} R^{-1}_{13} T_3 T_1 R_{13} T_2 R_{23} \\ &= R^{-1}_{23} R^{-1}_{13} T_3 (T_1 T_2) R_{13} R_{23} = R^{-1}_{23} R^{-1}_{13} T_3 R^{-1}_{12} T_2 T_1 R_{12} R_{13} R_{23} \\ &= R^{-1}_{23} R^{-1}_{13} R^{-1}_{12} T_3 T_2 T_1 R_{12} R_{13} R_{23}. \end{aligned} \tag{4.27}$$

(Note that $R_{ij} T_k = T_k R_{ij}$ if $i \neq j \neq k \neq i$ because R_{ij} only affects factors i and j where T_k is the identity.) We also have

$$\begin{aligned} (T_1T_2)T_3 &= R^{-1}_{12} T_2 T_1 R_{12} T_3 = R^{-1}_{12} T_2 (T_1 T_3) R_{12} = R^{-1}_{12} T_2 R^{-1}_{13} T_3 T_1 R_{13} R_{12} \\ &= R^{-1}_{12} R^{-1}_{13} (T_2 T_3) T_1 R_{13} R_{12} = R^{-1}_{12} R^{-1}_{13} R^{-1}_{23} T_3 T_2 R_{23} T_1 R_{13} R_{12} \\ &= R^{-1}_{12} R^{-1}_{13} R^{-1}_{23} T_3 T_2 T_1 R_{23} R_{13} R_{12}. \end{aligned} \tag{4.28}$$

The end products of (4.27) and (4.28) are the same proving the confluence conditions of the diamond lemma, [2], and the result follows. This argument: YBE $\Rightarrow$ confluence condition of diamond lemma has been observed before [6].

4.29. Comments on the Other Solutions of the YBE

The solutions consisting of one block of type I gives, as is easily checked, no relations at all among the t^i_j. The solutions consisting of one block with several components with mixed parameters λ_j give rise to a bialgebra $K\langle t\rangle/I(R)$ with nilpotent elements. Indeed if, say, $a \in C_1$ and $b \in C_2$ and $\lambda_1 \neq \lambda_2$, then by (4.11)

$$\lambda_1 t^a_b t^a_b = \lambda_2 t^a_b t^a_b, \tag{4.30}$$

so that $(t^a_b)^2 = 0$. These are, of course, perfectly good solutions of the YBE and as such are of potential use in, for example, the business of constructing link invariants (cf. Section 5 below) but the bialgebras they define are not quantum groups in the (more or less) accepted sense of the word. (There is no consensus and some authors equate the concepts Hopf algebra and quantum group; I would be inclined to reserve the phrase quantum group for a Hopf algebra that is a PBW algebra and is a deformation of the function algebra of a linear algebraic group.) Let me also remark that in spite of nilpotents, these bialgebras are still pretty nice in the sense that its defining rewriting rules (commutation relations) are confluent (so that it is easy to write down a basis and a version of Gröbner basis theory probably applies).

4.30. Quantum Groups

Let again R be a single block solution of the YBE with constant parameter λ_j defining a multiparameter quantum matrix algebra $M_q = K\langle t\rangle/I(R)$. As is shown in, e.g., [1], for the case of a single type II block there is an element d in M_q (a quantum determinant) such that the localization $M_q[d^{-1}]$ admits an antipode and thus becomes a Hopf algebra.

By the work of [6, 12, 13], cf. also [4], the fact that M_q comes from a solution of the YBE is useful in establishing such facts.

5. Yang–Baxter Operators and Link Invariants

For this section the Yang–Baxter equation takes the form

$$S_{12}S_{23}S_{12} = S_{23}S_{12}S_{23}. \tag{5.1}$$

If $S = (s^{ab}_{cd})$, then in terms of the entries of S, this works out as

$$s^{ab}_{kl}s^{lc}_{mw}s^{km}_{uv} = s^{ai}_{uk}s^{bc}_{ij}s^{kj}_{vw}. \tag{5.2}$$

There is a simple relation between (5.1) and the YBE (3.1): if $R = (r^{ab}_{cd})$ solves (3.1), then both

$$S = (s^{ab}_{cd}), \quad s^{ab}_{cd} = r^{ab}_{dc}, \qquad S' = (s'^{ab}_{cd}), \quad s'^{ab}_{cd} = r^{ba}_{cd} \tag{5.3}$$

solve (5.1) (and vice versa). Let's check that for S. Putting (5.3) in the LHS of (5.2) gives

$$r_{lk}^{ab} r_{wm}^{lc} r_{vu}^{km}, \tag{5.4}$$

which is the LHS of (3.2) with uvw replaced by wvu; now put (5.3) in the RHS of (5.2) to find

$$r_{ku}^{ai} r_{ji}^{bc} r_{wv}^{kj} = r_{ji}^{bc} r_{ku}^{ai} r_{wv}^{kj}, \tag{5.5}$$

which is the RHS of (3.2) also with uvw replaced by wvu. The proof for S' is as easy (except that now RHS and LHS switch).

5.6. DEFINITION ([22]). A Yang–Baxter operator consists of a quadruple (S, ν, α, β), where S is an $n^2 \times n^2$ matrix satisfying the YBE in the form (5.1), ν is an $n \times n$ matrix, and α, β are invertible scalars which are related to S by the conditions (5.7)–(5.9).

$$\nu \otimes \nu \text{ commutes with } S, \tag{5.7}$$

$$\mathrm{Tr}_2(S \circ (\nu \otimes \nu)) = \alpha\beta\nu, \tag{5.8}$$

$$\mathrm{Tr}_2(S^{-1} \circ (\nu \otimes \nu)) = \alpha^{-1}\beta\nu. \tag{5.9}$$

Here if $M = (m_{kl}^{ij})$ is an $n^2 \times n^2$ matrix (with the usual ordering $11, \ldots, 1n$; $21, \ldots, 2n; \ldots; n1, \ldots, nn$ of rows and columns), then $\mathrm{Tr}_2(M) = N$ is the $n \times n$ matrix with entries

$$n_j^i = m_{j1}^{i1} + \ldots + m_{jn}^{in}, \tag{5.10}$$

i.e. if M is written as an $n \times n$ matrix of $n \times n$ blocks, then N is constructed by replacing each block of M by its trace. If ν is invertible, then (5.8) and (5.9) are equivalent to

$$\mathrm{Tr}_2(S^{\pm 1} \circ (I_n \otimes \nu)) = \alpha^{\pm 1}\beta I_n \tag{5.11}$$

(where I_n is the $n \times n$ identity matrix).

Given a YB operator (S, ν, α, β), Turaev's formula

$$T_S(\xi) = \alpha^{-w(\xi)}\beta^{-m}\mathrm{Tr}(\rho_S(\xi) \circ \nu^{\otimes m}) \tag{5.12}$$

defines a link invariant. Here $\xi \in B_m$, the braid group on m letters, $w(\xi) = \Sigma\varepsilon_i$ if $\xi = \sigma_{i_1}^{\varepsilon_1} \ldots \sigma_{i_r}^{\varepsilon_r}$, where the σ_i are the standard generators of B_m, and ρ_S is the representation of the braid group (in $(K^n)^{\otimes m}$) defined by S, $\sigma_i \mapsto S_{ii+1}$; $T_S(\xi)$ is then independent of the particular braid that gives rise to a link ξ by closure of the braid.

Now, given the solutions of the YBE described in Section 3, it is natural to investigate whether these extend to Yang–Baxter operators in the sense of Turaev

(Definition 5.12), and, if so, what the resulting link and knot invariants bring. Here I report some preliminary results only. Further work is in progress.

5.13. *Remarks.* Both the constants α and β can be normalized to 1. Indeed if (S, ν, α, β) is a Yang–Baxter operator then $(\alpha^{-1}S, \beta^{-1}\nu, 1, 1)$ is another one. However, for the formulas below it is convenient to keep α (but β will always be 1). As Turaev observes, if ν is diagonal, then (5.8) implies that $\bar{S}\bar{\nu} = \bar{\alpha}$ where $\bar{S}$ is the $n \times n$ matrix $\bar{s}^i_j = s^{ij}_{ij}$, $\bar{\nu}$ is the column vector $(\nu_1, \ldots, \nu_n)^T$ and $\bar{\alpha}$ is the column vector $\alpha(1, 1, \ldots, 1)^T$. Thus, assuming ν is diagonal, it is unique if $\bar{S}$ is invertible.

5.14. THEOREM. *Let R be a solution of the YBE (as described in Theorem 3.35) consisting of a single block (with components $C_1, C_2, \ldots, C_p$ $(p \geqslant 2)$) with parameters y and z and let μ and λ be the two solutions of the equation $X^2 = yX + z$. Let $S = \tau R$ be the associated solution of (5.1), then S extends to a Yang–Baxter operator with the scalar α such that*

$$\alpha^2 = (-1)^{p-1}\lambda^{k_\lambda - k_\mu + 1}\mu^{k_\mu - k_\lambda + 1}, \tag{5.15}$$

where k_λ (resp., k_μ) is the number of components C_j with $\lambda_j = \lambda$ (resp., $\lambda_j = \mu$).

Proof. For the moment regard R, R^{-1} and S, S^{-1} as $n \times n$ matrices made up of blocks that are also $n \times n$ matrices. Observe that the diagonals of all the off-diagonal blocks are zero. Take $\nu = \operatorname{diag}(\nu_1, \ldots, \nu_n)$, the diagonal $n \times n$ matrix with diagonal entries $\nu_1, \ldots, \nu_n$. Because ν is diagonal and $s^{ab}_{cd} = 0$, unless $\{a, b\} = \{c, d\}$, (5.7) holds. It also follows (cf. (5.10)) that the conditions (5.8), (5.9) only involve the diagonal blocks of S and S^{-1}. As is easily checked, the inverse R^{-1} of R is also a solution of the YBE and has the same structure as R. One can easily verify that R^{-1} is equal to

$$\begin{array}{lll} (R^{-1})^{ab}_{ba} = \lambda^{-1} + \mu^{-1} & \text{if} & a < b, \\ (R^{-1})^{ab}_{ba} = (R^{ab}_{ba})^{-1} & \text{if} & a \simeq b, \\ (R^{-1})^{ab}_{ab} = z^{-1}x_{ba} & \text{if} & a < b, \\ (R^{-1})^{ab}_{ab} = x^{-1}_{ab} & \text{if} & a > b, \\ (R^{-1})^{aa}_{aa} = (R^{aa}_{aa})^{-1} & (= \lambda^{-1} \ (\text{resp.}, \mu^{-1})). & \end{array} \tag{5.16}$$

Indeed if $a < b$, $\{a, b\} \neq \{c, d\}$

$$(RR^{-1})^{ab}_{cd} = R^{ab}_{ij}(R^{-1})^{ij}_{cd} = R^{ab}_{ab}(R^{-1})^{ab}_{cd} + R^{ab}_{ba}(R^{-1})^{ba}_{cd} = 0.$$

Further, if $a < b$, $a = c$, $b = d$

$$(RR^{-1})^{ab}_{ab} = R^{ab}_{ab}(R^{-1})^{ab}_{ab} + R^{ab}_{ba}(R^{-1})^{ba}_{ab} = zx^{-1}_{ba}z^{-1}x_{ba} + 0 = 1$$

and if $a < b$, $a = d$, $b = c$

$$\begin{aligned} (RR^{-1})^{ab}_{ba} &= R^{ab}_{ab}(R^{-1})^{ab}_{ba} + R^{ab}_{ba}(R^{-1})^{ba}_{ba} \\ &= zx^{-1}_{ba}(\lambda^{-1} + \mu^{-1}) + (\lambda + \mu)x^{-1}_{ba} = 0 \end{aligned}$$

because $z = -\lambda\mu$.

The other cases $a \simeq b$, $a > b$ are even easier to check.

Switching λ and μ if necessary, we can assume that $\lambda_1 = \lambda$. Let the pattern of λ's and μ's be the following

$$\lambda_1 = \ldots = \lambda_{d_1} = \lambda; \qquad \lambda_{d_1+1} = \cdots = \lambda_{d_1+d_2} = \mu;$$

$$\lambda_{d_1+d_2+1} = \cdots = \lambda_{d_1+d_2+d_3} = \lambda; \ \ldots$$

Let r be the number of switches $(d_1, d_1+1), \ldots, (d_r, d_r+1)$, so that $\lambda_p = \lambda$ if r even and $\lambda_p = \mu$ if r is odd. We define a diagonal $p \times p$ matrix $T = \text{diag}(T_1, T_2, \ldots, T_p)$, where T_j is equal to the trace of ν with respect to the jth component, i.e. $T_j = \sum_{i \in C_j} \nu_i$.

It is now easy to see that Equations (5.8), (5.9) (with $\beta = 1$) amount to the following: (where the equations resulting from (5.8) constitute the upper block and those from (5.9) form the lower block. Here, as in the above, to follow the calculations, it is useful to keep the first example of (3.36) in front of one).

$$\begin{aligned}
&\lambda T_1 = \alpha \\
&\lambda T_2 + (\mu+\lambda)T_1 = \alpha \\
&\qquad\qquad \vdots \\
&\lambda T_{d_1-1} + (\mu+\lambda)(T_1 + \cdots + T_{d_1-2}) = \alpha \\
&\lambda T_{d_1} + (\mu+\lambda)(T_1 + \cdots + T_{d_1-1}) = \alpha \\
&\mu T_{d_1+1} + (\mu+\lambda)(T_1 + \cdots + T_{d_1}) = \alpha \\
&\mu T_{d_1+2} + (\mu+\lambda)(T_1 + \cdots + T_{d_1+1}) = \alpha \\
&\qquad\qquad \vdots \\
&\mu T_{d_1+d_2-1} + (\mu+\lambda)(T_1 + \cdots + T_{d_1+d_2-2}) = \alpha \\
&\mu T_{d_1+d_2} + (\mu+\lambda)(T_1 + \cdots + T_{d_1+d_2-1}) = \alpha \\
&\lambda T_{d_1+d_2+1} + (\mu+\lambda)(T_1 + \cdots + T_{d_1+d_2}) = \alpha \\
&\lambda T_{d_1+d_2+2} + (\mu+\lambda)(T_1 + \cdots + T_{d_1+d_2+1}) = \alpha \\
&\qquad\qquad \vdots \\
&\kappa T_p + (\mu+\lambda)(T_1 + \cdots + T_{p-1}) = \alpha
\end{aligned}$$

$$\begin{aligned}
&\frac{1}{\lambda}T_1 + (\lambda^{-1} + \mu^{-1})(T_2 + \cdots + T_p) = \frac{1}{\alpha} \\
&\frac{1}{\lambda}T_2 + (\lambda^{-1} + \mu^{-1})(T_3 + \cdots + T_p) = \frac{1}{\alpha} \\
&\qquad\qquad \vdots
\end{aligned}$$

$$\frac{1}{\lambda}T_{d_1-1} + (\lambda^{-1}+\mu^{-1})(T_{d_1}+\cdots+T_p) = \frac{1}{\alpha}$$
$$\frac{1}{\lambda}T_{d_1} + (\lambda^{-1}+\mu^{-1})(T_{d_1+1}+\cdots+T_p) = \frac{1}{\alpha}$$
$$\frac{1}{\mu}T_{d_1+1} + (\lambda^{-1}+\mu^{-1})(T_{d_1+2}+\cdots+T_p) = \frac{1}{\alpha}$$
$$\frac{1}{\mu}T_{d_1+2} + (\lambda^{-1}+\mu^{-1})(T_{d_1+3}+\cdots+T_p) = \frac{1}{\alpha}$$
$$\vdots$$
$$\frac{1}{\mu}T_{d_1+d_2-1} + (\lambda^{-1}+\mu^{-1})(T_{d_1+d_2}+\cdots+T_p) = \frac{1}{\alpha}$$
$$\frac{1}{\mu}T_{d_1+d_2} + (\lambda^{-1}+\mu^{-1})(T_{d_1+d_2+1}+\cdots+T_p) = \frac{1}{\alpha}$$
$$\frac{1}{\lambda}T_{d_1+d_2+1} + (\lambda^{-1}+\mu^{-1})(T_{d_1+d_2+2}+\cdots+T_p) = \frac{1}{\alpha}$$
$$\frac{1}{\lambda}T_{d_1+d_2+2} + (\lambda^{-1}+\mu^{-1})(T_{d_1+d_2+3}+\cdots+T_p) = \frac{1}{\alpha}$$
$$\vdots$$
$$\frac{1}{\kappa}T_p = \frac{1}{\alpha},$$

where $\kappa = \lambda$ (resp., μ) depending on whether r is even (resp., odd). Now observe that subtracting the $(i+1)$th from the ith equation in both the upper and lower blocks results in the same relation between T_{i+1} and T_i viz. $T_{i+1} = -\lambda^{-1}\mu T_i$, or $T_{i+1} = -\mu^{-1}\lambda T_i$, or $T_{i+1} = -T_i$. This results in the following recipe for the T's

$$T_1 = \lambda^{-1}\alpha,$$

$$T_i = \begin{cases} (-\lambda^{-1}\mu)T_{i-1} & \text{if } \lambda_i = \lambda = \lambda_{i-1}, \\ -T_{i-1} & \text{if } \lambda_i = \lambda, \lambda_{i-1} = \mu, \\ (-\mu^{-1}\lambda)T_{i-1} & \text{if } \lambda_i = \mu = \lambda_{i-1}, \\ -T_{i-1} & \text{if } \lambda_i = \mu, \lambda_{i-1} = \lambda, \end{cases} \tag{5.17}$$

$$T_p = \begin{cases} \lambda\alpha^{-1} & \text{if } r \text{ is even}, \\ \mu\alpha^{-1} & \text{if } r \text{ is odd}. \end{cases}$$

It follows that, depending on the number, r, of switches from λ to μ or vice versa

if r is even

$$T_p = (-1)^{p-1}\lambda^{k_\mu - k_\lambda + 1}\mu^{k_\lambda - k_\mu - 1}T_1, \quad T_1 = \lambda^{-1}\alpha,\ T_p = \lambda\alpha^{-1},$$

if r is odd

$$T_p = (-1)^{p-1}\lambda^{k_\mu - k_\lambda}\mu^{k_\lambda - k_\mu}T_1, \quad T_1 = \lambda^{-1}\alpha,\ T_p = \mu\alpha^{-1}, \tag{5.18}$$

where k_λ is the number of i's for which $\lambda_i = \lambda$ and k_μ the number of i's for which $\lambda_i = \mu$, $k_\lambda + k_\mu = p$. In both cases, it follows that

$$\alpha^2 = (-1)^{p-1}\lambda^{k_\lambda - k_\mu + 1}\mu^{k_\mu - k_\lambda + 1} \tag{5.19}$$

and for both α's solving (5.19) (taking, if necessary, a quadratic extension of K) (5.17) then specifies $T_1, \ldots, T_p$ such that (5.8), (5.9) are satisfied (with $\beta = 1$). This concludes the proof of the theorem.

5.20. *Remark.* Both choices for α in (5.19) give up to a sign the same link invariant, cf. [22, 3.3]. As for the uniqueness of the Yang–Baxter operator it is evident that the solution of $T_1, \ldots, T_p$ is unique, hence the solution of $\nu_1, \ldots, \nu_n$ is unique if and only if all components consist of one element, i.e. the block is of type II. This can also be seen from the fact that the matrix $\bar{S}$ satisfies $\bar{s}^i_j = s^{ij}_{ij} = r^{ji}_{ij} = \lambda_k$ if $i \simeq j$, y if $j < i$ and 0 if $i < j$, so it is invertible if and only if we are dealing with a type II block.

5.21. COROLLARY. *Let R be any solution of the YBE as described by Theorem* 3.35 *and $S = \tau R$ the corresponding solution of* (5.1). *Then S extends to a Yang–Baxter operator $(S, \nu, \alpha, 1)$ if any only if for all blocks*

$$\alpha^2 = (-1)^{p_i - 1}\lambda_i^{k_{\lambda_i} - k_{\mu_i} + 1}\mu_i^{k_{\mu_i} - k_{\lambda_i} + 1} \tag{5.22}$$

for a block with $p_i \geqslant 2$ components,

$$\alpha^2 = \lambda_i^2 \tag{5.23}$$

for a block with one component.

Proof. Take ν diagonal. From the form of S (and S^{-1} which has the same form), one easily sees that (5.8) and (5.9) only involve the separate blocks and the ν's with corresponding indices. It is trivial to check (5.23). Finally, (5.7) holds because $s^{ab}_{cd} = 0$ unless $\{a, b\} = \{c, d\}$ and ν is diagonal. □

The next result is perhaps a disappointment. With $\binom{n}{2}$ extra variables in an $n^2 \times n^2$ single type II block solution of (5.1) it might be hoped (even expected) that these will give some extra information when employed to define link invariants via Turaev's formula (5.12). This is not the case, and using both solutions λ and μ of $X^2 = yX + z$ (instead of just 1) for the $\rho_a = s^{aa}_{aa}$ also gives nothing new.

5.24. PROPOSITION. *Let S be a single type II block solution of* (5.1). *Let μ occur m times as a ρ_a, $m \leqslant \frac{1}{2}n$. Then the link invariant T_S defined by S by formula* (5.12) *using the extended YB operator $(S, \nu, \alpha, 1)$ defined by Theorem* 5.14 *is the same as the one defined by the single type II block solution S_1 of size $(n - 2m)^2 \times (n - 2m)^2$, $x_{ij} = 1 = z$ for all i, j, same y as S (i.e. it is one of the 'classical' A_s invariants of Turaev).*

Proof. It follows immediately from (5.12) that (S,ν,α,β) and $(\rho S,\nu,\rho\alpha,\beta)$ define the same link invariant. We can therefore assume $z=1$, i.e. $\lambda=q$, $\mu=-q^{-1}$. Then, by (5.15), $\alpha=\pm q^{n-2m}$. A simple check now shows that S satisfies the relation

$$S-S^{-1}=(q-q^{-1})I_{n^2} \tag{5.25}$$

and this also satisfied by S_1. It follows that the link invariants T and T_1 defined by S and S_1 (or $-S_1$ which does not matter by 5.20) both satisfy, [22], the same skein relation.

$$q^{n-2m}T_S(L_+)-q^{2m-n}T_S(L_-)=(q-q^{-1})T_S(L_0), \tag{5.26}$$

where L_+, L_-, and L_0 are three oriented links which are identical except for one crossing where they look, respectively, like

By repeated changing of + crossings to − crossings any link can be turned into an unlink. Thus, the value of T_S is uniquely determined by the skein relation (5.26) and its values on k-component unlinks. The latter are equal to $(\nu_1+\cdots+\nu_n)^k$. Finally one checks that

$$(\nu_1+\cdots+\nu_n)=(\bar\nu_1+\cdots+\bar\nu_{n-2m}),$$

where $(S_1,\bar\nu,\alpha,1)$ is the YB operator belonging to S_1. This is (with induction) seen as follows. If d_i is the shortest run of λ's or μ's, then if $i=1$, the pattern $d_2-d_1,d_3,\ldots,d_{r+1}$ gives the same trace value of ν as the original (because $\nu_{d_1+1}=-\nu_{d_1}$, $\nu_{d_1+i}=-\nu_{d_1-i+1}$, $i=1,\ldots,d_1$) and similarly if $i>1$, the pattern $d_1,\ldots,d_{i-1},d_{i+1},\ldots,d_{r+1}$ gives the same trace value of ν as the original. This proves the proposition.

5.27. *Remark.* This result (Proposition 5.24), illustrates the previous remark (cf. (3.37)) that the $\binom{n}{2}$ extra diagonal parameters in the general on type II block solution of the YBE, i.e. the x_{ij} and z, play in some sense a trivial role, while there is but one essential parameter, viz. y (or q). On the other hand, the corresponding quantum groups, the general $\binom{n}{2}+1$ parameter one, and the classical 1 parameter one are not isomorphic.

5.28. Invariants from Diagonal Solutions

On the other hand, perhaps surprisingly, the diagonal solutions of the YBE can give rise to nontrivial knot invariants. Take, for example, the $n = 2$, 2 blocks of size 1 solution:

$$R = \begin{pmatrix} x_{11} & & & \\ & zx_{21}^{-1} & & \\ & & x_{21} & \\ & & & x_{22} \end{pmatrix}, \qquad R^{-1} = \begin{pmatrix} x_{11}^{-1} & & & \\ & z^{-1}x_{21}^{-1} & & \\ & & x_{21}^{-1} & \\ & & & x_{22}^{-1} \end{pmatrix} \tag{5.29}$$

with corresponding solutions of (5.1)

$$S = \begin{pmatrix} x_{11} & & & \\ & 0 & x_{21} & \\ & zx_{21}^{-1} & 0 & \\ & & & x_{22} \end{pmatrix}, \qquad S^{-1} = \begin{pmatrix} x_{11}^{-1} & & & \\ & 0 & z^{-1}x_{21} & \\ & x_{21}^{-1} & 0 & \\ & & & x_{22}^{-1} \end{pmatrix} \tag{5.30}$$

This S, for $x_{11} = x_{22}$, extends to a Yang–Baxter operator (S, ν, α, β) with $\nu = I_2$, if $\alpha = x_{11} = x_{22}$, $\beta = 1$ and gives rise to a link invariant that takes the following values on the following links

L_0 L_1 L_2 L_3 L_4 (trefoil)

L_5 L_6 L_7 L_8

$$T(L_0) = 2, \qquad T(L_1) = 4, \qquad T(L_2) = 2 + 2\gamma,$$

$$T(L_3) = 2 + 2\gamma^2, \qquad T(L_4) = 2, \qquad T(L_5) = 2 + 2\gamma,$$

$$T(L_6) = 2(1+\gamma)^2, \qquad T(L_7) = 2 + 6\gamma^2, \qquad T(L_8) = 6 + 2\gamma^2. \tag{5.31}$$

Here $\gamma = r_{12}^{12} r_{21}^{21} = z$. Thus, this invariant counts components, can detect various ways in which components are linked but does not distinguish between, e.g., trefoil and unknot (L_0, and L_4; cf. also L_2 and L_5). The two size 1 blocks themselves give only the trivial invariant, thus this example shows conclusively that putting two blocks nontrivially together can definitely give nontrivial extra information.

5.32. *Remark.* The representations of the Braid group on k strings B_k defined by S and S_1 in Proposition 5.24 are different (even if $m = 0$), but this difference does not show up in the trace formula (5.12). This can also be seen directly in cases where there is no relation like (5.25), which is important in dealing with solutions S which do not consist of a single block. Indeed:

5.33. THEOREM. *R be an invertible $n^2 \times n^2$ matrix with diagonal entries x_{ij} and possibly nonzero diagonal entries $q_{ij} = r_{ij}^{ij}$, $i < j$, and no other nonzero entries. Let $S = \tau R$. Let $w = \sigma_{i_1}^{\varepsilon_1} \dots \sigma_{i_m}^{\varepsilon_m}$, $\varepsilon_i \in \{1, -1\}$ be an element of the braid group B_k of braids on k strings. Let $S_i = I_n^{\otimes i-1} \otimes S \otimes I_n^{\otimes k-i-1}$ and let $S_w = S_{i_1}^{\varepsilon_1} \dots S_{i_m}^{\varepsilon_m}$. Then the diagonal elements of S_w are Laurent polynomials in the q_{ij}, the x_{ii}, and the products $x_{ij}x_{ji} = z_{ij}$.*

Proof. The only off-diagonal elements of S are of the form

$$s_{ji}^{ij} = r_{ij}^{ij} = x_{ij}, \qquad s_{ij}^{ji} = r_{ji}^{ji} = x_{ji} = x_{ij}^{-1} z_{ij}. \tag{5.34}$$

The off-diagonal elements of R^{-1} are equal to $-q_{ij}x_{ji}^{-1}x_{ij}^{-1}$, $i < j$. It follows that the diagonal elements of $S^{-1} = R^{-1}\tau$ are of the form

$$x_{ii}^{-1}, \qquad -q_{ij}(x_{ij}x_{ij})^{-1} = -q_{ij}z_{ij}^{-1} \tag{5.35}$$

and that the off-diagonal elements of S^{-1} are of the form

$$(S^{-1})_{ji}^{ij} = x_{ji}^{-1} = z_{ij}^{-1}x_{ij}, \qquad (S^{-1})_{ij}^{ji} = x_{ij}^{-1}. \tag{5.36}$$

Now consider a diagonal element of S_w. Such an element is a sum of products of the form

$$t_{i_1(2)\dots i_n(2)}^{i_1(1)\dots i_n(1)} t_{i_1(3)\dots i_n(3)}^{i_1(2)\dots i_n(2)} \dots t_{i_1(m)\dots i_n(m)}^{i_1(m-1)\dots i_n(m-1)} \tag{5.37}$$

with $i_l(m) = i_l(1)$, $l = 1, \dots, n$, and $r_{i_1(l+1)\dots\dots i_n(l+1)}^{i_1(l)\dots i_n(l)}$ an element of $S_{i_l}^{\varepsilon_l}$. Because of (5.34)–(5.36) each product (5.37) is zero unless all the permutations

$$\begin{pmatrix} i_1(l) \dots i_n(l) \\ i_1(l+1) \dots i_n(l+1) \end{pmatrix}$$

are of the form identity or τ_k, where τ_k is the transposition $(k\ k+1)$ that interchanges the kth and $(k+1)$th entries and leaves all others in place. The identity permutations produce diagonal entries from S_{i_l} or $S_{i_l}^{-1}$ and by (5.34)–(5.36), these are of the desired form. The remaining permutations in (5.37) form a word ω in the $\tau_1, \ldots, \tau_{n-1}$ that is equal to the identity in the permutation group Π_n on n-letters. The relations between the generators $\tau_1, \ldots, \tau_{n-1}$ of Π_n are the following

$$\begin{aligned} &\tau_k^2 = 1, \\ &\tau_k\tau_{k+1}\tau_k\tau_{k+1}^{-1}\tau_k^{-1}\tau_{k+1}^{-1} = 1, \\ &\tau_k\tau_l\tau_k^{-1}\tau_l^{-1} = 1, \quad \text{if} \quad |k-l| \geqslant 2. \end{aligned} \tag{5.38}$$

It follows that somewhere in the word ω one of the three left hand sides of (5.38) occurs and by induction (in the length of ω) it follows that if suffices to check that in all three cases, the corresponding factors in (5.37) combine to give a monomial of the desired form. Observe that S and S^{-1} have the same off-diagonal entries except for a factor z_{ij}. Thus, replacing each S_l^{-1} with S_l only changes things by monomials in the z_{ij} and we may assume that all ε_l are 1.

In the first case we obtain a product

$$t_{\alpha_1 ba\alpha_2}^{\alpha_1 ab\alpha_2} t_{\alpha_1 ab\alpha_2}^{\alpha_1 ba\alpha_2}$$

which is equal to $x_{ab}x_{ba} = z_{ab}$. Here and below, the α_i stand for strings of indices that remain unchanged.

In the case of the second type of relation of (5.38) we obtain a product

$$t_{\alpha_1 bac\alpha_2}^{\alpha_1 abc\alpha_2} t_{\alpha_1 bca\alpha_2}^{\alpha_1 bac\alpha_2} t_{\alpha_1 cba\alpha_2}^{\alpha_1 bca\alpha_2} t_{\alpha_1 cab\alpha_2}^{\alpha_1 cba\alpha_2} t_{\alpha_1 acb\alpha_2}^{\alpha_1 cab\alpha_2} t_{\alpha_1 abc\alpha_2}^{\alpha_1 acb\alpha_2}$$

which is equal to $x_{ab}x_{ac}x_{bc}x_{ba}x_{ca}x_{cb} = z_{ab}z_{bc}z_{ac}$.

Finally, in the case of the third type of relation of (5.38), we obtain a product

$$t_{\alpha_1 ba\alpha_2 cd\alpha_3}^{\alpha_1 ab\alpha_2 cd\alpha_3} t_{\alpha_1 ba\alpha_2 dc\alpha_3}^{\alpha_1 ba\alpha_2 cd\alpha_3} t_{\alpha_1 ab\alpha_2 dc\alpha_3}^{\alpha_1 ba\alpha_2 dc\alpha_3} t_{\alpha_1 ab\alpha_2 cd\alpha_3}^{\alpha_1 ab\alpha_2 dc\alpha_3}$$

which is equal to $x_{ab}x_{cd}x_{ba}x_{dc} = z_{ab}z_{cd}$. This concludes the proof. □

5.39. COROLLARY. *Let R be any one of the solutions of the YBE described in Theorem* 3.35 *and suppose conditions* (5.22), (5.23) *of Corollary* 5.21 *hold* (*so that there is an YB operator* $(\tau R, \nu, \alpha, \beta)$). *Then the corresponding link invariant is a Laurent polynomial in the* λ_i, z_i, z_{ij}.

Proof. If there are no blocks of type I present this is an immediate corollary of Theorem 5.33. The presence of a block of type I changes very little (essentially on extra scalar multiple of the identity block in S) and the result remains true.

5.40. NEW INVARIANTS FROM MIXED SOLUTIONS

We already know from 5.28 that putting together several blocks (in a nontrivial way can give real extra information. In the case of $n = 4$ and 2 (different) type

II blocks of size 2 the resulting link invariant will be a Laurent polynomial in λ_1, λ_2, z_1, z_2, z_{12}. One of the z's, say z_1, can be normalized away (or absorbed into α which is the same thing) so that the result is a Laurent polynomial in four variables (with one nontrivial relation given by (5.22) between them and there does not seem to be any obvious way to write this polynomial in terms of known 'classical' ones. In particular, there is in general (e.g., for $\lambda_1 \neq \lambda_2$) no relation like (5.25). Just what this polynomial and all the other ones arising from Theorem 3.35 via Corollary 5.21 bring in terms of new invariants remains to be explored.

Appendix 1

Direct proof that the ideal generated by the elements (2.3), (2.4) *is a Hopf algebra ideal in* $K\langle t\rangle$.

Let I be the ideal in $K\langle t\rangle$ generated by the elements (2.3), (2.4). Under the comultiplication of $K\langle t\rangle$, we have

$$t_r^a t_r^b - q^{ab} t_r^b t_r^a \longmapsto t_{i_1}^a t_{i_2}^b \otimes t_r^{i_1} t_r^{i_2} - q^{ab} t_{j_1}^b t_{j_2}^a \otimes t_r^{j_1} t_r^{j_2}. \tag{A1.1}$$

First consider the terms on the right of (A1.1) with $i_1 = i_2$ and $j_1 = j_2$. These balance in pairs:

$$\begin{aligned} & t_i^a t_i^b \otimes t_r^i t_r^i - q^{ab} t_i^b t_i^a \otimes t_r^i t_r^i \\ & \quad = (t_i^a t_i^b - q^{ab} t_i^b t_i^a) \otimes t_r^i t_r^i \in I \otimes K\langle t\rangle. \end{aligned} \tag{A1.2}$$

The remaining terms on the right-hand side of (A1.1) are treated in groups of four $(i \neq j)$.

$$\begin{aligned} & t_i^a t_j^b \otimes t_r^i t_r^j - q^{ab} t_i^b t_j^a \otimes t_r^i t_r^j + t_j^a t_i^b \otimes t_r^j t_r^i - q^{ab} t_j^b t_i^a \otimes t_r^j t_r^i \\ & \quad \equiv (t_i^a t_j^b - q^{ab} t_i^b t_j^a + (q^{ij})^{-1} t_j^a t_i^b - (q^{ab})(q^{ij})^{-1} t_j^b t_i^q) \otimes t_r^i t_r^j \\ & \quad \equiv 0 \mod (I \otimes K\langle t\rangle + K\langle t\rangle \otimes I) \end{aligned} \tag{A1.3}$$

(where the first congruence is in fact $\mathrm{mod}(K\langle t\rangle \otimes I)$ and the second $\mathrm{mod}\, I \otimes (K\langle t\rangle))$.

The elements (2.4) are twice as complicated to treat. Under the comultiplication, (2.4) goes to

$$\begin{aligned} & t_{i_1}^a t_{i_2}^b \otimes t_r^{i_1} t_s^{i_2} - q^{ab} t_{j_1}^b t_{j_2}^a \otimes t_r^{j_1} t_s^{j_2} + \\ & \quad + (q^{rs})^{-1} t_{k_1}^a t_{k_2}^b \otimes t_s^{k_1} t_r^{k_2} - (q^{ab})(q^{rs})^{-1} t_{l_1}^b t_{l_2}^a \otimes t_s^{l_1} t_r^{l_2}. \end{aligned} \tag{A1.4}$$

The terms with $i_1 = i_2$ fit with those with $j_1 = j_2$ for the same value ($i_1 = i_2 = j_1 = j_2$):

$$t_i^a t_i^b \otimes t_r^i t_s^i - q^{ab} t_i^b t_i^a \otimes t_r^i t_s^i = (t_i^a t_i^b - q^{ab} t_i^b t_i^a) \otimes t_r^i t_s^i \in I \otimes K\langle t \rangle.$$

Similarly, the terms with $k_1 = k_2$ fit with those of $l_1 = l_2$ for the same value.

Recall that if $a = b$ the element (2.4) is zero. So $a \neq b$ in (A1.4). The remaining terms of (A1.4) are dealt with in groups of eight as follows:

$$\begin{aligned}
&t_i^a t_j^b \otimes t_r^i t_s^j + t_j^a t_i^b \otimes t_r^j t_s^i - q^{ab} t_i^b t_j^a \otimes t_r^i t_s^j - q^{ab} t_j^b t_i^a \otimes t_r^j t_s^i + \\
&\quad +(q^{rs})^{-1} t_i^a t_j^b \otimes t_s^i t_r^j + (q^{rs})^{-1} t_j^a t_i^b \otimes t_s^j t_r^i - \\
&\quad -(q^{ab})(q^{rs})^{-1} t_i^b t_j^a \otimes t_s^i t_r^j - (q^{ab})(q^{rs})^{-1} t_j^b t_i^a \otimes t_s^j t_r^i \\
&= (t_i^a t_j^b - q^{ab} t_i^b t_j^a + (q^{ij})^{-1} t_j^a t_i^b - (q^{ab})(q^{ij})^{-1} t_j^b t_i^a) \otimes t_r^i t_s^j - \\
&\quad -(q^{ij})^{-1} t_j^a t_i^b \otimes (t_r^i t_s^j - q^{ij} t_r^j t_s^i + (q^{rs})^{-1} t_s^i t_r^j - q^{ij} (q^{rs})^{-1} t_s^j t_r^i) + \\
&\quad +(q^{ab})(q^{ij})^{-1} t_j^b t_i^a \otimes (t_r^i t_s^j - q^{ij} t_r^j t_s^i + (q^{rs})^{-1} t_s^i t_r^j - q^{ij} (q^{rs})^{-1} t_s^j t_r^i) + \\
&\quad +(t_i^a t_j^b - q^{ab} t_i^b t_j^a + (q^{ij})^{-1} t_j^a t_i^b - (q^{ab})(q^{ij})^{-1} t_j^b t_i^a) \otimes (q^{rs})^{-1} t_s^i t_r^j,
\end{aligned}$$

which is in $I \otimes K\langle t \rangle + K\langle t \rangle \otimes I$. Above the RHS differs from the LHS only in regrouping and the insertion of the four terms

$$(q^{ij})^{-1} t_j^a t_i^b \otimes t_r^i t_s^j, \qquad q^{ab} (q^{ij})^{-1} t_j^b t_i^a \otimes t_r^i t_s^j,$$

$$(q^{ij})^{-1} (q^{rs})^{-1} t_j^a b_i^b \otimes t_s^i t_r^j, \qquad q^{ab} (q^{ij})^{-1} (q^{rs})^{-1} t_j^b t_i^a \otimes t_s^i t_r^j,$$

each both with a plus and a minus sign.

This proves that I_L is a bialgebra ideal. The proof for I_R is completely analogous.

Appendix 2

Derivation of the R-equations (R1)–(R7) *of Subsection* 3.6 *and proof that these are all equations.*

The general equation is (cf. (3.2))

$$r_{k_1 k_2}^{ab} r_{u k_3}^{k_1 c} r_{vw}^{k_2 k_3} = r_{l_1 l_2}^{bc} r_{l_3 w}^{a l_2} r_{uv}^{l_3 l_1}. \tag{A2.1}$$

By Lemma (3.4), we know that under the condition

$$r_{cd}^{ab} = 0 \quad \text{unless} \quad \{a, b\} = \{c, d\} \tag{A2.2}$$

both sides of (A2.1) are zero unless $\{a, b, c\} = \{u, v, w\}$.

CASE 1. $a = b = c = u = v = w$.
Then the LHS of (A2.1) is nonzero iff $k_1 = k_2 = k_3 = a$ and then is equal to

$(r^{aa}_{aa})^3$. Similarly, the RHS of (A2.1) is nonzero iff $l_1 = l_2 = l_3 = a$ and then it is also equal to $(r^{aa}_{aa})^3$. No extra equation results from this case.

CASE 2. $a \neq b \neq c \neq a$.
There are six subcases to be considerd, namely how the u, v, w match up with the a, b, c.

Subcase 2.1. $u = a,\ v = b,\ w = c$.
For a nonzero term on the LHS we need $k_1 = a,\ k_2 = b,\ k_3 = c$ giving a term $r^{ab}_{ab} r^{ac}_{ac} r^{bc}_{bc}$.
For a nonzero term on the RHS we need $l_1 = b,\ l_2 = c,\ l_3 = a$ giving a term $r^{bc}_{bc} r^{ac}_{ac} r^{ab}_{ab}$.
Thus, always LHS = RHS in this subcase and no extra equation results.

Subcase 2.2. $u = a,\ v = c,\ w = b$.
For a nonzero term on the LHS we need $k_1 = a,\ k_2 = b,\ k_3 = c$ giving a term $r^{ab}_{ab} r^{ac}_{ac} r^{bc}_{cb}$.
For a nonzero term on the RHS we need $l_1 = c,\ l_2 = b,\ l_3 = a$ giving a term $r^{bc}_{cb} r^{ab}_{ab} r^{ac}_{ac}$.
Thus, always LHS = RHS in this subcase and no extra equation results.

Subcase 2.3. $u = b,\ v = a,\ w = c$.
For a nonzero term on the left hand side we need $k_1 = b,\ k_2 = a,\ k_3 = c$ giving a term $r^{ab}_{ba} r^{bc}_{bc} r^{ac}_{ac}$.
For a nonzero term on the RHS we need $l_1 = b,\ l_2 = c,\ l_3 = a$ giving a term $r^{bc}_{bc} r^{ac}_{ac} r^{ab}_{ba}$.
Thus, always LHS = RHS in this subcase and no extra equation results. .

Subcase 2.4. $u = b,\ v = c,\ w = a$.
For a nonzero term on the LHS we need $k_1 = b,\ k_2 = a,\ k_3 = c$ giving a term $r^{ab}_{ba} r^{bc}_{bc} r^{ac}_{ca}$.
For a nonzero term on the RHS we need $l_1 = b,\ l_2 = c$ or $l_2 = c,\ l_1 = b$ and $l_3 = l_2$ giving the terms $r^{bc}_{bc} r^{ac}_{ca} r^{cb}_{bc}$ and $r^{bc}_{cb} r^{ab}_{ba} r^{bc}_{bc}$.
Thus, LHS = RHS in this subcase holds iff

$$r^{bc}_{bc}(r^{ab}_{ba} r^{ac}_{ca}) = r^{bc}_{bc}(r^{ac}_{ca} r^{cb}_{bc} + r^{bc}_{cb} r^{ab}_{ba}). \tag{R1}$$

Subcase 2.5. $u = c,\ v = a,\ w = b$.
For a nonzero term on the LHS we need $k_1 = a,\ k_2 = b$ or $k_1 = b,\ k_2 = a$ and $k_3 = k_1$ giving the terms $r^{ab}_{ab} r^{ac}_{ca} r^{ba}_{ab}$ and $r^{ab}_{ba} r^{bc}_{cb} r^{ab}_{ab}$.
For a nonzero term on the RHS we need $l_1 = c,\ l_2 = b,\ l_3 = a$ giving a term $r^{bc}_{cb} r^{ab}_{ab} r^{ac}_{ca}$.
Thus, LHS = RHS in this subcase holds iff

$$r^{ab}_{ab}(r^{ac}_{ca} r^{ba}_{ab} + r^{ab}_{ba} r^{bc}_{cb}) = r^{ab}_{ab}(r^{bc}_{cb} r^{ac}_{ca}). \tag{R2}$$

Subcase 2.6. $u = c,\ v = b,\ w = a$.
For a nonzero term on the LHS we need $k_1 = a$, $k_2 = b$ or $k_1 = b$, $k_2 = a$ and $k_3 = k_1$ giving the terms $r^{ab}_{ab}r^{ac}_{ca}r^{ba}_{ba}$ and $r^{ab}_{ba}r^{bc}_{cb}r^{ab}_{ba}$.
For a nonzero term on the RHS we need $l_1 = b$, $l_2 = c$ or $l_1 = c$, $l_2 = b$ and $l_3 = l_2$ giving the terms $r^{bc}_{bc}r^{ac}_{ca}r^{cb}_{cb}$ and $r^{bc}_{cb}r^{ab}_{ba}r^{bc}_{cb}$.
Thus, LHS = RHS in this subcase holds iff

$$r^{ab}_{ab}r^{ba}_{ba}r^{ac}_{ca} + r^{ab}_{ba}r^{avb}_{ba}r^{bc}_{cb} = r^{bc}_{bc}r^{cb}_{cb}r^{ac}_{ca} + r^{ab}_{ba}r^{ba}_{cb}r^{bc}_{cb}. \tag{R3}$$

CASE 3. $a = b \neq c$.
Again there are a number of subcases to consider depending on how the u, v, w match up with the a, b, c. The six possibilities a priori coincide in pairs giving three subcases.

Subcase 3.1. $u = v = a = b,\ w = c$.
For a nonzero term on the LHS we need $k_1 = a$, $k_2 = a$, $k_3 = c$ giving the term $r^{aa}_{aa}r^{ac}_{ac}r^{ac}_{ac}$.
For a nonzero term on the RHS we need $l_1 = a$, $l_2 = c$, $l_3 = a$ giving a term $r^{ac}_{ac}r^{ac}_{ac}r^{aa}_{aa}$.
Thus, always LHS = RHS in this subcase and no extra equation results.

Subcase 3.2. $u = w = a = b, v = c$.
For a nonzero term on the LHS we need $k_1 = k_2 = a$, $k_3 = c$ giving a term $r^{aa}_{aa}r^{ac}_{ac}r^{ac}_{ca}$.
For a nonzero term on the RHS we need $l_1 = a$, $l_2 = c$ or $l_1 = c$, $l_2 = a$ and $l_3 = l_2$ giving the terms $r^{ac}_{ac}r^{ac}_{ca}r^{ca}_{ac}$ and $r^{ac}_{ca}r^{aa}_{aa}r^{ac}_{ac}$.
Thus, LHS = RHS in this subcase iff

$$r^{ac}_{ac}r^{ac}_{ca}r^{ca}_{ac} = 0. \tag{R4}$$

Subcase 3.3. $u = c,\ v = w = a = b$.
For a nonzero term on the LHS we need $k_1 = k_2 = k_3 = a$ giving a term $r^{aa}_{aa}r^{ac}_{ca}r^{aa}_{aa}$.
For a nonzero term on the RHS we need $l_1 = a$, $l_2 = c$ or $l_1 = c$, $l_2 = a$ and $l_3 = l_2$ giving the terms $r^{ac}_{ac}r^{ac}_{ca}r^{ca}_{ca}$ and $r^{ac}_{ca}r^{aa}_{aa}r^{ac}_{ca}$.
Thus, LHS = RHS in this subcase iff

$$r^{aa}_{aa}r^{aa}_{aa}r^{ac}_{ca} = r^{aa}_{aa}r^{ac}_{ca}r^{ac}_{ca} + r^{ac}_{ac}r^{ca}_{ca}r^{ac}_{ca}. \tag{R5}$$

CASE 4. $a \neq b = c$.
As in case 3, there are three subcases to consider

Subcase 4.1. $u = a,\ v = w = b = c$.
For a nonzero term on the LHS we need $k_1 = a$, $k_2 = k_3 = b$ giving a term $r^{ab}_{ab}r^{ab}_{ab}r^{bb}_{bb}$.
For a nonzero term on the RHS we need $l_1 = l_2 = b$, $l_3 = a$ giving a term

$r^{bb}_{bb} r^{ab}_{ab} r^{ab}_{ab}$.
Thus, always LHS = RHS in this subcase and no extra equation results.

Subcase 4.2. $u = b = w = c$, $v = a$.
For a nonzero term on the LHS we need $k_1 = a$, $k_2 = b$ or $k_1 = b$, $k_2 = a$ and $k_3 = k_1$ giving the terms $r^{ab}_{ab} r^{ab}_{ba} r^{ba}_{ab}$ and $r^{ab}_{ba} r^{bb}_{bb} r^{ab}_{ab}$.
For a nonzero term on the RHS we need $l_1 = l_2 = b$, $l_3 = c$ giving a term $r^{bb}_{bb} r^{ab}_{ab} r^{ab}_{ba}$.
Thus, LHS = RHS in this subcase iff

$$r^{ab}_{ab} r^{ab}_{ba} r^{ba}_{ab} = 0$$

giving (R4) for the second time.

Subcase 4.3. $u = v = b = c$, $w = a$.
For a nonzero term on the LHS we need $k_1 = a$, $k_2 = b$ or $k_1 = b$, $k_2 = a$ and $k_3 = k_1$ giving the terms $r^{ab}_{ab} r^{ab}_{ba} r^{ba}_{ba}$ and $r^{ab}_{ba} r^{bb}_{bb} r^{ab}_{ba}$.
For a nonzero term on the RHS we need $l_1 = l_2 = l_3 = b$ giving a term $r^{bb}_{bb} r^{ab}_{ba} r^{bb}_{bb}$.
Thus, RHS = LHS in this subcase iff

$$r^{bb}_{bb} r^{bb}_{bb} r^{ab}_{ba} = r^{bb}_{bb} r^{ab}_{ba} r^{ab}_{ba} + r^{ab}_{ab} r^{ba}_{ba} r^{ab}_{ba}. \tag{R6}$$

Note that this is not the same equation as (R5) (also after changing b to a, a to c).

CASE 5. $a = c \neq b$.
As in Cases 3 and 4, there are three subcases to consider.

Subcase 5.1. $u = w = a = c$, $v = b$.
For a nonzero term in the LHS we need $k_1 = a$, $k_2 = b$ or $k_1 = b$, $k_2 = a$ and $k_3 = k_1$ giving the terms $r^{ab}_{ab} r^{aa}_{aa} r^{ba}_{ba}$ and $r^{ab}_{ba} r^{ba}_{ab} r^{ab}_{ba}$.
For a nonzero term on the RHS we need $l_1 = b$, $l_2 = a$ or $l_1 = a$, $l_2 = b$ and $l_3 = l_2$ giving the terms $r^{ba}_{ba} r^{aa}_{aa} r^{ba}_{ab}$ and $r^{ba}_{ab} r^{ab}_{ba} r^{ba}_{ab}$.
Thus, LHS = RHS in this subcase iff

$$r^{ab}_{ba} r^{ba}_{ab} r^{ab}_{ba} = r^{ab}_{ba} r^{ba}_{ab} r^{ba}_{ab}. \tag{R7}$$

Subcase 5.2. $u = v = a = c$, $w = b$.
For a nonzero term on the LHS we need $k_1 = a$, $k_2 = b$ or $k_1 = b$, $k_2 = a$ and $k_3 = k_1$ giving the terms $r^{ab}_{ab} r^{aa}_{aa} r^{ba}_{ab}$ and $r^{ab}_{ba} r^{ba}_{ab} r^{ab}_{ab}$.
For a nonzero term on the RHS we need $l_1 = a$, $l_2 = b$, $l_3 = a$ giving a term $r^{ba}_{ab} r^{ab}_{ab} r^{aa}_{aa}$.
Thus, LHS = RHS in this subcase iff

$$r^{ab}_{ab} r^{ab}_{ba} r^{ba}_{ab} = 0$$

giving (R4) for the third time.

Subcase 5.3. $u = b,\ v = w = a = c$.
For a nonzero term on the LHS we need $k_1 = a,\ k_2 = b$ or $k_1 = b,\ k_2 = k_3 = a$ giving a term $r^{ab}_{ba} r^{ba}_{ba} r^{aa}_{aa}$.
For a nonzero term on the RHS we need $l_1 = b,\ l_2 = a$ or $l_1 = a,\ l_2 = b$ and $l_3 = l_2$ giving the terms $r^{ba}_{ba} r^{aa}_{aa} r^{ab}_{ba}$ and $r^{ba}_{ab} r^{ab}_{ba} r^{ba}_{ba}$.
Thus, LHS = RHS in this subcase iff

$$r^{ba}_{ba} r^{ab}_{ba} r^{ba}_{ab} = 0$$

giving (R4) for the fourth time.

References

1. Artin, M., Schelter, W., and Tate, J.: Quantum deformations of GL_n, Preprint 1991.
2. Bergman, G. M.: The diamond lemma for ring theory, *Adv. Math.* **29** (1978), 178–218.
3. Dipper, R. and Donkin, S.: Quantum GL_n, *Proc. London Math. Soc.* **63** (1991), 162–211.
4. Drinfeld, V. G.: Quantum groups, in: *Encyclopaedia of Mathematics*, Vol. 7, Kluwer Acad. Publ., Dordrecht, 1991, pp. 408–410.
5. Jie Du, Parshall, B., and Jian-Pan Wang: Two parameter quantum linear groups and the hyperbolic invariance of q-Schur algebras, Preprint 1990, to appear, in *J. London Math. Soc.*
6. Faddeev, L. D., Reshetikhin, N. Yu., and Takhtadzhyan, L. A.: Quantized Lie groups and Lie algebras, *Leningrad Math. J.* **1** (1990), 491–513.
7. Gerstenhaber, M. and Schack, S. D.: Bialgebra cohomology, deformations and quantum groups, *Proc. Nat. Acad. Sci. USA* **87** (1990), 478–481.
8. Gurevich, D. I.: Algebraic aspects of the quantum Yang–Baxter equation, *Leningrad Math. J.* **2** (1991), 801–828.
9. Hazewinkel, M.: On quadratic Hopf algebras, unpublished notes, 10 March 1990.
10. Hazewinkel, M.: Introductory recommendations for the study of Hopf algebras in mathematics and physics, *CWI Quarterly* **4**(1) (1991), 3–26.
11. Kulish, P. P.: The two parameter quantum group and gauge transformation, *Zap. Nauchn. Sem. LOMI* **180** (1990), 89–93.
12. Lyabashenko, V. V.: Hopf algebras and vector symmetries, *Russian Math. Survey* **41**(5) (1986), 153–154.
13. Lyabashenko, V. V.: Vector symmetries, Preprint, Math. Inst., Univ. of Stockholm, 1987.
14. Maltsinotis, H.: Groupes quantiques et structures différentielles, Preprint.
15. Manin, Yu. I.: Multiparameter quantum deformation of the general linear supergroup, *Comm. Math. Phys.* **123** (1989), 163–175.
16. Manin, Yu. I.: Quantum groups and noncommutative geometry, CRM, Univ. de Montréal, 1988.
17. Reshetikhin, N.: Multiparameter quantum groups and twisted quasi-triangular Hopf algebras, *Lett. Math. Physics* **20** (1990), 331–335.
18. Sudbery, A.: Consistent multiparameters quantization of $\mathrm{GL}(n)$, Preprint, Dept. Math., Univ. of York, 1990.
19. Sudbery, A.: Quantum matrix groups determined by quadratic coordinate algebras, Preprint, Dept. Math., Univ. of York, Sept. 1990.
20. Takeuchi, M.: A two parameter quantization of GL_n, *Proc. Japan Acad., Ser. A* **66** (1990), 112–114.
21. Takeuchi, M.: Matrix bialgebras and quantum groups, *Israel J. Math.* **72** (1990), 232–251.
22. Turaev, V. G.: The Yang–Baxter equation and invariants of links, *Invent. Math.* **92** (1988), 527–553.

Acta Applicandae Mathematicae **41:** 99–121, 1995.

Infinite-Dimensional Flag Manifolds in Integrable Systems

G. F. HELMINCK
Department of Mathematics, Universiteit Twente, Enschede, The Netherlands.
e-mail: helminck@math.utwente.nl

and

A. G. HELMINCK
Department of Mathematics, North Carolina State University, Raleigh, NC 27695-8205, U.S.A.
e-mail: loek@math.ncsu.edu

(Received: 28 February 1994)

Abstract. In this paper, we present several instances where infinite-dimensional flag varieties and their holomorphic line bundles play a role in integrable systems. As such, we give the correspondence between flag varieties and Darboux transformations for the KP hierarchy and the nth KdV hierarchy. We construct solutions of the nth MKdV hierarchy from the space of periodic flags and we treat the geometric interpretation of the Miura transform. Finally, we show how the group extension connected with these line bundles shows up at integrable deformations of linear systems on $\mathbb{P}^1(\mathbb{C})$.

Mathematics Subject Classifications (1991): 22E65, 14M15, 35Q58, 43A80, 17B65.

Key words: infinite-dimensional flag varieties, modified equations, Darboux transforms, integrable deformations.

1. Introduction

Let H be a finite-dimensional complex Hilbert space. Then a flag $V = (V(i))$ in H is a chain of subspaces in H,

$$\{0\} = V(0) \subsetneq V(1) \subsetneq \cdots \subsetneq V(m) = H.$$

If we put $\underline{d} = (d_1, \ldots, d_m)$, with $d_i = \dim(V(i))$ and we write $\mathcal{F}_{\underline{d}}$ for the collection of all flags $V = (V(i))$ in H such that $\dim V(i) = d_i$, then $\mathcal{F}_{\underline{d}}$ is a compact Kähler manifold. The flag variety of type $\underline{d}$. The flag varieties $\mathcal{F}_{\underline{d}}$ are the main ingredients in field theory for the definition of the fundamental complex manifolds of twistor geometry, see [22]. If $m = 2$, then $\mathcal{F}_{\underline{d}}$ is simply the collection of all subspaces of H of dimension d_1 and is, as such, better known as the Grassmann manifold $\mathrm{Gr}(d_1, n - d_1)$. These manifolds are the natural scene for Ricatti type equations and play, as such, an important role in control theory, see [9].

The manifolds $\mathcal{F}_{\underline{d}}$ are homogeneous spaces for both $\mathrm{Gl}(H)$ and $\mathrm{U}(H)$. By adding additional requirements to the flags under consideration, one obtains flag varieties corresponding to other groups.

Holomorphic line bundles over the varieties $\mathcal{F}_{\underline{d}}$ lead to a geometric realization of the irreducible unitary representations of $\mathrm{U}(H)$ as the natural action of $\mathrm{U}(H)$ on the holomorphic sections of these line bundles. An important example from mathematical physics, where these bundles are crucial is the Penrose transform. For details and generalizations of this theme, we refer to [2]

In the second section of this paper, we give the geometric structure of the flag varieties and their holomorphic line bundles as they are needed in the context of integrable systems.

The third section describes the systems of equations and transformations that will be fit into the geometric framework.

In the fourth section, we recall first some results for the KP hierarchy. Next we show how Darboux transformations can be seen geometrically and we conclude with the construction of solutions for the nth MKdV hierarchy and the interpretation of the Miura transformation.

The final section describes the role the group extension related to the line bundles plays at integrable deformations of a linear system on $\mathbb{P}^1(\mathbb{C})$.

2. Flag Varieties

2.1. The Type of Flags

Like with infinite-dimensional vector spaces, there is a wide range of manifold structures that can be considered for infinite-dimensional flag varieties. Since we like to discuss analytic properties of our varieties and apply them in analytic situations, we will not take the algebraic set-up from [13], for example. We will also not try to generalize the Sato Grassmanian, see, e.g., [12] and [5], or the one from [26]. Our choice will be a natural generalization of the Grassmann manifolds in [20]. Let H be a separable complex Hilbert space with inner product $\langle \cdot \mid \cdot \rangle$. Since we work in a topological context, it is reasonable to consider merely chains $F = (F(i))$ in H,

$$\{0\} = F(0) \subsetneq F(1) \subsetneq \cdots \subsetneq F(m) = H,$$

where all the $F(i)$ are closed subspaces of H. If we put for all i, $1 \leqslant i \leqslant m$,

$$F_i := F(i-1)^{\perp} \cap F(i),$$

then we see that each flag $F = (F(i))$ corresponds precisely to an orthogonal decomposition of H

$$H = F_1 \oplus \cdots \oplus F_m.$$

We will use both notations $F = (F(i))$ and $F = (F_i)$ to describe a flag. Our starting point is a given orthogonal decomposition

$$H = H_1 \oplus \cdots \oplus H_m \quad \text{with } H_i \perp H_j \text{ for } i \neq j. \tag{2.2.1}$$

The flag corresponding to the (H_i), we denote by $F^{(0)}$ and we call it the *basic flag*. We will take together all flags 'similar' to the basic flag and make this into some kind of manifold. First, we present two examples of decompositions that occur naturally at modelling certain equations from mathematical physics. We start with what will be the leading example during the rest of this paper

EXAMPLE 2.1. Let H be the Hilbert space

$$L^2(S^1, \mathbb{C}) = \left\{ \sum_{n\in\mathbb{Z}} a_n z^n,\ a_n \in \mathbb{C},\ \sum_{n\in\mathbb{Z}} |a_n|^2 < \infty \right\}$$

with the inner product

$$\left\langle \sum_{n\in\mathbb{Z}} a_n z^n \mid \sum_{n\in\mathbb{Z}} b_n z^n \right\rangle = \sum_{n\in\mathbb{Z}} a_n \bar{b}_n.$$

Let $\underline{s} = (s_1, \ldots, s_{m-1})$, where $s_i \in \mathbb{Z}$ and $s_{i+1} < s_i$, then we take the basic flag defined by

$$H(i) = \left\{ \sum_{n\geqslant s_i} a_n z^n \in H \right\} \quad \text{for } i,\ 1 \leqslant i \leqslant m-1 \text{ and } H(m) = H.$$

We will see later on that the flag manifold corresponding to this flag is the 'basis' of the modified equations.

EXAMPLE 2.2. In quantum field theory the states of the system are vectors in a Fock space. In the fermionic case, this space is built up from the splitting of a basic Hilbert space into positive and negative energy states, e.g., at the Dirac theory for a one dimensional particle of mass $m \geqslant 0$, one considers, see [3], the Dirac Hamiltonian

$$D = \begin{pmatrix} -i & 0 \\ 0 & -i \end{pmatrix} \frac{\mathrm{d}}{\mathrm{d}x} + \begin{pmatrix} 0 & m \\ m & 0 \end{pmatrix}.$$

It acts on the Hilbert space $H = L^2(\mathbb{R})^2$ and the relevant decomposition of H for the Fock space representation is $H = H_+ \oplus H_-$, where H_+ is the subspace of H corresponding to the positive spectrum of D and H_- the one corresponding to the negative spectrum of D. In general, see [16], Dirac operators are associated to a number of geometric data, like a spin bundle over an oriented Riemannian manifold, another vector bundle over this manifold and a connection. A similar splitting of the square-integrable extended Dirac spinors is the starting point for the construction of a representation of an extension of the group of gauge transformations, see [16].

In the finite-dimensional case, it was sufficient to fix the dimension of each F_i and then to take all flags of the same size. This idea is too vague in this context and requires adaptation. For each i, $1 \leqslant i \leqslant m$, let p_i denote the orthogonal projection from H onto H_i. Then we introduce

DEFINITION 2.3. The *flag variety* $\mathcal{F}$ corresponding to the decomposition (2.2.1) is the collection of flags $F = \{F_1, \ldots, F_m\}$ in H, satisfying $\dim(F_i) = \dim(H_i)$ and for all i and j with $j \neq i$, the orthogonal projection $p_j\colon F_i \to H_j$ is a Hilbert–Schmidt operator.

If H and $\underline{s}$ are as in Example 2.1, then we write $\mathcal{F} = \mathcal{F}(\underline{s})$. The flag variety $\mathcal{F}$ is a homogeneous space for an analogue adapted to this situation of the general linear group. The Banach structure of this group follows directly from that of its Lie algebra. This requires a

Notation 2.4. If g belongs to $\mathcal{B}(H)$, the space of bounded linear operators from H to H, then $g = (g_{ij})$, $1 \leqslant i \leqslant m$, and $1 \leqslant j \leqslant m$, denotes the decomposition of g w.r.t. the $\{H_i \mid 1 \leqslant i \leqslant m\}$. That is to say $g_{ij} = p_i \circ g \mid H_j$.

Now we come to the Lie algebra of the analogue of the general linear group.

DEFINITION 2.5. A *restricted endomorphism* of H is a $u = (u_{ij})$ in $\mathcal{B}(H)$ such that u_{ij} is a Hilbert–Schmidt operator for all $i \neq j$. We denote the space of all restricted endomorphisms of H by $\mathcal{B}_{\text{res}}(H)$.

The algebra $\mathcal{B}_{\text{res}}(H)$ becomes a Banach algebra if we equip it with the norm $\|\cdot\|_2$ defined by

$$\|u\|_2 = \|u\| + \sum_{i \neq j} \|u_{ij}\|_{\mathcal{HS}}.$$

where $\|\cdot\|_{\mathcal{HS}}$ denotes the Hilbert–Schmidt norm. The Lie group corresponding to $\mathcal{B}_{\text{res}}(H)$ is the following group.

DEFINITION 2.6. The *restricted linear group*, $\mathrm{Gl}_{\text{res}}(H)$, is the group of invertible elements in $\mathcal{B}_{\text{res}}(H)$. As such, it is a natural Banach Lie group.

One easily shows that $\mathrm{Gl}_{\text{res}}(H)$ can be described as

$$\mathrm{Gl}_{\text{res}}(H) = \left\{ g = (g_{ij}) \in \mathrm{Gl}(H) \;\middle|\; \begin{array}{l} g_{ii} \text{ is a Fredholm operator for all } i \\ g_{ij} \in \mathcal{HS}(H_j, H_i), \text{ for } i \neq j. \end{array} \right\}$$

Moreover, $\mathrm{Gl}_{\text{res}}(H)$ acts on $\mathcal{F}$ through its natural action on subspaces of H and this action is transitive, see [8]. Hence, we have $\mathcal{F} = \mathrm{Gl}_{\text{res}}(H)/P$, where P is the parabolic subgroup stabilizing the basic flag.

Let τ: $\mathrm{Gl}_{\mathrm{res}}(H) \to \mathcal{F}$ be the projection $\tau(g) = g \cdot F^{(0)}$. From the definition of the parabolic group P, one sees directly that the Lie algebra of P is given by

$$L(P) = \{g \mid g = (g_{ij}) \in \mathcal{B}_{\mathrm{res}}(H),\ g_{ij} = 0 \ \text{ for all } \ i > j\}$$

and that a complement of $L(P)$ in $\mathcal{B}_{\mathrm{res}}(H)$ is the Hilbert space $(E, \|\cdot\|_{\mathcal{HS}})$ with

$$E = \bigoplus_{\substack{1 \leqslant j \leqslant m-1 \\ i>j}} \mathcal{HS}(H_j, H_i).$$

From [8], we know then that the homogeneous space $\mathcal{F} = \mathrm{Gl}_{\mathrm{res}}(H)/P$ carries an analytic E-manifold structure for which τ is a submersion and for which the natural action of $\mathrm{Gl}_{\mathrm{res}}(H)$ on $\mathcal{F}$ is analytic.

It is shown in [8] that the group of unitary operators in $\mathrm{Gl}_{\mathrm{res}}(H)$ acts transitively on $\mathcal{F}$. Hence, to make $\mathcal{F}$ into a Hermitian manifold, it suffices to give on the tangent space E of the basic flag $F^{(0)}$ a Hermitian form, which is invariant under the stabilizer of $F^{(0)}$. Consider on $E \times E$ the form $H(\cdot \mid \cdot)$ given by

$$H(X \mid Y) = H((X_{ij})|(Y_{ij})) = \mathrm{Trace}\left(\sum_{\substack{i=2 \\ j<i}}^{m} X_{ij} Y_{ij}^*\right)$$

Let $\omega(X, Y)$ be the imaginary part of $H(X \mid Y)$. One verifies directly that ω is a nondegenerate bilinear form. By direct computation or by using the Lie algebra 2-cocycles for $\mathcal{B}_{\mathrm{res}}(H)$ from [8], one shows that ω is closed: $\mathrm{d}\omega = 0$. Hence, $\mathcal{F}$ is a symplectic manifold and even a Kähler manifold, another property that carries over from the finite-dimensional $F_{\underline{d}}$.

Remark 2.7. For simplicity, we restrict ourselves in this paper to finite chains of subspaces. However, one could just as well consider infinite chains of closed subspaces. They also arise naturally in the context of integrable systems, see [6] and in an implicit form [7].

2.2. PROPERTIES OF $\mathcal{F}$

As in the Grassmanian case, one shows that the connected component of $\mathcal{F}$ containing the basic flag equals the orbit of the connected component $\mathrm{Gl}_{\mathrm{res}}^{(0)}(H)$, i.e.

$$\begin{aligned} \mathcal{F}^{(0)} &= \{g \cdot \mathcal{F}^{(0)} \mid g \in \mathrm{Gl}_{\mathrm{res}}^{(0)}(H)\} \\ &= \{g \cdot F^{(0)} \mid g = (g_{ij}),\ \text{index } (g_{ii}) = 0 \ \text{ for all } \ i\} \end{aligned}$$

It may look that all these flag varieties are homogeneous spaces for different groups. However, from the basic properties of Fredholm operators one deduces

PROPOSITION 2.8. *Let G_2 be the subgroup of* $\mathrm{Gl}_{\mathrm{res}}(H)$ *with the induced topology, defined by*

$$G_2 = \{g \mid g \in \mathrm{Gl}_{\mathrm{res}}(H),\ g - \mathrm{Id} \in \mathcal{HS}(H)\}.$$

Then G_2 acts transitively on all the connected components of the $\mathcal{F}$.

In the rest of this subsection, we will take H and the basic flag as in Example 2.1. For that example, we also need a description of the other connected components. Also here there holds

PROPOSITION 2.9. *The connected components of $\mathcal{F}(\underline{s})$ are given by*

$$\mathcal{F}^{(k)}(\underline{s}) = \{g \cdot F^{(0)} \mid \mathit{index}\ (g_{11}) = k\}, \quad k \in \mathbb{Z}.$$

For a proof, we refer to [8]. An important operator in $\mathrm{Gl}_{\mathrm{res}}(H)$ is the shift operator Λ defined by

$$\Lambda\Big(\sum_{n\in\mathbb{Z}} a_n z^n\Big) = \sum_{n\in\mathbb{Z}} a_n z^{n+1}.$$

As Λ_{11} has index -1, we see that for each l and k in $\mathbb{Z}$, Λ^l maps $\mathcal{F}^{(k)}(\underline{s})$ onto $\mathcal{F}^{(k-l)}(\underline{s})$. Taking this into account, we get the following result:

For each strictly decreasing sequence $\underline{s} = (s_1, \ldots, s_{m-1})$ and each $l \in \mathbb{Z}$, we have $\mathcal{F}(\underline{s}) = \mathcal{F}(\underline{t})$, where $\underline{t} = (t_1, \ldots, t_{m-1})$ is defined by $t_i = s_i + l$ for all i. (2.2.2)

Finally, we introduce a group of commuting flows on $\mathcal{F}(\underline{s})$ that plays a central role in the sequel. Let U be a connected neighbourhood of S^1 in $\mathbb{C}$ and let $\Gamma(U)$ be the collection of all nonzero holomorphic maps $\gamma\colon U \to \mathbb{C}^*$. In a natural way $\Gamma(U)$ is a group. If $U_1 \supset U_2$, then we get by restriction an embedding of $\Gamma(U_1)$ into $\Gamma(U_2)$. We write Γ for the inverse limit of the $\{\Gamma(U)\}$. Each $\gamma \in \Gamma$ has a Fourier series $\sum_{i\in\mathbb{Z}} \gamma_i z^i$, $\gamma_i \in \mathbb{C}$, that converges uniformly on some neighbourhood of S^1. Clearly, the multiplication with γ, gives the element $\sum_{i\in\mathbb{Z}} \gamma_i \Lambda^i$ of $B_{\mathrm{res}}(H)$ and this defines a continuous injective group homomorphism from Γ to $\mathrm{Gl}_{\mathrm{res}}(H)$. The closure of the image of Γ in $\mathrm{Gl}_{\mathrm{res}}(H)$ is a maximal Abelian subgroup of $\mathrm{Gl}_{\mathrm{res}}(H)$. In Γ we consider the following subgroups

$$\begin{aligned}
\Gamma_+ &= \Big\{\gamma \mid \gamma \in \Gamma, \gamma = \exp\Big(\sum_{i=1}^{\infty} t_i z^i\Big)\Big\},\\
\Gamma_- &= \Big\{\gamma \mid \gamma = \sum_{j\leqslant 0} \gamma_j z^j \in \Gamma\Big\}, \quad \text{and}\\
\Delta &= \{z^k \mid k \in \mathbb{Z}\}.
\end{aligned}$$

Then we have, see [10],

PROPOSITION 2.10. *The group* Γ *is the direct product of these* 3 *groups, i.e.* $\Gamma = \Gamma_+ \Delta \Gamma_-$.

This decomposition will play a role at the geometric description of the KP hierarchy.

2.3. Holomorphic line bundles over $\mathcal{F}^{(0)}$

If one wants to construct an analogue of the holomorphic line bundles over finite-dimensional flag varieties, one needs a description of $\mathcal{F}^{(0)}$ as a homogeneous space of a smaller group, for which certain 'minors' exist. Consider

$$\mathfrak{G} = \left\{ g \mid g = (g_{ij}) \in \mathrm{Gl}(H) \begin{array}{l} g_{ii} - \mathrm{Id} \text{ is trace-class} \\ g_{ij} \in \mathcal{HS}(H_j, H_i) \text{ for } i \neq j. \end{array} \right\}.$$

Then one verifies directly that $\mathfrak{G}$ is a group. We give it the topology based on

$$\bigoplus_{i \neq j} \mathcal{HS}(H_j, H_i) \oplus \bigoplus_{i=1}^{m} \mathcal{N}(H_i),$$

where $\mathcal{N}(H_i)$ is the space of trace class operators on H_i, equipped with the trace norm. The group $\mathfrak{G}$ is chosen such that for each i, $1 \leqslant i \leqslant m$, and for each $g = (g_{ij})$ in $\mathfrak{G}$ the minor

$$\det \begin{pmatrix} g_{11} & \cdots & g_{1i} \\ \cdots & \vdots & \cdots \\ g_{i1} & \cdots & g_{ii} \end{pmatrix}$$

exists. The natural embedding of $\mathfrak{G}$ into $\mathrm{Gl}_{\mathrm{res}}(H)$ is continuous. Moreover, it can be shown, see [8], that $\mathfrak{G}$ acts transitively on $\mathcal{F}^{(0)}$. Hence, we have that $\mathcal{F}^{(0)}$ is isomorphic to $\mathfrak{G}/\mathfrak{T}$, with $\mathfrak{T}$ the stabilizer of $F^{(0)}$ in $\mathfrak{G}$, i.e.

$$\mathfrak{T} = \left\{ t = \begin{pmatrix} t_{ii} & & t_{1m} \\ 0 & \cdots & \\ \vdots & \cdots & \\ 0 & \cdots 0 & t_{mm} \end{pmatrix}, \ t \in \mathfrak{G} \right\}.$$

For each $\underline{k} = (k_1, \cdots, k_m) \in \mathbb{Z}^m$, we define the analytic characters $\psi_{\underline{k}}\colon \mathfrak{T} \to \mathbb{C}^*$ by

$$\psi_{\underline{k}}(t) = \det(t_{11})^{k_1} \cdots \det(t_{mm})^{k_m}.$$

To each $\psi_{\underline{k}}$ there is associated a holomorphic line bundle $L(\underline{k})$ over $\mathcal{F}^{(0)}$. It can be defined as follows: consider on the space $\mathfrak{G} \times \mathbb{C}$ the equivalence relation

$$(g_1, \lambda_1) \sim (g_2, \lambda_2) \iff g_1 = g_2 \circ t \quad \text{with } t \in \mathfrak{T} \quad \text{and } \lambda_2 = \lambda_1 \psi_{\underline{k}}(t).$$

The space $\mathfrak{G} \times \mathbb{C}$ modulo this equivalence relation is $L(\underline{k})$. We denote the class of the pair $(g, \lambda), g \in \mathfrak{G}, \lambda \in \mathbb{C}$, by $[g, \lambda]$.

Remark 2.11. If we take $m = 2, \dim(H_1) = \dim(H_2) = \infty$, $k = (-1,0)$, resp., $k = (1,0)$, then the line bundle $L((-1,0))$, resp., $L((1,0))$, are the determinant bundle and its dual introduced in [13].

The group $\mathfrak{G}$ acts naturally on $L(\underline{k})$ by left translations. However, e.g., the group of flows Γ_+ from Subsection 2.2 is not contained in $\mathfrak{G}$ and one would like to let it act nevertheless. In general, one cannot lift the natural action of $\mathrm{Gl}_{\mathrm{res}}^{(0)}(H)$ on $\mathcal{F}^{(0)}$ to an action of the whole group on $L(\underline{k})$ and one is forced to pass to an extension G of $\mathrm{Gl}_{\mathrm{res}}^{(0)}(H)$. If D is the subgroup of $\mathrm{Gl}_{\mathrm{res}}^{(0)}(H)$, given by

$$D = \{g = (g_{ij}) \in \mathrm{Gl}_{\mathrm{res}}^{(0)}(H),\ g_{ij} = 0 \text{ if } i \neq j\},$$

then G is defined by

$$G = \{(g,d) \mid g \in \mathrm{Gl}_{\mathrm{res}}^{(0)}(H),\ d \in D,\ gd^{-1} \in \mathfrak{G}\}.$$

The parabolic subgroup P of $\mathrm{Gl}_{\mathrm{res}}^{(0)}(H)$ embeds as follows into G:

$$p = (p_{ij}) \longmapsto \left(p, \begin{pmatrix} p_{11} & & 0 \\ & \cdots & \\ 0 & & p_{mm} \end{pmatrix}\right).$$

The group G acts on $L(\underline{k})$ by

$$(g,d) \cdot [g_1, \lambda_1] = [g g_1 d^{-1}, \lambda_1].$$

In particular, there is thus a lifting of the action of Γ^+ on $\mathcal{F}^{(0)}$ to one on $L(\underline{k})$. This action of G leads to a natural representation of G on the holomorphic sections of $L(\underline{k})$. These representations are the analogues of the finite-dimensional irreducible representations of $\mathrm{Gl}_n(\mathbb{C})$.

3. The Equations

3.1. The KP hierarchy

We first recall the Lax form of this system of equations. One starts out with some commutative ring R of functions depending of the variables $\{x_1, t_1, t_2, t_3, \ldots\}$. Next, one assumes that $\partial := \partial/\partial x$ and all the $\partial_k := \partial/\partial t_k$ form derivations of R. To R and ∂ is associated the ring $R[\xi, \xi^{-1}]$ of pseudodifferential operators, whose elements are expressions

$$\sum_{i=-\infty}^{N} a_i \xi^i \quad \text{with } a_i \in R.$$

Addition and multiplication in $R[\xi,\xi^{-1}]$ are defined by

$$\sum_i a_i\xi^i + \sum_i b_i\xi^i = \sum_i (a_i+b_i)\xi^i,$$

$$\left(\sum_i a_i\xi^i\right)\cdot\left(\sum_j b_j\xi^j\right) = \sum_{\substack{i,j\\ m\geqslant 0}} \binom{i}{m} a_i\partial^m(b_j)\xi^{i+j-m}.$$

If $P=\sum_j p_j\xi^j$ belongs to $R[\xi,\xi^{-1}]$, then we denote its differential operator part $\sum_{j\geqslant 0} p_j\xi^j$ by P_+ and we write P_- for $\sum_{j<0} p_j\xi^j$.

In $R[\xi,\xi^{-1}]$, one considers so-called *Lax operators*, i.e. operators L of the form

$$L=\xi+\sum_{j>0} u_{j+1}\xi^{-j}. \tag{3.3.1}$$

Examples of Lax operators can be obtained by the *dressing procedure*: take a $K=\xi^n+\sum_{j<n} k_j\xi^j$, then K is invertible in $R[\xi,\xi^{-1}]$ and $L=K\xi K^{-1}$ is a Lax operator. All Lax operators can be obtained in this way if and only if ∂ is surjective. An important class of examples is the set of roots of differential operators. For, if $\mathcal{L}\in R[\xi]$ has the form

$$\mathcal{L}=\xi^n+\sum_{j=2}^{n} l_j\xi^{n-j}. \tag{3.3.2}$$

Then there exist a unique Lax operator L with $L^n=\mathcal{L}$ and this L is also denoted as $\mathcal{L}^{1/n}$.

We extend the derivations ∂ and $\{\partial_k \mid k\geqslant 1\}$ from R to derivations of $R[\xi,\xi^{-1}]$ by letting them act coefficientwise. Now we can introduce the KP hierarchy in Lax form

DEFINITION 3.1. The *Kadomtsev–Petviashvili hierarchy* (KP) is the following system of equations for a Lax operator L:

$$\partial_k(L)=[(L^k)_+,L] \quad \text{for all } k\geqslant 1. \tag{3.3.3}$$

A Lax operator that satisfies Equations (3.3.3) is called a *solution* of the KP-hierarchy. For $k=1$, we see that this equation says that $\partial=\partial_1$ on the coefficients of L. Therefore, we may just as well assume $x=t_1$ and this will be done from now on. If $L=\mathcal{L}^{\frac{1}{n}}$, with $\mathcal{L}$ as in (3.3.2), it is not difficult to show that Equations (3.3.3) are equivalent to

$$\partial_k(\mathcal{L})=[(\mathcal{L}^{k/n})_+,\mathcal{L}] \quad \text{for all } k\geqslant 1. \tag{3.3.4}$$

For $n=2$, $\mathcal{L}=\xi^2+2u$ and Equation (3.3.4) for $k=3$ is then equivalent to the KdV equation for u. Therefore, one calls Equations (3.3.4) for $n=2$, the

KdV hierarchy and for general n, the nth KdV hierarchy. The next subsection describes some linear equations related to (3.3.3).

3.2. The Linearization

Let L be a Lax operator. Equations (3.3.3) can be seen as the compatibility of the equations

$$L \cdot \psi = z\psi \quad \text{and} \quad \partial_k(\psi) = (L^k)_+ \cdot \psi \text{ for all } k \geqslant 1 \tag{3.3.5}$$

and Equations (3.3.5) are called the *linearization of the KP hierarchy*. In order to be able to obtain (3.3.3) from (3.3.5), one must first give a context in which operators like L and the $(L^k)_+$ act on 'functions' ψ. Further, we should be able to speak of $\partial_k(\psi)$ and, finally the manipulations with the equations should hold. This context is an appropriate $R[\xi, \xi^{-1}]$-module. To justify its form, we consider the trivial solution of the KP hierarchy $L = \xi$. In that case

$$g(z) = \exp\left(\sum_{k=1}^{\infty} t_k z^k\right)$$

is a solution of (3.3.5). In viewing the general solutions of the KP hierarchy as 'perturbations' of the trivial solution, one takes the ψ as 'perturbations' of $g(z)$. Thus one comes to consider

$$M(g(z)) = \left\{ f(z) \cdot g(z) \mid f(z) = \sum_{i=-\infty}^{N} a_i z^i \text{ with } a_i \in R \right\}.$$

The product $f(z) \cdot g(z)$ is a formal one, for in general the product of these 2 series in z makes no sense. $M(g(z))$ becomes an $R[\xi, \xi^{-1}]$-module by linear extension of

$$\left(\sum_{i \leqslant N} a_i z^i\right) \cdot g(z) + \left(\sum_{i \leqslant N} b_i z^i\right) \cdot g(z) = \left(\sum_{i \leqslant N} (a_i + b_i) z^i\right) \cdot g(z),$$

$$p\xi^j \cdot \left\{ \left(\sum a_i z^i\right) \cdot g(z) \right\} = \left(\sum_{i,\ m \geqslant 0} \binom{j}{m} p\partial^m(a_i) z^{i+j-m} \right) \cdot g(z).$$

In particular, one sees that ξ acts as differentiating the formal product w.r.t. the variable x. We let ∂_k act in the same fashion on $M(g(z))$:

$$\partial_k\left(\left(\sum a_i z^i\right) g(z)\right) = \left(\sum_i \partial_k(a_i) z^i + \sum_i a_i z^{i+k}\right) g(z).$$

Note that it makes sense now to consider Equations (3.3.5) inside $M(g(z))$. Another important observation is that $M(g(z))$ is a free $R[\xi, \xi^{-1}]$-module with generator $g(z)$, for we have

$$\left(\sum_i a_i \xi^i\right) \cdot g(z) = \left(\sum_i a_i z^i\right) \cdot g(z).$$

Thus one can translate relations in $M(g(z))$ to equations in $R[\xi, \xi^{-1}]$. In $M(g(z))$ we are interested in elements of a special form

DEFINITION 3.2. The element ψ in $M(g(z))$ is called of *type* z^n if we have $\psi = (z^n + \sum_{j<n} a_j z^j)g(z)$.

So, for each element ψ of type z^n there is a unique operator $K_\psi = \xi^n +$ 'lower order in ξ' such that $\psi = K_\psi \cdot g(z)$. The crucial notion is now

DEFINITION 3.3. An element ψ in $M(g(z))$ of type z^n is called a *Baker function* of type z^n for the Lax operator L if Equations (3.3.5) hold.

Remark 3.4. In the literature, one can also find the names wavefunction and Baker–Akhieser function.

The main result is now

THEOREM 3.5. *If L is a Lax operator and ψ is a Baker function of type z^n for L, then L is a solution of the KP hierarchy and $L = K_\psi \xi K_\psi^{-1}$.*

3.3. Darboux Transformations

If the Lax operator L is a solution of the KP hierarchy, one might wonder of simple transformations that yield out of L new solutions to the KP hierarchy. A typical example of such a transformation could be conjugation with an element

$$K = a_n \xi^n + \text{'lower order in } \xi\text{'}$$

with a_n invertible in R. One directly computes that $\widetilde{L} = KLK^{-1}$ is a Lax operator without constant term iff $\partial_1(a_n) = 0$, but then we can just as well assume $a_n = 1$. In particular, we are interested in differential operators

$$P = \sum_{i=0}^{n} p_i \xi^i = \xi^n + \sum_{i<n} p_i \xi^i,$$

such that $\widetilde{L} = PLP^{-1}$ is a solution of the KP hierarchy. If ψ is a Baker function of type z^m, then $\widetilde{\psi} = P \cdot \psi$ is the candidate Baker function for $\widetilde{L}$ and it will be a solution of the KP hierarchy, if we can show for all $k \geqslant 1$

$$\begin{aligned} \partial_k(\widetilde{\psi}) &= \{\partial_k(P)P^{-1} + PB_kP^{-1}\}\widetilde{\psi} \\ &= \{\partial_k(P)P^{-1} + PB_kP^{-1}\}_+\widetilde{\psi}. \end{aligned}$$

In fact, it even suffices to show for each $k \geqslant 1$, that there is a C_k in $R[\xi]$ such that

$$\partial_k(\widetilde{\psi}) = C_k \cdot \widetilde{\psi}.$$

This equation implies namely directly that

$$C_k = (\widetilde{L})^k_+ = (\partial_k(P)P^{-1} + PB_kP^{-1})_+.$$

Likewise, one can put for solutions L of the KP hierarchy that are of the form $L = \mathcal{L}^{1/m}$, with

$$\mathcal{L} = \sum_{j=0}^{m} l_j\xi^j = \xi^m + \sum_{j=0}^{m-2} l_j\xi^j,$$

the question which differential operators $P = \xi^n + \sum_{i<n} P_i\xi^i$ yield by the transformation $\mathcal{L} \to P\mathcal{L}P^{-1} = \widetilde{\mathcal{L}}$, again an operator $\widetilde{\mathcal{L}}$ in $R[\xi]$. In that case, $\widetilde{\mathcal{L}}$ is clearly of the same form as $\mathcal{L}$. We call the transformation $\mathcal{L} \to \widetilde{\mathcal{L}}$ a *Darboux transformation of order* n. Consider, namely, the special case of the Schrödinger operator $\mathcal{L} = \xi^2 - u$. In 1882, G. Darboux associated with $\mathcal{L}$ another Schrödinger operator

$$\widetilde{\mathcal{L}} = \xi^2 - \left(u - 2\frac{\partial(\psi)}{\psi}\right) \quad \text{with } \psi \neq 0,\ \mathcal{L}\psi = c\psi.$$

It was shown in [4] that this Darboux–Bäcklund transformation $\mathcal{L} \to \widetilde{\mathcal{L}}$ is simply the conjugation of $\mathcal{L}$ with $A = \psi\xi\psi^{-1} = \xi - (\partial(\psi)/\psi)$.

3.4. Miura transformations

In this subsection, we follow [25]. To explain the name of this subsection, we again consider the Schrödinger operator $\mathcal{L} = \xi^2 - u$. Consider a splitting of $\mathcal{L}$ into linear factors,

$$\mathcal{L} = (\xi - v)(\xi + v). \tag{3.3.6}$$

One computes directly that u is related to v by the so-called Miura transformation

$$u = \partial(v) - v^2. \tag{3.3.7}$$

Miura showed that it had the following remarkable property: if v satisfies the modified KdV equation

$$\frac{\partial v}{\partial t} = \partial^3(v) - 6v^2\partial(v), \tag{3.3.8}$$

then u satisfied the KdV equation. This last equation is equivalent to the first nontrivial equation in the KdV hierarchy

$$\frac{\partial \mathcal{L}}{\partial t} = [(\mathcal{L}^{3/2})_+, \mathcal{L}]. \tag{3.3.9}$$

The modified KdV or shortly MKdV equation, has also a description in Lax form, see [14]. Take the matrix differential operator

$$\underline{\mathcal{L}} = \begin{pmatrix} 0 & 1 \\ 1 & 0 \end{pmatrix} \xi + \begin{pmatrix} 0 & -v \\ v & 0 \end{pmatrix}. \tag{3.3.10}$$

Then a direct computation shows

$$\mathcal{L}^2 = \begin{pmatrix} 1 & 0 \\ 0 & 1 \end{pmatrix} \xi^2 + \begin{pmatrix} \partial(v) - v^2 & 0 \\ 0 & -\partial(v) - v^2 \end{pmatrix} = \begin{pmatrix} 1 & 0 \\ 0 & 1 \end{pmatrix} \xi^2 + \begin{pmatrix} u & 0 \\ 0 & \tilde{u} \end{pmatrix}.$$

Let Q now be the operator in $M_2(R)[\xi, \xi^{-1}]$ defined by

$$Q = (\underline{\mathcal{L}}^2)^{1/2} = \begin{pmatrix} (\xi^2 + u)^{1/2} & \\ 0 & (\xi^2 + \tilde{u})^{1/2} \end{pmatrix}.$$

One directly verifies then that the Lax equation

$$\frac{\partial \underline{\mathcal{L}}}{\partial t} = [4(Q^{3/2})_+, \underline{\mathcal{L}}] \tag{3.3.11}$$

is equivalent to the MKdV equation. This Lax equation is again the simplest nontrivial equation for a hierarchy, the so-called *MKdV hierarchy*. This hierarchy cannot only be introduced for the Schrödinger operator, but also for higher order differential operators. Consider thereto, in $M_n(R)[\xi]$ the operator

$$\underline{\mathcal{L}} = \begin{pmatrix} 0 & \cdots & 0 & 1 \\ 1 & \ddots & & 0 \\ \vdots & \ddots & & \vdots \\ 0 & \cdots & 1 & 0 \end{pmatrix} \xi + \begin{pmatrix} 0 & \cdots & 0 & v_{n-1} \\ v_0 & \ddots & & 0 \\ 0 & \ddots & & \vdots \\ \vdots & & \ddots & \vdots \\ 0 & \cdots 0 & v_{n-2} & 0 \end{pmatrix}, \tag{3.3.12}$$

with $\sum_{i=0}^{n-1} v_i = 0$ Then one computes directly that

LEMMA 3.6. *We have*

$$\underline{\mathcal{L}}^n = \begin{pmatrix} \mathcal{L}_0 & & 0 \\ & \ddots & \\ 0 & & \mathcal{L}_{n-1} \end{pmatrix},$$

where $\mathcal{L}_i$ *in* $R[\xi]$ *is given by*

$$\mathcal{L}_i = (\xi + v_{n+i-1})(\xi + v_{n+i-2}) \cdots (\xi + v_i)$$

and the subscripts in the $\mathcal{L}_i$ *are read* mod n.

The conditions $\sum_{i=0}^{n-1} v_i = 0$ implies that the coefficient of ξ^{n-1} in each $\mathcal{L}_i$ is zero. If we write

$$\mathcal{L}_0 = (\xi + v_{n-1})(\xi + v_{n-2}) \cdots (\xi + v_0) = \xi^n + \sum_{i=0}^{n-2} u_i \xi^i,$$

then the $\{u_i\}$ are polynomials in the $\{\partial^s(v_j) \mid s \geqslant 0,\ 0 \leqslant j \leqslant n-1\}$. The transformation

$$\{v_i\} \longmapsto \{u_i\} \tag{3.3.13}$$

we call again the *Miura transformation.* To $\underline{\mathcal{L}}$ we associate the operator

$$Q = \begin{pmatrix} \mathcal{L}_0^{1/n} & & 0 \\ & \ddots & \\ 0 & & \mathcal{L}_{n-1}^{1/n} \end{pmatrix} \tag{3.3.14}$$

in $M_n(R)[\xi, \xi^{-1}]$. For each $k \geqslant 1$, we consider the Lax equation

$$\frac{\partial}{\partial_{t_k}} \underline{\mathcal{L}} = [(Q^k)_+, \underline{\mathcal{L}}]. \tag{3.3.15}$$

We call this the kth *equation* of the nth *MKdV hierarchy* and the operator $\underline{\mathcal{L}}$ that satisfies all these equations is called a solution of this hierarchy. Since (3.3.15) also holds for arbitrary powers of a solution, we see that we have

PROPOSITION 3.7. *The Miura transformation maps solutions of the* nth *MKdV hierarchy to solutions of the* nth *KdV hierarchy.*

4. Solutions

4.1. Solutions of the KP hierarchy

Here we recall the basic facts from [21] for the construction of solutions of the KP hierarchy such as we need them. One considers the case $m = 2$, i.e. $\mathcal{F}(\underline{s})$ is the Grassmann manifold. Because of (2.2.2), we may assume that $\underline{s} = (\underline{0}) = (0)$. Take any $W \in \mathcal{F}(\underline{0})$ and consider

$$\Gamma_W = \{\gamma \in \Gamma_+ \mid \Lambda^k \circ p_1 \circ \Lambda^{-k}\colon\ \gamma^{-1}W \to \Lambda^k H_1 \ \text{ is a bijection}\}$$

Then we have

LEMMA 4.1. *The set* Γ_W *is nonempty and open in* Γ_+.

Let R be the ring of analytic functions on Γ_W. For each $\gamma \in \Gamma_W$, we define

$$\begin{aligned}\hat{\psi}_W(\gamma) &= (\Lambda^k \circ p_1 \circ \Lambda^{-k})^{-1}(z^k) \in \gamma^{-1}W \\ &= z^k + \sum_{i<k} a_i(\gamma) z^i\end{aligned}$$

and

$$\psi_W(\gamma) = \gamma(\hat{\psi}_W(\gamma)) \in W.$$

Then ψ_W is by construction a function of type z^k in $M(g(\lambda))$ and the central result is

THEOREM 4.2. *If we write $\psi_W = K_W \cdot g(z)$, with $K_W \in R[\xi, \xi^{-1}]$, then ψ_W is a Baker function of type z^k for $L_W = K_W \xi K_W^{-1}$.*

From the fact that

$$L_W = K_W \xi K_W^{-1} = K_W \Big(a_0 + \sum_{i<0} a_i \xi^i\Big) \xi \Big(a_0 + \sum_{i<0} a_i \xi^i\Big)^{-1} K_W^{-1},$$

if $\partial(a_i) = 0$ for all i and the actual form of the action of Γ_-, one obtains moreover

THEOREM 4.3. (i) *For each γ in Γ_- and $\delta \in \Delta$, we have $L_{\gamma\delta W} = L_W$.*

(ii) *Let $\overline{\Gamma\Delta}$ be the closure of the image of $\Gamma_-\Delta$ in* $\mathrm{Gl}_{\mathrm{res}}(H)$. *The set of solutions of the KP hierarchy obtained in this way can be identified with $\overline{\Gamma_-\Delta} \backslash \mathrm{Gl}_{\mathrm{res}}(H)/P(\underline{0})$.*

That Δ acts trivially on the solutions shows that each connected component of $\mathcal{F}(\underline{0})$ renders the same set of solutions of the KP hierarchy. On the other hand, the triviality of the Γ_- action tells you that the commuting flows from Γ_+ give you practically a maximal set of independent directions.

One can also characterize geometrically the solutions from Theorem 4.2 that correspond to solutions of the nth KdV hierarchy.

THEOREM 4.4. *For each $W \in \mathcal{F}(\underline{0})$ the following properties are equivalent*

(i) $(L_W^n)_+ = L_W^n$.
(ii) $\Lambda^n W \subset W$.

Proof. (i) $\Rightarrow$ (ii). If $(L_W^n)_+ = L_W^n$, then we have

$$\partial_n(\psi_W)(\gamma) = z^n \psi_W(\gamma) \quad \text{for all } \gamma \in \Gamma_W.$$

Since the $\{\psi_W(\gamma) | \gamma \in \Gamma_W\}$ are dense in W and the left-hand side of this equation belongs to W, we get $\Lambda^n W \subset W$.

(ii) ⇒ (i). Assume $W \in \mathcal{F}(0)^{(-k)}$. Then we have

$$\begin{aligned}\partial_n(\psi_W) &= \{\partial_n(K_W)K_W^{-1} + L_W^n\} \cdot \psi_W \\ &= z^n \cdot \psi_W + \partial_n(K_W)K_W^{-1} \cdot \psi_W = (L_W^n)_+ \cdot \psi_w.\end{aligned}$$

If $\wedge^n W \subset W$, then we know that $z^n \psi_W(\gamma) \in W$. This implies $\partial_n(K_w) \cdot g(z) \in W \cap (\gamma(z^k H_1)^\perp) = \{0\}$ and, hence, $(L_W^n)\psi_W = (L_W^n)_+ \psi_W$. This gives the required equality. □

4.2. Darboux transforms

We start with the fundamental observation that shows how flags make their appearance in this framework.

THEOREM 4.5. *Let W_1 and W_2 be elements of $\mathcal{F}(\underline{0})$ belonging to respectively $\mathcal{F}(\underline{0})^{(m)}$ and $\mathcal{F}(\underline{0})^{(m+n)}$, with $n \geqslant 1$. Then the following 2 points are equivalent:*

(i) *There is a polynomial P in $R[\xi]$ of order n such that $P \cdot \psi_{W_2} = \psi_{W_1}$.*
(ii) $W_1 \subset W_2$.

Proof. (i) ⇒ (ii) for each γ in $\Gamma_{W_1} \cap \Gamma_{W_2}$ the vector $(P \cdot \psi_{W_2})(\gamma)$ belongs by construction to W_2. Since the vectors

$$\{\psi_{W_1}(\gamma) \mid \gamma \in \Gamma_{W_1} \cap \Gamma_{W_2}\}$$

are dense in W_1, we see that the equality in (i) implies that $W_1 \subset W_2$.

(ii) ⇒ (i) As we have seen at the construction of the wavefunctions we have for each $i \geqslant 0$.

$$\xi^i \cdot \psi_{W_2}(\gamma) = \left\{z^{-m-n+i} + \sum_{j<-m-n+i} a_j(\gamma) z^j\right\} g(z)$$

and

$$\psi_{W_1}(\gamma) = \left\{z^{-m} + \sum_{l<-m} b_l(\gamma) z^l\right\} g(z).$$

Hence, we can always find a polynomial P in $R[\xi]$ of order n such that

$$P \cdot \psi_{W_2}(\gamma) - \psi_{W_1}(\gamma) = \left\{\sum_{r<-m-n} c_r(\gamma) z^r\right\} g(z).$$

If $W_1 \subset W_2$, then the left-hand side of this equation belongs to W_2 and to the space $\gamma((z^{-m-n} H_1)^\perp)$. By construction, this has to be zero and we obtain the desired equality. □

Take $\underline{s} = (0, -n)$ and consider the manifold $\mathcal{F}(\underline{s})$. The foregoing theorem shows that the map

$$(W_1, W_2) \longrightarrow (L_{W_1}, L_{W_2})$$

assigns to each point in $\mathcal{F}(\underline{s})$ a pair of solutions of the KP hierarchy that are coupled by a Darboux transformation of order n, i.e.

$$L_{W_1} = PL_{W_2}P^{-1} \quad \text{with } P \in R[\xi] \text{ of order } n.$$

By combining this with Theorem 4.3, we get analogously

COROLLARY 4.6. *Consider the collection of pairs $(L, \widetilde{L})$, where L and $\widetilde{L}$ are solutions of the KP hierarchy as constructed in Theorem 4.2 and such that L and $\widetilde{L}$ are coupled by $L = P\widetilde{L}P^{-1}$, with $P \in R[\xi]$ of order n. Then this set can be identified with*

$$\overline{\Gamma_-\Delta} \setminus \mathrm{Gl}_{\mathrm{res}}(H)/P(\underline{s}).$$

Next we consider the subvariety $\mathcal{F}(\underline{0})_m$ in $\mathcal{F}(\underline{0})$, consisting of all W in $\mathcal{F}(\underline{0})$ satisfying

$$\Lambda^m W \subset W.$$

Since the points of $\mathcal{F}(\underline{0})_m$ yield exactly the mth roots of differential operators, we look for Darboux transforms that are also of this form. For $n \leqslant m$ and $\underline{s} = (0, -n)$, consider the subvariety $\mathcal{F}_m(\underline{s})$ of $\mathcal{F}(\underline{s})$ given by

$$\mathcal{F}_m(\underline{s}) = \{(\widetilde{W}, W) \in \mathcal{F}(\underline{s}),\ \Lambda^m W \subset \widetilde{W} \subset W\}.$$

Since $\Lambda^m \widetilde{W} \subset \Lambda^m W \subset \widetilde{W}$, we see that for each pair $(\widetilde{W}, W)$ in $\mathcal{F}_m(\underline{s})$ both requirements hold, i.e.

$$L^m_{\widetilde{W}} = (L^m_{\widetilde{W}})_+ \quad \text{and} \quad L_{\widetilde{W}} = PL_W P^{-1},$$

with $P \in R[\xi]$ of order n.

4.3. Solutions of the nth MKdV Hierarchy

In this subsection we consider the class of so-called periodic flags that is an intersection of the flag varieties $\mathcal{F}_m(\underline{s})$ from Subsection 4.2. Again we start with a W in $\mathcal{F}(\underline{0})_n$. A *periodic flag* associated to W is a chain

$$z^n(W) = W_n \subsetneq W_{n-1} \subsetneq \cdots \subsetneq W_0 = W.$$

If $\underline{s} = (s_1, s_1 - 1, \ldots, s_1 - n + 1)$, then the collection of all these flags is a subvariety $\mathcal{F}_{\mathrm{per}}(\underline{s})$ of $\mathcal{F}(\underline{s})$. According to Theorem 4.5 the Baker functions of the $\{W_i\}$ satisfy

$$\psi_{W_{i+1}} = (\xi + v_i)\psi_{W_i} \quad \text{for all } 1 \leqslant i \leqslant n-1.$$

If we use the operator $\underline{\mathcal{L}}$ in $M_n(R)[\xi]$ from Subsection 3.4 and the fact that $\psi_{W_m} = z^n \psi_{W_0}$, these relations can be written as

$$\mathcal{L}\begin{pmatrix} \psi_{W_0} \\ \vdots \\ \psi_{W_{n-1}} \end{pmatrix} = \begin{pmatrix} z^n \psi_{W_0} \\ \psi_{W_1} \\ \vdots \\ \psi_{W_{n-1}} \end{pmatrix}.$$

By iterating this action, we get the form

$$\underline{\mathcal{L}}^n \begin{pmatrix} \psi_{W_0} \\ \vdots \\ \psi_{W_{n-1}} \end{pmatrix} = \begin{pmatrix} \mathcal{L}_0 & & 0 \\ & \cdots & \\ 0 & & \mathcal{L}_{n-1} \end{pmatrix} \begin{pmatrix} \psi_{W_0} \\ \vdots \\ \psi_{W_{n-1}} \end{pmatrix} = z^n \begin{pmatrix} \psi_{W_0} \\ \vdots \\ \psi_{W_{n-1}} \end{pmatrix}.$$

In particular, this implies that for all i, $0 \leqslant i \leqslant n-1$, $L^n_{W_i} = \mathcal{L}_i$ and that $\underline{\mathcal{L}}^n$ satisfies the equations

$$\frac{\partial}{\partial t_k} \underline{\mathcal{L}}^n = [(Q^k)_+, \underline{\mathcal{L}}^n], \quad k \geqslant 1,$$

where $Q = (\underline{\mathcal{L}}^n)^{1/n} \in M_n(R)[\xi, \xi^{-1}]$ as in Subsection 3.4. We want to show now the following result.

THEOREM 4.7. *The operator $\underline{\mathcal{L}}$ constructed above is a solution of the mth MKdV hierarchy, i.e. for all $k \geqslant 1$ it satisfies*

$$\frac{\partial \underline{\mathcal{L}}}{\partial t_k} = [(Q^k)_+, \underline{\mathcal{L}}]$$

Proof. Because of Proposition 3.3 in [16], it suffices to show that $\underline{\mathcal{L}}$ and Q commute. We have that $[\underline{\mathcal{L}}, \underline{\mathcal{L}}^n] = [\underline{\mathcal{L}}, Q^n] = 0$ and, since $\mathrm{ad}(\underline{\mathcal{L}})$ is a derivation, we get

$$[\underline{\mathcal{L}}, Q^n] = \sum_{i=0}^{n-1} Q^i [\underline{\mathcal{L}}, Q] \cdot Q^{n-i-1} = 0.$$

If $[\underline{\mathcal{L}}, Q] = \alpha \xi^s +$ 'lower order', then there must hold

$$[\underline{\mathcal{L}}, Q^n] = n\alpha \xi^{n+s-1} + \text{'lower order'} = 0.$$

This shows $[\mathcal{L}, Q] = 0$. □

If $W \in \mathcal{F}(\underline{0})_n$, then we have seen that each $(W_i) \in \mathcal{F}_{\mathrm{per}}(\underline{s})$, with $W_0 = W$, gives you a decomposition of L^n_W into linear factors

$$L^n_W = \xi^n + \sum_{i=0}^{n-2} u_i \xi^i = (\xi + v_{n-1}) \cdots (\xi + v_0).$$

Because of Theorem 4.7, we may say that the Miura transformation $\{v_i\} \to \{u_i\}$ corresponds geometrically to the natural projection $(W_i) \to W_0$ of $\mathcal{F}_{\text{per}}(\underline{s})$ onto $\mathcal{F}(\underline{0})_n$.

Remark 4.8. The periodic flags also turn up in a natural way in algebraic geometric situations, see, e.g., [1] and [20].

Remark 4.9. The KP hierarchy and its modified versions can also be formulated in the so-called Hirota bilinear form. The equations are then expressed in quadratic relations for a finite number of so-called τ-functions. The geometric interpretation of these functions requires the transition to the bundle Det*, see [21]. For the geometric description of the relations, we refer to [8].

5. Deformations

5.1. Deformations of Meromorphic Equations on $\mathbb{P}^1(\mathbb{C})$

It is a well-known fact, see, e.g., [11], that deformations of linear meromorphic differential equations on $\mathbb{P}^1(\mathbb{C})$ may lead to solutions of nonlinear differential equations that possess interesting features, like the Painlévé property. We start with a description of the setting. Consider a linear differential equation on $\mathbb{P}^1(\mathbb{C})$ that is meromorphic over $Y_0 = \{a_1^0, \ldots, a_m^0, \infty\}$. By this, we mean the set of data

1.3. (i) a holomorphic vector bundle E^0 over $\mathbb{P}^1(\mathbb{C})$.

(ii) An integrable connection ∇^0 of the bundle $E^0 \mid \mathbb{P}^1(\mathbb{C}) - Y_0$ that is meromorphic over Y_0.

If x is the meromorphic function that identifies $\mathbb{P}^1(\mathbb{C})$ with $\mathbb{C} \cup \{\infty\}$, then the connection form Ω^0 of ∇^0 can be written as

$$\Omega^0 = \left\{ \sum_{1 \leqslant k \leqslant m} \left\{ \sum_{l \geqslant 1} \frac{A_{kl}^0}{(x - a_k^0)^l} \right\} + \sum_{l \geqslant 0} A_{\infty l}^0 x^l \right\} \mathrm{d}x,$$

where the A_{kl}^0 and $A_{\infty l}^0$ are complex $n \times n$ matrices that are zero for sufficiently large 1.

Now we are interested in deformations that move the poles $\{a_i^0\}$ of Ω^0 in a holomorphic way inside $\mathbb{C}$. For simplicity, we assume that the parameter space of the deformation, shortly called *deformation space*, is a connected complex variety T. The way the poles are moved is determined by a set of holomorphic functions $a_i\colon T \to \mathbb{C}$, $1 \leqslant i \leqslant m$, the so-called *deformation functions*. Furthermore, there has to be a base point t_0 in T such that $a_i(t_0) = a_i^0$ for all i. To avoid a-priori topological obstructions, we will restrict ourselves to deformations for which the poles never coincide, i.e. the deformation functions satisfy

$$a_i(t) \neq a_j(t) \quad \text{for all } t \text{ in } T \text{ and all } i \neq j.$$

Let X be $\mathbb{P}^1(\mathbb{C}) \times T$. We can introduce a smooth codimension one subvariety Y of X by

$$Y = Y_1 \cup \cdots \cup Y_m \cup Y_\infty$$

with

$$Y_i = \{(x,t) \mid (x,t) \in X,\ x = a_i(t)\} \quad \text{and} \quad Y_\infty = \{(\infty,t) \mid (\infty,t) \in X\}.$$

The object one is interested in, is then

DEFINITION 5.1. An *integrable deformation* (E, ∇) of the pair (E^0, ∇^0) with deformation space T, deformation functions $\{a_i\}$ and base point t_0 in T consists of

(i) a holomorphic vector bundle E over $X = \mathbb{P}^1(\mathbb{C}) \times T$ of rank n,
(ii) an integrable connection ∇ of $E \mid X - Y$, that is meromorphic over Y and that is such that the restriction of (E, ∇) to $\mathbb{P}^1(\mathbb{C}) \times \{t_0\}$ is isomorphic to (E^0, ∇^0).

Given the deformation space, the deformation functions $\{a_i\}$ and the base point $\{t_0\}$, a relevant question is if there exists an integrable deformation with these deformation data. Even if there is no fusion of the poles, one might hit at topological obstructions. Let i: $\mathbb{P}^1(\mathbb{C}) \to T$ be the embedding $x \mapsto (x, t_0)$. It induces a natural map i^*: $\pi_1(\mathbb{P}^1(\mathbb{C}) - Y_0) \to \pi_1(X - Y)$. If there exists an integrable deformation (E, ∇) of (E^0, ∇^0), then the representation ρ of $\pi_1(\mathbb{P}^1(\mathbb{C}) - \{a_1^0, \ldots, a_m^0, \infty\})$ corresponding to (E^0, ∇^0) has to factorize through i^*. Consider the projection p_2: $X - Y \to T$, given by $p_2((x,t)) = t$. The fiber over t is equal to $\mathbb{C} - \{a_1(t), \ldots, a_m(t)\}$. Thus, $(X - Y, T, p_2)$ is a fiber bundle and we have the long exact sequence

$$\begin{aligned}\pi_2(T) &\longrightarrow \pi_1(\mathbb{P}^1(\mathbb{C}) - \{a_1^0, \ldots, a_m^0, \infty\}) \xrightarrow{i^*} \pi_1(X - Y) \\ &\longrightarrow \pi_1(T) \longrightarrow 1.\end{aligned}$$

Now ρ has to be trivial on the image of $\pi_2(T)$ and should be extendable to $\pi_1(X - Y)$. Such problems do not occur if $\pi_2(T) = \pi_1(T) = \{1\}$. In that case the monodromy is preserved by the deformation. An example of such a deformation space occurred in [5] and is given by

EXAMPLE 5.2. Let Z be the space

$$Z = \mathbb{C}^m - \bigcup_{i \neq j} D_{ij}, \quad \text{where } D_{ij} = \{(x_k) \in \mathbb{C}^m,\ x_i = x_j\}.$$

As deformation space we take the universal covering space $\widetilde{Z}$ of Z and for a_i: $\widetilde{Z} \to \mathbb{C}$ we take the composition of the natural projection π: $\widetilde{Z} \to Z$ with the projection on the ith coordinate.

Remark 5.3. From the construction one sees directly that in a neighbourhood of the singular point $(a_j^0, t_0) \in Y_j$, the singular part of the connection form Ω has w.r.t. any trivializing basis of local sections of E, the form

$$\sum_{l \geqslant 1} \frac{B_{jl}(t)}{(x - a_j(t))^l} \,\mathrm{d}(x - a_j),$$

where all the B_{jl} are holomorphic. By applying a proper coordinate transformation, one may moreover assume

$$B_{jl}(t_0) = A_{jl}(t_0) \quad \text{for all } l.$$

Since we considered integrable deformations, the functions B_{jl} satisfy the nonlinear compatibility conditions and those are the nonlinear equations we referred to at the beginning, e.g., if ∇^0 has a logarithmic pole over Y_0, that is $A_{kl}^0 = 0$ for all $l > 1$ and $A_{\infty l}^0 = 0$ for all $l \geqslant 0$, then the B_{j1} satisfy the so-called 'Schlesinger equations'

$$\mathrm{d}B_{i1} = -\sum_{j \neq i} [B_{i1}, B_{j1}] \frac{\mathrm{d}(a_i - a_j)}{a_i - a_j}.$$

For the space $\widetilde{Z}$ with the deformation functions $\{a_i\}$ as in the foregoing example we had no topological obstructions and this turns out to be sufficient, for following carefully the lines of proof in [15], on can show

THEOREM 5.4. *For each pair (E^0, ∇^0) there exists an integrable deformation (E, ∇) of (E^0, ∇^0) with $\widetilde{Z}$ as deformation space and the $\{a_i\}$ as deformation functions.*

Let (E, ∇) be the integrable deformation from Theorem 5.4 and assume E^0 was a trivial vector bundle. Then we consider as in [15], the set

$$\Theta = \{t \in \widetilde{Z} \mid E|_{\mathbb{P}^1(\mathbb{C}) \times \{t\}} \text{ is nontrivial}\}$$

Since all holomorphic line bundles over Z are trivial, one can show

PROPOSITION 5.5. *There is a holomorphic $\underline{\tau}$: $\widetilde{Z} \to \mathbb{C}$ such that*

$$\Theta = \{t \in \widetilde{Z},\ \underline{\tau}(t) = 0\}.$$

Choose open balls D_1 and D_2 in $\mathbb{P}^1(\mathbb{C})$ such that $D_1 \cup D_2 = \mathbb{P}^1(\mathbb{C})$ and that $E \mid D_1 \times \widetilde{Z}$ and $E \mid D_2 \times \widetilde{Z}$ are trivial. The comparison of trivializing sections of these two bundles, gives you a holomorphic transfer funtion

$$S\colon\ D_1 \cap D_2 \times \widetilde{Z} \longrightarrow \mathrm{Gl}_n(\mathbb{C}).$$

Let H be $L^2(S^1, \mathbb{C}^n)$ with the decomposition $H = H_+ \oplus H_+^{\perp}$, where

$$H_+ = \left\{ \sum_{i \geqslant 0} b_i x^i \in H,\ b_i \in \mathbb{C}^n \right\}.$$

Then S determines a holomorphic family of operators $\{\mathbb{S}(t) \mid t \in \widetilde{Z}\}$ in $\mathrm{Gl}_{\mathrm{res}}(H)$, such that $\mathbb{S}(t)$ is in the big cell for all $t \in \widetilde{Z} - \Theta$. Locally one can fit the $\mathbb{S}(t)$ into the group G and this suffices to prove that

PROPOSITION 5.6. (i) *There exist holomorphic maps* $S_+\colon D_2 \times \{\widetilde{Z} - \Theta\} \to \mathrm{Gl}_n(\mathbb{C})$ *and* $S_-\colon D_1 \times \{\widetilde{Z} - \Theta\} \to \mathrm{Gl}_n(\mathbb{C})$ *with* $S_-(\infty, t) = \mathrm{Id}$ *such that* $S = S_-^{-1} S_+$.

(ii) *The maps* S_+ *and* S_- *are meromorphic over* $D_2 \times \Theta$, *resp.* $D_1 \times \Theta$.

In the case that ∇^0 has a logarithmic pole at infinity this proposition is used to show that the connection form of $\nabla \mid D_i \times \widetilde{Z} - Y \cap \{D_i \times \widetilde{Z}\}$ w.r.t. suitable trivializing basis extends meromorphically to $D_i \times \widetilde{Z}$. Thus, one obtains the following result.

THEOREM 5.7. (i) *Assume that* E^0 *is trivial and that* ∇^0 *has a logarithmic pole at infinity. Let* (E, ∇) *be the integrable deformation from* 5.4. *Then there is a neighbourhood* U *of* t_0 *such that* $E \mid \mathbb{P}^1(\mathbb{C}) \times U$ *is trivial and a basis of sections* $\{h_i\}$ *such that the connection form* Ω *w.r.t. the* $\{h_i\}$ *looks like*

$$\Omega = \sum_{i=1}^{m} \sum_{l \geqslant 1} \frac{B_{il}(t)}{(x - a_i(t))^l} d(x - a_i),$$

with $B_{il}(t_0) = A_{il}^0$ *for all* i *and* l.

(ii) *These solutions* $\{B_{il}\}$ *of the integrability equations extend holomorphically to* $\widetilde{Z} - \Theta$ *and are meromorphic on* $\widetilde{Z}$.

Remark 5.8. In the case that ∇^0 has logarithmic poles over Y_0, Malgrange has shown in [15] that there is a holomorphic $\tau\colon \widetilde{Z} \to \mathbb{C}$ with zero-set Θ such that $\omega = d\tau/\tau$ is the differential form presented in [11]. This τ-function can also be interpreted as the determinant of an associated Cauchy–Riemann operator on the spin bundle over $\mathbb{P}^1(\mathbb{C})$, see [17].

Remark 5.9. The bundle Det* plays also an important role at monodromy preserving deformations of Dirac equations, see [18] and [19].

References

1. Adler, M., Haine, L., and van Moerbeke, P.: Limit matrices for the Toda flow and periodic flags for loop groups, Preprint.

2. Baston, R. J. and Eastwood, M. G.: *The Penrose Transform*, Clarendon Press, Oxford, 1989.
3. Carey, A. L. and Ruysenaars, S. N. M.: On fermion gauge groups, current algebras and Kac–Moody algebras, *Acta Appl. Math.* **10** (1987), 1–86.
4. Ehlers, F. and Knörrer, H.: An algebro-geometric interpretation of the Bäcklund transformation for the Korteweg–de Vries equation, *Comment. Math. Helv.* **57** (1982), 1–10.
5. Fukuma, M., Kawai, H., and Nakayama, R.: Infinite-dimensional Grassmannian structure of two-dimensional quantum gravity, *Comm. Math. Phys.* **143** (1992), 371–403.
6. Haine, L. and Horozov, E.: Toda orbits of Laguerre polynomials and representations of the Virasoro algebra, *Bull. Sci. Math., 2e Serie* **117** (1993), 485–518.
7. Helminck, G.: A geometric construction of solutions of the Toda lattice hierarchy, in *Lecture Notes in Phys.* 424, Springer-Verlag, New York, pp. 91–101.
8. Helminck, G. F. and Helminck, A. G.: The structure of Hilbert flag varieties, *Publ. RIMS, Kyoto Univ.* **30**(3) (1994), 410–441.
9. Hazewinkel, M. and Martin, C. F.: Representations of the symmetric group the specialization order, Systems and Grassmann manifolds, *Enseign. Math.* **29** (1983), 53–87.
10. Helminck, G. F. and Post, G. F.: The geometry of differential difference equations, *Indag. Math. N.S.* **5**(4) 411–438.
11. Jimbo, M., Muire, T., and Ueno, K.: Monodromy preserving deformations of linear ordinary differential equations with rational coefficients 1. General theory and τ-functions, *Physics 2D* **402** (1981), 306–352.
12. Kawamoto, N., Narukawa, Y., Tsychiya, A., and Yamada, Y.: Geometric Realization of Conformal Field Theory on Riemann Surfaces, *Comm. Math. Phys.* **116** (1988), 247–308.
13. Kac, V. G. and Peterson, D. H.: Infinite flag varieties and conjugacy theorems, *Proc. Nat. Acad. Sci. USA* **18** (1983), 1778–1782.
14. Kupershmidt, B. A. and Wilson, G.: Modifying Lax equations and the second Hamiltonian structure, *Invent. Math.* **62** (1981), 403–436.
15. Malgrange, B.: *Sur les déformations isomonodromiques 1. Singularités réguliers*, Progress in Mathematics, Vol. 37, Birkhäuser, Boston, 1983.
16. Mickelsson, J.: *Current Algebras and Groups*, Plenum Monographs in Nonlinear Physics, Plenum Press, New York, 1989.
17. Palmer, J.: Determinants of Cauchy–Riemann operators as τ-functions, *Acta Appl. Math.* **18**(3) (1990), 199–223.
18. Palmer, J.: Tau functions for the Dirac operator in the Euclidean plane, *Pacific J. Math.* **160**(2) (1993), 259–342.
19. Palmer, J., Beatty, M., and Tracy, C. A.: Tau functions for the Dirac operator on the Poincaré disk, Preprint.
20. Pressley, A. and Segal, G.: *Loop Groups*, Clarendon Press, Oxford 1986.
21. Segal, G. and Wilson, G.: Loop groups and equations of KdV-type, *Publ. Math. IHES* **61** (1985), 5–65.
22. Ward, R. S. and Wells, R. O., Jr.: *Twistor Geometry and Field Theory*, Cambridge Monographs on Mathematical Physics, University Press, Cambridge, 1991.
23. Wilson, G.: Commuting flows and conservation laws for Lax equations, *Math. Proc. Cambridge Philos. Soc.* **86** (1979), 131–143.
24. Wilson, G.: Habillage et fonctions τ, *C.R. Acad. Sci., Paris, Sér. I. Math.* **299** (1984), 587–590.
25. Wilson, G.: Loop groups and equations of KdV type II, Flag manifolds and the modified equations, Preprint Math. Inst., Oxford, 1983.
26. Witten, W.: Quantum field theory, Grassmannian and algebraic curves, *Comm. Math. Phys.* **113** (1988), 539–600.

Acta Applicandae Mathematicae **41:** 123–134, 1995.

Computation by Computer of Lie Superalgebra Homology and Cohomology

N. v.d. HIJLIGENBERG* and G. F. POST
Department of Applied Mathematics, University of Twente, P.O. Box 217, 7500 AE Enschede, The Netherlands

(Received: 28 February 1994)

Abstract. In this paper, we introduce a package to compute homology and cohomology spaces of Lie superalgebras. We describe most of its features and the implementation in REDUCE.

Mathematics Subject Classifications (1991): 17B55, 17B56, 17B70, 18G35, 69J99.

Key words: homology, cohomology, Lie superalgebra, computer algebra.

In this paper, we present a package that is designed to compute homology and cohomology of Lie superalgebras. Since a lot of the properties of Lie superalgebras (such as, e.g., the existence of central extensions, deformations, and exterior derivations) can be expressed in terms of certain cohomology classes, this Homology Package can be used to investigate such properties. This has been the main motivation for developing the package. Since an ordinary Lie algebra can be considered to be a Lie superalgebra with an odd part equal to zero, the package can also be used to compute homology and cohomology of Lie algebras.

1. (Co)Homology of Lie Superalgebras

We assume that the reader is familiar with superspaces, Lie superalgebras and representations of Lie superalgebras. We recall the definition of homology and cohomology of Lie superalgebras and we set some notations. For a more detailed description, we refer to [3, 6, 9].

For a superspace V, V_0 will denote its even part and V_1 its odd part. By $|v|$ we denote the $\mathbb{Z}_2$-degree of the element v in V, which is implicity assumed to be homogeneous. By V^* we denote the linear dual of V. All superspaces are assumed to be vector spaces over the field of complex numbers $\mathbb{C}$.

* Current address: CWI, Dept. AM, Kruislaan 413, 1098 SJ Amsterdam.

Let L be a Lie superalgebra and M an L-module. The $\mathbb{Z}$-graded superspace $C.(L;M)$ is defined as

$$C.(L;M) = \bigoplus_{q\in\mathbb{Z}} C_q(L;M) \quad \text{with} \quad C_q(L;M) = M \otimes \wedge^q L.$$

Here $\wedge^q L$ is the superspace of $\mathbb{Z}_2$-graded q-alternating tensors on L, i.e. $(\wedge L)_i$ is spanned by

$$x_1 \wedge x_2 \wedge \cdots \wedge x_q \quad (x_j \in L)$$

satisfying $\sum_{j=1}^{q} |x_j| = i$ and

$$\begin{aligned} & x_1 \wedge x_2 \wedge \cdots x_j \wedge x_{j+1} \wedge \cdots \wedge x_q \\ & \quad = -(-1)^{|x_j||x_{j+1}|} x_1 \wedge \cdots \wedge x_{j+1} \wedge x_j \wedge \cdots \wedge x_q. \end{aligned}$$

The boundary operator ∂ is the even linear operator of $\mathbb{Z}$-degree -1 on $C.(L;M)$ given by

$$\begin{aligned} & \partial(v \otimes x_1 \wedge \cdots \wedge x_q) \\ & \quad = \sum_{1\leqslant i<j\leqslant q} (-1)^{i+j+1+(|x_i|+|x_j|)k_i+|x_j|k_{i,j}} v \otimes [x_i, x_j] \times \\ & \qquad \times x_1 \wedge \cdots \hat{x}_i \cdots \hat{x}_j \cdots \wedge x_q + \\ & \qquad + \sum_{i=1}^{q} (-i)^{i+1+|x_i|(|v|+k_i)} (x_i \cdot v) \otimes x_1 \wedge \cdots \hat{x}_i \cdots \wedge x_q \\ & \qquad (x_i \in L;\ v \in M) \end{aligned} \tag{1}$$

with $k_i = \sum_{j=1}^{i-1} |x_j|$ and $k_{i,j} = \sum_{l=i+1}^{j-1} |x_l|$, the $\cdot$ denotes the action of L on M. The caret means that the corresponding argument is to be omitted. Usually we write ∂_q for the restriction of ∂ to $C_q(L;M)$. Since $\partial^2 = 0$ (i.e. $\partial_{q-1} \circ \partial_q = 0$ for all q in $\mathbb{Z}$) $(C.(L;M), \partial)$ is a chain complex and we write $B_q(L;M)$ for the space of q-boundaries, $Z_q(L;M)$ for the space of q-cycles and $H_q(L;M)$ for the space of q-homology classes. Usually one speaks of the homology of L with coefficients in M.

The cochain complex $(C^{\cdot}(L;M), d)$ is defined by

$$C^{\cdot}(L;M) = \bigoplus_{q\in\mathbb{Z}} C^q(L;M) \quad \text{with} \quad C^q(L;M) = \mathrm{Hom}(\wedge^q L; M),$$

and the coboundary operator d is the even linear operator of $\mathbb{Z}$-degree $+1$ satisfying

$$\begin{aligned} & d(f)(x_1 \wedge \cdots \wedge x_{q+1}) \\ & \quad = \sum_{1\leqslant i<j\leqslant q+1} (-1)^{i+j+(|x_i|+|x_j|)k_i+|x_j|k_{i,j}} \times \end{aligned}$$

$$\times f\left([x_i, x_j] \wedge x_1 \wedge \cdots \hat{x}_i \cdots \hat{x}_j \cdots \wedge x_{q+1}\right) +$$
$$+\sum_{j=1}^{q+1}(-1)^{j+1+|x_j|(k_j+|f|)} x_j \cdot f(x_1 \wedge \cdots \hat{x}_j \cdots \wedge x_{q+1})$$
$$(x_i \in L;\ f \in C^q(L;M)), \tag{2}$$

where k_i and $k_{i,j}$ are as previously defined. We write d_q for the restriction of d to $C^q(L;M)$. It is indeed a cochain complex since $d^2 = 0$ (i.e. $d_{q+1} \circ d_q = 0$ for all q in $\mathbb{Z}$). By $B^q(L;M)$ we denote the space of q-coboundaries, by $Z^q(L;M)$ the space of q-cocycles and by $H^q(L;M)$ the space of q-cohomology classes. This cohomology is called the cohomology of L with coefficients in M.

We end this section by making two remarks. The first concerns the notations of the complexes in case $M = \mathbb{C}$, i.e. the (one-dimensional) trivial representation; we simply write $C_{\cdot}(L)$ and $C^{\cdot}(L)$ instead of $C_{\cdot}(L;\mathbb{C})$ and $C^{\cdot}(L;\mathbb{C})$. Secondly we note that both complexes are a direct sum of the even and the odd subcomplexes, this is due to the fact that d and ∂ are even operators. This is an useful remark from the computational point of view.

2. The Homology Package

The source code of the Homology Package is written in REDUCE, for documention on REDUCE we refer to [2]. We implemented the boundary and the coboundary operator by fruitfully using the 'TOOLS Package for REDUCE' (see [7] and [8]). This package enables one to define and make use of multilinear operators in an elegant manner. In this section we discuss most of the features that are present in the Homology Package and we give some insight in the implementation.

2.1. THE INPUTS: THE STRUCTURE OF L AND M

Let L be a finite-dimensional Lie superalgebra with a homogeneous basis $\{x_i\}_{i \in I^L}$, where

$$I^L = \{-\dim L_1, -\dim L_1 + 1, \ldots, -1, 1, 2, \ldots, \dim L_0\}.$$

The elements x_i with $i < 0$ are odd and the elements x_i with $i > 0$ are even. We represent x_i by the operator element $x(i)$. Similarly, let $\{v_i\}_{i \in I^M}$ be a homogeneous basis of the finite-dimensional L-module M, we represent v_j by the operator element $v(j)$. The dimensions of L and M are set by the procedures *put_dimension* and *put_module_dimension*. The first one puts on the property list of x an *evenpart*, this is the dotted pair $(1 \cdot \dim L_0)$, and an *oddpart* equal to $(-\dim L_1 \cdot -1)$. The second one puts on the property list of v the properties *dimeven* and *dimodd*, these are equal to $\dim M_0$ and $\dim M_1$, respectively. The algebraic procedure *lie* is defined to present the commutator of L, i.e.

$$lie(i,j) = \sum_{k \in I^L} c_{i,j}^k x(k) \quad (i,j \in I^L),$$

where $c_{i,j}^k$ are the structure constants of L w.r.t. the basis $\{x_i\}_{i \in I^L}$. Similarly, the algebraic procedure *act* is defined by

$$act(i,j) = \sum_{k \in I^M} a_{i,j}^k v(k) \quad (i \in I^L; j \in I^M),$$

where the coefficients $a_{i,j}^k$ denote the structure constants of the representation w.r.t. the bases $\{x_i\}_{i \in I^L}$ and $\{v_j\}_{j \in I^M}$.

We remark, that in case of the trivial representation, one does not need to define the module structure. This will be clear from the implementation of ∂ and d.

2.2. THE IMPLEMENTATION OF ∂

As a basis of $\wedge^q L = C_q(L)$ we choose the monomials

$$x_{i_1} \wedge x_{i_2} \wedge \cdots \wedge x_{i_q} \quad (i_j \in I^L)$$

such that $i_j \leqslant i_{j+1}$ and $i_j = i_{j+1}$ implies $i_j < 0$. Let I_q^L be the set of all multi-indices $I = (i_1, i_2, \ldots, i_q)$ of length q satisfying these conditions. We simply write x_I to denote the monomial corresponding to $I \in I_q^L$ and represent it by the operator element $ext(i_1, i_2, \ldots, i_q)$. The elements $v_i \otimes x_I$ $(i \in I^M; I \in I_q^L)$ constitute a basis of $C_q(L; M)$. The tensor product is implemented by the infix operator @ which is given a higher precedence than *times*. On the property list of @ we put the *simpfn simp_multilinear*, the *oplist* $(v\ ext)$ and the *resimp_fn simpiden*. By consequence, @ is simplified as being an algebraic operator that acts multilinear w.r.t. the algebraic operators v and *ext*. Hence, the monomial $v_i \otimes x_I$ is represented as

$$v(i)@\ ext(i_1, i_2, \ldots, i_q)$$

In order to compute

$$\partial_q \left(\sum_{\substack{i \in I^M \\ I \in I_q^L}} \alpha_{i,I} v_i \otimes x_I \right) \quad (\alpha_{i,I} \in \mathbb{C}),$$

we make use of the linearity of the boundary operator. ∂ is represented by *dint*, on its property list we put the *simpfn simp_multilinear* and the *oplist* (@ v *ext*).

Hence, *dint* acts multilinear w.r.t. the operators @, v and *ext*. As *resimp_fn* we put *simp_d_interior*, this procedure takes care of the action of ∂ on monomials $v_i \otimes x_I$ according to the formula

$$\begin{aligned} \partial_q(v_i \otimes x_I) &= \sum_{1\leqslant s<t\leqslant q} (-1)^{\hat\sigma(i_s)(s+1)+\hat\sigma(i_t)t} \sum_{j\in I^L} c^j_{i_s,i_t} v_i \otimes x_j \wedge x_{i_1} \cdots \hat{x}_{i_s} \cdots \hat{x}_{i_t} \cdots x_{i_q} + \\ &+ \sum_{1\leqslant s\leqslant q} (-1)^{\sigma(i_s)\sigma(i)+\hat\sigma(i_s)(s+1)} \sum_{j\in I^M} a^j_{i_s,i} v_j \otimes x_{i_1} \wedge \cdots \hat{x}_{i_s} \cdots \wedge x_{i_q}, \end{aligned} \qquad (3)$$

where $\sigma\colon \mathbb{Z} \to \mathbb{Z}_2$ is defined by $\sigma(i) = 0$ for $i \geqslant 0$ and $\sigma(i) = 1$ for $i < 0$ and $\hat\sigma\colon \mathbb{Z} \to \mathbb{Z}_2$ is given by $\hat\sigma(i) = 1 + \sigma(i)$. The resulting monomials are reordered such that a linear combination of well-ordered (i.e. basis) monomials is delivered. The procedure *simp_d_interior* is organized in such a way that it can act on three types of elements:

(1) $v(i)@\,ext(i_1, i_2, \ldots, i_q)$	$(q > 0)$
(2) $v(i)$	(the case $q = 0$)
(3) $ext(i_1, i_2, \ldots, i_q)$	(the trivial module)

From formula (3), we learn that a quick access to the structure constants of L and M is of vital importance for the efficiency of the operator *dint*. For that reason we compute all structure constants during the initializing procedure called *initialize_vectors* and store them in the nested vectors *d_interior_vector* and *module_action_vector*. Since we use negative indices to indicate odd elements, these vectors are shifted over the odd dimensions of L and M. The element of *d_interior_vector* corresponding to the entry $(i, j) \in I^L_2$ is a list containing all lists $(c^k_{i,j}\ k)$ with $c^k_{i,j} \neq 0$. The *module_action_vector* has a similar structure.

2.3. THE IMPLEMENTATION OF d

As basis of $(\wedge^q L)^* = C^q(L)$ we take $\{x^*_I\}_{I\in I^L_q}$, where x^*_I is the element dual to x_I, i.e.

$$\langle x^*_I, x_J\rangle = \delta_{I,J} \quad (I, J \in I^L_q).$$

We use the same representation for x^*_I as for x_I, this will turn out to be useful in dealing with the Laplacian (see Section 2.8). By identifying $C^q(L; M)$ and $M \otimes (\wedge^q L)^*$ we consider the elements $v_i \otimes x^*_I$ $(i \in I^m; I \in I^L_q)$ to constitute a basis of $C^q(L; M)$. The monomial $v_i \otimes x^*_I$ is represented as $v(i)@ext(i_1, i_2, \ldots, i_q)$.

The implementation of the coboundary operator is similar to the one of the boundary operator. So d is represented by *dext* which is declared to be multilinear w.r.t. @, v and *ext*. Its *resimp_fn simp_d_exterior* takes care of the action on monomials $v_i \otimes x^*_I$.

From the formula

$$d_1(x_k^*) = \sum_{(i,j)\in I_2^L} c_{i,j}^k (x_i \wedge x_j)^*$$

we see that, in case of the coboundary operator, access to $c_{i,j}^k$ with k as input is needed. Therefore, the vector *d_exterior_vector* is constructed by the procedure *initialize_vectors*, to the entry k corresponds a list containing all lists $(i\ j\ c_{i,j}^k)$ with $(i,j) \in I_2^L$ and $c_{i,j}^k \neq 0$.

By setting the switch *d_interior_calculation* (*d_exterior_calculation*) on, one enforces the *d_interior_vector* (*d_exterior_vector*) to be made by the procedure *initialize_vectors*. By default both switches are on.

2.4. Generating a basis of $C_q(L; M)$

Generating a basis of $C_q(L; M)$ simply boils down to (recursively) generating all multi-indices $I \in I_q^L$ and joining them with all indices $i \in I^M$. In general, in computing $H_q(L; M)$ one encounters computations on vector spaces of large dimensions, even though L and M are rather 'small'. The splitting of the complex into an odd and an even part does not solve this problem. In many cases, however, there exist additional compatible gradings on L and M that constitute a so-called multigrading. Since the boundary operator is homogeneous of degree zero w.r.t. the induced multigrading on $C.(L; M)$ (see, e.g., [1], [3]), the chain complex can be split into homogeneous subcomplexes. We shall write $C_q^{(\alpha)}(L; M)$ to denote the homogeneous subspace of $C_q(L; M)$ of degree α. With regard to the multigrading we impose the following conditions:

- the bases of L and M are homogeneous. We put the degrees of the basis elements x_i and v_i by *put_degree* and *put_module_degree*.
- the degree is given by a list. This list can be of arbitrary length greater or equal than one.
- the first component of the multigrading, which we will call the total-degree, is a $\mathbb{Z}$-grading (it may be 0 for all elements). The basis $\{x_i\}_{i\in I^L}$ of L is ordered increasingly w.r.t. the total-degree, i.e.

$$\begin{aligned} &\text{total-degree } (x_i) \geqslant \text{total-degree } (x_{i-1}), \quad 2 \leqslant i \leqslant \dim L_0, \\ &\text{total-degree } (x_i) \geqslant \text{total-degree } (x_{i+1}), \quad -\dim L_1 \leqslant i \leqslant -2. \end{aligned}$$

  Moreover, the maximal and minimal total-degrees are put on the property list of x as *max_degree* and *min_degree* respectively.
- the other components of the multigrading are by default assumed to be $\mathbb{Z}$-gradings as well. By redefining *add_rest_degree* and *subtr_rest_degree*, these procedures take care of addition and subtraction of the degrees with the exception of the first component, one can introduce gradings of a completely different nature.

The procedure *make_complex(q,degree)* generates a basis of $C_q^{(degree)}(L)$. This enumeration process is very efficient due to the ordering of the basis w.r.t. the total-degree. For this reason the properties *min_degree* and *max_degree* are set. They enable us to cut of certain branches in the recursion process at an early stage. The other components of the *degree* are simply used to select, from all generated elements with the correct total-degree, the elements with the prescribed *degree*. The basis of $C_q^{(degree)}(L)$ is stored by means of the algebraic operator *ccom*: the dimension is put into *ccom(q,0)* and the operator elements *ccom(q,i)* represent the basis elements ($1 \leqslant i \leqslant$ *ccom(q,0)*). By using *make_even_complex* (*make_odd_complex*) one can generate the even (odd) part of this superspace. The *degree* can be just a beginning part from the multigrading, so we can compute bases of homogeneous spaces with respect to partial degrees. For instance, by taking the empty list for *degree* the procedure *make_complex* will generate a basis of $C_q(L)$. The procedure *make_module_complex(q,degree)* generates a basis of $C_q^{(degree)}(L;M)$ by simply considering all basis elements of M and using the alternating tensorproduct to fit the *degree*. Also in this case the algebraic operator *ccom* is used to represent the basis.

2.5. GENERATING A BASIS OF $C^q(L;M)$

The multigrading splits up the cochain complex into homogeneous subcomplexes, for which we will use simular notations. Since the representation of x_I^* is equal to the representation of x_I, *make_complex* and *make_module_complex* also generate bases of $C^q_{(\alpha}(L)$ and $C^q_{(\alpha)}(L;M)$. The difference is that the degree of x_I^* is opposite to the degree of x_I. The switch *cohomology* takes care of this minus sign, i.e. if the switch is on (off) a basis of the space of cochains (chains) is generated.

2.6. THE COMPUTATION OF $H_q(L;M)$

We discuss the construction of a basis of $H_q^{(\alpha)}(L;M)$, this involves the computation of the image of ∂_{q+1} and the kernel of ∂_q restricted to $C_{\cdot}^{(\alpha)}(L;M)$:

$$C_{q-1}^{(\alpha)}(L;M) \xleftarrow{\partial_q} C_q^{(\alpha)}(L;M) \xleftarrow{\partial_{q+1}} C_{q+1}^{(\alpha)}(L;M).$$

A basis of $B_q^{(\alpha)}(L;M)$, this is the image of ∂_{q+1}, is computed by means of *comp_boundary(qfrom, qto)* with *qfrom*$= q+1$ and *qto*$= q$ (it is assumed that a basis of $C_{q+1}^{(\alpha)}(L;M)$ has been constructed) according to the following algorithm. For $i :=$ 1: *ccom(qfrom,0)* we compute *expression* = *dint ccom(qfrom, i)*; if *expression* $\neq 0$ then add the *expression* as a new basis element of $B^q_{(\alpha)}(L;M)$. Afterwards an arbitrary term is chosen and solved from the equation *expression* =

0. Solving these equations ensures that the computations are performed modulo previously constructed boundaries and, hence, we obtain linear independent boundaries. The basis is stored by means of the algebraic operator *bcom* (same convention as for *ccom*) and the list *solved_list* contains all terms that have been solved during this process.

A basis of $H_q^{(\alpha)}(L;M)$ is computed by *comp_homology(qfrom,qto)* with *qfrom*$= q$ and *qto*$= q-1$, bases of $C_q^{(\alpha)}(L;M)$ and $B_q^{(\alpha)}(L;M)$ are assumed to be present. We compute a basis of the kernel of ∂_q on $C_q^{(\alpha)}(L;M)/B_q^{(\alpha)}(L;M)$ as follows. Solve the set of linear equations for the coefficients *repr_hom_coeff(i)* given by

$$\sum_{ccom(qfrom,i)\notin solved\_list} repr\_hom\_coeff(i) * dint\ ccom(qfrom,i) = 0$$

and compute the coefficients of *repr_hom_coeff* in

$$\sum_{ccom(qfrom,i)\notin solved\_list} repr\_hom\_coeff(i) * ccom(qfrom,i).$$

These coefficients form a basis of $H_q^{(\alpha)}(L;M)$, we store them by using the algebraic operator *hcom*. We note that this basis is a basis of a subspace in $Z_q^{(\alpha)}(L;M)$ which is mapped isomorphically to $H_q^{(\alpha)}(L;M)$ by the canonical projection π_q: $Z_q(L;M) \to H_q(L;M)$. This is in fact what we mean if we speak of a basis of $H_q^{(\alpha)}(L;M)$.

It is evident that this construction is noncanonical; the result depends on the choice of elements that were put on the *solved_list*. This choice is made on account of the internal ordering of the basis elements in REDUCE. There are methods to obtain a more intrinsic choice of a basis of $H_q^{(\alpha)}(L;M)$, see, for instance, Subsections 2.8 and 2.9. Finally, we note that by generating a basis of $C_q^{(\alpha)}(L)$ instead of $C_q^{(\alpha)}(L;M)$ one can compute bases of $B_q^{(\alpha)}(L)$ and $H_q^{(\alpha)}(L)$ by using the same procedures.

2.7. The computation of $H^q(L;M)$

The computation of bases of $B^q_{(\alpha)}(L;M)$ and $H^q_{(\alpha)}(L;M)$ is handled completely analogous, just replace *dint* by *dext*. It is implemented by the same procedures *comp_boundary* and *comp_homology*. Since the $\mathbb{Z}$-degree of the coboundary operator d is equal to $+1$ one needs to set *qto*$=$*qfrom*$+1$ instead of *qto*$=$*qfrom*-1.

2.8. The Laplacian

One way of obtaining a 'natural' basis of $H_q(L)$ is to make use of the Laplacian. We will shortly explain the construction of this operator. For all q in $\mathbb{Z}$ we define an even Hermitian positive definite from $(\ ,\ )_q$ on $C_q(L)$ by

$$(x_I, x_J)_q = \delta_{I,J} \quad (I, J \in I_q^L).$$

This form is semi-linear in the first and linear in the second argument. The semi-linear mapping ϕ_q: $C_q(L) \to C^q(L)$ is given by

$$\langle \phi_q(x), y \rangle = (x, y)_q \quad (x, y \in C_q(L)),$$

in particular we have $\phi_q(x_I) = x_I^*$ for all $I \in I_q^L$. The Laplacian Δ is the even linear operator of $\mathbb{Z}$-degree 0 on $C.(L)$ defined by

$$\Delta_q = \hat{d}_{q-1} \circ \partial_q + \partial_{q+1} \circ \hat{d}_q \quad \text{with}$$

$$\hat{d}_q = \phi_{q+1}^{-1} \circ d_q \circ \phi_q \quad (q \in \mathbb{Z}),$$

where Δ_q is the restriction of Δ to $C_q(L)$. Note that Δ_q is self-adjoint on $C_q(L)$ w.r.t. $(\ ,\ )_q$. The kernel of Δ_q is a subspace in $Z_q(L)$ which is mapped isomorphically onto $H_q(L)$ by π_q (see, e.g., [1] and [3]). Since our representation of x_I^* is equal to the representation of x_I, the implementation of the Laplacian is simply given by *Laplacian(el)= dint dext el + dext dint el.* The procedure *comp_kernel_Laplacian(q)* computes a basis of the kernel of Δ_q (and, hence, a basis of $H_q(L)$) by using the property:

$$kernel(\Delta_q) = kernel(\partial_q) \cap kernel(\hat{d}_q).$$

It is assumed that a basis of $C_q(L)$ (or of $C_q^{(\alpha)}(L)$) has been constructed. The result is given by means of the operator *hcom.*

2.9. The representation $C.(L; M)$

The representation of L on M and the adjoint representation of L induce a representation of L on $C.(L; M)$. On $C_q(L; M)$ this action is given by

$$\begin{aligned} &x \cdot (v \otimes x_1 \wedge \cdots \wedge x_q) \\ &\quad = (x \cdot v) \otimes (x_1 \wedge \cdots \wedge x_q) + \\ &\qquad + (-1)^{|x||v|} \sum_{j=1}^{q} (-1)^{|x|k_j} v \otimes x_1 \wedge \cdots [x, x_j] \cdots \wedge x_q, \end{aligned}$$

with

$$k_j = \sum_{l=1}^{j-1} |x_l| \quad (x, x_i \in L;\ v \in M).$$

Due to the fact that the action $x\cdot$ ($x \in L$) commutes with the boundary operator, it is a useful tool for homology computations. We implemented it by the operator *dot*, which is declared to be multilinear w.r.t. the operators *x, v, ext* and @. The *resimp_fn simp_dot* takes care of the actual work, i.e. the action of an element x_i ($i \in I$) on a monomial $v_i \otimes x_I$ ($j \in I^m$, $I \in I_q^L$).

The superspaces $B_q(L;M)$ and $Z_q(L;M)$ are L-submodules of $C_q(L;M)$; the action of L on the quotient module $H_q(L;M) = Z_q(L;M)/B_q(L;M)$ is trivial (see, e.g., [1] and [3]). By means of the operator *dot* we can investigate if there exists an L-invariant subspace in $Z_q(L;M)$ complementary to $B_q(L;M)$. If so, a basis of this space yields a natural basis of $H_q(L;M)$ in terms of L-representations.

Another application of *dot* is the use of inner gradings. Suppose there exists an element x in L_0 such that

$$L = \bigoplus_\alpha L_{(\alpha)} \quad \text{with } L_{(\alpha)} = \{y \in L \mid x \cdot y = [x,y] = \alpha y\},$$

$$M = \bigoplus_\alpha M_{(\alpha)} \quad \text{with } M_{(\alpha)} = \{m \in M \mid x \cdot m = \alpha m\}.$$

Such gradings of L and M are called inner. For the induced grading on $C.(L;M)$ we have $H_q(L;M) = H_q^{(0)}(L;M)$ (see, e.g., [1]), i.e. we can confine ourselves to computing the homology of the subcomplex $C.^{(0)}(L;M)$. This subcomplex can be computed by using *dot*.

Often, it is more interesting to consider a representation of another Lie superalgebra $\tilde{L}$ on L and M. Let us denote such representations by $\cdot_L$ and $\cdot_M$. We assume that they are compatible, i.e.

$$x \cdot_M (y \cdot m) = (x \cdot_L y) \cdot m + (-1)^{|x||y|} y \cdot (x \cdot_M m)$$

$$(x \in \tilde{L};\ y \in L;\ m \in M).$$

Due to the compatibility the induced representation of $\tilde{L}$ on $C.(L;M)$ commutes with the boundary operator. Hence, $Z_q(L;M)$ and $B_q(L;M)$ are $\tilde{L}$-submodules of $C_q(L;M)$ and $H_q(L;M)$ is the quotient module. A basis of a $\tilde{L}$-submodule of $Z_q(L;M)$ complementary to $B_q(L;M)$ would be a natural basis of $H_q(L;M)$ as $\tilde{L}$-module. We consider the following example. Let K be a $\mathbb{Z}$-graded Lie superalgebra and M a K-module. Define $\tilde{L} = K_{(0)}$ and $L = \bigoplus_{i>0} K_{(i)}$, then the compatiblity condition is obviously satisfied for $\cdot_M = \cdot$ (of K) and $\cdot_L = [\ ,\ ]$ (in K).

With the previous example in mind, we implemented the action of $\tilde{L}$ by means of shifting the operator elements $x(i)$ in such a way that $x(i)$ represents $\tilde{x}_i$ for $i \in I^{\tilde{L}}$, $x_{i-\dim \tilde{L}_0}$ for $\dim \tilde{L}_0 < i \leqslant \dim \tilde{L}_0 + \dim L_0$ and $x_{i+\dim \tilde{L}_1}$ for $-(\dim L_1 + \dim \tilde{L}_1) \leqslant i < -\dim \tilde{L}_1$. Note that $\{\tilde{x}_i\}_{i \in I^{\tilde{L}}}$ is a homogeneous basis w.r.t. the $\mathbb{Z}_2$-grading of $\tilde{L}$. According to this shift, the procedures *lie* and *act* need to be adjusted such that they represent the commutators of $\tilde{L}$ and L (*lie*) the action of $\tilde{L}$ on L (*lie*) and the actions of $\tilde{L}$ and L on M (*act*). To distinguish the bases of $\tilde{L}$ and L, we put on the property list of x the *evenpart* $((\dim \tilde{L}_0 + 1) \cdot (\dim \tilde{L}_0 + \dim L_0))$ and the *oddpart* $((-\dim L_1 - \dim \tilde{L}_1) \cdot (-\dim \tilde{L}_1 - 1))$. These properties are used by *dint* and *dext* to ensure that (co)homology of L (and not of $\tilde{L} \oplus L$) is computed. With this implementation *dot* can be used to compute a natural basis of $H_q(L; M)$ w.r.t. the representation of $\tilde{L}$.

By defining a multigrading on $\tilde{L}$ compatible with the multigradings on L and M (*put_degree* sets the degrees of the basis elements, they are assumed to be homogeneous), we can compute a natural basis of $H_q^{(\alpha)}(L; M)$ w.r.t. the action of $\tilde{L}$. Finally, we remark that *dot* can also be used to compute $H_q(L)$ instead of $H_q(L; M)$ in a natural manner.

2.10. The representation $C^{\cdot}(L; M)$

It is evident that, similar to the reasoning from the preceding section, the induced action of L (or $\tilde{L}$) on $C^{\cdot}(L; M)$ gives rise to a natural basis of $H^q(L; M)$. This action is implemented by the procedure *dual_dot*. This procedure uses the *dual_action_vector*. This nested vector contains the structure constants of the representation of L (and $\tilde{L}$) on L^* w.r.t. the bases $\{x_i\}_{i \in I^L}$ ($\{\tilde{x}_i\}_{i \in I^{\tilde{L}}}$) and $\{x_I^*\}_{i \in I^L}$. Its structure is similar to that of the *module_action_vector*. By putting on the switch *dual_action* one enforces the *dual_action_vector* to be made by *initialize_vectors*.

3. Applications of the Homology Package

The homology package can be used for several constructions that are based on homology or cohomology computations. In this section, we mention two of them: the construction of formal deformations and of a minimal set of generators and defining relations.

The infinitesional part of a formal deformation of a Lie superalgebra L is a homogeneous element of $H^2(L; L)$, in case of an even (odd) element we speak of an even (odd) formal deformation of L. For $S(0, n)$, this is the Lie superalgebra consisting of divergence free vector fields on a superspace with 0 even and n odd coordinates, we computed $H^2(S(0, n); S(0, n))$ for $2 \leqslant n \leqslant 4$. In order to perform this computation, we made use of a $\mathbb{Z} \times \mathbb{Z}^n$-grading in combination with several inner gradings. We found that $\dim H^2_{(\alpha)}(S(0, n); S(0, n)) = 1$ for $\alpha = (n; 1, 1, \ldots, 1)$ and zero for all other degrees. Since $B^2_{(\alpha)}(S(0, n);\ S(0, n)) = 0$, the basis of $H^2_{(\alpha)}(S(0, n);\ S(0, n))$ was automatically unique. Due to this, we were able to generalize the result to all values of n, i.e. we proved that $\dim H^2(S(0, n);\ S(0, n)) \geqslant 1$ for all $n \geqslant 2$. Furthermore, it appeared that the constructed infinitesimal part defines a linear deformation of $S(0, n)$ which is even (odd) for n even (odd). For the details we refer to [5] in which we discuss similar results for the Lie superalgebra of Hamiltonian vector fields $H(0,n)$.

For a nilpotent Lie superalgebra L a basis of $H_1(L)$ reflects a minimal set of generators of L and a basis of $H_2(L)$ reflects a minimal set of defining relations. We computed these for the graded maximal nilpotent parts of the classical Lie superalgebras of vector fields $W(m,n)$, $S(m,n)$, $H(k,n)$, and $K(k,n)$. For a description of these infinite-dimensional Lie superalgebras we refer to [1]. These Lie superalgebras are $\mathbb{Z}$-graded such that $L = \bigoplus_{p\geqslant -1} L_{(p)}$. The graded maximal nilpotent part is defined by $\hat{L} = \bigoplus_{p>0} L_{(p)}$. We presented the homology spaces $H_1^{(p)}(\hat{L})$ and $H_2^{(p)}(\hat{L})$ as module over $L_{(0)}$; for the case $n = 0$ we refer to [4] and the general case can be found in [3]. We note that $W(m,n)_{(0)} = \mathrm{gl}(m,n)$, $S(m,n)_{(0)} = \mathrm{sl}(m,n)$, $H(k,n)_{(0)} = \mathrm{osp}(2k,n)$ and $K(k,n) = \mathrm{osp}(2k,n) \times \mathbb{C}$. We were able to perform these calculations for infinite-dimensional Lie superalgebras due to the fact that the $\mathbb{Z}$-grading gives rise to finite-dimensional subcomplexes. So again, the use of a grading is essential in the computations.

References

1. Fuchs, D. B.: *Cohomology of Infinite-Dimensional Lie Algebras*, Plenum, New York, 1986.
2. Hearn, A. C.: *REDUCE User's Manual Version 3.4*, RAND, Santa Monica, 1991.
3. v.d. Hijligenberg, N.: Computations and applications of Lie superalgebra cohomology, PhD Thesis, University of Twente, The Netherlands, 1993.
4. v.d. Hijligenberg, N. and Post, G. F.: Defining relations for Lie algebras of vector fields, *Indag. Math. N.S.* **2** (1991), 207.
5. v.d. Hijligenberg, N., Kotchetkov, Y., and Post, G. F.: Deformations of $S(0,n)$ and $H(0,n)$, *Internat. J. Algebra Comput.* **3**(1) (1993), 57.
6. Leites, D. A.: Cohomology of Lie superalgebras, *Func. Anal. Appl.* **9** (1975), 75.
7. Roelofs, G. H. M.: The TOOLS-package for REDUCE, Memorandum 942, University of Twente, The Netherlands, 1991.
8. Roelofs, G. H. M. and Gragert, P. K. H.: Implementation of multilinear operators in REDUCE and applications in mathematics, in *Proc. 1991 Internat. Symp. on Symbolic and Algebraic Computation*, 1991, 390.
9. Tripathy, K. C. and Patra, M. K.: Cohomology theory and deformations of $\mathbb{Z}_2$-graded Lie algebras, *J. Math. Phys.* **31** (1990), 2822.

Acta Applicandae Mathematicae **41:** 135–144, 1995.

Conservation Laws and the Variational Bicomplex for Second-Order Scalar Hyperbolic Equations in the Plane

IAN M. ANDERSON
Department of Mathematics, Utah State University, Logan, UT 84322–3900, U.S.A.

and

NIKY KAMRAN
Department of Mathematics, McGill University, Montreal, Quebec H3A 2K6, Canada

(Received: 15 October 1993)

Abstract. In this paper, we announce several new results concerning the cohomology of the variational bicomplex for a second-order scalar hyperbolic equation in the plane. These cohomology groups are represented by the conservation laws, and certain form-valued generalizations, for the equation. Our methods are based upon the introduction of an adapted coframe for the the variational bicomplex which is constructed by generalizing the classical Laplace transformation used to integrate certain linear hyperbolic equations in the plane.

Mathematics Subject Classifications (1991): 58G16, 35A30, 35L65.

Key words: variational bicomplex, hyperbolic second-order equations, conservation laws.

The horizons of modern geometrical methods in differential equations have been widened considerably over the past ten to fifteen years with the infusion of new, sophisticated ideas and techniques from topology, homological and commutative algebra, and differential geometry. Still, the classical literature in this subject and the problems addressed therein, beginning with the works of S. Lie and G. Darboux and continuing through to E. Cartan, continue to motivate many current developments and to provide a rich, stimulating source of examples and case studies. The purpose of this paper is to announce, without proof, some new theorems on the cohomology of the variational bicomplex and on the existence and characterization of conservation laws for second-order hyperbolic partial differential equations in the plane. The integration of such equations was extensively studied prior to the turn of the century and much of our work is inspired by results found in Goursat's superb treatise [8]. We refer the reader to [2] and [3] for the proofs of our theorems as well as for the corresponding results for second-order elliptic and parabolic equations in the plane. For second-order scalar parabolic equations in the plane, we refer to the recent papers of Bryant and Griffiths ([4]

and [5]) who have applied their theory of characteristic cohomology theory for exterior differential systems to classify such equations according to their number of classical conservation laws.

Consider the second-order partial differential equation

$$F(x, y, u, u_x, u_y, u_{xx}, u_{xy}, u_{yy}) = 0. \tag{1}$$

Let $J^2(E)$ denote the second-order jet bundle over the trivial bundle E: $\mathbf{R}^2 \times \mathbf{R} \to \mathbf{R}^2$. The coordinates for E are π: $(x, y, u) \to (x, y)$ and the coordinates for $J^2(E)$ are $(x, y, u, u_x, u_y, u_{xx}, u_{xy}, u_{yy})$. Let $\mathcal{R}^2$ be an open, connected and contractible subset of the locus defined by Equation (1). We assume that $\mathcal{R}^2$ fibers over an open subset U of the (x, y) plane and that $\mathcal{R}^2$ is a smooth 7-dimensional submanifold of $J^2(E)$. We can then construct, from the derivatives of (1), the infinite prolongation manifold

$$\iota\colon\ \mathcal{R}^\infty \longrightarrow J^\infty(E)$$

as a submanifold of the infinite jet bundle $J^\infty(E)$ of E, with projection map π_U^∞: $\mathcal{R}^\infty \to U$. Let $\mathcal{C}(\mathcal{R}^\infty)$ denote the pull-back of the contact ideal on $J^\infty(E)$ to $\mathcal{R}^\infty$ and let $\mathcal{R}$ denote the triple

$$\mathcal{R} = \{\mathcal{R}^\infty,\ \pi_U^\infty,\ \mathcal{C}(\mathcal{R}^\infty)\}.$$

Classical local solutions to (1) are in bijective correspondence with solutions to $\mathcal{R}$, that is, sections σ of π_U^∞: $\mathcal{R}^\infty \to U$ which satisfy $\sigma^*[\mathcal{C}(\mathcal{R}^\infty)] = 0$. The *variational bicomplex* for $\mathcal{R} = \{\mathcal{R}^\infty, \pi_U^\infty, \mathcal{C}(\mathcal{R}^\infty)\}$ is the pull-back of the free variational bicomplex $(\Omega^{*,*}(J^\infty(E)), d_H, d_V)$ (see [1, 10, 11]) to $\mathcal{R}^\infty$. We write the variational bicomplex for $\mathcal{R}$ as

$$\begin{array}{ccccccccc}
 & & & & \uparrow d_V & & \uparrow d_V & & \uparrow d_V \\
 & & 0 & \longrightarrow & \Omega^{0,3}(\mathcal{R}^\infty) & \xrightarrow{d_H} & \Omega^{1,3}(\mathcal{R}^\infty) & \xrightarrow{d_H} & \Omega^{2,3}(\mathcal{R}^\infty) \\
 & & & & \uparrow d_V & & \uparrow d_V & & \uparrow d_V \\
 & & 0 & \longrightarrow & \Omega^{0,2}(\mathcal{R}^\infty) & \xrightarrow{d_H} & \Omega^{1,2}(\mathcal{R}^\infty) & \xrightarrow{d_H} & \Omega^{2,2}(\mathcal{R}^\infty) \\
 & & & & \uparrow d_V & & \uparrow d_V & & \uparrow d_V \\
 & & 0 & \longrightarrow & \Omega^{0,1}(\mathcal{R}^\infty) & \xrightarrow{d_H} & \Omega^{1,1}(\mathcal{R}^\infty) & \xrightarrow{d_H} & \Omega^{2,1}(\mathcal{R}^\infty) \\
 & & & & \uparrow d_V & & \uparrow d_V & & \uparrow d_V \\
0 & \longrightarrow & \mathbf{R} & \longrightarrow & \Omega^{0,0}(\mathcal{R}^\infty) & \xrightarrow{d_H} & \Omega^{1,0}(\mathcal{R}^\infty) & \xrightarrow{d_H} & \Omega^{2,0}(\mathcal{R}^\infty)\ .
\end{array}$$

When (1) is written in the form

$$u_{xx} + f(x, y, u, u_x, u_y, u_{xy}, u_{yy}) = 0, \tag{2}$$

we can explicitly describe the variational bicomplex on $\mathcal{R}^\infty$ as follows. Natural coordinates for $\mathcal{R}^\infty$ are

$$(x, y, u, u_x, u_y, u_{xy}, u_{yy}, \ldots, u_{xy^{k-1}}, u_{y^k}, \ldots).$$

The *total* derivative operators D_y and D_x on $\mathcal{R}^\infty$ are given by

$$D_y = \frac{\partial}{\partial y} + u_y \frac{\partial}{\partial u} + u_{xy} \frac{\partial}{\partial u_x} + u_{yy} \frac{\partial}{\partial u_y} + u_{xyy} \frac{\partial}{\partial u_{xy}} + \cdots$$

and

$$D_x = \frac{\partial}{\partial y} + u_x \frac{\partial}{\partial u} - f \frac{\partial}{\partial u_x} + u_{xy} \frac{\partial}{\partial u_y} - (D_y f) \frac{\partial}{\partial u_{xy}} + \cdots$$

and a basis for the contact ideal on $\mathcal{R}^\infty$ is

$$\{\theta, \theta_x, \theta_y, \theta_{xy}, \theta_{yy}, \ldots, \theta_{xy^{k-1}}, \theta_{y^k}, \ldots\}, \tag{3}$$

where

$$\theta = \mathrm{d}u - u_x\, \mathrm{d}x - u_y\, \mathrm{d}y,$$

$$\theta_{xy^{k-1}} = \mathrm{d}u_{xy^{k-1}} + ((D_y)^{k-1} f) \mathrm{d}x - u_{xy^{k-1}}\, \mathrm{d}y, \quad \text{and}$$

$$\theta_{y^k} = \mathrm{d}u_{y^k} - u_{xy^{k-1}}\, \mathrm{d}x - u_{y^{k+1}}\, \mathrm{d}y.$$

The basis given by $\mathrm{d}x$, $\mathrm{d}y$ and the forms (3) is called the coordinate coframe on $\mathcal{R}^\infty$.

A differential form $\alpha \in \Omega^{0,s}(\mathcal{R}^\infty)$ is an s form on $\mathcal{R}^\infty$ in the exterior algebra generated by the contact forms (3). Forms $\beta \in \Omega^{1,s}(\mathcal{R}^\infty)$ and $\gamma \in \Omega^{2,s}(\mathcal{R}^\infty)$ can be written as

$$\beta = A \wedge \mathrm{d}x + B \wedge \mathrm{d}y \quad \text{and} \quad \gamma = C \wedge \mathrm{d}x \wedge \mathrm{d}y,$$

where A, B, and C belong to $\Omega^{0,s}(\mathcal{R}^\infty)$. The d_H and d_V structure equations for this coordinate coframe are

$$\mathrm{d}_H(g) = (D_x g)\, \mathrm{d}x + (D_y g)\, \mathrm{d}y,$$

and

$$\mathrm{d}_V(g) = \frac{\partial g}{\partial u}\theta + \frac{\partial g}{\partial u_x}\theta_x + \frac{\partial g}{\partial u_y}\theta_y + \frac{\partial g}{\partial u_{xy}}\theta_{xy} + \frac{\partial g}{\partial u_{yy}}\theta_{yy} + \cdots,$$

where $g = g(x, y, u, u_x, u_y, u_{xy}, u_{yy}, \ldots)$ is a function on $\mathcal{R}^\infty$, and

$$\mathrm{d}_H(\mathrm{d}x) = \mathrm{d}_H(\mathrm{d}y) = 0, \qquad \mathrm{d}_H(\theta) = \mathrm{d}x \wedge \theta_x + \mathrm{d}y \wedge \theta_y,$$

$$\mathrm{d}_H(\theta_{xy^{k-1}}) = -\mathrm{d}x \wedge \mathrm{d}_V((D_y)^{k-1} f) + \mathrm{d}y \wedge \theta_{xy^k},$$

$$d_H(\theta_{y^k}) = dx \wedge \theta_{xy^k} + dy \wedge \theta_{y^{k+1}},$$

and

$$d_V(dx) = d_V(dy) = d_V(\theta) = d_V(\theta_{xy^{k-1}}) = d_V(\theta_{y^k}) = 0.$$

The spectral sequence associated to the bicomplex $(\Omega^{*,*}(\mathcal{R}^\infty), d_H, d_V)$ obtained from the filtration by vertical degree is Vinogradov's $\mathcal{C}$ spectral sequence [12] for Equation (1). The E_1 term of this spectral sequence is just the d_H cohomology of the variational bicomplex, namely,

$$E_1^{r,s} = H^r(\Omega^{*,s}(\mathcal{R}^\infty), d_H) \stackrel{\text{def}}{=} H^{r,s}(\mathcal{R}^\infty).$$

For $r = 0$, it follows from a general theorem of Vinogradov [12] that

$$H^{0,s}(\mathcal{R}^\infty) = \begin{cases} \mathbf{R} & \text{for } s = 0, \\ 0 & \text{for } s \geqslant 1. \end{cases}$$

Therefore the only E_1 terms of interest are

$$E_1^{1,s} = H^{1,s}(\mathcal{R}^\infty) \quad \text{and} \quad E_1^{2,s} = H^{2,s}(\mathcal{R}^\infty).$$

Let

$$\omega = dx \wedge M + dy \wedge N$$

be a cohomology representative in $H^{1,s}(\mathcal{R}^\infty)$, so that

$$d_H \omega = 0.$$

For $s = 0$, M and N are smooth functions on $\mathcal{R}^\infty$ and ω is a *classical conservation law*. For $s \geqslant 1$, M and N are contact $(0, s)$ forms on $\mathcal{R}^\infty$ and ω is called a *form-valued conservation law.* The conservation law ω is said to be trivial if it represents the zero class, that is, if there is a $(0, s)$ form η such that $\omega = d_H\eta$. We focus our study of the variational bicomplex for second-order scalar equations in the plane on the classification of classical and form-valued conservation laws.

It is not difficult to prove (see [1, 2]) that if (1) is derivable from a variational principle then there is always a nontrivial form-valued conservation law of type $(1, 2)$.

The hyperbolicity of Equation (1) means that the inequality

$$\frac{\partial F}{\partial u_{xx}} \frac{\partial F}{\partial u_{yy}} - \frac{1}{4}\left(\frac{\partial F}{\partial u_{xy}}\right)^2 < 0$$

is satisfied everywhere on $\mathcal{R}^2$. Consequently, we have the following factorization on $\mathcal{R}^2$

$$\kappa\left(\frac{\partial F}{\partial u_{xx}}\lambda^2 - \frac{\partial F}{\partial u_{xy}}\lambda\mu + \frac{\partial F}{\partial u_{yy}}\mu^2\right) = (m_x\lambda - m_y\mu)(n_x\lambda - n_y\mu),$$

for some smooth real-valued functions κ, m_x, m_y, n_x, n_y satisfying

$$\kappa \neq 0 \quad \text{and} \quad m_x n_y - m_y n_x \neq 0.$$

The two *characteristic vector field systems* for the hyperbolic Equation (1) are generated by the two characteristic total vector fields

$$X = m_x D_x + m_y D_y \quad \text{and} \quad Y = n_x D_x + n_y D_y.$$

Of particular interest to us are Equations (1) for which the total vector fields X and Y may be scaled so as to commute. This is notably the case when there exist nonconstant functions I and J such that

$$X(I) = 0 \quad \text{and} \quad Y(J) = 0.$$

Then it is easy to see that the characteristic vector fields

$$\widetilde{X} = \frac{1}{X(J)} X \quad \text{and} \quad \widetilde{Y} = \frac{1}{Y(I)} Y$$

commute.

We let σ and τ be the type $(1,0)$ forms dual to X and Y,

$$\sigma(X) = 1, \qquad \sigma(Y) = 0, \qquad \tau(X) = 0, \qquad \tau(Y) = 1.$$

If $\omega \in \Omega^{r,s}(\mathcal{R}^\infty)$ is any type (r,s) form, then

$$d_H \omega = \sigma \wedge X(\omega) + \tau \wedge Y(\omega),$$

where, for any total vector field Z on $\mathcal{R}^\infty$, we have defined

$$Z(\omega) = Z \lrcorner\, d_H \omega + d_H(Z \lrcorner\, \omega).$$

Restricted to $\mathcal{R}^\infty$, the contact forms θ, θ_x θ_y, θ_{xx}, θ_{xy} and θ_{yy} are not independent but are related by the equation $d_V F = 0$, that is,

$$\frac{\partial F}{\partial u_{xx}} \theta_{xx} + \frac{\partial F}{\partial u_{xy}} \theta_{xy} + \frac{\partial F}{\partial u_{yy}} \theta_{yy} + \frac{\partial F}{\partial u_x} \theta_x + \frac{\partial F}{\partial u_y} \theta_y + \frac{\partial F}{\partial u} \theta = 0. \tag{4}$$

We call this equation the *universal linearization* of (1). Equation (4) plays a fundamental role in the study of the variational bicomplex for (1). In terms of the characteristic vector fields X and Y, the universal linearization becomes

$$XY(\Theta) + A\,X(\Theta) + B\,Y(\Theta) + C\,\Theta = 0, \tag{5}$$

where A, B and C are certain combinations of κ, m_x, m_y, n_x, n_y, F and their derivatives and where Θ is nonzero multiple of θ. This freedom to scale θ can often be used to simplify the coefficients A, B, and C in the universal linearization (5).

The next crucial step in our study of the variational bicomplex for (1) is to construct a coframe on $\mathcal{R}^\infty$ which is adapted to the universal linearization (5). This coframe, which we call the Laplace adapted coframe and which we denote by

$$\{\sigma, \tau, \Theta, \xi_1, \eta_1, \xi_2, \eta_2, \ldots, \xi_k, \eta_k, \ldots\},$$

is obtained by applying a formal generalization of the classical *Laplace transformation* (see, for example, Darboux [6]) for second order linear hyperbolic equations to the universal linearization (5). For example, under the assumption that the characteristics X and Y commute, the contact one forms ξ_1 and η_1 are defined by

$$\xi_1 = X(\Theta) + B\,\Theta \quad \text{and} \quad \eta_1 = Y(\Theta) + A\,\Theta. \tag{6}$$

Then, by virtue of the universal linearization (5), these forms satisfy

$$Y(\xi_1) + A\,\xi_1 - K_0\Theta = 0 \quad \text{and} \quad X(\eta_1) + B\,\eta_1 - H_0\Theta = 0, \tag{7}$$

where the functions H_0 and K_0 are defined by

$$H_0 = X(A) + AB - C \quad \text{and} \quad K_0 = Y(B) + AB - C.$$

These functions generalize to the case of the fully nonlinear Equation (1) the classical Laplace invariants for the linear hyperbolic equations [6]. Provided $H_0 \neq 0$ and $K_0 \neq 0$, it can be shown that ξ_1 and η_1 satisfy linear relations similar to (5) and the Laplace transforms (6) can be applied to ξ_1 and η_1, in place of Θ, to define the next contact forms ξ_2 and η_2 of our adapted coframe. Thus, in the process of constructing the Laplace adapted coframe, we also introduce the two sequences of invariants

$$H_0, H_1, H_2, \ldots \quad \text{and} \quad K_0, K_1, K_2, \ldots$$

which we call the *generalized Laplace invariants* of the hyperbolic Equation (1). In particular, the next Laplace invariant H_1 is given by

$$H_1 = X(A_1) + A_1B_1 - C_1,$$

where

$$A_1 = A - \frac{Y(H_0)}{H_0}, \qquad B_1 = B,$$

$$C_1 = C - X(A) + Y(B) - B\frac{Y(H_0)}{H_0}.$$

Under the infinite prolongation of a classical first-order contact transformation $\Phi\colon J^1(E) \to J^1(E')$, we know that the characteristic vector fields X and Y and the contact one form Θ transform according to

$$X' = mX, \qquad Y' = nY, \qquad \Theta' = l\Theta,$$

where m, n and l are nonzero functions on $\mathcal{R}^\infty$, and we can prove that

$$H_i' = mnH_i \quad \text{and} \quad K_j' = mnK_j$$

and, also provided $H_{i-1} \neq 0$ and $K_{j-1} \neq 0$,

$$\eta_i' = n^i l\,\eta_i \quad \text{and} \quad \xi_j' = m^j l\,\xi_j.$$

Thus any homogeneous relationship amongst the generalized Laplace invariants is a contact invariant condition on Equation (1).

To the extent that the generalized Laplace invariants do not vanish, we obtain structure equations

$$\mathrm{d}_H(\eta_i) \equiv H_{i-1}\,\sigma \wedge \eta_{i-1} + \tau \wedge \eta_{i+1} \bmod \{\eta_i\}$$

and

$$\mathrm{d}_H(\xi_j) \equiv \sigma \wedge \xi_{j+1} + K_{j-1}\,\tau \wedge \xi_{j-1} \bmod \{\xi_j\}.$$

The usefulness of the Laplace adapted coframe lies largely in the simplicity of these structure equations and their invariance under classical contact transformations.

A contact form $\omega \in \Omega^{0,s}(\mathcal{R}^\infty)$ is called an X characteristic invariant form if $X(\omega) = 0$. Such forms naturally arise in computations of the d_H cohomology of the variational bicomplex for $\mathcal{R}$.

THEOREM 1. *If the generalized Laplace invariants $H_0, H_1, H_2, \ldots$ are all nonzero, then there are no X invariant contact forms.*

Proofs for Theorems 1–5 can be found in [1]. Theorems 6–9 are established in [2].

The generalized Laplace invariants play an important role in the geometric integrability properties of second order hyperbolic partial differential equations. Recall that (1) is said to be *Darboux integrable* if there exist two functionally independent X invariant functions and two functionally independent Y invariant functions, say

$$X(I) = X(\tilde{I}) = 0 \quad \text{and} \quad X(J) = X(\tilde{J}) = 0.$$

Consequently, for Darboux integrable equations, the characteristic vector fields can always be scaled so as to commute. The general solution to a Darboux integrable equation may be obtained from the solution of a completely integrable Frobenius system, that is, Darboux integrable equations can be solved by integrating ordinary differential equations. If X and Y commute, if I_1, I_2 and I_3 are X invariant functions such that $Y(I_1) = 1$ and $I_3 = Y(I_2)$, then it is not difficult to show that the contact form

$$\omega = \mathrm{d}_V I_2 - I_2 \mathrm{d}_V I_1$$

is X invariant. This observation, together with Theorem 1 and similar results for Y characteristic invariant forms, proves the following theorem.

THEOREM 2. *If a hyperbolic equation is Darboux integrable, then there exist integers $p \geqslant 0$ and $q \geqslant 0$ such that $H_p = 0$ and $K_q = 0$.*

The generalized Laplace invariants also arise in the solution to the inverse problem of the calculus of variations for (1).

THEOREM 3. *If Equation (1) is equivalent to the Euler–Lagrange equations for a Lagrangian first order $L = L(x, y, u, u_x, u_y)$, then $H_0 = K_0$.*

We now turn to the issue of conservation laws. We begin by making the simple but important observation that if α and β are any X and Y invariant $(0, s)$ forms, that is, $X(\alpha) = 0$ and $Y(\beta) = 0$, then the $(1, s)$ form

$$\omega = \sigma \wedge \beta + \tau \wedge \alpha \tag{8}$$

is d_H closed and defines a (possibly trivial) conservation law of type $(1, s)$ for $\mathcal{R}^\infty$. Starting from this observation, we can prove the following.

THEOREM 4. *Let $\mathcal{R}$ be a hyperbolic equation with commuting characteristics. Suppose that $H_p = 0$ and that there exists at least one nonzero X invariant contact $(0, 1)$ form. Then $\mathcal{R}$ admits infinitely many nontrivial contact form-valued conservation laws of type $(1, s)$, for all $s \geqslant 2$.*

THEOREM 5. *If $\mathcal{R}$ is Darboux integrable then, for all $s \geqslant 0$, $H^{1,s}(\mathcal{R})$ is infinite dimensional.*

We now state some partial converses to Theorem 4. These are based upon the following theorem which gives a normal form for any form-valued conservation law.

THEOREM 6. *Let ω be a type $(1, s)$, $s \geqslant 1$, conservation law for a hyperbolic equation $\mathcal{R}$ with commuting characteristics X and Y. Then ω is cohomologous to a unique conservation law $\widetilde{\omega}$ of the form*

$$\widetilde{\omega} = \sigma \wedge \Theta \wedge \chi + \tau \wedge \eta_1 \wedge \rho,$$

where the $(0, s-1)$ forms χ and ρ satisfy

$$X(\rho) = B\rho + \chi \quad \text{and} \quad Y(\chi) = H_0\rho + A\chi.$$

COROLLARY 7. *Assume that the characteristics X and Y of $\mathcal{R}$ commute. Then, for all $s \geqslant 1$, the vector space of type $(1, s-1)$ forms ρ which satisfy the equation*

$$XY(\rho) - AX(\rho) - BY(\rho) - (X(A) + Y(B) - C)\,\rho = 0 \tag{9}$$

surjects onto the vector space of nontrivial $(1, s)$ conservation laws $H^{1,s}(\mathcal{R}^\infty)$.

Equation (9) can be viewed as the analogue, for contact-form valued conservation laws, of Olver's [9] characteristic equation for a classical conservation law. It is important to note that (9) is the formal adjoint of the universal linearization (5) and that it can be studied in its own right by means of the generalized Laplace transform. The relations between the Laplace invariants of (5) and those of (9) are the same as those described by Forsyth [7] in the classical context. A careful analysis of (9) leads us to the following conclusions.

THEOREM 8. *Let $\mathcal{R}$ be a hyperbolic equation with commuting characteristics. If $\mathcal{R}$ admits a nontrivial conservation law of type* $(1, s)$, *for* $s \geqslant 3$, *then either* $H_p = 0$ *or* $K_q = 0$ *for some* $p \geqslant 0$ *or* $q \geqslant 0$.

Theorem 6, Corollary 7 and Theorem 8 can be extended to the case where the characteristics X and Y may not commute.

THEOREM 9. *If a hyperbolic equation is Darboux integrable, then every conservation law of type* $(1, s)$, *for* $s \geqslant 3$, *is cohomologous to one of the type* (8).

Theorems 8 and 9 are false for $s = 2$ although our methods do provide considerable information on conservation laws of type (1, 2).

To illustrate these results, we consider the Liouville equation $u_{xy} = \mathrm{e}^u$. We hasten to add the cautionary remark that this simple equation reflects none of the complexities which are inherent in the general theory of Darboux integrable equations to which our results apply. The characteristic vector fields are $X = D_x$ and $Y = D_y$, the form-valued linearization is $XY(\theta) - \mathrm{e}^u\,\theta = 0$, the first elements of the Laplace adapted coframe are just $\xi_1 = \theta_x$ and $\eta_1 = \theta_y$, and the initial Laplace invariants are $H_0 = K_0 = \mathrm{e}^u$.

We find that η_1 and ξ_1 satisfy

$$XY(\eta_1) - u_y X(\eta_1) - \mathrm{e}^u \eta_1 = 0 \quad \text{and} \quad XY(\xi_1) - u_x Y(\xi_1) - \mathrm{e}^u \xi_1 = 0,$$

so that the next Laplace invariants are $H_1 = K_1 = 0$. The characteristic X and Y invariant functions are

$$I_1 = y, \qquad I_2 = u_{yy} - \tfrac{1}{2}u_y^2, \qquad I_3 = D_y I_2, \qquad I_4 = D_y I_3, \ \ldots$$

and

$$J_1 = x, \qquad J_2 = u_{xx} - \tfrac{1}{2}u_x^2, \qquad J_3 = D_x J_2, \qquad J_4 = D_x J_3, \ \ldots .$$

Every X invariant type $(0, s)$ contact form α is a sum of terms

$$f\,\mathrm{d}_V I_{k_1} \wedge \mathrm{d}_V I_{k_2} \cdots \wedge \mathrm{d}_V I_{k_s},$$

where f is a function of the invariants I_1, I_2, $\ldots$ and every Y invariant type $(0, s)$ contact form β is a sum of terms

$$g\,\mathrm{d}_V J_{k_1} \wedge \mathrm{d}_V J_{k_2} \cdots \wedge \mathrm{d}_V J_{k_s},$$

where g is a function of the invariants J_1, J_2, ... By Theorem 9, this characterizes all the type $(1, s)$, $s \geqslant 3$, conservation laws for Liouville's equation. Every type (1, 2) conservation law is also of this form with the one exception being

$$\omega_0 = \theta \wedge (\theta_y \wedge dy - \theta_x \wedge dx).$$

The sine-Gordon equation

$$u_{xy} = \sin u$$

has infinitely many classical type $(1, 0)$ conservation laws. But it is not integrable by the method of Darboux and it has no form-valued conservation laws of type $(1, s)$, for $s \geqslant 3$.

Acknowledgements

This research is supported by grants DMS-9100674 and OGP-0105490 from the National Science Foundation and the Natural Sciences and Engineering Research Council of Canada.

References

1. Anderson, M.: Introduction to the variational bicomplex, in M. Gotay, J. Marsden and V. Moncrief (eds), *Mathematical Aspects of Classical Field Theory*, Contemporary Mathematics 132, Amer. Math Soc., Providence, 1992, pp. 51–73.
2. Anderson, Ian M. and Kamran, N.: The variational bicomplex for second order scalar partial differential equations in the plane, Centre de recherches mathématiques, Technical report, September 1994.
3. Anderson, Ian M. and Kamran, N.: The variational bicomplex for hyperbolic second order scalar partial differential equations in the plane (submitted May 1995).
4. Bryant, R. L. and Griffiths, P. A.: Characteristic cohomology of differential systems, I: General theory, Duke University, Mathematics Preprint Series, January, 1993.
5. Bryant, R. L. and Griffiths, P. A.: Characteristic cohomology of differential systems, II: Conservation laws for a class of parabolic equations, Duke University, Mathematics Preprint Series, January, 1993.
6. Darboux, G.: *Leçons sur la théorie générale des suraces et les applications géométriques du calcul infinitésimal*, Gauthier-Villars, Paris, 1896.
7. Forsyth, A.: *Theory of Differential Equations*, Vol. 6, Dover, New York, 1959.
8. Goursat, E.: *Leçon sur l'intégration des équations aux dérivées partielles du second ordre á deux variables indépendantes*, Tome 2, Hermann, Paris, 1896.
9. Olver, P. J.: *Applications of Lie Groups to Differential Equations*, Springer, New York, 1986.
10. Tsujishita, T.: On variation bicomplexes associated to differential equations, *Osaka J. Math.* **19** (1982), 311–363.
11. Tsujishita, T.: Formal geometry of systems of differential equations, *Sugaku Exposition* **2** (1989), 1–40.
12. Vinogradov, A. M.: The $\mathcal{C}$-spectral sequence, Lagrangian formalism and conservation laws, I, II, *J. Math. Anal. Appl.* **100** (1984), 1–129.

Acta Applicandae Mathematicae **41:** 145–152, 1995.

On the $\mathcal{C}$-Spectral Sequence for Systems of Evolution Equations

N. G. KHOR'KOVA*
Department of Applied Mathematics, Moscow State Technical University, ul. 2-aja Baumanskaja 5, 107005 Moscow, Russia

(Received: 30 November 1993)

Abstract. The groups $E_1^{2,n-1}(\mathcal{Y}_\infty)$ of Vinogradov's $\mathcal{C}$-spectral sequence for determined systems of evolution equations are considered. Presentation of these groups useful in practical computations is obtained. The group $E_1^{2,1}(\mathcal{Y}_\infty)$ is calculated for a system of Schrödinger type equations.

Mathematics Subject Classifications (1991): 58H10, 35Q53, 58G35.

Key words: nonlinear differential equations, $\mathcal{C}$-differential operators, the $\mathcal{C}$-spectral sequence.

Introduction

In [4]–[5] we considered the groups $E_1^{p,n-1}(\mathcal{Y}_\infty)$, $p \geqslant 2$, of Vinogradov's $\mathcal{C}$-spectral sequence for differential equations in a one-dimensional linear bundle, that is we restricted ourselves to these equations, and systems of differential equations were eliminated from our considerations. Now we generalize the results of [4]–[5] concerning evolution equations for determined systems. Namely, we obtain a representation of the group $E_1^{2,n-1}(\mathcal{Y}_\infty)$ for the systems under consideration and then using it calculate the group $E_1^{2,1}(\mathcal{Y}_\infty)$ for a system of Schrödinger-type equations.

In what follows, we use notations from [1]–[3]. Namely,

- $J^k(\pi)$, $0 \leqslant k \leqslant \infty$, is the space of k-jets of a fiber bundle $\pi\colon E_\pi^{n+m} \to M^n$,
- $\pi_k\colon J^k(\pi) \to M^n$, $\pi_{k,s}\colon J^k(\pi) \to J^s(\pi)$, $k > s$, are natural projections,
- $F_k(\pi) = C^\infty(J^k(\pi))$ and $F(\pi) = \lim\operatorname{dir}_{k\to\infty} F_k(\pi)$ are rings of smooth functions on $J^k(\pi)$ and $J^\infty(\pi)$, respectively,
- $F_k(\pi,\xi) = \Gamma(\pi_k^*(\xi))$, $F(\pi,\xi) = \lim\operatorname{dir}_{k\to\infty} F_k(\pi,\xi)$, where $\pi\colon E_\pi^{n+m} \to M^n$ and $\xi\colon E_\xi^{n+m} \to M^n$ are bundles,
- $\mathcal{Y}_\infty$ is the infinite prolongation of a differential equation $\mathcal{Y} = \{F = 0\}$, $F \in F_k(\pi,\xi)$,
- $J_{\mathcal{Y}} = \{f \in F(\pi) \mid f|_{\mathcal{Y}_\infty} = 0\}$,
- $[Q] = Q/Q \cdot J_{\mathcal{Y}}$ is the restriction of an $F(\pi)$-module Q to $\mathcal{Y}_\infty$,

* Correspondence to: N. G. Khor'kova, Volokolamskoe sh. 16b, korp. 2, Apt. 45, 125080 Moscow, Russia.

- $\mathcal{Y}_s = \pi_{\infty,s}(\mathcal{Y}_\infty)$, $F_s(\mathcal{Y}) = F(\pi)|_{\mathcal{Y}_s}$, $F(\mathcal{Y}_\infty) = \lim\operatorname{dir}_{k\to\infty} F_s(\mathcal{Y})$,
- $\mathcal{C}\mathrm{Diff}(P, Q)$ is the module of $\mathcal{C}$-differential operators acting from the $F(\pi)$ ($F(\mathcal{Y}_\infty)$)-module P to the $F(\pi)$ ($F(\mathcal{Y}_\infty)$)-module Q,
- $\bar{\mathbf{\Delta}}$ is a restriction of the operator $\mathbf{\Delta} \in \mathcal{C}\mathrm{Diff}(P, Q)$ to $\mathcal{Y}_\infty$ (P and Q are $F(\pi)$-modules),
- $x_1, \ldots, x_n, p^j_\sigma$, $1 \leqslant j \leqslant m$, $\sigma = (i_1, \ldots, i_s)$, $1 \leqslant i_s \leqslant n$, $0 \leqslant s < \infty$, is a canonical coordinate system in $J^\infty(\pi)$,
- D_i is the total derivative with respect to x_i,
- $\mathcal{C}D(\pi) = \{\sum_i a_i D_i \mid a_i \in F(\pi)\}$ is the Cartan distribution on $J^\infty(\pi)$,
- $D_\mathcal{C}(\pi)$ is the Lie algebra of vector fields S on $J^\infty(\pi)$ such that $[S, \mathcal{C}D(\pi)] \subset \mathcal{C}D(\pi)$,
- $\varkappa = D_\mathcal{C}(\pi)/\mathcal{C}D(\pi)$,
- l_F is the universal linearization operator,
- $\widehat{P} = \mathrm{Hom}(P, \Lambda^n)$,
- $\mathbf{\Delta}^*$ is the adjoint operator.

1. Basic Constructions [1]

1.1. Let $\mathcal{Y}$ be an equation and $\mathcal{C}\Lambda^1(\mathcal{Y}_\infty)$ be the module of 1-forms vanishing on the Cartan distribution. Consider in the algebra $\Lambda^*(\mathcal{Y}_\infty)$ the ideal $\mathcal{C}\Lambda^*(\mathcal{Y}_\infty)$ generated by $\mathcal{C}\Lambda^1(\mathcal{Y}_\infty)$. This ideal and all its powers $\mathcal{C}^k\Lambda^*(\mathcal{Y}_\infty)$ are stable with respect to the operator d. Thus, in the de Rham complex $\{\Lambda^*(\mathcal{Y}_\infty), d\}$ on $\mathcal{Y}_\infty$, the filtration

$$\Lambda^*(\mathcal{Y}_\infty) = \mathcal{C}^0\Lambda^*(\mathcal{Y}_\infty) \supset \mathcal{C}^1\Lambda^*(\mathcal{Y}_\infty) \supset \cdots \supset \mathcal{C}^k\Lambda^*(\mathcal{Y}_\infty) \supset \cdots$$

arises.

The spectral sequence $\{E_r^{p,q}(\mathcal{Y}_\infty), d_r^{p,q}\}$ determined by this filtration is called *the $\mathcal{C}$-spectral sequence* for $\mathcal{Y}_\infty$.

1.2. An equation $\mathcal{Y} = \{F = 0\} \subset J^k(\pi)$, $F \in F_k(\pi, \xi)$ is called regular, if

$$J_\mathcal{Y} = \mathcal{C}\mathrm{Diff}(P, F(\pi))(F),$$

where $P = F(\pi, \xi)$.

It can be easily seen that a determined system of evolution equations is regular.

1.3. Let $\mathcal{Y} = \{F = 0\}$ be a regular equation. Then

$$E_1^{2,n-1}(\mathcal{Y}_\infty) = G/H,$$

where G is the group of all operators $\mathbf{\Delta} \in \mathcal{C}\mathrm{Diff}([\varkappa], [\widehat{P}])$ such that

$$\bar{l}_F^* \circ \mathbf{\Delta} = \mathbf{\Delta}^* \circ \bar{l}_F,$$

while $H = \{\mathbf{\nabla} \circ \bar{l}_F \mid \mathbf{\nabla} \in \mathcal{C}\mathrm{Diff}([P], [\widehat{P}]),\ \mathbf{\nabla} = \mathbf{\nabla}^*\}$.

2. A Description of the Group $E_1^{2,n-1}(\mathcal{Y}_\infty)$ for Systems of Evolution Equations

In this section, we prove a generalization of Theorem 2.2.1 from [5] for a determined system of evolution equations.

THEOREM 2.1. *Let $\mathcal{Y}$ be a determined system of evolution equations*

$$\begin{aligned} u_t &= f(x, u, \dots, u_\sigma), \qquad u = (u^1, \dots, u^m), \\ x &= (x_1, \dots, x_n), \qquad t = x_n, \quad n \notin \sigma, \\ f &= (f_1, \dots, f_m), \quad f \in F_k(\pi). \end{aligned} \tag{2.1}$$

Then the group $E_1^{2,n-1}(\mathcal{Y}_\infty)$ is isomorphic to the linear space of all matrix $\mathcal{C}$-differential operators $\mathbf{\Delta} = (\Delta_{ij})$,

$$\Delta_{ij} = \sum_{i_1,\dots,i_{n-1}} a^{ij}_{i_1\cdots i_{n-1}} \overline{D}_1^{i_1} \circ \dots \circ \overline{D}_{n-1}^{i_{n-1}} \colon F(\mathcal{Y}_\infty) \longrightarrow F(\mathcal{Y}_\infty),$$

$$a^{ij}_{i_1\cdots i_{n-1}} \in F(\mathcal{Y}_\infty),$$

such that

$$\bar{l}_F^* \circ \mathbf{\Delta} = \mathbf{\Delta}^* \circ \bar{l}_F, \tag{2.2}$$

where $F = p_n - f(x, p_\emptyset, \dots, p_\sigma)$.

Proof. If $\mathcal{Y}$ is a system of evolution equations, then any scalar $\mathcal{C}$-differential operator $\mathbf{\Delta}\colon F(\mathcal{Y}_\infty) \to F(\mathcal{Y}_\infty)$ can be locally represented as

$$\mathbf{\Delta} = \sum_{k=0}^{N} A_k \circ \overline{D}_t^k, \quad A_N \neq 0, \tag{2.3}$$

where

$$A_k = \sum_{i_1,\dots,i_{n-1}} a_{k,i_1\cdots i_{n-1}} \overline{D}_1^{i_1} \circ \dots \circ \overline{D}_{n-1}^{i_{n-1}}, \quad a_{k,i_1\cdots i_{n-1}} \in F(\mathcal{Y}_\infty). \tag{2.4}$$

We say that N is the *order* of $\mathbf{\Delta}$ with respect to t and write $N = \mathrm{ord}_t \mathbf{\Delta}$.

If an operator $\mathbf{\Delta} = (\Delta_{ij})$ generates an element of the group $E_1^{2,n-1}(\mathcal{Y}_\infty)$, then it satisfies Equation (2.2), which for Equation (2.1) has the form

$$(-\mathrm{Id} \circ \overline{D}_t - \bar{l}_f^*) \circ \mathbf{\Delta} = \mathbf{\Delta}^* \circ (\mathrm{Id} \circ \overline{D}_t - \bar{l}_f),$$

or the following system of equations

$$-\overline{D}_t \circ \Delta_{jk} - \sum_{i=1}^{n} l_{ij}^* \circ \Delta_{ik} = \Delta_{kj}^* \circ \overline{D}_t - \sum_{i=1}^{n} \Delta_{ij}^* \circ l_{ik}, \tag{2.5$_{jk}$}$$

where $\bar{l}_f = (l_{ij})$, $j, k = 1, \ldots, n$.

Now let us represent each operator Δ_{ij} in the form (2.3)–(2.4):

$$\Delta_{ij} = \sum_{k=0}^{N_{ij}} A_k^{ij} \circ \overline{D}_t^k, \qquad A_{N_{ij}}^{ij} \neq 0, \quad N_{ij} = \mathrm{ord}_t \Delta_{ij}$$

and let $N_{\alpha\beta} = \max_{i,j} N_{ij} \geqslant 1$.

Equating to zero the coefficient of $\overline{D}_t^{N_{\alpha\beta}+1}$ in Equation ($2.5_{\alpha\beta}$), we get $N_{\alpha\beta} = N_{\beta\alpha}$ and

$$A_{N_{\alpha\beta}}^{\alpha\beta} = (-1)^{N_{\alpha\beta}+1} (A_{N_{\alpha\beta}}^{\beta\alpha})^*. \tag{2.6}$$

Our goal now is to show that the operator $\mathbf{\Delta}$ can be represented in the form

$$\mathbf{\Delta} = \mathbf{\nabla} \circ \bar{l}_F + \mathbf{\Delta}', \tag{2.7}$$

where

$$\mathbf{\nabla}^* = \mathbf{\nabla}, \qquad \mathbf{\Delta}' = (\Delta'_{ij}),$$
$$\mathrm{ord}_t \Delta'_{ij} \leqslant \mathrm{ord}_t \Delta_{ij}, \quad \text{if } (ij) \neq (\alpha\beta), \qquad \mathrm{ord}_t \Delta'_{\alpha\beta} < \mathrm{ord}_t \Delta_{\alpha\beta}. \tag{2.8}$$

Then, using the induction over $\mathrm{ord}_t \Delta_{ij}$, one can easily prove that

$$\mathbf{\Delta} = \widetilde{\mathbf{\nabla}} \circ \bar{l}_F + \tilde{\mathbf{\Delta}},$$

where $\widetilde{\mathbf{\nabla}}^* = \widetilde{\mathbf{\nabla}}$ and $\mathrm{ord}_t \tilde{\Delta}_{ij} = 0$ for all i, j. Since the operators $\mathbf{\Delta}$ and $\tilde{\mathbf{\Delta}}$ generate the same element of the group $E_1^{2,n-1}(\mathcal{Y}_\infty)$, we shall get our statement.

So we look for representation (2.7) for $\mathbf{\Delta}$.

Introduce the operator $\hat{\mathbf{\Delta}} = (\hat{\Delta}_{ij})$ with the components

$$\hat{\Delta}_{\alpha\beta} = A_{N_{\alpha\beta}}^{\alpha\beta} \circ \overline{D}_t^{N_{\alpha\beta}-1}, \qquad \hat{\Delta}_{\beta\alpha} = A_{N_{\beta\alpha}}^{\beta\alpha} \circ \overline{D}_t^{N_{\alpha\beta}-1},$$
$$\hat{\Delta}_{ij} = 0 \quad \text{for } (ij) \neq (\alpha\beta) \text{ or } (\beta\alpha).$$

Then we have representation (2.7) for $\mathbf{\Delta}$ with

$$\mathbf{\nabla} = \frac{\hat{\mathbf{\Delta}} + \hat{\mathbf{\Delta}}^*}{2}, \qquad \mathbf{\Delta}' = \frac{\hat{\mathbf{\Delta}} - \hat{\mathbf{\Delta}}^*}{2} \circ \bar{l}_F + \mathbf{\Delta} - \hat{\mathbf{\Delta}} \circ \bar{l}_F.$$

It can be easily checked that, by virtue of (2.6) and the form of $\hat{\mathbf{\Delta}}$ and $\bar{l}_F$, the operator $\mathbf{\Delta}'$ satisfies (2.8).

The theorem is proved.

COROLLARY. *Let $\mathcal{Y}$ be a determined system of evolution Equations* (2.1). *Then the group $E_1^{2,n-1}(\mathcal{Y}_\infty)$ is isomorphic to the linear space of all matrix C-differential operators $\mathbf{\Delta} = (\Delta_{ij})$,*

$$\Delta_{ij} = \sum_{i_1,\ldots,i_{n-1}} a_{i_1\cdots i_{n-1}}^{ij} \overline{D}_1^{i_1} \circ \cdots \circ \overline{D}_{n-1}^{i_{n-1}} \colon F(\mathcal{Y}_\infty) \longrightarrow F(\mathcal{Y}_\infty),$$

$$a_{i_1\ldots i_{n-1}}^{ij} \in F(\mathcal{Y}_\infty),$$

such that

$$\mathbf{\Delta}^* = -\mathbf{\Delta}, \tag{2.9}$$
$$\overline{D}_t(\mathbf{\Delta}) + \mathbf{\Delta} \circ \bar{l}_f + \bar{l}_f^* \circ \mathbf{\Delta} = 0. \tag{2.10}$$

where $\overline{D}_t(\mathbf{\Delta}) = [\mathrm{Id} \circ \overline{D}_t, \mathbf{\Delta}]$.

3. An Example

In this section to demonstrate a method of computation of the group $E_1^{2,n-1}(\mathcal{Y}_\infty)$, based on Theorem 2.1 we find the group $E_1^{2,1}(\mathcal{Y}_\infty)$ for a system of Schrödinger-type equations (cf. [6]).

THEOREM 3.1. *Let $\mathcal{Y}$ be the following system of evolution equations*

$$\begin{aligned} u_t &= \lambda u_{xx} - 2uu_xv, \\ v_t &= -\lambda v_{xx} - 2uvv_x, \quad \lambda \in \mathbb{R}. \end{aligned} \tag{3.1}$$

Then the group $E_1^{2,1}(\mathcal{Y}_\infty)$ is isomorphic to the one-dimensional linear space generated by the operator

$$\mathbf{\Delta} = \begin{pmatrix} 0 & 1 \\ -1 & 0 \end{pmatrix}.$$

Proof. Let $\mathbf{\Delta}$ generates an element of the group $E_1^{2,1}(\mathcal{Y}_\infty)$. Then it satisfies (2.9)–(2.10). From (2.9) we obtain

$$\mathbf{\Delta} = \begin{pmatrix} A & B \\ -B^* & C \end{pmatrix}, \qquad A^* = -A, \quad C^* = -C.$$

For system (3.1), operator $\bar{l}_f$ has the form

$$\bar{l}_f = \begin{pmatrix} \Box & -2uu_x \\ -2vv_x & -\Box^* \end{pmatrix},$$

where $\Box = \lambda D_x^2 - 2uvD_x - 2u_xv$, $D_x = \overline{D}_x$.

Then Equation (2.10) is equivalent to the following system

$$\overline{D}_t(A) + A \circ \Box + \Box^* \circ A + 2vv_xB^* - B \circ 2vv_x = 0, \tag{3.2}$$
$$\overline{D}_t(B) - B \circ \Box^* + \Box^* \circ B - 2vv_xC - A \circ 2uu_x = 0, \tag{3.3}$$
$$\overline{D}_t(C) - C \circ \Box^* - \Box \circ C - 2uu_xB + B^* \circ 2uu_x = 0. \tag{3.4}$$

Suppose $A \neq 0$ or $C \neq 0$. Then we obtain from (3.2) or (3.4)

$$\mathrm{ord}_x A + 2 = \mathrm{ord}_x(vv_xB^* - B \circ vv_x) \leqslant \mathrm{ord}_x B$$

or

$$\mathrm{ord}_x C + 2 = \mathrm{ord}_x(uu_xB^* - B \circ uu_x) \leqslant \mathrm{ord}_x B.$$

Let

$$B = \sum_{k=0}^{N+2} b_k D_x^k, \qquad b_{N+2} \neq 0, \quad N \geqslant 0.$$

Then $\mathrm{ord}_x A \leqslant N$, $\mathrm{ord}_x C \leqslant N$.

Equating to zero the coefficients of D_x^{N+3} and D_x^{N+2} in (3.3), we get

$$b_{N+2} = \beta(t), \qquad b_{N+1} = \frac{1}{2\lambda}(2\beta(N+2)uv - \beta' x) + \gamma(t).$$

Equating to zero the coefficients of D_x^{N+2} and D_x^{N+1} in (3.2), (3.4), we obtain

$$\begin{aligned}
a_N &= (1-(-1)^N)\frac{\beta}{\lambda}vv_1, \\
a_{N-1} &= \frac{1}{\lambda}\{\beta(N+2)D(vv_1) + (1+(-1)^N)vv_1 b_{N+1}\} - D(a_N), \\
c_N &= ((-1)^N - 1)\frac{\beta}{\lambda}uu_1, \\
c_{N-1} &= \frac{1}{\lambda}\{(-1)^N\beta(N+2)D(uu_1) + ((-1)^{N+1} - 1)uu_1 b_{N+1}\} - D(c_N).
\end{aligned}$$

Further, it can be easily proved by induction that

$$\begin{aligned}
a_j &\equiv \beta\tilde{a}_j vv_{N-j+1} \bmod F_{N-j}(\mathcal{Y}_\infty), \\
c_j &\equiv \beta\tilde{c}_j uu_{N-j+1} \bmod F_{N-j}(\mathcal{Y}_\infty), \\
b_j &\equiv \beta\tilde{b}_j uv_{N-j+1} + \beta\hat{b}_j vu_{N-j+1} \bmod F_{N-j}(\mathcal{Y}_\infty), \\
j &= 0, 1, \ldots, N,
\end{aligned}$$

where $\tilde{a}_j$, $\tilde{c}_j$, $\tilde{b}_j$, $\hat{b}_j$ are constants, while $\phi \equiv \psi \bmod F_k(\mathcal{Y}_\infty)$ means $\phi - \psi \in F_k(\mathcal{Y}_\infty)$.

Equating to zero the coefficient of D_x^0 in (3.4), we get

$$\overline{D}_t(c_0) - \lambda D_x^2(c_0) + 2(-1)^{N+2}\beta uu_{N+3} \equiv 0 \mod F_{N+2}(\mathcal{Y}_\infty).$$

Hence, $\beta = 0$. But this contradicts assumptions $A \neq 0$, $C \neq 0$.

Thus, we have $A = C = 0$. Solving system (3.2)–(3.4) in this case, one can easily get

$$\boldsymbol{\Delta} = \begin{pmatrix} 0 & a \\ -a & 0 \end{pmatrix},$$

where $a = \mathrm{const}$.

Theorem 3.1 is proved.

4. Some Applications

Equation (2.2) shows that the group $E_1^{2,n-1}$ consists of operators taking generating functions of symmetries to elements of $\ker \bar{l}_F^*$. It is known (see [1], [3]) that the kernel of the adjoint universal linearization operator contains generating functions of all conservation laws of a given equation and that to find all conservation laws for an equation $\mathcal{Y} = \{F = 0\}$ it is necessary to solve the equation $\bar{l}_F^*(\psi) = 0$. If there exists a nontrivial element $\mathbf{\Delta} \in E_1^{2,n-1}(\mathcal{Y}_\infty)$, then there exists a linear map of linear spaces

$$\mathbf{\Delta}\colon \ker \bar{l}_F \longrightarrow \ker \bar{l}_F^*.$$

In the case of system (3.1) this map is an isomorphism of spaces $\ker \bar{l}_F$ and $\ker \bar{l}_F^*$. Hence, if symmetry algebra $\operatorname{Sym}\mathcal{Y} \approx \ker \bar{l}_F$ for the system under consideration had been found we need not solve the equation $\bar{l}_F^*(\psi) = 0$ to find conservation laws: any vector-function $\varphi \in \ker \bar{l}_F^*$ has the form $\psi = \mathbf{\Delta}(\varphi)$ for some $\varphi \in \ker \bar{l}_F$. Thus, having in mind the problems of computation of symmetry algebra and the group of conservation laws for a given equation we may look for some element of the group $E_1^{2,n-1}(\mathcal{Y}_\infty)$. The latter may be easier than to solve the equation $\bar{l}_F^*(\psi) = 0$.

To illustrate this idea let us consider the following system $\mathcal{Y}$ of evolution equations

$$\begin{aligned} u_t &= \lambda u_{xx} + g(u, v, u_x, v_x), \\ v_t &= \mu v_{xx} + h(u, v, u_x, v_x), \quad \lambda, \mu \in \mathbb{R}. \end{aligned} \tag{4.1}$$

One can look for an element $\mathbf{\Delta} \in E_1^{2,n-1}(\mathcal{Y}_\infty)$ of the form

$$\mathbf{\Delta} = \begin{pmatrix} a_{11} & a_{12} \\ a_{21} & a_{22} \end{pmatrix},$$

where $a_{ij} \in \mathbb{R}$ and $a_{11}a_{22} - a_{12}a_{21} \neq 0$. But by the virtue of Equation (2.9) one may restrict oneself by the operator

$$\mathbf{\Delta} = \begin{pmatrix} 0 & 1 \\ -1 & 0 \end{pmatrix}.$$

Let us prove the following statement

PROPOSITION 4.1. *Let $\mathcal{Y}$ be system* (4.1). *Then*

$$\mathbf{\Delta} = \begin{pmatrix} 0 & 1 \\ -1 & 0 \end{pmatrix} \in E_1^{2,1}(\mathcal{Y}_\infty)$$

if and only if equation $\mathcal{Y}$ has the form

$$\begin{aligned} u_t &= \lambda u_{xx} + a(u,v)u_x + b_v(u,v), \\ v_t &= -\lambda v_{xx} + a(u,v)v_x - b_u(u,v), \quad \lambda \in \mathbb{R}. \end{aligned} \tag{4.2}$$

Proof. Equation (2.10) for system (4.1) and

$$\Delta = \begin{pmatrix} 0 & 1 \\ -1 & 0 \end{pmatrix}$$

has the form

$$(l_h^u)^* = l_h^u, \qquad (l_g^v)^* = l_g^v, \qquad (\mu + \lambda)D^2 + (l_g^u)^* + l_h^v = 0, \tag{4.3}$$

where l_f^u for $f = f(u, v, u_x, v_x)$ denotes the operator $l_f^u = f_{u_x}D + f_u$, while $l_f^v = f_{v_x}D + f_v$.

One can easily get from (4.3) the following

$$\mu + \lambda = 0, \qquad g = g(u, v, u_x), \qquad h = h(u, v, v_x).$$

Then from the last equation in (4.3) we obtain the final form of g and h.

Hence, we have also the following statement

PROPOSITION 4.2. *Let $\mathcal{Y}$ be system* (4.2). *Then operator*

$$\Delta = \begin{pmatrix} 0 & 1 \\ -1 & 0 \end{pmatrix}$$

generates an isomorphism of linear spaces $\ker \bar{l}_F$ *and* $\ker \bar{l}_F^*$.

References

1. Vinogradov, A. M.: The C-spectral sequence, Lagrangian formalism and conservation laws, *J. Math. Anal. Appl.* **100**(3) (1984), 1–129.
2. Krasil'shchik, I. S., Lychagin, V. V., and Vinogradov, A. M.: *Geometry of Jet Spaces and Nonlinear Partial Differential Equations*, Gordon and Breach, New York, 1986.
3. Vinogradov, A. M.: Local symmetries and conservation laws, *Acta Appl. Math.* **2**(1) (1984), 21–78.
4. Khor'kova, N. G.: On the C-spectral sequence of evolution equations, *Math. Notes* **49**(6) (1991), 145–147.
5. Khor'kova, N. G.: On the C-spectral sequence of differential equations, *Differential Geom. Appl.* **3** (1993), 219–243.
6. Bogolubov, N. N. Jr, *et al.*: Nonlinear model of Schrödinger type: Conservation laws, Hamiltonian structure and complete integrability, *Theoret. Mech. Math. Phys.* **65**(2) (1985), 271–284 (in Russian).

Acta Applicandae Mathematicae **41:** 153–165, 1995.

Exact Gerstenhaber Algebras and Lie Bialgebroids

Y. KOSMANN-SCHWARZBACH
Centre de Mathématiques, URA 169 du CNRS, Ecole Polytechnique, F-91128 Palaiseau Cedex, France. E-mail: yks@orphee.polytechnique.fr

(Received: 2 May 1994)

Abstract. We show that to any Poisson manifold and, more generally, to any triangular Lie bialgebroid in the sense of Mackenzie and Xu, there correspond two differential Gerstenhaber algebras in duality, one of which is canonically equipped with an operator generating the graded Lie algebra bracket, i.e. with the structure of a Batalin–Vilkovisky algebra.

Mathematics Subject Classifications (1991): 17B70, 17B81, 17B66, 53C15, 58F05.

Key words: Poisson manifolds, Lie algebroids, Lie bialgebroids, Lie pseudo-algebras, Schouten brackets, graded Lie algebras, Gerstenhaber algebras, Batalin–Vilkovisky algebras, topological field theories, string theory.

Recent papers by Lian and Zuckerman [18] and by Getzler [9], and several preprints on string theory [23, 29], make extensive use of algebraic structures which have previously appeared in various contexts. We wish both to point out important earlier references and to establish a connection between these structures and the theory of Lie algebroids due to Pradines [24] (see [19]) and, more specifically, the new concept of Lie bialgebroid introduced by Mackenzie and Xu [20].

The notion of a Gerstenhaber algebra goes back to Gerstenhaber's work on the cohomology rings of algebras in 1963 [6] (see [7, 8]), while the notion of a generating operator for a Gerstenhaber algebra was introduced in 1985 by Koszul [15] in his study of the graded Lie algebra of multivectors on a manifold.

DEFINITION. Let $(\mathcal{A} = \bigoplus_{i\in\mathbf{Z}} \mathcal{A}^i, \wedge)$ be a graded commutative associative algebra. A graded Lie algebra structure on $\mathcal{A} = \bigoplus_{i\in\mathbf{Z}} \mathcal{A}^{(i)}$, where $\mathcal{A}^{(i)} = \mathcal{A}^{i+1}$, is called a Gerstenhaber algebra bracket if, for each a in $\mathcal{A}^{(i)}$, $[a, .]$ is a derivation of degree i of $(\mathcal{A} = \bigoplus_{i\in\mathbf{Z}} \mathcal{A}^i, \wedge)$.

Moreover, an operator ∂ of degree -1 is said to generate the Gerstenhaber algebra bracket if, for all $a \in \mathcal{A}^{|a|}$ and $b \in \mathcal{A}$,

$$[a, b] = (-1)^{|a|}(\partial(a \wedge b) - \partial a \wedge b - (-1)^{|a|} a \wedge \partial b)\,.$$

When there exists a generating operator of square 0 for the bracket, the Gerstenhaber algebra is called exact.

The biderivation property of the bracket is called the (graded) Leibniz rule. Koszul showed that if ∂ is a differential operator of order at most 2, i.e. if the trilinear mapping Φ^3_∂, defined by formula (1.3) of [15], vanishes, then the graded Leibniz rule for the associated bracket is satisfied. Koszul also showed that if $\partial^2 = 0$, then the associated bracket satisfies the Jacobi identity, and ∂ is a derivation of $(\mathcal{A}, [\, ,\,])$.

So a *Gerstenhaber algebra* is a triple denoted by, e.g., $(\mathcal{A}, \wedge, [\, ,\,])$, and an *exact Gerstenhaber algebra* is a quadruple denoted by, e.g., $(\mathcal{A}, \wedge, [\, ,\,], \partial)$, where ∂ is an operator of degree -1, and of square 0, while $[\, ,\,]$ measures the extent to which ∂ fails to be a derivation of $(\mathcal{A}, \wedge)$.

For short, we shall sometimes call the Gerstenhaber algebras defined above G-algebras, following [7, 8]. What we have called exact Gerstenhaber algebras are *coboundary Gerstenhaber algebras* in the sense of Lian and Zuckerman [18], where we require the generating operator to be of square 0.

What Getzler calls *braid algebras* in [9] are Gerstenhaber algebras with a differential, i.e. a derivation of degree 1 and square 0 of the graded commutative associative structure of the G-algebra, so we shall call them *differential Gerstenhaber algebras*. What Getzler calls *Batalin–Vilkovisky algebras* are braid algebras which are exact as G-algebras, i.e. *differential exact Gerstenhaber algebras*. (But Getzler's convention for the gradings is actually the opposite of the usual one, which we adopt here.) So, a differential Gerstenhaber algebra will be denoted by a quadruple, e.g., $(\mathcal{A}, \wedge, \mathrm{d}, [\, ,\,])$, where $(\mathcal{A}, \wedge, [\, ,\,])$ is a Gerstenhaber algebra and d is a derivation of degree 1 and square 0 of the graded commutative associative algebra $(\mathcal{A}, \wedge)$, while a differential exact Gerstenhaber algebra will be denoted by a quintuple, e.g., $(\mathcal{A}, \wedge, \mathrm{d}, [\, ,\,], \partial)$.

1. The Gerstenhaber Algebra of a Lie Algebroid

It is a well-known fact that, given a Lie algebroid A, the algebra of sections of $\bigwedge A$, $\Gamma(\bigwedge A)$, equipped with the exterior product together with the generalized Schouten bracket, forms a Gerstenhaber algebra (see Kosmann-Schwarzbach and Magri [13, 14], where the term Schouten algebra was used instead of Gerstenhaber algebra, and Mackenzie and Xu [20].) The generalized Schouten bracket is defined as the unique extension $[\, ;\,]$ of the Lie algebroid bracket such that

(i) $[Q, Q'] = -(-1)^{qq'}[Q', Q]$, for $Q \in \Gamma(\bigwedge^{q+1} A)$, $Q' \in \Gamma(\bigwedge^{q'+1} A)$,
(ii) $[X, f] = a(X)f$, for $X \in \Gamma(\bigwedge^1 A)$, $f \in \Gamma(\bigwedge^0 A)$, where the mapping $a\colon \Gamma(A) \to \Gamma(TM)$ is the anchor of the Lie algebroid A, with base M, and
(iii) for $Q \in \Gamma(\bigwedge^{q+1} A)$, $[Q, .]$ is a derivation of degree q of the graded algebra $(\Gamma(\bigwedge A) = \bigoplus_{p\geqslant 0} \Gamma(\bigwedge^p A), \wedge)$.

(In our convention, the exterior algebra of a module F, which we denote by $\bigwedge F$ is the sum of its exterior powers, not the sum of the exterior powers of its dual, F^*.)

Moreover, $\Gamma(\bigwedge(A^*))$ is equipped with a derivation of square 0 of its graded commutative associative stucture, denoted by d, and called, by analogy with Example 1.2 below, the de Rham differential of forms. The differential d is defined by a formula identical to the Cartan formula for the de Rham differential,

$$\begin{aligned} &\mathrm{d}\alpha(x_0, \dots, x_p) \\ &= \sum_{i=0}^{p}(-1)^i a(x_i)(\alpha(x_0, \dots, \hat{x}_i, \dots, x_p)) + \\ &+ \sum_{0 \leqslant i<j \leqslant p} (-1)^{i+j}\alpha([x_i, x_j], x_0, \dots, \hat{x}_i, \dots, \hat{x}_j, \dots, x_p), \end{aligned}$$

where α is a p-form, a is the anchor of the Lie algebroid A, and $[\,,\,]$ is the Lie algebroid bracket of sections of A. The Lie derivative of forms with respect to an element x of $\Gamma(A)$ is the derivation $L_x = [i_x, \mathrm{d}]$, where we have denoted the graded commutator by $[\,,\,]$.

More generally, one can consider the algebraic version of Lie algebroids, variously called Palais pairs [8], differential Lie algebras, Lie–Rinehart algebras, pseudo-Lie algebras, Elie Cartan spaces, Lie–Cartan pairs, etc., see [22] and, e.g., [13, 10]. Given a Palais pair (H_0, H_1), the exterior algebra of H_1 over the ring H_0 is a Gerstenhaber algebra. This fact is clearly stated in Section 6.3 of [13] (see also the 'note added in proof' in that article which refers to an unpublished manuscript of B. Kostant and S. Sternberg). It appears as Theorem 5 in Gerstenhaber and Schack [8]. See Krasilshchik [16, 17] for the foundations of the algebraic theory of the Schouten bracket, and for further developments.

EXAMPLE 1.1. Any finite-dimensional Lie algebra $\mathfrak{g}$ is a Lie algebroid with trivial base manifold. The Gerstenhaber algebra bracket on $\bigwedge \mathfrak{g}$ is the *algebraic Schouten bracket* (see [3, 12, 15, 25]). The corresponding derivation $\mathrm{d} = \mathrm{d}_\mu$ of $\bigwedge(\mathfrak{g}^*)$ is the cohomology operator of the Lie algebra $\mathfrak{g}$ acting on scalar-valued cochains on $\mathfrak{g}$. The restriction to $\mathfrak{g}^*$ of the derivation d_μ of $\bigwedge(\mathfrak{g}^*)$ is the transpose of the Lie algebra bracket, while its restriction to $\bigwedge^0(\mathfrak{g}^*)$ vanishes.

In fact, this Gerstenhaber algebra is exact. It is generated by the operator $\partial = \partial_\mu$ which is the transpose of d_μ, i.e. the Lie algebra homology operator defined by

$$\begin{aligned} &\partial(x_1 \wedge \cdots \wedge x_p) \\ &= \sum_{1 \leqslant i<j \leqslant p} (-1)^{i+j}[x_i, x_j] \wedge x_1 \wedge \cdots \wedge \hat{x}_i \wedge \cdots \wedge \hat{x}_j \wedge \cdots \wedge x_p, \end{aligned}$$

for $x_1, \cdots, x_p \in \mathfrak{g}$. Thus

$$[q, q'] = (-1)^{|q|}(\partial(q \wedge q') - \partial q \wedge q' - (-1)^{|q|} q \wedge \partial q'),$$

for $q, q' \in \bigwedge \mathfrak{g}$. (See Koszul [15, p. 261], and Lian and Zuckermann [18, p. 644].)

EXAMPLE 1.2. The tangent bundle of a smooth manifold M is a Lie algebroid with respect to the Lie bracket of vector fields. The Gerstenhaber algebra bracket on $\bigwedge(M) = \Gamma(\bigwedge(TM))$ is the Schouten bracket of fields of multivectors. In this case, the differential d is the usual de Rham differential of differential forms.

If, in particular, the manifold M is a Lie group G, the left- (resp., right-) invariant fields of multivectors on G constitute a Gerstenhaber subalgebra of $\bigwedge(G)$ which is isomorphic (resp., antiisomorphic) to $\bigwedge \mathfrak{g}$ equipped with the algebraic Schouten bracket, and the Lie algebra cohomology operator d_μ is the restriction to invariant forms of the de Rham differential of differential forms on G.

EXAMPLE 1.3. Any Poisson structure P on a manifold M defines a Lie algebroid structure on the dual T^*M of TM. In fact, the cotangent bundle is a Lie algebroid when equipped with the Lie bracket of differential 1-forms $[\,,\,]_P$,

$$[\alpha, \beta]_P = L_{P\alpha}\beta - L_{P\beta}\alpha - \mathrm{d}(P(\alpha, \beta)), \tag{1}$$

defined independently by Magri and Morosi [21], Gelfand and Dorfman [5] and Karasev [11], and interpreted as a Lie algebroid bracket by Coste, Dazord and Weinstein [2, 28].

Remark. Our convention here is $\langle \beta, P\alpha \rangle = P(\alpha, \beta)$, which is the opposite of the convention adopted in [13], so the bracket defined above, $[\,,\,]_P$, is opposite to the one defined in [13].

Under our present conventions,

$$[\alpha, f\beta]_P = f[\alpha, \beta]_P + (L_{P\alpha}f)\beta, \qquad [\mathrm{d}f, \mathrm{d}g]_P = \mathrm{d}\{f, g\}$$

and

$$P([\alpha, \beta]_P) = [P\alpha, P\beta].$$

The associated G-algebra is the exterior algebra of all differential forms $\bigwedge^*(M) = \Gamma(\bigwedge(T^*M))$, equipped with the Koszul bracket defined in [15]. The fact that the Koszul bracket is a G-algebra bracket was shown by Koszul (formula (1.7)). The fact that it is the generalized Schouten bracket associated with the Lie algebroid structure of the cotangent bundle of a Poisson manifold was shown in Section 6.5 of [13]. Also see Krasil'shchik [16, 17] and Vaisman [26, Section 4.6].

In the case of the Lie algebroid T^*M of a Poisson manifold, the associated differential on $\Gamma(\bigwedge(TM))$ is the Lichnerowicz–Poisson differential on fields of multivectors, i.e.

$$\mathrm{d}_P = [P, .].$$

This property was proved, independently and in various contexts, by Bhaskara and Viswanath [1], Kosmann-Schwarzbach and Magri [13] and Huebschmann [10].

EXAMPLE 1.4. Any *Nijenhuis tensor* N on a manifold M, a field of (1,1)-tensors with vanishing Nijenhuis torsion, defines a deformed Lie algebroid structure on TM and therefore a deformed G-algebra structure on $\bigwedge(M)$. (See [13, 14], and Vaisman [27] for further developments.) This deformed Lie algebroid structure is compatible with the usual one (the sum of the Lie brackets is a Lie bracket) and this implies that the sum of the usual G-bracket and the deformed one on $\bigwedge(M)$ is itself a G-bracket, i.e. the deformed G-bracket gives rise to an infinitesimal deformation of the usual G-bracket.

2. The Dual Differential Exact G-Algebras of a Poisson Manifold

We have just recalled that both the tangent and cotangent bundles of a Poisson manifold are Lie algebroids, and that therefore $\bigwedge(M)$ and $\bigwedge^*(M)$, the spaces of sections of their exterior algebras, are Gerstenhaber algebras. In fact, they are exact G-algebras and the following proposition summarizes the numerous relations that hold in this case.

PROPOSITION 2.1. *Let (M, P) be a Poisson manifold. Then*

(i) $(\bigwedge^*(M), \wedge, \mathrm{d}, [\,,\,]_P, \partial_P)$ *is a differential exact G-algebra, where* d *is the de Rham differential, and* $\partial_P = [i_P, \mathrm{d}]$ *is the Poisson homology operator of Koszul. The associated bracket,*

$$[\alpha, \beta]_P = (-1)^{|\alpha|}(\partial_P(\alpha \wedge \beta) - \partial_P\alpha \wedge \beta - (-1)^{|\alpha|}\alpha \wedge \partial_P\beta),$$

is the Koszul bracket of differential forms on a Poisson manifold.

(ii) *On* $\bigwedge(M)$, *there exists a differential operator,* ∂, *which generates the Schouten bracket in the sense of Koszul, i.e.*

$$[Q, Q'] = (-1)^{|Q|}(\partial(Q \wedge Q') - \partial Q \wedge Q' - (-1)^{|Q|}Q \wedge \partial Q'),$$

and, when ∂ *is chosen to be of square 0, then* $(\bigwedge(M), \wedge, \mathrm{d}_P, [\,,\,], \partial)$ *is a differential exact G-algebra, where* d_P *is the Lichnerowicz–Poisson differential,* $\mathrm{d}_P = [P, .]$. *If, in particular,* M *is symplectic, with* $\Omega = P^{-1}$, *then we may set* $\partial = [i_\Omega, \mathrm{d}_P]$.

All the proofs are in Koszul [15]. See also Kosmann-Schwarzbach and Magri [13], Sections 3.1, 6.3, and 6.5.

3. The Dual Differential G-Algebras of a Lie Bialgebroid

Lie bialgebroids introduced by Mackenzie and Xu [20] are a beautiful generalization of both the Poisson manifolds and the Lie bialgebras. In their definition, a Lie bialgebroid is a pair (A, A^*) of Lie algebroids in duality, where the Lie brackets satisfy a compatibility condition which can be expressed in terms of the differential d_* on $\Gamma(\bigwedge A)$ defined by the Lie algebroid structure of A^* and the G-algebra bracket $[\,,\,]$ on $\Gamma(\bigwedge A)$ defined by the Lie algebroid structure of A,

$$d_*[x,y] = [d_*x, y] + [x, d_*y]\,, \tag{2}$$

for all x and y in $\Gamma(A)$. (This is clearly equivalent to condition (16) in [20].)

PROPOSITION 3.1. *In a Lie bialgebroid* (A, A^*), *the following relations hold for all* $f, g \in \Gamma(\bigwedge^0 A) = C^\infty(M)$, $x, y \in \Gamma(A)$, $\xi, \eta \in \Gamma(A^*)$,

$$d_*[x,f] = [d_*x, f] + [x, d_*f], \tag{3}$$

$$L_{d_*f}x + L^*_{df}x = 0, \tag{3'}$$

$$\langle d_*f, dg\rangle + \langle df, d_*g\rangle = 0, \tag{4}$$

$$L_{d_*f}\xi + L^*_{df}\xi = 0, \tag{3'*}$$

$$d[\xi, f]_* = [d\xi, f]_* + [\xi, df]_*, \tag{3*}$$

$$d[\xi,\eta]_* = [d\xi,\eta]_* + [\xi, d\eta]_*. \tag{2*}$$

Proof. Relation (3) follows from (2), and the identity

$$\begin{aligned} &d_*[x, fy] - [d_*x, fy] - [x, d_*(fy)] - f(d_*[x,y] - [d_*x,y] - [x, d_*y]) \\ &\quad = (d_*[x,f] - [d_*x, f] - [x, d_*f]) \wedge y\,, \end{aligned}$$

which is proved using the fact that d_* is a derivation of degree 1 of $(\Gamma(\bigwedge A), \wedge)$.

Relation (3) can be rewritten as

$$[d_*f, x] + d_*[x,f] - [d_*x, f] = 0$$

or

$$L_{d_*f}x + d_*i_{df}x + i_{df}d_*x = 0\,,$$

which is (3′). (We have used the relations $[a,f] = -i_{df}a$ for $a \in \Gamma(\bigwedge^2 A)$, and $L^*_y = [i_y, d_*]$).

Relation (4) follows from (3′) and the identity

$$L_{d_*f}(gx) + L^*_{df}(gx) - g(L_{d_*f}x + L^*_{df}x) = (L_{d_*f}g + L^*_{df}g)x\,.$$

We then obtain (3′∗) from (4), (3′) and the identity

$$(L_{d_*f} + L^*_{df})\langle \xi, x\rangle = \langle L_{d_*f}\xi + L^*_{df}\xi, x\rangle + \langle \xi, L_{d_*f}x + L^*_{df}x\rangle\,.$$

Now (3∗) is equivalent to (3′∗) just as (3) is equivalent to (3′).

Let us now prove (2∗). We first remark that, since

$$(L_x L^*_\xi - L^*_{L_x\xi})f = \langle \xi, [x, \mathrm{d}_* f]\rangle$$

and

$$(L^*_\xi L_x - L_{L^*_\xi x})f = \langle [\xi, \mathrm{d}f]_*, x\rangle\,,$$

we obtain, using (4),

$$([L_x, L^*_\xi] - L^*_{L_x\xi} + L_{L^*_\xi x})f = L_{\mathrm{d}_*\langle\xi,x\rangle} f\,,$$

and therefore,

$$([L_x, L^*_\xi] - L^*_{L_x\xi} + L_{L^*_\xi x})\langle\eta, y\rangle + ([L_y, L^*_\eta] - L^*_{L_y\eta} + L_{L^*_\eta y})\langle\xi, x\rangle = 0\,.$$

Now a direct computation, similar to that in [20], shows that

$$\begin{aligned}
&(L_x \mathrm{d}_* y - L_y \mathrm{d}_* x - \mathrm{d}_*[x,y])(\xi,\eta) - (L^*_\xi \mathrm{d}\eta - L^*_\eta \mathrm{d}\xi - \mathrm{d}[\xi,\eta]_*)(x,y)\\
&\quad = ([L_x, L^*_\xi] - L^*_{L_x\xi} + L_{L^*_\xi x})\langle\eta, y\rangle + ([L_y, L^*_\eta] - L^*_{L_y\eta} + L_{L^*_\eta y})\langle\xi, x\rangle -\\
&\qquad -([L_x, L^*_\eta] - L^*_{L_x\eta} + L_{L^*_\eta x})\langle\xi, y\rangle - ([L_y, L^*_\xi] - L^*_{L_y\xi} + L_{L^*_\xi y})\langle\eta, x\rangle\,,
\end{aligned}$$

and, therefore, this quantity vanishes identically. Thus, (2) implies (2∗).

Let us remark that the proof of (2∗) for exact forms is immediate,

$$\mathrm{d}[\mathrm{d}f, \mathrm{d}g]_* = \mathrm{d}(L^*_{\mathrm{d}f}\mathrm{d}g) = -\mathrm{d}(L_{\mathrm{d}_* f}\mathrm{d}g) = 0\,,$$

since d commutes with L_x and d is of square 0. Combining this fact with (3∗), we obtain a one-line proof of (2∗) when $\Gamma(A^*)$ is generated by the image of d.

COROLLARY 3.2. *In a Lie bialgebroid* (A, A^*), d_* *is a derivation of the graded Lie algebra* $(\Gamma(\bigwedge A), [\,,\,])$, *and* d *is a derivation of* $(\Gamma(\bigwedge A^*), [\,,\,]_*)$.

Proof. The statement for d_* follows from (2) and (3) and the Leibniz rule for $[\,,\,]$, and similarly for d.

We can now state a proposition which yields alternate definitions of Lie bialgebroids and proves the selfduality of this notion.

PROPOSITION 3.3. *Let* (A, A^*) *be a pair of Lie algebroids in duality. The following properties are equivalent*:

(i) (A, A^*) *is a Lie bialgebroid,*
(ii) d_* *is a derivation of* $(\Gamma(\bigwedge A), [\,,\,])$,
(iii) d *is a derivation of* $(\Gamma(\bigwedge A^*), [\,,\,]_*)$,
(iv) (A^*, A) *is a Lie bialgebroid.*

Proof. In view of Corollary 3.2, we know that in a Lie bialgebroid, d_* is a derivation of $[\,,\,]$ and d is a derivation of $[\,,\,]_*$. Since these properties clearly imply (2) and (2∗), the proposition follows.

Thus we have given an alternate proof of Mackenzie and Xu's Theorem 3.10 [20] asserting that the notion of a Lie bialgebroid is *selfdual.*

We can now show that a Poisson structure can be defined canonically on the base manifold M of any Lie bialgebroid. For f and $g \in \Gamma(\bigwedge^0 A) = C^\infty(M)$, let us set

$$\{f, g\}_{(A,A^*)} = \langle \mathrm{d}f, \mathrm{d}_* g\rangle. \tag{5}$$

By (4), this bracket is skew-symmetric.

PROPOSITION 3.4. *The bracket* $\{\,,\,\}_{(A,A^*)}$ *satisfies*

$$\mathrm{d}\{f, g\}_{(A,A^*)} = [\mathrm{d}f, \mathrm{d}g]_* \tag{6}$$

and

$$\mathrm{d}_*\{f, g\}_{(A,A^*)} = -[\mathrm{d}_* f, \mathrm{d}_* g], \tag{6*}$$

and is a Poisson bracket on the base manifold M.

Proof. By definition, $\langle \mathrm{d}f, \mathrm{d}_* g\rangle = [\mathrm{d}f, g]_* = [\mathrm{d}_* g, f]$, thus (6) (resp., (6∗)) follows from (3∗) (resp., (3)) together with $\mathrm{d}^2 = 0$ (resp., $(\mathrm{d}_*)^2 = 0$):

$$\begin{aligned} \mathrm{d}\{f, g\}_{(A,A^*)} &= \mathrm{d}[\mathrm{d}f, g]_* = [\mathrm{d}f, \mathrm{d}g]_*\,, \\ \mathrm{d}\{f, g\}_{(A,A^*)} &= \mathrm{d}_*[\mathrm{d}_* g, f] = [\mathrm{d}_* g, \mathrm{d}_* f] = -[\mathrm{d}_* f, \mathrm{d}_* g]\,. \end{aligned}$$

As is usual in the Hamiltonian formalism, let us set $X_f = -\mathrm{d}_* f$, and let us write $\{\,,\,\}$ for $\{\,,\,\}_{(A,A^*)}$. Then $\{f, g\} = X_f \cdot g$, and relation (6∗) becomes $[X_f, X_g] = X_{\{f,g\}}$. Thus, the following relations on sums over cyclic permutations hold,

$$\oint \{\{f_1, f_2\}, f_3\} = \oint (X_{\{f_1,f_2\}} \cdot f_3) = \oint [X_{f_1}, X_{f_2}] \cdot f_3 = 0\,,$$

which proves the last assertion.

In addition, let d_M be the de Rham differential of functions on M, and let $[\,,\,]_{(A,A^*)}$ be the Lie algebroid bracket on T^*M defined by the above Poisson structure on M. Then

$$\mathrm{d}_M\{f, g\}_{(A,A^*)} = [\mathrm{d}_M f, \mathrm{d}_M g]_{(A,A^*)}\,. \tag{7}$$

The results of this section can now be expressed in terms of G-algebras.

PROPOSITION 3.5. *If* (A, A^*) *is a Lie bialgebroid in the sense of Mackenzie and Xu, then* $(\Gamma(\bigwedge A), \wedge, \mathrm{d}_*, [\,,\,])$ *and* $(\Gamma(\bigwedge A^*), \wedge, \mathrm{d}, [\,,\,]_*)$ *are differential G-algebras. Moreover,* d_* *is a derivation of* $[\,,\,]$ *and* d *is a derivation of* $[\,,\,]_*$.

EXAMPLE 3.1. If $\mathfrak{g}$ is a finite-dimensional Lie bialgebra [3, 4], then the pair $(\mathfrak{g}, \mathfrak{g}^*)$ is a bialgebroid [20], and $(\bigwedge \mathfrak{g}, \wedge, \mathrm{d}_*, [\,,\,])$ and $(\bigwedge \mathfrak{g}^*, \wedge, \mathrm{d}, [\,,\,]_*)$ are differential G-algebras, where $\mathrm{d}\colon \bigwedge \mathfrak{g}^* \to \bigwedge \mathfrak{g}^*$ (resp., $\mathrm{d}_*\colon \bigwedge \mathfrak{g} \to \bigwedge \mathfrak{g}$) is the cohomology operator of $\mathfrak{g}$ (resp., $\mathfrak{g}^*$) acting on scalar-valued cochains on $\mathfrak{g}$ (resp., $\mathfrak{g}^*$), and $[\,,\,]$ (resp.,$[\,,\,]_*$) is the algebraic Schouten bracket defined by the Lie algebra structure of $\mathfrak{g}$ (resp., $\mathfrak{g}^*$). Here $\mathrm{d}f = \mathrm{d}_*f = 0$, for $f \in \bigwedge^0 \mathfrak{g}$. Relation (2) is the cocycle condition, which states that the cobracket (the restriction of d_* to $\mathfrak{g}$), $\gamma\colon \mathfrak{g} \to \bigwedge^2 \mathfrak{g}$, is a 1-cocycle of $\mathfrak{g}$ with values in $\bigwedge^2 \mathfrak{g}$ with respect to the adjoint action, while (2∗) is the dual condition, which is known to be equivalent to the former [3]. For a direct proof of the fact that d (resp., d_*) is a derivation of $[\,,\,]_*$ (resp., $[\,,\,]$), see [12].

EXAMPLE 3.2. If (M, P) is a Poisson manifold, we know that both TM and T^*M are Lie algebroids. Mackenzie and Xu showed that together they constitute a Lie bialgebroid. In fact, we know that in this case $\mathrm{d}_* = \mathrm{d}_P = [P, .]$ is the Lichnerowicz–Poisson differential on fields of multivectors. Together with the graded Jacobi identity for the Schouten bracket, this yields an immediate proof of the fact that d_* is a derivation of $[\,,\,]$ and, therefore, that (TM, T^*M) is a Lie bialgebroid. Thus Poisson manifolds give rise to pairs of differential G-algebras in duality.

All the relations in Proposition 3.1 are well-known properties of Poisson manifolds. For instance, relation (3′) can be written

$$L_{\mathrm{d}_P f} x + L_{P(\mathrm{d}f)} x = 0\,,$$

which is true because $\mathrm{d}_P f = [P, f] = -P(\mathrm{d}f)$, while relation (4) means that

$$\langle P\mathrm{d}f, \mathrm{d}g \rangle + \langle \mathrm{d}f, P\mathrm{d}g \rangle = 0\,.$$

In this case, the Poisson bracket $\{\,,\,\}_{(TM,T^*M)}$ coincides with the original Poisson bracket, since

$$\{f, g\}_{(TM,T^*M)} = \langle \mathrm{d}f, P\mathrm{d}g \rangle = P(\mathrm{d}f, \mathrm{d}g)\,.$$

Formula (6) reduces to a well-known property of the Koszul bracket, which is the Lie algebroid bracket on T^*M, while formula (6∗) states that the mapping $\mathrm{d}_P\colon f \in C^\infty(M) \to X_f^P \in \Gamma(TM)$, which carries a function f into the Hamiltonian vector field $X_f^P = -\mathrm{d}_P f = P(\mathrm{d}f)$, is a Lie algebra homomorphism, and clearly, (7) reduces to (6).

Because there exists a section P of $\bigwedge^2(TM)$ such that $[P, P] = 0$, and $\mathrm{d}_* = [P, .]$, the Lie algebroid (TM, T^*M) is an example of what Mackenzie and Xu [20] called a *triangular Lie bialgebroid,* a situation which we shall now study.

4. The Batalin–Vilkovisky Algebra of a Triangular Lie Bialgebroid

The triangular Lie bialgebroids generalize both the Lie bialgebroids of Poisson manifolds and the triangular Lie bialgebras (see Drinfeld [4]), whence their name.

Let A be a Lie algebroid, with anchor a. As above, we denote the Schouten bracket on $\Gamma(\bigwedge A)$ by $[\,,\,]$, and the differential on $\Gamma(\bigwedge A^*)$ by d.

Let P be a section of $\bigwedge^2 A$ which we identify with a map $P\colon A^* \to A$, and let us assume that $[P, P] = 0$. We call P a 'Poisson bivector'. Then A^* is a Lie algebroid when it is equipped with the anchor $a \circ P$, and the Lie bracket is defined by a formula identical to formula (1) for the Lie bracket of differential 1-forms on a Poisson manifold. Mackenzie and Xu [20] showed that, in fact, (A, A^*) is a Lie bialgebroid which they called *triangular.*

We denote the Gerstenhaber algebra bracket on $\Gamma(\bigwedge A^*)$ by $[\,,\,]_P = [\,,\,]_*$, and the differential on $\Gamma(\bigwedge A)$ by $\mathrm{d}_P = \mathrm{d}_*$.

PROPOSITION 4.1. *The differential* d_P *satisfies* $\mathrm{d}_P = [P, .\,]$. *The operator* $\partial_P = [i_P, \mathrm{d}]$ *generates* $[\,,\,]_P$,

$$[\alpha, \beta]_P = (-1)^{|\alpha|}(\partial_P(\alpha \wedge \beta) - \partial_P\alpha \wedge \beta - (-1)^{|\alpha|}\alpha \wedge \partial_P\beta) \tag{8}$$

and ∂_P *is a derivation of* $[\,,\,]_P$.

Proof. In fact, (8) holds for α and β of degree 0 or 1, and therefore for all forms.

Moreover, it follows from the fact that P is a morphism of Lie algebroids from A^* to A (see [20]) that P extends to a morphism of graded Lie algebras from $(\Gamma(\bigwedge A^*), [\,,\,]_P)$ to $(\Gamma(\bigwedge A), [\,,\,])$.

Thus if A is a Lie algebroid with a 'Poisson bivector', then (A, A^*) is a Lie bialgebroid and $\Gamma(\bigwedge A)$ and $\Gamma(\bigwedge A^*)$ are both differential Gerstenhaber algebras, $\Gamma(\bigwedge A^*)$ being exact.

We now describe the generalization of the case of symplectic manifolds to the Lie algebroid setting.

PROPOSITION 4.2. *If* P *is invertible, with inverse* $P^{-1} = \Omega\colon A \to A^*$, *then* $\mathrm{d}\Omega = 0$ *and the differential* d *satisfies* $\mathrm{d} = [\Omega, .\,]_P$. *We set* $\partial = [i_\Omega, \mathrm{d}_P]$. *Then* ∂ *generates* $[\,,\,]$,

$$[a, b] = (-1)^{|a|}(\partial(a \wedge b) - \partial a \wedge b - (-1)^{|a|}a \wedge \partial b),$$

and ∂ *is a derivation of* $[\,,\,]$.

Proof. If P is Poisson and invertible, then $d\Omega = 0$ since $d\Omega(P\alpha, P\beta, P\gamma)$ is proportional to $[P, P](\alpha, \beta, \gamma)$. The other statements follow from the fact that they are true for elements of degree 0 or 1 and from the derivation properties.

Thus if A is a Lie algebroid with a 'symplectic structure', i.e. if there exists an invertible map $\Omega\colon A \to A^*$ of Lie algebroids such that $d\Omega = 0$, then (A, A^*) is a Lie bialgebroid and $\Gamma(\bigwedge A)$ and $\Gamma(\bigwedge A^*)$ are isomorphic differential exact Gerstenhaber algebras.

EXAMPLE 4.1. Let $\mathfrak{g}$ be a triangular Lie bialgebra, defined by $r \in \bigwedge^2 \mathfrak{g}$, a skew-symmetric solution of the classical Yang–Baxter equation, $[r, r] = 0$. Such an element of $\bigwedge^2 \mathfrak{g}$ is called a *classical triangular r-matrix*. Then $d_* = d_\mu r$ is the transpose of a Lie bracket on $\mathfrak{g}^*$, and $(\mathfrak{g}, \mathfrak{g}^*)$ is a triangular Lie bialgebroid.

EXAMPLE 4.2. For any Poisson manifold (M, P), the pair (TM, T^*M) is a triangular Lie bialgebroid. We remark that in this case both G-algebras $\bigwedge(M)$ and $\bigwedge^*(M)$ are exact. However, on $\bigwedge^*(M)$ the generating operator ∂_P is canonically defined, while that on $\bigwedge(M)$ depends on the choice of a torsionless connection or a volume element on M. (See Koszul [15].)

In conclusion, we see that a Lie algebroid gives rise to a G-algebra, a Lie bialgebroid gives rise to a pair of differential G-algebras in duality, and a triangular Lie bialgebroid gives rise to a pair of differential G-algebras in duality, one of which is exact, i.e. is a Batalin–Vilkovisky algebra. In the case of the Lie bialgebroid (TM, T^*M) of a Poisson manifold M, both G-algebras are exact, and the differential structure of the manifold on the one hand, and the Poisson structure on the other hand play dual roles. However, by the preceding remark, there remains an asymmetry between the role of the differential structure of the manifold and that of the Poisson structure: the G-algebra $\bigwedge^*(M)$ is canonically equipped with the structure of a Batalin–Vilkovisky algebra, while $\bigwedge(M)$ is not.

5. The Linear Case

The linear case was treated by Koszul [15]. Let $\mathfrak{g}$ be a Lie algebra. Then $\mathfrak{g}^*$ is a Poisson manifold with the linear Poisson structure of Lie, Berezin, Kirillov, Kostant and Souriau. Thus both $\Gamma(\bigwedge(T\mathfrak{g}^*))$ and $\Gamma(\bigwedge(T^*\mathfrak{g}^*))$ are Gerstenhaber algebras. We remark that in restriction to the vector space $\mathfrak{g} \otimes \bigwedge \mathfrak{g}^*$ of linear fields of multivectors on $\mathfrak{g}^*$, the Schouten bracket reduces to the Nijenhuis–Richardson bracket, a fact which is true on any vector space (see [12, Section 2.6]).

Because $\mathfrak{g}^*$ is a Poisson manifold, $\Gamma(\bigwedge(T^*\mathfrak{g}^*)) = C^\infty(\mathfrak{g}^*) \otimes \bigwedge \mathfrak{g}$ is a differential exact Gerstenhaber algebra, where the differential, d, is the de Rham differential of differential forms, and the graded Lie algebra structure is the Koszul bracket associated with the linear Poisson structure P of $\mathfrak{g}^*$. The Koszul bracket is generated by the operator, $\partial_P = [i_P, \mathrm{d}]$. In restriction to the constant differential forms on $\mathfrak{g}^*$, we recover the algebraic Schouten bracket of $\bigwedge \mathfrak{g}$ and the restriction of the operator ∂_P is the Lie algebra homology operator ∂_μ described in Example 1.1. Thus, $(\bigwedge \mathfrak{g}, \wedge, 0, [\, ,\,], \partial_\mu)$ is a Batalin–Vilkovisky subalgebra of $\Gamma(\bigwedge(T^*\mathfrak{g}^*))$.

On $\Gamma(\bigwedge(T\mathfrak{g}^*)) = C^\infty(\mathfrak{g}^*) \otimes \bigwedge \mathfrak{g}^*$, the differential d_P is the Lie algebra cohomology operator acting on cochains on $\mathfrak{g}$ with values in $C^\infty(\mathfrak{g}^*)$, with respect to the coadjoint action (see Koszul [15]). In fact, this space is a Batalin–Vilkovisky algebra $(\Gamma(\bigwedge(T\mathfrak{g}^*)), \wedge, \mathrm{d}_P, [\, ,\,], D)$, where an operator generating the Schouten bracket of fields of multivectors on $\mathfrak{g}^*$ is

$$D = -\sum_i \frac{\partial}{\partial \xi_i} \frac{\partial}{\partial x_i},$$

the operator which was introduced by Koszul (who cites Elie Cartan!) and also appears in the approach of Batalin and Vilkovisky to the quantization of string field theory. (See, e.g., [18, 23, 29].) Thus, the algebraic structures arising in topological field theories were in fact developped earlier in differential geometry.

Acknowledgement

I am very grateful for the stimulating atmosphere of the Centre Emile Borel where this article was written.

References

1. Bhaskara, K. H. and Viswanath, K.: Calculus on Poisson manifolds, *Bull. London Math. Soc.* **20** (1988), 68–72.
2. Coste, A., Dazord, P., and Weinstein, A.: Groupoïdes symplectiques, *Publ. Dép. Math. Univ. Lyon* I, 2A (1987).
3. Drinfeld, V. G.: Hamiltonian structures on Lie groups, Lie bialgebras and the geometric meaning of the classical Yang–Baxter equation, *Soviet. Math. Dokl.* **27**(1) (1983), 68–71. (In Russian, *Dokl. Akad. Nauk SSSR* **268**(2) (1983).)
4. Drinfeld, V. G.: Quantum groups, *Proc. Internat. Congr. Math.* (*Berkeley, 1986*), Vol. 1, Amer. Math. Soc., Providence, 1987, pp. 798–820.
5. Gelfand, I. M. and Dorfman, I. Ya.: Hamiltonian operators and the classical Yang–Baxter equation, *Funct. Anal. Appl.* **16**(4) (1982), 241–248.
6. Gerstenhaber, M.: The cohomology structure of an associative ring, *Ann. Math.* **78** (1963), 267–288.
7. Gerstenhaber, M. and Schack, S. D.: Algebraic cohomology and deformation theory, in M. Hazewinkel and M. Gerstenhaber (eds), *Deformation Theory of Algebras and Structures and Applications*, Kluwer, Dordrecht, 1988, pp. 11–264.
8. Gerstenhaber, M. and Schack, S. D.: Algebras, bialgebras, quantum groups and algebraic deformations, in M. Gerstenhaber and J. Stasheff (eds), *Deformation Theory and Quantum Groups with Applications to Mathematical Physics*, Contemporary Mathematics 134, Amer. Math. Soc., Providence, 1992, pp. 51–92.
9. Getzler, E.: Batalin–Vilkovisky algebras and two-dimensional topological field theories, *Comm. Math. Phys.* **159** (1994), 265–285.

10. Huebschmann, J.: Poisson cohomology and quantization, *J. Reine Angew. Math.* **408** (1990), 57–113.
11. Karasev, M. V.: Analogues of the objects of Lie group theory for nonlinear Poisson brackets, *Math. USSR Izv.* **28**(3) (1987), 497–527. (In Russian, *Izvestyia* **50** (1986).)
12. Kosmann-Schwarzbach, Y.: Jacobian quasi-bialgebras and quasi-Poisson Lie groups, in M. Gotay, J. E. Marsden, and V. Moncrief (eds), *Mathematical Aspects of Classical Field Theory*, Contemporary Mathematics 132, Amer. Math. Soc., Providence, 1992, pp. 459–489.
13. Kosmann-Schwarzbach, Y. and Magri, F.: Poisson–Nijenhuis structures, *Ann. Inst. Henri Poincaré A* **53**(1) (1990), 35–81.
14. Kosmann-Schwarzbach, Y., and Magri, F.: Dualization and deformation of Lie brackets on Poisson manifolds, in J. Janyška and D. Krupka (eds), *Differential Geometry and its Applications* (Brno, 1989), World Scientific, Singapore, 1990, pp. 79–84.
15. Koszul, J.-L.: Crochet de Schouten–Nijenhuis et cohomologie, in *'Elie Cartan et les mathématiques d'aujourd'hui', Astérisque, n° hors série*, Soc. Math. Fr., 1985, pp. 257–271.
16. Krasilshchik, I.: Schouten brackets and canonical algebras, in *Global Analysis III*, Lecture Notes Math. 1334, Springer-Verlag, Berlin, 1988, pp. 79–110.
17. Krasilshchik, I.: Supercanonical algebras and Schouten brackets, *Mat. Zametki* **49** (1991), 70–76.
18. Lian, B. H. and Zuckerman, G. J.: New perspectives on the BRST-algebraic structure of string theory, *Comm. Math. Phys.* **154** (1993), 613–646.
19. Mackenzie, K.: *Lie Groupoids and Lie Algebroids in Differential Geometry*, London Math. Soc. Lect. Notes Series 124, Cambridge University Press, Cambridge, 1987.
20. Mackenzie, K. C. H. and Ping Xu: Lie bialgebroids and Poisson groupoids, *Duke Math. J.* **73** (1994), 415–452.
21. Magri, F. and Morosi, C.: A geometrical characterization of integrable Hamiltonian systems through the theory of Poisson–Nijenhuis manifolds, Quaderno S 19 (1984), University of Milan.
22. Palais, R. S.: *The Cohomology of Lie Rings*, Proc. Symp. Pure Math. 3, Amer. Math. Soc., Providence, 1961, pp. 130–137.
23. Penkava, M. and Schwarz, A.: On some algebraic structures arising in string theory, Preprint hep-th/912071.
24. Pradines, J.: Théorie de Lie pour les groupoïdes différentiables. Calcul différentiel dans la catégorie des groupoïdes infinitésimaux, *C.R. Acad. Sci. Paris, Série A* **264** (1967), 245–248.
25. Roger, C.: Algèbres de Lie graduées et quantification, in P. Donato *et al.* (eds), *Symplectic Geometry and Mathematical Physics*, Progress in Mathematics 99, Birkhäuser, Boston, 1991, pp. 374–421.
26. Vaisman, I.: *Lectures on the Geometry of Poisson Manifolds*, Progress in Mathematics 118, Birkhäuser, Boston, 1994.
27. Vaisman, I.: Poisson–Nijenhuis structures revisited, *Rendiconti Sem. Mat. Torino* **52** (1994).
28. Weinstein, A.: Some remarks on dressing transformations, *J. Fac. Sci. Univ. Tokyo, IA, Math.* **35** (1988), 163–167.
29. Zwiebach, B.: Closed string theory: an introduction, Preprint hep-th/9305026.

Acta Applicandae Mathematicae **41:** 167–191, 1995.

Graded Differential Equations and Their Deformations: A Computational Theory for Recursion Operators

I. S. KRASIL'SHCHIK
*Moscow Institute for Municipal Economy and Civil Engineering, Moscow, Russia**

and

P. H. M. KERSTEN
University of Twente, Department of Applied Mathematics, PO Box 217, 7500 AE Enschede, The Netherlands

(Received: 28 February 1994)

Abstract. An algebraic model for nonlinear partial differential equations (PDE) in the category of n-graded modules is constructed. Based on the notion of the graded Frölicher–Nijenhuis bracket, cohomological invariants $H^*_\nabla(\mathcal{A})$ are related to each object $(\mathcal{A}, \nabla)$ of the theory. Within this framework, $H^0_\nabla(\mathcal{A})$ generalizes the Lie algebra of symmetries for PDE's, while $H^1_\nabla(\mathcal{A})$ are identified with equivalence classes of infinitesimal deformations. It is shown that elements of a certain part of $H^1_\nabla(\mathcal{A})$ can be interpreted as recursion operators for the object $(\mathcal{A}, \nabla)$, i.e. operators giving rise to infinite series of symmetries. Explicit formulas for computing recursion operators are deduced. The general theory is illustrated by a particular example of a graded differential equation, i.e. the Super KdV equation.

Mathematics Subject Classifications (1991): 58F07, 58G07, 58H10, 58H15, 58G37, 58A50, 35Q53, 35Q55, 35Q58, 58G35, 16W55.

Key words: Graded differential equations, Frölicher–Nijenhuis bracket, cohomologies, deformations, symmetries, recursion operators, coverings, integrable systems, super KdV equation.

Introduction

In his book, P. Olver wrote: "The deduction of the form of the recursion operator (if it exists) requires a certain amount of inspired guesswork ..." (see [12, p. 315]). What we present here is a theory making computations of recursion operators a straightforward algorithmic procedure.

Our approach is based on three components of modern invariant theory of partial differential equations: the geometry of PDE [6], essential algebraic structures resulting from (or parallel to) this geometry [4, 16], and nonlocal theory of PDE [8]. For its implementation, no external structures (e.g., Hamiltonian ones [10]) are needed. To be more particular, it means the following.

* 1 Tverskoy-Yamskoy per. 14, Apt. 45, 125047 Moscow, Russia.

The main point of our interests is the infinite prolongation $\mathcal{E}^{(\infty)}$ of a differential equation $\mathcal{E}$ considered as a fibre bundle $\pi_\infty\colon \mathcal{E}^{(\infty)} \to M$ and endowed with a natural flat connection, the Cartan connection. We present an algebraic model of this construction and generalize it to the case of $\mathbb{Z}$ (or $\mathbb{Z}_2$) polygraded algebras. The objects obtained, together with corresponding morphisms, form a category which we call a category of graded differential equations, or GDE category (cf. A. M. Vinogradov's [15]).

For the objects of GDE category a kind of differential geometrical calculus has evolved which is based on the notions of the (graded) de Rham differential and of a contraction (interior product) and included graded counterparts of Lie derivatives, the Frölicher–Nijenhuis, and the Richardson–Nijenhuis brackets. In particular, the Frölicher–Nijenhuis bracket is used for constructing a cohomology theory $H^* = \sum_{i\geqslant 0} H^i$ in GDE which seems to be a natural language in dealing with symmetries and recursion operators.

Remaining in the algebraic setting we construct a functor $\mathcal{T}$: GDE $\Rightarrow$ GDE, such that for any object $\mathcal{F} \in$ Ob (GDE) there exists an embedding

$$H^*(\mathcal{F}) \subset H^1(\mathcal{T}(\mathcal{F})). \qquad (0.1)$$

This construction reduces computations of $H^*(\mathcal{F})$ to those of a new object $\mathcal{T}(\mathcal{F})$ and is used in particular applications.

We return to the geometrical framework and define graded extensions of differential equations as the objects of GDE of a special type. For these objects H^0 happens to coincide with the Lie algebra of higher symmetries, while contraction of H^0 into H^1 gives H^0 again, thus identifying elements of H^1 with recursion operators. On the other hand, elements of H^1 are classes of infinitesimal deformations of the equation structure (see [4, 7]; also [1] for the general theory), which provides another point of view at recursion operators. We also show that the Lie algebra structure of symmetries is strongly related to the triviality of H^2.

Using embedding (0.1) we treat recursion operators of an equation $\mathcal{E}$ as symmetries of special type of a new object, $\mathcal{T}(\mathcal{E})$, and derive corresponding computational relations. *This is the crucial step to making a search for recursion operators strictly algorithmical* (see above). But this step is not the last one.

Usually (except for some special cases), when solving the equations for recursion operators, one finds trivial solutions only. To get nontrivial ones, it needs to extend the set of variables with new, nonlocal variables, i.e. to consider a covering over initial differential equation [8]. In this new setting one can find nontrivial solutions, but in terms of so-called *shadows*. We offer two solutions for the problem of recovering recursion operators from there shadows. Namely, following [3] we consider a functor $\mathcal{K}$: GDE $\Rightarrow \widetilde{\text{GDE}}$ of killing horizontal cohomologies (here $\widetilde{\text{GDE}}$ denotes the category derived from GDE by considering classes of isomorphic objects). We show that of $\mathcal{F} \in$ Ob (GDE) and $\mathcal{R}_s$ is a shadow of a recursion operator for $\mathcal{F}$ then:

(a) for any summetry $\varphi \in H^0(\mathcal{F})$ there exists a symmetry $\mathcal{R}\varphi \in H^0(\mathcal{K}^\infty\mathcal{F})$, such that $\mathcal{R}_s\varphi$ is its shadow,
(b) there exists a recursion operator $\mathcal{R} \in H^1((\mathcal{T} \circ \mathcal{K})^\infty\mathcal{F})$, such that $\mathcal{R}_s$ is its shadow.

The notation $(\cdot)^\infty$ above means the infinite power of corresponding functors.

The efficiency of our approach was once demonstrated in [7], where recursion operators for several classical integrable systems wer rediscovered. Here we demonstrate it by computing a recursion operator for the super KdV equation (see [2, 11]).

1. Preliminaries

Here we briefly describe the initial algebraic setting needed below. See [5, 18] for the details.

Let $\mathbb{F}$ be a field of the characteristic 0 and A be an associative commutative $\mathbb{F}$-algebra with a unit $1 \in A$ (in geometrical applications $\mathbb{F} = \mathbb{R}$ and $A = C^\infty(M)$ for some smooth manifold M). We fix a $\mathbb{Z}^n$-graded associative commutative unitary A-algebra $\mathcal{A} = \sum_{\alpha\in\mathbb{Z}^n} \mathcal{A}_\alpha$, such that $\mathcal{A}_\mathbf{0} = A, \mathbf{0} = (0, \dots, 0)$. It means, in particular, that

$$\mathcal{A}_\alpha \cdot \mathcal{A}_\beta \subset \mathcal{A}_{\alpha+\beta}, \quad \alpha, \beta \in \mathbb{Z}^n \tag{1.1}$$

and

$$ab = (-1)^{a\cdot b}ba, \quad a, b \in \mathcal{A},$$

where a, b are homogeneous elements and $a \cdot b$ denotes the scalar product of their gradings $\mathrm{gr}(a)$ and $\mathrm{gr}(b)$.

In what follows we consider the category $GM_n(\mathcal{A})$ of $\mathbb{Z}^n$-graded left $\mathcal{A}$-modules $\mathcal{P} = \sum_{\alpha\in\mathbb{Z}^n} \mathcal{P}_\alpha$ such that all $\mathcal{P}_\alpha$ (including $\mathcal{A}_\alpha$ as well) are filterd by projective A-modules of finite type. The morphisms in question are graded morphisms respecting filtrations. Each $\mathcal{P}$ is also considered as a right $\mathcal{A}$-module by setting $pa = (-1)^{a\cdot p}ap$ for homogeneous $a \in \mathcal{A},\ p \in \mathcal{P}$.

We consider the functors D_s: $GM_n(\mathcal{A}) \Rightarrow GM_n(\mathcal{A}), s = 0, 1, \dots,$ which put into correspondence to a module $\mathcal{P}$ the module of graded $\mathcal{P}$-valued s-polyderivations of $\mathcal{A}$. Then the module

$$D_*(\mathcal{A}) = \sum_{s=0} D_s(\mathcal{A}) \tag{1.2}$$

carries the structure of $\mathbb{Z}^{n+1}$-graded commutative algebra with respect to the (graded) wedge product

$$X \wedge Y = (-1)^{X\cdot Y + X_1\cdot Y_1} Y \wedge X, \quad X, Y \in D_*(\mathcal{A}),$$

where $X \cdot Y$ is the scalar product of the n-gradings inherited from $\mathcal{A}$, while X_1, Y_1 are $(n+1)$-st gradings arising from (1.2).

The functors D_s are representable in $GM_n(\mathcal{A})$, i.e. there exist objects Λ^s, such that

$$D_s(\mathcal{P}) = \hom_{\mathcal{A}}(\Lambda^s, \mathcal{P}) \tag{1.3}$$

for any $\mathcal{P}$. Elements of Λ^s are called (graded) forms over $\mathcal{A}$. it can be shown that:

(i) there exists a uniquely defined derivation

$$\mathrm{d}\colon\ \mathcal{A} \longrightarrow \Lambda^1 \tag{1.4}$$

of the grading **0**, such that any $X \in D_1(\mathcal{P})$ is represented as $X = f_X \circ \mathrm{d}$ for some $f_X \in \hom_{\mathcal{A}}(\Lambda^1, \mathcal{P})$;

(ii) Λ^1 as an $\mathcal{A}$-module is generated by the image of d;

(iii) for any $s \geqslant 0$ one has

$$\Lambda^s = \Lambda^1 \wedge \cdots \wedge \Lambda^1, \quad s \text{ times}, \tag{1.5}$$

where the wedge is understood in the graded sense.

Due to (1.5),

$$\Lambda^* = \sum_{s \geqslant 0} \Lambda^s \tag{1.6}$$

forms a commutative $(n+1)$-graded algebra:

$$\omega \wedge \theta = (-1)^{\omega \cdot \theta + \omega_1 \theta_1} \theta \wedge \omega,$$

where, as before, $\omega \cdot \theta$ comes from the n-grading inherited from $\mathcal{A}$, while ω_1, θ_1 are defined by (1.6).

Derivation (1.4) is extended up to a complex

$$0 \longrightarrow \mathcal{A} \longrightarrow \Lambda^1 \longrightarrow \cdots \longrightarrow \Lambda^s \xrightarrow{\ \mathrm{d}\ } \Lambda^{s+1} \longrightarrow \cdots, \tag{1.7}$$

where the differential d is defined by

$$\mathrm{d}(\mathrm{d}a) = 0, \quad a \in \mathcal{A},$$

and

$$\mathrm{d}(\omega \wedge \theta) = \mathrm{d}\omega \wedge \theta + (-1)^{\omega_1} \omega \wedge \mathrm{d}\theta.$$

Complex (1.7) is called the de Rham complex for the algebra $\mathcal{A}$.

In what follows, we deal with the algebras $\mathcal{A}$ satisfying all previous conditions and such that Λ^1 is an object of $GM_n(\mathcal{A})$. We call such algebras *geometrical* ones (cf. [6]). For geometrical algebras one has

$$D_s(\mathcal{P}) = \mathcal{P} \otimes D_1(\mathcal{A}) \wedge \cdots \wedge D_1(\mathcal{A}). \tag{1.8}$$

We also use, when it is needed, a common reduction from $\mathbb{Z}^n$- to $\mathbb{Z}_2^n$-graded objects.

Remark 1.1. A polyderivation $X \in D_s(\mathcal{P})$, $s > 1$, can be considered as a derivation of $\mathcal{A}$ taking its values in $D_{s-1}(\mathcal{P})$ and satisfying natural skew-symmetric conditions (see [5]). For a decomposable $X = Y \wedge Z$ one has

$$Xa = Y \wedge Za + (-1)^{Za+Z_1} Ya \wedge Z, \quad a \in \mathcal{A}. \tag{1.9}$$

Using (1.8) and (1.9) one can derive an explicit expression for any $X = p \otimes X_1 \wedge \cdots \wedge X_s$, $p \in \mathcal{P}$, $X_i \in D_1(\mathcal{P})$.

2. Calculus of Brackets

Let $\mathcal{A}$ be an algebra satisfying the conditions of the previous section. In the module $\Lambda^* \otimes_{\mathcal{A}} D_1(\mathcal{A})$, we construct two graded brackets, those of Richardson–Nijenhuis (see also [9]) and Frölicher–Nijenhuis, and derive their main properties. The construction is based on the notions of an inner product (contraction) and of a Lie derivative adapted to the n-graded case. For a general bracket construction (in the nongraded setting), we also refer to [17].

A contraction $i \equiv \rfloor$: $D_r(\mathcal{A}) \otimes_{\mathcal{A}} \Lambda^s \to \Lambda^{s-r}$ is defined as follows. Let $X \in D_r(\mathcal{A})$ and $\omega \in \Lambda^s$. Then we set:

1. if $X = a \in D_0(\mathcal{A}) = \mathcal{A}$, then $i_a\omega = a\omega$;
2. $i_X(\omega) = 0$ for $r > s$;
3. $i_X(\omega) = X(\omega)$ for $r = s$, where $X(\omega) \in \mathcal{A}$ is understood due to (1.3);
4. for $r < s$ we define inductively

$$i_X(\mathrm{d}a \wedge \omega) = i_{X_a}\omega + (-1)^{X\cdot a + X_1} \mathrm{d}a \wedge i_X\omega,$$

where $a \in \mathcal{A}$ and Xa is understood due to Remark 1.1.

Consider now a tensor product $\Lambda^* \otimes_{\mathcal{A}} D_*(\mathcal{A})$ and extend the operations of wedge product and contraction. Let $\Omega, \Theta \in \Lambda^* \otimes_{\mathcal{A}} D_*(\mathcal{A})$ be two decomposable homogeneous elements: $\Omega = \omega \otimes X$, $\Theta = \theta \otimes Y$. Then we set

$$i_\Omega \Theta \equiv \Omega \rfloor \Theta = \omega \wedge i_X\theta \otimes Y, \tag{2.1}$$

where $i_X\theta$ was defined above, and

$$\Omega \wedge \Theta = (-1)^{X\cdot\theta + X_1\theta_1} \omega \wedge \theta \otimes X \wedge Y, \tag{2.2}$$

where wedge products in the right-hand side of (2.2) come from Λ^* and $D_*(\mathcal{A})$, respectively.

Define an $(n+1)$-grading in $\Lambda^* \otimes D_*(\mathcal{A})$ by

$$\mathrm{gr}(\Omega) = (\mathrm{gr}(\omega) + \mathrm{gr}(X),\ w_1 - X_1). \tag{2.3}$$

for $\Omega = \omega \otimes X \in \Lambda^* \otimes_{\mathcal{A}} D_*(\mathcal{A})$.

Remark 2.1. Grading (2.3) defines $(n+1)$-gradings in all the submodules $\Lambda^* \otimes_{\mathcal{A}} D_r(\mathcal{A})$. These gradings are shifted in the $(n+1)$st component by r with respect to the 'original' grading $\mathrm{gr}(\omega \otimes X) = (\mathrm{gr}(\omega) + \mathrm{gr}(X), \omega_1)$. Below we mainly deal with the module $\Lambda^* \otimes_{\mathcal{A}} D_1(\mathcal{A})$ and in all notations express this shift explicitly.

PROPOSITION 2.2. (i) *$\Lambda^* \otimes_{\mathcal{A}} D_*(\mathcal{A})$ is an $(n+1)$-graded commutative algebra over Λ^* with respect to grading* (2.3) *and multiplication* (2.2).

(ii) *For any $\Omega \in \Lambda^* \otimes_{\mathcal{A}} D_r(\mathcal{A})$ the contraction i_Ω: $\Lambda^* \otimes_{\mathcal{A}} D_*(\mathcal{A}) \to \Lambda^* \otimes_{\mathcal{A}} D_*(\mathcal{A})$ is an $(n+1)$-graded linear differential operator of the order r.*

(iii) *In particular, for $r = 1$ and any $\rho \in \Lambda^*$, $\Theta \in \Lambda^* \otimes_{\mathcal{A}} D_*(\mathcal{A})$ one has*

$$i_\Omega(\rho \wedge \Theta) = i_\Omega(\rho) \wedge \Theta + (-1)^{\Omega\cdot\rho + (\Omega_1+1)\rho_1} \rho \wedge i_\Omega \Theta. \tag{2.4}$$

(iv) *For any $\Omega, \Theta \in \Lambda^* \otimes D_1(\mathcal{A})$ one has*

$$i_\Omega \circ i_\Theta = i_{\Omega \rfloor \Theta} + i_{\Omega \wedge \Theta}. \tag{2.5}$$

From (2.5) it follows that for any $\Omega, \Theta \in \Lambda^* \otimes_{\mathcal{A}} D_1(\mathcal{A})$ the graded commutator $[i_\Omega, i_\Theta]$ is again an operator of the form i_Ξ for some $\Xi \in \Lambda^* \otimes_{\mathcal{A}} D_1(\mathcal{A})$. We denote Ξ by $[\![\Omega, \Theta]\!]^R$ and call it *the graded Richardson–Nijenhuis bracket* of Ω and Θ.

PROPOSITION 2.3. *Let $\Omega, \Theta \in \Lambda^* \otimes_{\mathcal{A}} D_1(\mathcal{A})$. Then*

(i) $$[\![\Omega, \Theta]\!]^R = i_\Omega \Theta - (-1)^{\Omega\cdot\Theta + (\Omega_1+1)(\Theta_1+1)} i_\Theta \Omega. \tag{2.6}$$

(ii) $$[\![\Omega, \Theta]\!]^R + (-1)^{\Omega\cdot\Theta + (\Omega_1+1)(\Theta_1+1)} [\![\Theta, \Omega]\!]^R = 0. \tag{2.7}$$

(iii) *If $\Xi \in \Lambda^* \otimes_{\mathcal{A}} D_1(\mathcal{A})$, then*

$$\oint (-1)^{\Theta\cdot(\Omega+\Xi) + (\Theta_1+1)(\Omega_1+\Xi_1)} [\![[\![\Omega, \Theta]\!]^R, \Xi]\!]^R = 0, \tag{2.8}$$

where $\oint$ denotes the sum of cyclic permutations.

(iv) *If $\rho \in \Lambda^*$, then*

$$[\![\Omega, \rho \wedge \Theta]\!]^R = i_\Omega \rho \wedge \Theta + (-1)^{\Omega\cdot\rho + (\Omega_1+1)\rho_1} \wedge [\![\Omega, \Theta]\!]^R. \tag{2.9}$$

From (2.7) and (2.8) one can see that $[\![\cdot,\cdot]\!]^R$ defines an $(n+1)$-graded Lie algebra structure in $\Lambda^* \otimes_{\mathcal{A}} D_1(\mathcal{A})$ with respect to grading (2.3). Formula (2.9) shows that the operator $\partial_\Omega^R = [\![\Omega,\cdot]\!]$ is a derivation of the Λ^*-module $\Lambda^* \otimes_{\mathcal{A}} D_1(\mathcal{A})$.

Now, for any $\Omega \in \Lambda^* \otimes_{\mathcal{A}} D_1(\mathcal{A})$ we define *a Lie derivative* L_Ω: $\Lambda^* \to \Lambda^*$ by

$$L_\Omega = [i_\Omega, \mathrm{d}] = i_\Omega \circ \mathrm{d} + (-1)^{\Omega_1} \mathrm{d} \circ i_\Omega. \tag{2.10}$$

The basic properties of Lie derivatives are expressed by

PROPOSITION 2.4. *Let* $\Omega \in \Lambda^* \otimes_{\mathcal{A}} D_1(\mathcal{A})$, $\rho, \theta \in \Lambda^*$. *Then*

(i) $$L_\Omega(\rho \wedge \theta) = L_\Omega(\rho) \wedge \theta + (-1)^{\Omega\cdot\rho + \Omega_1\rho_1} \rho \wedge L_\Omega \theta. \tag{2.11}$$

(ii) $$L_{\rho\wedge\Omega} = \rho \wedge L_\Omega + (-1)^{\rho_1 + \Omega_1} \mathrm{d}\rho \wedge i_\Omega. \tag{2.12}$$

(iii) $$[L_\Omega, d] = 0. \tag{2.13}$$

(iv) *For any* $\Theta \in \Lambda^* \otimes_{\mathcal{A}} D_1(\mathcal{A})$ *the graded commutator* $[L_\Omega, L_\Theta]$ *is of form* L_Ξ *for some uniquely defined* $\Xi \in \Lambda^* \otimes_{\mathcal{A}} D_1(\mathcal{A})$.

The element Ξ defined in Proposition 2.4(iv) is called *the Frölicher–Nijenhuis bracket* of Ω and Θ and is denoted by $[\![\Omega,\Theta]\!]$. From Propositions 2.2 and 2.4 it follows that if $\Omega = \omega \otimes X$, $\Theta = \theta \otimes Y \in \Lambda^* \otimes_{\mathcal{A}} D_1(\mathcal{A})$ are homogeneous decomposable elements, then

$$\begin{aligned}[\![\Omega,\Theta]\!] &= (-1)^{X\cdot\theta} \omega \wedge \theta \otimes [X,Y] + L_\Omega \Theta \otimes Y - \\ &\quad -(-1)^{\Omega\cdot\Theta + \Omega_1\Theta_1} L_\Theta(\Omega) \otimes X,\end{aligned} \tag{2.14}$$

where $[X,Y]$ is the commutator of derivations X and Y.

From the definitions and equalities (2.4), (2.11)–(2.14) one has the following properties of $[\![\cdot,\cdot]\!]$.

PROPOSITION 2.5. *Let* $\Omega, \Theta, \Xi \in \Lambda^* \otimes_{\mathcal{A}} D_1(\mathcal{A})$. *Then*

(i) $$[\![\Omega,\Theta]\!] + (-1)^{\Omega\cdot\Theta + \Omega_1\cdot\Theta_1} [\![\Theta,\Omega]\!] = 0. \tag{2.15}$$

(ii) $$\oint (-1)^{\Theta\cdot(\Omega+\Xi) + \Theta_1(\Omega_1+\Xi_1)} [\![\Omega, [\![\Theta,\Xi]\!]]\!] = 0. \tag{2.16}$$

(iii) $$i_{[\![\Omega,\Theta]\!]} = [L_\Omega, i_\Theta] + (-1)^{\Omega\cdot\Theta + \Omega_1(\Theta_1+1)} L_{\Theta\rfloor\Omega}. \tag{2.17}$$

$$\begin{aligned}
&i_\Xi[\![\Omega,\Theta]\!] \\
\text{(iv)}\qquad &= [\![i_\Xi\Omega,\Theta]\!] + (-1)^{\Omega\cdot\Xi+\Omega_1(\Xi_1+1)}[\![\Omega,i_\Xi\Theta]\!]+ \\
&+(-1)^{\Omega_1} i_{[\![\Xi,\Omega]\!]}\Theta - (-1)^{\Omega\cdot\Theta+(\Omega_1+1)\Theta_1} i_{[\![\Xi,\Theta]\!]}\Omega.
\end{aligned}\tag{2.18}$$

(v) *If* $\rho \in \Lambda^*$, *then*

$$\begin{aligned}
&[\![\Omega,\rho\wedge\Theta]\!] \\
&= L_\Omega(\rho)\wedge\Theta - (-1)^{\Omega\cdot(\Theta+\rho)+(\Omega_1+1)(\Theta_1+\rho_1)} d\rho\wedge i_\Theta\Omega + \\
&+(-1)^{\Omega\cdot\rho+\Omega_1\rho_1}\rho\wedge[\![\Omega,\Theta]\!].
\end{aligned}\tag{2.19}$$

Equalities (2.15), (2.16) show that $\Lambda^*\otimes_{\mathcal{A}} D_1(\mathcal{A})$ is an $(n+1)$-graded Lie algebra with respect to the Frölicher–Nijenhuis bracket in the original, nonshifted grading (see Remark 2.1).

Consider an element $U\in\Lambda^1\otimes_{\mathcal{A}} D_*(\mathcal{A})$. We say it to be *integrable* if $U\cdot U \in 2\mathbb{Z}$ and $[\![U,U]\!]=0$. From (2.16) it follows that for any integrable U the operator

$$\partial_U = [\![U,\cdot]\!]\colon\ \Lambda^s\otimes_{\mathcal{A}} D_1(\mathcal{A}) \longrightarrow \Lambda^{s+1}\otimes_{\mathcal{A}} D_1(\mathcal{A}) \tag{2.20}$$

is a differential, i.e. $\partial_U\circ\partial_U=0$. Hence, one has cohomology modules

$$H_U^*(\mathcal{A}) = \sum_{s\geqslant 0} H_U^s(\mathcal{A}) = \ker\ \partial_U/\ \mathrm{Im}\ \partial_U. \tag{2.21}$$

We shall deal with these cohomologies in a more concrete situation.

3. Categories with Connections and Cohomologies

Fix a commutative associative unitary $\mathbb{F}$-algebra A and consider an n-graded algebra $\mathcal{A}$ satisfying the conditions of Section 1. Denote by $D_1(A,\mathcal{A})$ an $\mathcal{A}$-module of $\mathcal{A}$ valued derivations of A.

DEFINITION 3.1. A homomorphism $\nabla\colon\ D_1(A,\mathcal{A})\to D_1(\mathcal{A})$ is called *a connection* in the graded algebra $\mathcal{A}$ if

$$\nabla(\mu)|_A = \mu \tag{3.1}$$

for any $\mu\in D_1(A,\mathcal{A})$, where for a derivation $X\in D_1(\mathcal{A})$ by $X|_A\in D_1(A,\mathcal{A})$ its restriction onto A is denoted

Let $\mu,\nu\in D_1(A,\mathcal{A})$. Define

$$R_\nabla(\mu,\nu) = [\nabla(\mu),\nabla(\nu)] - \nabla(\nabla(\mu)\circ\nu - (-1)^{\mu\cdot\nu}\nabla(\nu)\circ\mu). \tag{3.2}$$

It is easy to show that (3.2) defines a homomorphism

$$R_\nabla \in \hom_{\mathcal{A}}(D_1(A,\mathcal{A})\wedge D_1(A,\mathcal{A}), D_1(\mathcal{A})).$$

It is called *the curvature form* of the connection ∇. A connection is called *flat*, if its curvature vanishes.

Consider a category $\mathrm{Grc}_n(A)$ whose *objects* are pairs $(\mathcal{A}, \nabla)$, where $\mathcal{A}$ is an n-graded geometrical A algebra (see Section 1) and ∇ is a flat connection in $\mathcal{A}$. Let $\mathcal{O}^1 = (\mathcal{A}^1, \nabla^1)$, $\mathcal{O}^2 = (\mathcal{A}^2, \nabla^2)$ be two objects. An A-homomorphism $\varphi\colon \mathcal{A}^1 \to \mathcal{A}^2$ is said to be *a morphism* in $\mathrm{Grc}_n(A)$, if it respects gradings and filtrations in $\mathcal{A}^1$, $\mathcal{A}^2$ and for any $\mu \in D_1\,(A, \mathcal{A})$ one has a commutative diagram

$$
\begin{array}{ccc}
\mathcal{A}^1 & \xrightarrow{\ \varphi\ } & \mathcal{A}^2 \\
{\scriptstyle \nabla^1(\mu)}\big\downarrow & & \big\downarrow{\scriptstyle \nabla^2(\varphi\circ\mu)} \\
\mathcal{A}^1 & \xrightarrow{\ \varphi\ } & \mathcal{A}^2
\end{array}
$$

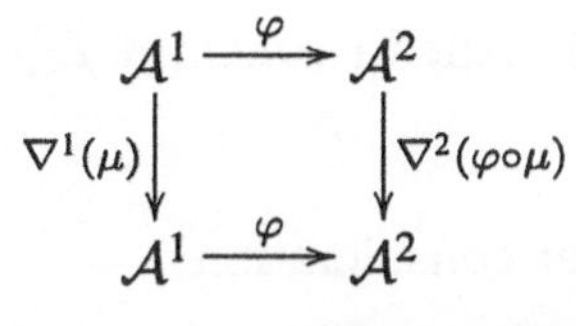

Due to the natural embedding $\mathbb{Z}^n \hookrightarrow \mathbb{Z}^m$, $n < m$, where $\mathbb{Z}^n$ is mapped onto first n summands of $\mathbb{Z}^m$, one can consider $\mathrm{Grc}_n(A)$ as a subcategory of $\mathrm{Grc}_m(A)$. It enables us to consider the category $\mathrm{Grc}_\infty(A)$ of infinitely graded objects, which includes all categories $\mathrm{Grc}_n(A)$, $n = 0, 1, \dots$.

Given an object $\mathcal{O} = (\mathcal{A}, \nabla)$, we shall also deal with a category $\mathrm{Grc}_n(\mathcal{O})$ whose objects are morphisms

$$
\varphi\colon \mathcal{O} \longrightarrow \mathcal{O}', \qquad \mathcal{O}' \in \mathrm{Ob}(\mathrm{Grc}_n(A)),\ \varphi \in \mathrm{Mor}(\mathcal{O}, \mathcal{O}'),
$$

and morphisms consist of commutative diagrams

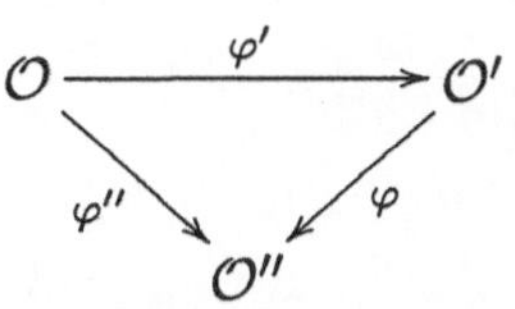

where $\varphi \in \mathrm{Mor}(\mathcal{O}', \mathcal{O}'')$.

Two objects $\mathcal{O}'$ and $\mathcal{O}''$ are said to be *equivalent* if they are isomorphic in $\mathrm{Grc}_n(A)$. They are called *$\mathcal{O}$-equivalent*, if they are isomorphic in $\mathrm{Grc}_n(\mathcal{O})$. Obviously, classes of equivalent objects form categories denoted by $\widetilde{\mathrm{Grc}}_n(A)$ and $\widetilde{\mathrm{Grc}}_n(\mathcal{O})$, respectively.

Let now $\mathcal{O} = (\mathcal{A}, \nabla)$ be an object of $\mathrm{Grc}_n(A)$. Define a homomorphism

$$
U_{\mathcal{O}} \in \mathrm{hom}\,(D_1(\mathcal{A}), D_1(\mathcal{A}))
$$

by

$$
U_{\mathcal{O}}(X) = X - \nabla(X|_A), \qquad X \in D_1(\mathcal{A}). \tag{3.3}
$$

Due to the restrictions posed on algebras $\mathcal{A}$ in Section 1, $U_{\mathcal{O}}$ can be considered as an element of $\Lambda^1(\mathcal{A}) \otimes_{\mathcal{A}} D_1(\mathcal{A})$; this element is called *a connection form* of the object $\mathcal{O}$. From the definitions one has

THEOREM 3.2. *Let $\mathcal{O} = (\mathcal{A}, \Delta)$ be an object of* $\mathrm{Grc}_n(A)$. *Then for any $X, Y \in D_1(\mathcal{A})$ the following identity holds*

$$i_Y(i_X U_{\mathcal{O}}) \equiv U_{\mathcal{O}}(X, Y) = (-1)^{X \cdot Y} 2R_\nabla(X|_A, Y|_A).$$

Since, by definition, $\mathrm{gr}(U_{\mathcal{O}}) = (\mathbf{0}, 1)$, it follows that the connection form of any object is integrable and the cohomology theory (2.20), (2.21) can be related to this object. To our purpose, a subtheory of this theory will be useful.

DEFINITION 3.3. An element $\Omega \in \Lambda^* \otimes_{\mathcal{A}} D_1(\mathcal{A})$ is called *vertical*, if $L_\Omega a = 0$ for any $a \in A \subset \mathcal{A} = \Lambda^\circ(\mathcal{A})$.

Let $(\cdot)^v$ denote the vertical part of modules under consideration

PROPOSITION 3.4. *Let $\mathcal{O} = (\mathcal{A}, \Lambda)$ be an object of* $\mathrm{Grc}_n(A)$. *Then*

(i) $(\Lambda^*(\mathcal{A}) \otimes_{\mathcal{A}} D_1(\mathcal{A}))^v = \Lambda^*(\mathcal{A}) \otimes_{\mathcal{A}} D_1^v(\mathcal{A})$.
(ii) $\Omega \in \Lambda^*(\mathcal{A}) \otimes_{\mathcal{A}} D_1^v(\mathcal{A})$ *iff* $i_\Omega U_{\mathcal{O}} = \Omega$.
(iii) $U_{\mathcal{O}} \in \Lambda^1(\mathcal{A}) \otimes_{\mathcal{A}} D_1^v(\mathcal{A})$.
(iv) $[\![\Lambda^r(\mathcal{A}) \otimes_{\mathcal{A}} D_1^v(\mathcal{A}), \Lambda^s(\mathcal{A}) \otimes_{\mathcal{A}} D_1^v(\mathcal{A})]\!] \subset \Lambda^{r+s}(\mathcal{A}) \otimes_{\mathcal{A}} D_1^v(\mathcal{A})$.
(v) $(\Lambda^r(\mathcal{A}) \otimes_{\mathcal{A}} D_1^v(\mathcal{A})) \rfloor (\Lambda^s(\mathcal{A}) \otimes_{\mathcal{A}} D_1^v(\mathcal{A})) \subset \Lambda^{r+s-1}(\mathcal{A}) \otimes_{\mathcal{A}} D_1^v(\mathcal{A})$.

In particular, differential (2.20) can be restricted up to

$$\partial_{\mathcal{O}}\colon \Lambda^s \otimes_{\mathcal{A}} D_1^v(\mathcal{A}) \longrightarrow \Lambda^{s+1} \otimes_{\mathcal{A}} D_1^v(\mathcal{A}), \tag{3.4}$$

and one gets cohomologies

$$H^*(\mathcal{O}) \equiv H^*_\nabla(\mathcal{A}) = \ker \partial_{\mathcal{O}} / \mathrm{Im}\, \partial_{\mathcal{O}}.$$

From Propositions 2.5 and 3.4 one gets the following properties of differential (3.4).

PROPOSITION 3.5. *Let $\mathcal{O} = (\mathcal{A}, \nabla)$ be an object of* $\mathrm{Grc}_n(A)$ *and $\rho \in \Lambda^* = \Lambda^*(\mathcal{A})$; $\Omega, \Theta \in \Lambda^* \otimes_{\mathcal{A}} D_1^v(\mathcal{A})$. Let $L_\nabla = L_{U_{\mathcal{O}}}$. Then*

$$\text{(i)} \quad \partial_{\mathcal{O}}(\rho \wedge \Omega) = (L_\nabla - d)\rho \wedge \Omega + (-1)^{\rho_1} \rho \wedge \partial_{\mathcal{O}} \Omega. \tag{3.5}$$

$$\text{(ii)} \quad [L_\nabla, i_\Omega] = i_{\partial_{\mathcal{O}} \Omega} + (-1)^{\Omega_1} L_\Omega. \tag{3.6}$$

$$\text{(iii)} \quad [i_\Omega, \partial_{\mathcal{O}}] = (-1)^{\Omega_1} i_{\partial_{\mathcal{O}} \Omega}. \tag{3.7}$$

$$\text{(iv)} \quad \partial_{\mathcal{O}} [\![\Omega, \Theta]\!] = [\![\partial_{\mathcal{O}} \Omega, \Theta]\!] + (-1)^{\Omega_1} [\![\Omega, \partial_{\mathcal{O}} \Theta]\!]. \tag{3.8}$$

Equalities (3.7) and (3.8) show that the cohomology module $H^*(\mathcal{O})$ inherits the operations $\rfloor$, $[\![\cdot,\cdot]\!]$, and $[\![\cdot,\cdot]\!]^R$. We use this fact in the next section.

To finish this section, we outline a general construction allowing us to generate new objects of $\mathrm{Grc}_n(A)$ from the existing ones.

Let $\mathcal{O} = (\mathcal{A}, \nabla)$ be an object of $\mathrm{Grc}_n(A)$ and consider a set $\mathcal{V} = \{v^\alpha\}$ of formal variables with the prescribed gradings $\mathrm{gr}(v^\alpha) \in \mathbb{Z}^m$. Consider a commutative ℓ-graded algebra $\mathcal{A}^\circ = \mathcal{A}[\mathcal{V}]$ freely generated over $\mathcal{A}$ by $\{v^\alpha\}$, where $\ell = \max(m,n)$, and define an $\mathcal{A}^\circ$-algebra $\mathcal{A}^1$ as follows:

(i) $\mathcal{A}^1$ is generated by the symbols of the form $[Z,a]$, where $Z \in \operatorname{Im}\nabla \subset D_1(\mathcal{A}), a \in \mathcal{A}^\circ$;
(ii) $\mathrm{gr}([Z,a]) = \mathrm{gr}(Z) + \mathrm{gr}(a)$;
(iii) Generators of $\mathcal{A}^1$ satisfy the relations

$$\begin{aligned}&[Z,a] = Z(a), \text{ if } a \in \mathcal{A},\\ &[Z,ab] = Z(a)b + (-1)^{Z\cdot a} aZ(b), a,b \in \mathcal{A}^\circ,\\ &[Z,a] \text{ is } \mathcal{A}\text{-linear with respect ot the first argument}\\ &\text{and } \mathbb{F}\text{-linear with respect to the second one.}\end{aligned} \tag{3.9}$$

Thus to any $X \in D_1(A,\mathcal{A})$ one can put into correspondence a derivation $\nabla^1 X \in D_1(\mathcal{A}^\circ, \mathcal{A}^1)$ by setting

$$\nabla^1 X(a) = [\nabla X, a], \quad a \in \mathcal{A}^\circ. \tag{3.10}$$

This correspondence is extendable up to an $\mathcal{A}^1$-linear map

$$\nabla^1\colon D_1(A,\mathcal{A}^1) \longrightarrow D_1(\mathcal{A}^\circ, \mathcal{A}^1). \tag{3.11}$$

Defining now an $\mathcal{A}^1$-algebra $\mathcal{A}^2$ by the generators $[Z,a], Z \in \operatorname{Im}\nabla^1, a \in \mathcal{A}^1$, and by the relations similar to (3.9) one gets an $\mathcal{A}^2$-linear map

$$\nabla^2\colon D_1(A,\mathcal{A}^2) \longrightarrow D_1(\mathcal{A}^1, \mathcal{A}^2),$$

etc. At the i-th step we shall have a $\mathbb{Z}^\ell$-graded $\mathcal{A}^{i-1}$-algebra $\mathcal{A}^i$, $\mathcal{A}^{i-1} \subset \mathcal{A}^i$, together with an $\mathcal{A}^i$-module homomorphism

$$\nabla^i\colon D_1(A,\mathcal{A}^i) \longrightarrow D_1(\mathcal{A}^{i-1}, \mathcal{A}^i).$$

Setting

$$\mathcal{A}_\mathcal{V} = \bigcup_{i \geqslant 0} \mathcal{A}^i \tag{3.12}$$

we get at the infinity

$$\nabla_\mathcal{V}\colon D_1(A,\mathcal{A}_\mathcal{V}) \longrightarrow D_1(\mathcal{A}_\mathcal{V}). \tag{3.13}$$

PROPOSITION 3.6. *The map $\nabla_\mathcal{V}$ from* (3.13) *is a flat connection in the $\mathbb{Z}^\ell$-graded algebra $\mathcal{A}_\mathcal{V}$.*

The object $\mathcal{O}_{\mathcal{V}} = (\mathcal{A}_{\mathcal{V}}, \nabla_{\mathcal{V}})$ is called *a free $\mathcal{V}$-extension of the object* $\mathcal{O} = (\mathcal{A}, \nabla)$.

4. Deformations, Symmetries and Recursion Operators in Categories with Connections

Referring the reader to [1, 4, 6] for motivations, we give the following definition

DEFINITION 4.1. Let $\mathcal{O} = (\mathcal{A}, \nabla)$ be an object of $\mathrm{Grc}_n(A)$. Then

(i) The elements of $H^0(\mathcal{O})$ are called *symmetries* of the object $\mathcal{O}$. We also use the notation $H^0(\mathcal{O}) = \mathrm{sym}_\nabla \mathcal{A} = \mathrm{sym}\, \mathcal{O}$.
(ii) The elements of $H^1(\mathcal{O})$ are called *(infinitesimal) deformations* of $\mathcal{O}$.

The nature of the elements of $\mathrm{sym}_\nabla \mathcal{A}$ is clarified by

PROPOSITION 4.2. *A derivation $X \in D_1^v(\mathcal{A})$ is a symmetry of an object $\mathcal{O} = (\mathcal{A}, \nabla)$ iff*

$$[X, \nabla(\mu)] = 0$$

for any $\mu \in D_1(A) \subset D_1(A, \mathcal{A})$.

From the results of Section 3, one gets

THEOREM 4.3. *Let $\mathcal{O} = (\mathcal{A}, \nabla)$ be an object of* $\mathrm{Grc}_n(A)$. *Then*

(i) $\mathrm{sym}_\nabla \mathcal{A}$ *is an n-graded Lie algebra over k with respect to* $[\cdot,\cdot]$.
(ii) *A correspondence $X \mapsto [\![X, \cdot]\!]$ determines a representation of* $\mathrm{sym}_\nabla \mathcal{A}$ *in $H^s(\mathcal{O})$ for any $s \geqslant 0$.*
(iii) *$H^1(\mathcal{O})$ is an associative n-graded algebra with respect to contraction operation.*
(iv) *A correspondence $\mathbb{R}\colon H^1(\mathcal{O}) \to \mathrm{End}_k H^0(\mathcal{O})$ defined by*

$$\mathbb{R}_\Omega(X) = i_X(\Omega), \quad X \in H^0(\mathcal{O}),\ \Omega \in H^1(\mathcal{O}), \tag{4.1}$$

is a representation of this algebra in $\mathrm{sym}_\nabla \mathcal{A}$.

Thus, given a symmetry, one can construct new elements of $H^*(\mathcal{O})$ and with a given deformation it is possible to generate new symmetries. In the latter case, though, a certain part of deformations will act trivially on symmetries by (4.1). To eliminate these trivial actions, we go deeper into the structure of $H^*(\mathcal{O})$.

Consider the first summand at the right-hand side of (3.5) and define

$$\mathrm{d}_h = \mathrm{d} - L_\nabla\colon \Lambda^* \longrightarrow \Lambda^*. \tag{4.2}$$

Thus, (3.5) can be rewritten as

$$\partial_{\mathcal{O}}(\rho \wedge \Omega) = -\mathrm{d}_h \rho \wedge \Omega + (-1)^{\rho_1} \rho \wedge \partial_{\mathcal{O}} \Omega. \tag{4.3}$$

Since, by definition,

$$2L_\nabla \circ L_\nabla = L_{[\![U_\mathcal{O},U_\mathcal{O}]\!]} = 0$$

and due to (2.13) $[L_\nabla, \mathrm{d}] = 0$, one has $\mathrm{d}_h \circ \mathrm{d}_h = 0$. We call d_h *the horizontal differential* for the object $\mathcal{O}$ and denote corresponding cohomology by

$$H_h^*(\mathcal{O}) = \sum_{s \geqslant 0} H_h^s(\mathcal{O}).$$

Definition 4.2 gives one a direct sum decomposition

$$\Lambda^1 = \nabla\Lambda^1 \oplus \Lambda_h^1 \tag{4.4}$$

where $\nabla\Lambda^1$ and Λ_h^1 are $\mathcal{A}$-submodules generated by $L_\nabla(\mathcal{A})$ and $\mathrm{d}_h(\mathcal{A})$, respectively. From (4.4), we get that $\Lambda^s = \sum_{p+q=s} \Lambda^{p,q}$, where $\Lambda^{p,q} = \Lambda^{p,q}(\mathcal{A})$ consists of the forms

$$\rho = aL_\nabla a_1 \wedge \cdots \wedge L_\nabla a_p \wedge \mathrm{d}_h a_{p+1} \wedge \cdots \wedge \mathrm{d}_h a_{p+q}, \quad a, a_1, \ldots, a_{p+q} \in \mathcal{A}.$$

THEOREM 4.4. *Let* $\mathcal{O} = (\mathcal{A}, \nabla)$ *be an object of* $\mathrm{Grc}_n(A)$. *Then*

(i) $H^*(\mathcal{O})$ *is an* $H_h^*(\mathcal{O})$*-module.*
(ii) *A triple* $(\Lambda^*(\mathcal{A}), \mathrm{d}_h, L_\nabla)$ *forms a bicomplex, i.e.*

$$\mathrm{d}_h \circ \mathrm{d}_h = L_\nabla \circ L_\nabla = [\mathrm{d}_h, L_\nabla] = 0. \tag{4.5}$$

(iii) *For any* $p, q \geqslant 0$ *one has*

$$\mathrm{d}_h(\Lambda^{p,q}) \subset \Lambda^{p,q+1}, \tag{4.6}$$
$$L_\nabla(\Lambda^{p,q}) \subset \Lambda^{p+1,q}, \tag{4.7}$$
$$\partial_\mathcal{O}(\Lambda^{p,q} \otimes_\mathcal{A} D_1^v(\mathcal{A}) \subset \Lambda^{p,q+1} \otimes_\mathcal{A} D_1^v(\mathcal{A}). \tag{4.8}$$

(iv) *The modules* $\Lambda^{p,q} \otimes_\mathcal{A} D_1^v(\mathcal{A})$ *can be described as*

$$\begin{aligned}\Lambda^{p,q} &\otimes_\mathcal{A} D_1^v(\mathcal{A}) \\ &= \{\Omega \in \Lambda^s \otimes_\mathcal{A} D_1^v(\mathcal{A}) \mid X_1 \rfloor \cdots \rfloor X_{p+1} \rfloor \Omega = 0 \\ &\quad X_i \in D_1^v(\mathcal{A}), Y_1 \rfloor \cdots \rfloor Y_{q+1} \rfloor \Omega = 0,\ Y_j \in \mathrm{Im}\nabla,\ p+q=s\}\end{aligned} \tag{4.9}$$

Remark 4.5. The first spectral sequence associated with bicomplex (4.5) converges to the de Rham cohomology of $\mathcal{A}$ and is a natural generalization of A. M. Vinogradov's $\mathcal{C}$-spectral sequence [14].

From (4.6) and (4.8) it follows that cohomologies $H_h^*(\mathcal{O})$ and $H^*(\mathcal{O})$ can be split into direct sums

$$H_h^*(\mathcal{O}) = \sum_{p,q \geqslant 0} H_h^{p,q}(\mathcal{O})$$

and

$$H^*(\mathcal{O}) = \sum_{p,q\geqslant 0} H^{p,q}(\mathcal{O}) \equiv \sum_{p,q\geqslant 0} H^{p,q}_{\nabla}(\mathcal{A}),$$

where

$$H^{p,q}_h(\mathcal{O}) = \frac{\ker(\mathrm{d}_h\colon\ \Lambda^{p,q} \to \Lambda^{p,q+1})}{\operatorname{Im}\,(\mathrm{d}_h\colon\ \Lambda^{p,q-1} \to \Lambda^{p,q})}$$

and

$$H^{p,q}_{\nabla}(\mathcal{O}) = \frac{\ker(\partial_{\mathcal{O}}\colon\ \Lambda^{p,q} \otimes_{\mathcal{A}} D^v_1(\mathcal{A}) \to \Lambda^{p,q+1} \otimes_{\mathcal{A}} D^v_1(\mathcal{A}))}{\operatorname{Im}(\partial_{\mathcal{O}}\colon\ \Lambda^{p,q-1} \otimes_{\mathcal{A}} D^v_1(\mathcal{A}) \to \Lambda^{p,q} \otimes_{\mathcal{A}} D^v_1(\mathcal{A}))}.$$

Note that $H^{0,*}_h(\mathcal{O}) = \sum_{q\geqslant 0} H^{0,q}_h(\mathcal{O})$ is called *horizontal de Rham cohomology* of the object $\mathcal{O}$. Coming back to representation (4.1) note also that due to (4.9) it is trivial for the elements of $H^{0,1}(\mathcal{O})$. Therefore, we restrict ourselves to the modules

$$H^{p,0}_{\nabla}(\mathcal{A}) = \ker\,(\partial_{\mathcal{O}}\colon\ \Lambda^{p,0} \otimes_{\mathcal{A}} D^v_1(\mathcal{A}) \to \Lambda^{p,1} \otimes_{\mathcal{A}} D^v_1(\mathcal{A}))$$

and give the following

DEFINITION 4.6. Elements of $H^{1,0}_{\nabla}(\mathcal{A})$ are called *recursion operators* for the object $\mathcal{O} = (\mathcal{A}, \nabla)$. Their action on symmetries of $\mathcal{O}$ is given by (4.1). We also use the notation $H^{1,0}_{\nabla}(\mathcal{A}) = \operatorname{rec}_{\nabla}\mathcal{A} = \operatorname{rec}\mathcal{O}$.

PROPOSITION 4.7. *Recursion operators form a subalgebra in the algebra* $H^1_{\nabla}(\mathcal{O})$ (*see Theorem* 4.3 (iii)). *Moreover, this subalgebra is unitary with* $U_{\mathcal{O}}$ *as a unit.*

Consider an object $\mathcal{O} = (\mathcal{A}, \nabla)$ and the $(n+1)$-graded algebra $\Lambda^{*,0}(\mathcal{A}) := \mathcal{T}(\mathcal{A})$. Conditions imposed on $\mathcal{A}$ in Section 1 imply that any derivation $\mu \in D_1(A, \mathcal{T}(\mathcal{A}))$ is of the form $\mu = \sum_\alpha \rho_\alpha \otimes \mu_\alpha$, where $\rho_\alpha \in \mathcal{T}(\mathcal{A})$ and $\mu_\alpha \in D_1(A, \mathcal{A})$. Let $\rho \in \mathcal{T}(\mathcal{A})$ and set

$$\mathcal{T}\nabla(\mu)(\rho) = \sum_\alpha \rho_\alpha \wedge L_{\nabla(\mu_\alpha)}(\rho). \tag{4.10}$$

Due to the $\mathcal{A}$-linearity of $\nabla, \mathcal{T}\nabla$ is well-defined. Let now $X \in \operatorname{Im}\nabla$. Then, obviously, $[\![X, U_{\mathcal{O}}]\!] = 0$ and from (2.17) it follows that $i_X\mathcal{T}\nabla(\mu)(\rho) = 0$, or, due to (4.9), $\mathcal{T}\nabla(\mu)(\rho) \in \mathcal{T}(\mathcal{A})$.

By definition, the map $\mathcal{T}\nabla\colon\ D_1(A, \mathcal{T}(\mathcal{A}))$ is $\mathcal{T}(\mathcal{A})$-linear, i.e. is a connection and it is easy to show that $R_{\mathcal{T}\nabla} = 0$.

Note that for any $\Phi \in H^{*,0}_{\nabla}(\mathcal{A})$, the corresponding Lie derivative $L_\Phi\colon\ \Lambda^* \to \Lambda^*$ preserves $\Lambda^{*,0} = \mathcal{T}(\mathcal{A}) \subset \Lambda^*$ and for any $\mu \in D_1(A)$ one has

$$[L_\Phi, \mathcal{T}\nabla(\mu)] = [L_\Phi, L_{\nabla(\mu)}] = L_{[\![\Phi, \nabla(\mu)]\!]} = 0. \tag{4.11}$$

In fact, applying (2.18) with

$$\Xi = \Phi, \Omega = U_{\mathcal{O}}, \Theta = \nabla(\mu),$$

one gets $[\![\Phi, \nabla(\mu)]\!] = [\![\Phi, \nabla(\mu)]\!] \rfloor U_{\mathcal{O}}$ and applying the same identity with

$$\Xi = \nabla(\mu), \Omega = U_{\mathcal{O}}, \Theta = \Phi,$$

one gets $[\![\Phi, \nabla(\mu)]\!] \rfloor U_{\mathcal{O}} = 0$. Hence, from Proposition 4.2 and (4.11) it follows that L_Φ is a symmetry of the object $(\mathcal{T}(\mathcal{A}), \mathcal{T}\nabla)$.

THEOREM 4.8. *Let* $\mathcal{O} = (\mathcal{A}, \nabla)$ *be an object of* $\mathrm{Grc}_n(A)$ *and* $\mathcal{T}\mathcal{O} := (\mathcal{T}(\mathcal{A}), \mathcal{T}\nabla)$. *Then*

(i) *The correspondence* $\mathcal{O} \Rightarrow \mathcal{T}\mathcal{O}$ *defines a covariant functor*

$$\mathcal{T}\colon \mathrm{Grc}_n(A) \Longrightarrow \mathrm{Grc}_{n+1}(A). \tag{4.12}$$

(ii) *For any object* $\mathcal{O}$ *ther exists a natural embedding*

$$L\colon H^{*,0}(\mathcal{O}) \longrightarrow H^{1,0}(\mathcal{T}\mathcal{O}) = \mathrm{sym}_{\mathcal{T}\nabla}\mathcal{T}(\mathcal{A}) \tag{4.13}$$

defined by the Lie derivative.

(iii) *If* $\mathcal{O}', \mathcal{O}''$ *are equivalent (resp.,* $\mathcal{O}$*-equivalent) objects, then* $\mathcal{T}\mathcal{O}', \mathcal{T}\mathcal{O}''$ *are equivalent (resp.,* $\mathcal{T}\mathcal{O}$*-equivalent) as well.*

We call $\mathcal{T}$ *the functor of superlinearization.*

Remark 4.9. The definition of the functor $\mathcal{T}$ can be guessed from the paper [13].

In what follows, we shall need another functor defined on the category $\mathrm{Grc}_n(A)$ and generalizing the construction of [3]. Consider an object $\mathcal{O} = (\mathcal{A}, \nabla)$ and its first horizontal cohomology $H_h^{0,1}(\mathcal{O})$. Let $S \subset H_h^{0,1}(\mathcal{O})$ be an $\mathbb{F}$-subspace and $\{w_\alpha\}$ be its basis. For each class $w_\alpha \in H_h^{0,1}(\mathcal{O})$, consider its representative $\omega_\alpha \in \Lambda^{0,1}(\mathcal{A})$. Let $\mathcal{A}[S]$ be an n-graded commutative algebra freely generated over $\mathcal{A}$ by $\{w_\alpha\}$ with $\mathrm{gr}(w_\alpha) = \mathrm{gr}(\omega_\alpha)$. As in the previous case, any derivation $\mu \in D_1(A, \mathcal{A}[S])$ is of the form $\mu = \sum_\beta \varphi_\beta \otimes \mu_\beta$, where $\varphi_\beta \in \mathcal{A}[S]$ and $\mu_\beta \in D_1(A, \mathcal{A})$.

Let $a \in \mathcal{A}$ and set

$$\mathcal{K}\nabla(\mu)(aw_\alpha) = \sum_\beta \varphi_\beta\Big(\nabla(\mu_\beta)(a)w_\alpha + (-1)^{a\cdot\mu_\beta} a(\nabla(\mu_\beta)\rfloor\omega_\alpha)\Big) \tag{4.14}$$

From the $\mathcal{A}$-linearity of ∇ and from (2.17) it follows that $\mathcal{K}\nabla(\mu)$ is well-defined. Using the graded Leibniz rule, (4.14) can be extended up to a derivation of $\mathcal{A}[S]$ and one can prove that the correspondence $\mathcal{K}\nabla\colon D_1(A, \mathcal{A}[S]) \to D_1(\mathcal{A}[S])$ is a flat connection in $\mathcal{A}[S]$.

THEOREM 4.10. *Let $\mathcal{O} = (\mathcal{A}, \nabla)$ be an object of* $\mathrm{Grc}_n(A)$, $H_h^{0,1}(\mathcal{O})$ *be its first horizontal cohomology and $S \subset H_h^{0,1}(\mathcal{O})$ be a k-subspace.*
Then

(i) *The equivalence class of the object $\mathcal{K}_S(\mathcal{O}) = (\mathcal{A}[S], \mathcal{K}\nabla)$ in* $\mathrm{Grc}_n(\mathcal{O})$ *(see Section 3) is dependent neither on the choice of a basis $\{w_\alpha\}$ in S, nor on the choice of representative forms $\omega_\alpha \in \Lambda^{0,1}(\mathcal{A})$.*
(ii) *In particular, for $S = H_h^{0,1}(\mathcal{O})$, the correspondence*

$$\mathcal{K} = \mathcal{K}_S\colon \mathrm{Grc}_n(A) \longrightarrow \widetilde{\mathrm{Grc}_n(A)}$$

is a covariant functor.
(iii) *For any $w_\alpha \in H_h^{0,1}(\mathcal{O})$ with a representative $\omega_\alpha \in \Lambda^{0,1}(\mathcal{A})$, one has*

$$\omega_\alpha = \mathrm{d}_k w_\alpha \tag{4.15}$$

in $\mathcal{K}(\mathcal{O})$.

We call $\mathcal{K}$ the killing functor.

Here we finish with a purely algebraic part of the theory and in what follows, consider what geometrical specifics of differential equations add to the general picture.

5. Graded Differential Equations

Let M be a smooth manifold and $\pi\colon E \to M$ be a locally trivial smooth vector bundle over M. Consider the manifold of k-jets and corresponding projection $\pi_k\colon J^k(\pi) \to M$. Let $\mathcal{E} \subset J^k(\pi)$ be a differential equation of the order k and $\mathcal{E}^\infty \subset J^\infty(\pi)$ be its infinite prolongation. We consider the case when $\mathcal{E}$ is *fomally integrable*, i.e. when the projection $\pi_\infty\colon \mathcal{E}^\infty \to M$ is a fibre bundle. Then in this bundle there exists an integrable connection, *the Cartan connection*, which sends a vector field $X \in D_1(M)$ to a field $\mathcal{C}X \in D_1(\mathcal{E})$ on $\mathcal{E}^\infty$. This construction is commonly known as *a total derivative*: one takes a vector field X on the base M, extends it to the total derivative on $J^\infty(\pi)$ and the restriction of the latter onto $\mathcal{E}^{(\infty)}$ gives the field $\mathcal{C}X$ (see [6] for details).

Let $\mathcal{F} = \mathcal{F}(\mathcal{E})$ be the algebra of smooth functions on $\mathcal{E}^{(\infty)}$ filtered by subalgebras $\mathcal{F}_k$ of smooth functions of the kth prolongations of $\mathcal{E}, k = 0, 1, \ldots$. We formally set $\mathcal{F}_{-\infty} = C^\infty(M) \subset \mathcal{F}$. Obviously, the pair $(\mathcal{F}, \mathcal{C})$ is an object of $\mathrm{Grc}_0(\mathcal{F}_{-\infty})$. In particular, for the empty equation $\mathcal{E} = J^0(\pi)$, one gets the algebra $\mathcal{F}(\pi) = \bigcup_{k \geqslant 0} \mathcal{F}_k(\pi)$, where $\mathcal{F}_k(\pi) = C^\infty(J^k(\pi))$.

Consider a set V of graded formal variables $v^\alpha, \mathrm{gr}(v^\alpha) \in \mathbb{Z}^b$, and the free V-extension $(\mathcal{F}_V(\pi), \mathcal{C}_V)$ of $(\mathcal{F}(\pi), \mathcal{C})$ (see Section 3). The algebra $\mathcal{F}_V(\pi)$ is bifiltered: the first filtration is inherited from $\mathcal{F}(\pi)$, the second one is defined by (3.12). Denote by $\mathcal{F}_{i,j,V}(\pi) \subset \mathcal{F}_V(\pi)$ the set of elements of bifiltration (i,j) and define

$$\mathcal{F}_{k,V}(\pi) = \sum_{\max(i,j) \leqslant k} \mathcal{F}_{i,j,V}(\pi). \tag{5.1}$$

Obviously, $\{\mathcal{F}_{k,V}(\pi)\}$ forms a filtration in $\mathcal{F}_V(\pi)$. Let I be a graded ideal in $\mathcal{F}_V(\pi)$, filtered with respect to (5.1) and such that $\mathcal{C}_V(X)I \subset I$ for any $X \in D_1(\mathcal{F}_{-\infty}, \mathcal{F}_V(\pi))$. Then $\mathcal{C}_V$ determines a connection $\mathcal{C}_I$ in the quotient algebra $\mathcal{F}_I = \mathcal{F}_V(\pi)/I$.

DEFINITION 5.1. The object $(\mathcal{F}_I, \mathcal{C}_I)$ is called an ℓ-graded differential equation in the bundle $\pi\colon E \to M$ with graded variables $V = \{v^\alpha\}$. A subcategory of $\mathrm{Grc}_\ell(\mathcal{F}_{-\infty})$ formed by the objects of such a type is denoted by $\mathrm{GDE}_\ell(\pi)$. A category of all graded equations in π is denoted by $\mathrm{GDE}_\infty(\pi)$.

EXAMPLE 5.2. If $\mathcal{E} \subset J^k(\pi)$ is a differential equation, then $\mathcal{T}(\mathcal{F}, \mathcal{C})$ is a $\mathbb{Z}_2$-graded equation (see Theorem 4.9).

Now let $\mathcal{E} \subset J^k(\pi)$ be a differential equation and $\mathcal{E}^g = (\mathcal{F}_I, \mathcal{C}_I)$ be a graded equation with graded variables $V = \{v^\alpha\}$. The algebra $\mathcal{F}(\mathcal{E})$ of smooth functions on $\mathcal{E}^{(\infty)}$ can be represented as a quotient algebra $\mathcal{F}(\pi)/I_\mathcal{E}$, where $I_\mathcal{E}$ is the ideal of functions vanishing on $\mathcal{E}^{(\infty)}$. Let

$$\varphi_\mathcal{E}\colon \mathcal{F}(\pi) \longrightarrow \mathcal{F}(\mathcal{E}) = \mathcal{F}(\pi)/I_\mathcal{E}$$

and

$$\varphi_I\colon \mathcal{F}_V(\pi) \longrightarrow \mathcal{F}_I = \mathcal{F}_V(\pi)/I$$

be canonical projections. Consider a $\mathcal{C}_V$-stable graded filtered ideal (V) generated in $\mathcal{F}_V(\pi)$ by V. Then $\mathcal{F}(\pi) = \mathcal{F}_V(\pi)/(V)$. Let $\varphi_V\colon \mathcal{F}_V(\pi) \to \mathcal{F}(\pi)$ be the corresponding projection. Obviously, $\varphi_\mathcal{E}$, φ_I, and φ_V are morphisms in the category $\mathrm{Grc}_\infty(\mathcal{F}_{-\infty})$.

DEFINITION 5.3. Let $\mathcal{E} \subset J^k(\pi)$ be a differential equation and $\mathcal{O} = (\mathcal{F}_I, \mathcal{C}_I)$ be a graded equation in π. Then $\mathcal{O}$ is said to be a graded extension of $\mathcal{E}$ if there exists an epimorphism $\varphi\colon \mathcal{F}_I \to \mathcal{F}(\mathcal{E})$ such that

(i) φ is a morphism in $\mathrm{Grc}_\infty(\mathcal{F}_{-\infty})$;
(ii) the diagram

$$\begin{array}{ccc} \mathcal{F}_V(\pi) & \xrightarrow{\varphi_I} & \mathcal{F}_I \\ {\scriptstyle\varphi_V}\downarrow & & \downarrow{\scriptstyle\varphi} \\ \mathcal{F}(\pi) & \xrightarrow{\varphi_\mathcal{E}} & \mathcal{F}(\mathcal{E}) \end{array}$$

is commutative. The category of all graded extensions of $\mathcal{E}$ is denoted by $\mathrm{GDE}_\infty(\mathcal{E})$.

Finally, similar to Section 3, for each object $\mathcal{O}$ of $\mathrm{GDE}_\infty(\mathcal{E})$ we consider a category whose objects are morphisms

$$\varphi\colon \mathcal{O} \to \mathcal{O}', \mathcal{O}' \in \mathrm{Ob}(\mathrm{GDE}_\infty(\mathcal{E})), \quad \varphi \in \mathrm{Mor}(\mathcal{O}, \mathcal{O}'),$$

and whose morphisms are commutative diagrams

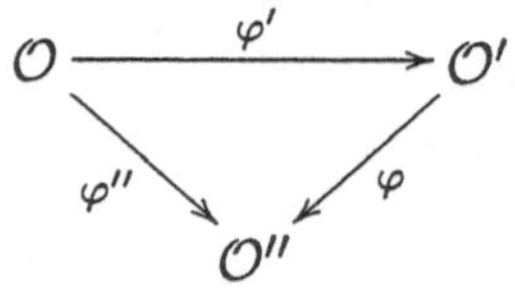

,

$\varphi \in \mathrm{Mor}(\mathcal{O}', \mathcal{O}'')$. Its subcategory consisting of monomorphisms is denoted by $\mathrm{Cov}(\mathcal{O})$ and objects of $\mathrm{Cov}(\mathcal{O})$ are called *coverings* over $\mathcal{O}$.

EXAMPLE 5.4. Due to Theorem 4.11, for any object $\mathcal{O}$ of $\mathrm{GDE}_\infty(\mathcal{E})$ and any subspace $S \subset H_h^{0,1}(\mathcal{O})$, the object $\mathcal{K}_S(\mathcal{O})$ is a class of equivalent coverings over $\mathcal{O}$ (where equivalence is understood as the $\mathcal{O}$-equivalence in the sense of Section 3).

From now on, we confine ourselves to the objects of $\mathrm{GDE}_\ell(\mathcal{E})$ which we call *localized*. It means that a neighbourhood $\mathcal{U} \subset M$ with local coordinates $(x_1, \ldots, x_n)$, $n = \dim M$, is chosen in such a way that the bundle $\pi\colon E \to M$ is trivial over $\mathcal{U}$. Let $(e_1, \ldots, e_m)$ be a basis of local sections of π over $\mathcal{U}, m = \dim \pi$. Then any section of π over $\mathcal{U}$ is expressed as $u = u^1 e_1 + \cdots + u^m e_m$ and the functions

$$\{u_\sigma^j\}, \quad j = 1, \ldots, m, \quad \sigma = (\sigma_1, \ldots, \sigma_\pi), \qquad |\sigma| = \sigma_1 + \cdots + \sigma_N \leqslant k,$$

together with $x_1, \ldots, x_n$ are natural coordinates in $\pi_k^{-1}(\mathcal{U}) \subset J^k(\pi)$, while the whole set (x_i, u_σ^j), $|\sigma| < \infty$, forms a coordinate system in $\pi_\infty^{-1}(\mathcal{U}) \subset J^\infty(\pi)$. The functions u_σ^j are detemined by characteristic relations

$$u_\sigma^j = \frac{\partial^{|\sigma|} u^j}{\partial x_1^{\sigma_1} \ldots \partial x_n^{\sigma_n}}$$

for any local section u. In this setting, any equation $\mathcal{E} \subset J^k(\pi)$ can be represented as a system

$$\begin{aligned} &F^1(x_1, \ldots, x_n, \ldots, u_\sigma^j, \ldots) = 0, \\ &\ldots \\ &F^r(x_1, \ldots, x_n, \ldots, u_\sigma^j, \ldots) = 0, \end{aligned} \tag{5.3}$$

where $F^1, \ldots, F^r$ are smooth functions, $j = 1, \ldots, m$, $|\sigma| \leqslant k$, k being the order of $\mathcal{E}$. Graded differential equations in π are locally described by the relations

$$\begin{aligned} &G^1(x_1, \ldots, x_n, \ldots, u_\sigma^j, \ldots, v_\tau^\alpha, \ldots) = 0, \\ &\ldots \\ &G^s(x_1, \ldots, x_n, \ldots, u_\sigma^j, \ldots, v_\tau^\alpha, \ldots) = 0. \end{aligned} \tag{5.4}$$

Here $\tau = (\tau_1, \dots, \tau_n)$ is a multi-index and, due to (3.9), variables v^α_τ are defined by the induction:

$$v^\alpha_{(0,\dots,0)} = v^\alpha, \qquad v^\alpha_{\tau+1_i} = [D_i, v^\alpha_{(\tau_1,\dots,\tau_n)}],$$

where $D_i = \mathcal{C}(\partial/\partial x_i)$ is the ith total derivative on $J^\infty(\pi)$ and $\tau + 1_i = (\tau_1, \dots, \tau_i + 1, \dots, \tau_n)$.

Equation (5.4) is a graded extension of (5.3), if it reduces to (5.3), when all v^α_τ vanish.

In the following general considerations, we shall not distinguish between variables u^j and v^α and shall represent graded equations in the form

$$\begin{aligned} &G^1(x_1, \dots, x_n, \dots, v^\alpha_\tau, \dots) = 0, \\ &\dots \\ &G^s(x_1, \dots, x_n, \dots, v^\alpha_\tau, \dots) = 0, \end{aligned} \tag{5.5}$$

where v^α in (5.5) stand both for u^j and 'old' v^α. Denote the graded equation associated with (5.5) by $\mathcal{Y} = (\mathcal{F}(\mathcal{Y}), \mathcal{C})$ and retain notation D_i for $\mathcal{C}(\partial/\partial_{x_i})$ on $\mathcal{Y}$, $i = 1, \dots, n$. Let also $\mathcal{J} = (\mathcal{F}(\mathcal{J}), \mathcal{C})$ be the empty equation associated with the variables $\{v^\alpha\}$. The following result is a graded generalization of the classical one (see [6]).

THEOREM 5.5. (i) *A symmetry algebra* $\mathrm{sym}_\mathcal{C}\mathcal{J}$ *consists of evolutionary derivations*

$$\ni_{\widetilde{\omega}} = \sum_{\alpha,\tau} D_\tau(\widetilde{\omega}^\alpha)\partial/\partial v^\alpha_\tau, \tag{5.6}$$

where $\widetilde{\omega} = (\widetilde{\omega}^1, \dots, \widetilde{\omega}^\alpha, \dots)$, $\widetilde{\omega}^\alpha \in \mathcal{F}(\mathcal{J})$, *and* $D_\tau = D_1^{\tau_1} \circ \dots \circ D_n^{\tau_n}$.

(ii) *If* $\ni_\omega$ *can be restricted onto* $\mathcal{Y}$, *then this restriction is uniquely determined by the restrictions* $\{\omega^\alpha\}$ *of* $\{\widetilde{\omega}^\alpha\}$ *onto* $\mathcal{Y}$:

$$\ni_{\widetilde{\omega}} |_\mathcal{Y} := \ni_\omega \tag{5.7}$$

(iii) $\mathrm{sym}_\mathcal{C}\mathcal{Y}$ *consists of derivations of the form (5.7).*

(iv) *A differential operator* $\tilde{\ell}_\mathcal{Y}$ *acting on* $\widetilde{\omega} = \{\widetilde{\omega}^\alpha\}$, $\widetilde{\omega}^\alpha \in \mathcal{F}(\mathcal{J})$, *by*

$$(\tilde{\ell}_\mathcal{Y}\widetilde{\omega})^\beta = \sum_{\alpha,\tau} D_\tau(\widetilde{\omega}^\alpha)\frac{\partial G^\beta}{\partial v^\alpha_\tau} \tag{5.8}$$

is restrictable onto $\mathcal{Y}$. *Denote this restriction by* $\ell_\mathcal{Y}$.

(v) *Let* $\omega = \{\omega^\alpha\}$, $\omega^\alpha \in \mathcal{F}(\mathcal{Y})$. *Then* ω *determines a derivation of the form* (5.7) *iff*

$$\ell_\mathcal{Y}\omega = 0. \tag{5.9}$$

In other words,

$$\mathrm{sym}_\mathcal{C}\mathcal{Y} = \ker \ell_\mathcal{Y}. \tag{5.10}$$

The last theorem, in combination with Theorem 4.9, solves the problem of computation of the modules $H_{\mathcal{C}}^{p,0}(\mathcal{Y})$, $\mathcal{Y} = \mathcal{F}(\mathcal{Y})$, $\mathcal{C}$ being a graded differential equation. Namely, let $\omega \in \Lambda^{p,0}(\mathcal{Y})$ and $\Delta\colon \mathcal{F}(\mathcal{Y}) \to \mathcal{F}(\mathcal{Y})$ be a linear differential operator of the form

$$\Delta(\varphi) = \sum_{\tau} D_{\tau}(\varphi)\Delta_{\tau}, \quad \Delta_{\tau} \in \mathcal{F}(\mathcal{Y}), \tag{5.11}$$

where $D_{\tau} = D_1^{\tau_1} \circ \cdots \circ D_n^{\tau_n}$ and D_i is the restriction of corresponding total derivative onto $\mathcal{Y}$. Since $D_i \rfloor \rho = 0$ for any $i = 1, \ldots, n$ and $\rho \in \Lambda^{p,0}(\mathcal{Y})$, one gets a well-defined action

$$\Delta^{[p]}\colon \Lambda^{p,0}(\mathcal{Y}) \longrightarrow \Lambda^{p,0}(\mathcal{Y})$$

by setting

$$\Delta^{[p]}(\rho) = \sum_{\tau} L_{\tau}(\rho)\Delta_{\tau}, \tag{5.12}$$

where

$$L_{\tau}(\rho) = (\underbrace{L_{D_1} \circ \cdots \circ L_{D_1}}_{\tau_1 \text{ times}} \circ \cdots \circ \underbrace{L_{D_n} \circ \cdots \circ L_{D_n}}_{\tau_n \text{ times}})\rho. \tag{5.13}$$

Then, using (4.13) and definition (5.11)–(5.12), one gets the following

THEOREM 5.6. *Let $\mathcal{Y} = (\mathcal{F}(\mathcal{Y}), \mathcal{C})$ be a graded differential equation and $\ell_{\mathcal{Y}}$ be defined by Theorem* 5.5(iv). *Let $\omega = (\omega^1, \ldots, \omega^{\alpha}, \ldots)$, $\omega_{\alpha} \in \Lambda^{p,0}(\mathcal{Y})$. Then $H_{\mathcal{C}}^{p,0}(\mathcal{Y})$ consists of the elements $\ni_{\omega}$ (see* (5.6)–(5.7)), *such that*

$$\ell_{\mathcal{Y}}^{[p]}\omega = 0. \tag{5.14}$$

In other words,

$$H_{\mathcal{C}}^{p,0}(\mathcal{Y}) = \ker \ell_{\mathcal{Y}}^{[p]}. \tag{5.15}$$

From here until the end of the paper, we confine ourselves with the case of determined systems of evolution equations in two independent variables, x and t, i.e. with the equations of the form

$$\begin{aligned} v_t^1 &= f^1(x, v^1, \ldots, v^s, \ldots, v_k^1, \ldots, v_k^s), \\ &\cdots \\ v_t^s &= f^s(x, v^1, \ldots, v^s, \ldots, v_k^1, \ldots, v_k^s), \end{aligned} \tag{5.16}$$

where x, t stand for x_1 and x_2, respectively, and v_i^{α} denotes $v_{(0,i)}^{\alpha}$.

In this case $\Lambda^{*,0}(\mathcal{Y})$, as an $\mathcal{F}(\mathcal{Y})$-algebra, is generated by the forms

$$\omega_r^{\alpha} = L_{\mathcal{C}}(v_r^{\alpha}) = \mathrm{d}v_r^{\alpha} - v_{r+1}^{\alpha}\mathrm{d}x - D_x^r(f^{\alpha})\,\mathrm{d}t, \tag{5.17}$$

where

$$D_x = \frac{\partial}{\partial x} + \sum_{\alpha,i} v_{i+1}^{\alpha} \frac{\partial}{\partial v_i^{\alpha}}.$$

Since we are interested in symmetries and recursion operators, we reformulate Theorem 5.6 as

THEOREM 5.7. *Let $\mathcal{Y} = (\mathcal{F}(\mathcal{Y}), \mathcal{C})$ be an equation of the form* (5.16). *Then:*

(i) *Any symmetry $X \in \operatorname{sym}\mathcal{Y}$ is represented as*

$$X = \ni_{\varphi} = \sum_{i,\alpha} D_x^i(\varphi^{\alpha}) \frac{\partial}{\partial v_i^{\alpha}},$$

where $\varphi = (\varphi^1, \ldots, \varphi^s)$, $\varphi^{\alpha} \in \mathcal{F}(\mathcal{Y})$, and

$$\begin{aligned} &D_t(\varphi^1) = \textstyle\sum_{i,\alpha} D_x^i(\varphi^{\alpha}) \frac{\partial f^1}{\partial v_i^{\alpha}}, \\ &\ldots \\ &D_t(\varphi^s) = \textstyle\sum_{i,\alpha} D_x^i(\varphi^{\alpha}) \frac{\partial f^s}{\partial v_i^{\alpha}} \end{aligned} \tag{5.18}$$

(ii) *Any recursion operator $\Omega \in \operatorname{rec}\mathcal{Y}$ is represented as*

$$\Omega = \ni_{\omega} = \sum_{i,\alpha} D_x^i(\omega^{\alpha}) \frac{\partial}{\partial v_i^{\alpha}},$$

where

$$\omega = (\omega^1, \ldots, \omega^s), \qquad \omega^{\alpha} = \sum_{r,\beta} \omega_r^{\beta} \varphi_{\beta}^{\alpha,r}, \quad \varphi_{\beta}^{\alpha,r} \in \mathcal{F}(\mathcal{Y}),$$

ω_r^{β} are given by (5.18), *and*

$$\begin{aligned} &D_t(\omega^1) = \textstyle\sum_{i,\alpha} D_x^i(\omega^{\alpha}) \frac{\partial f^1}{\partial v_i^{\alpha}}, \\ &\ldots \\ &D_s(\omega^s) = \textstyle\sum_{i,\alpha} D_x^i(\omega^{\alpha}) \frac{\partial f^s}{\partial v_i^{\alpha}} \end{aligned} \tag{5.19}$$

(iii) *If $X = \ni_{\varphi}$ is a symmetry and $\Omega = \ni_{\omega}$ is a recursion operator, then $X \rfloor \Omega$ is a symmetry of the form $\ni_{\psi}$, where $\psi = (\psi^1, \ldots, \psi^s)$ and*

$$\psi^{\alpha} = \sum_{r,\beta} D_x^r(\varphi^{\beta}) \varphi_{\beta}^{\alpha,r}, \quad \alpha = 1, \ldots, s. \tag{5.20}$$

The last theorem, in principle, silves the problem of constructing recursion operators for an arbitrary evolution equation. Nevertheless, in practice, the situation is more complicated: if one tries to solve Equations (5.19), the only solutions (with rare exceptions) will be trivial ones (i.e. $\mathcal{O}$ and $U_{\mathcal{C}}$). Nontrivial solutions arise when one passes to specially chosen coverings of the initial equation.

Let $\mathcal{Y}$ be an equation of the form (5.20) and $S \subset H_h^{0,1}(\mathcal{Y})$ be a subspace in the first horizontal cohomology module of $\mathcal{Y}$ (see Theorem 4.11). Note that in this case, elements of $H_h^{0,1}(\mathcal{Y})$ coincide with the classes of equivalent conservation laws for $\mathcal{Y}$. Consider a covering $\mathcal{Y}_S = \mathcal{K}_S(\mathcal{Y})$ with new (nonlocal) variables $\{w^\alpha\}$ corresponding to some forms $\gamma^1, \dots, \gamma^\alpha, \dots$ which determine a basis in S. Denote by D_x^S and D_t^S extensions of the total derivatives onto $\mathcal{K}_S(\mathcal{Y})$.

THEOREM 5.8. *A derivation Ω lies in $H^{p,0}_{\mathcal{KC}}(\mathcal{Y}_S)$ if and only if it is of the form*

$$\Omega = \ni_\omega^S + \sum_\alpha \rho^\alpha \frac{\partial}{\partial w^\alpha},$$

where $\omega, \rho^\alpha \in \Lambda^{p,0}(\mathcal{Y}_S)$ and

$$\ell_S^{[p]}(\omega) = 0, \tag{5.21}$$

$$\ni_\omega^S(\gamma^1) = d_h \rho^1,$$

$$\dots$$

$$\ni_\omega^s(\gamma^\alpha) = d_h \rho^\alpha, \tag{5.22}$$

$$\dots$$

Here the operators $\ni_\omega^S$ and $\ell_S^{[p]}$ are constructed from $\ni_\omega$ and $\ell_{\mathcal{Y}}^{[p]}$ by changing D_x and D_t to D_x^S and D_t^S, respectively.

We call the solutions of (5.22) *the p-shadows* of elements from $H^{p,0}_{\mathcal{KC}}(\mathcal{Y}_S)$. Our last theorem concerns the recovering of real solutions of (5.22) from their shadows and is a generalization of the main result from [3].

THEOREM 5.9. *Let $\mathcal{Y}$ be an equation of the form* (5.16). *Then*:

(i) *A shadow ω extends to a solution of* (5.22) *if and only if $\ni_\omega^s(\gamma^\alpha)$ determine d_h-trivial cohomology classes in $H_h^{p,1}(\mathcal{Y}_S)$.*
(ii) *If $X \in \mathrm{sym}\,\mathcal{Y}_S$ and let Ω be a shadow of a recursion operator. Then $X \rfloor \Omega$ is a shadow of a symmetry.*
(iii) *Let $\mathcal{K}^\infty\mathcal{Y} = \lim_{n\to\infty}\mathcal{K}^n\mathcal{Y}$. Then any shadow of a symmetry can be recovered up to a symmetry in $\mathcal{K}^\infty\mathcal{Y}$. Hence, if X is a symmetry and Ω is a shadow of a recursion operator, then $X \rfloor \Omega$ can be recovered up to a symmetry in $\mathcal{K}^\infty\mathcal{Y}$.*
(iv) *Let $(\mathcal{K}\circ\mathcal{T})^\infty\mathcal{Y} = \lim_{n\to\infty}(\mathcal{K}\circ\mathcal{R})^n\mathcal{Y}$. Then all p-shadows can be recovered up to elements from $H^{p,0}_{(\mathcal{K}\circ\mathcal{T})^\infty\mathcal{C}}((\mathcal{K}\circ\mathcal{T})^\infty\mathcal{Y})$.*

The theorem remains valid if, instead of a graded equation $\mathcal{Y}$, one considers a covering $\mathcal{K}_T(\mathcal{Y})$, $T \subset H_h^{0,1}(\mathcal{Y})$. In other words, it can be applied to any sequence of coverings $\mathcal{Y} = \mathcal{Y}_0, \mathcal{Y}_1, \ldots, \mathcal{Y}_r$, where $\mathcal{Y}_r = \mathcal{K}_{S_r}(\mathcal{Y}_{r-1})$, $S_r \subset H_h^{0,1}(\mathcal{Y}_{r-1})$.

We finish this paper with

EXAMPLE 5.10 (super KdV equation). Consider the super KdV equation

$$\begin{aligned} u_t &= -u_3 + 6uu_1 - 3\vartheta\vartheta_2, \\ \vartheta_t &= -\vartheta_3 + 3\vartheta_1 u + 3\vartheta u_1 \end{aligned} \tag{5.23}$$

(see [11]), where u is the even variable and ϑ is the odd one.

Then the forms

$$\begin{aligned} \gamma_0^0 &= u\mathrm{d}x - (u_2 + 3u^2 - \vartheta\vartheta_1)\mathrm{d}t, \\ \gamma_0^1 &= \vartheta\mathrm{d}x - (\vartheta_2 - 3\vartheta u)\mathrm{d}t \end{aligned}$$

are d_h-closed and define a two-dimensional subspace S_1 in $H_h^{0,1}(\mathcal{Y})$. Let w_0^0 and w_0^1 be corresponding nonlocal variables in $\mathcal{Y}_1 = \mathcal{K}_{s_1}(\mathcal{Y})$. The forms

$$\begin{aligned} \gamma_1^0 &= \vartheta w_0^1\mathrm{d}x - ((\vartheta_2 - 3\vartheta u)w_0^1 - 2\vartheta_1\vartheta)\mathrm{d}t, \\ \gamma_1^1 &= \vartheta w_0^0\mathrm{d}x - ((\vartheta_2 - 3\vartheta u)w_0^0 + \vartheta u_1 - V_1 u)\mathrm{d}t \end{aligned}$$

are d_h-closed in $\mathcal{Y}_1$ and define a two-dimensional subspace S_2 in $H_h^{0,1}(\mathcal{Y}_1)$. Denote by w_1^0 and w_1^1 corresponding nonlocal variables in $\mathcal{Y}_2 = \mathcal{K}_{S_2}(\mathcal{Y}_1)$. We also use the notations

$$\omega_\alpha = L_C u_\alpha, \qquad \theta_\beta = L_C v_\beta$$

and

$$\rho_\beta^\alpha = L_C w_\beta^\alpha.$$

Then solving (5.21), in this case one can see that

$$\begin{aligned} \Omega = {} & \{\omega_2 - 4\omega_0 u - 2\theta v + \theta_0 v_1 + \\ & + \rho_0^1(w_0^1 + w_0^0 v_1 + v_2 - uv) - \rho_1^1 v_1 - 2\rho_0^0 u_1 + \rho_1^0 u_1\}\partial/\partial u + \\ & + \{\theta_2 - 2\theta_0 u - 2\omega_0 v - \rho_0^1(w_0^1 v^1 - w_0^0 u - u_1) - \\ & - \rho_0^0 v_1 + \rho_1^0 v_1 - \rho_1^1 u\}\partial/\partial V \end{aligned} \tag{5.24}$$

is a shadow of a recursion operator in $\mathcal{Y}_2$.

If

$$X = \ni_{\varphi,\psi} = \sum_{\alpha,\beta}\left(D_x^\alpha(\varphi)\frac{\partial}{\partial u_\alpha} + D_x^\beta(\psi)\frac{\partial}{\partial v_\beta}\right) \tag{5.25}$$

is a shadow of a symmetry, then the action of Ω on X is given by contraction of (5.25) into (5.24). A more conventional expression of the action of recursion operator (5.24) on symmetries of (5.23) can be deduced from the equalities

$$\begin{aligned}
&\ni_{\varphi,\psi} \rfloor \omega_\alpha = D_x^\alpha(\varphi), \qquad \ni_{\varphi,\psi} \rfloor \theta_\beta = D_x^\beta(\psi),\\
&\ni_{\varphi,\psi} \rfloor \rho_0^0 = D_x^{-1}(\varphi), \qquad \ni_{\varphi,\psi} \rfloor \rho_0^1 = D_x^{-1}(\psi),\\
&\ni_{\varphi,\psi} \rfloor \rho_1^0 = D_x^{-1}(\psi w_0^1 + v D_x^{-1}(\psi)),\\
&\ni_{\varphi,\psi} \rfloor \rho_1^1 = D_x^{-1}(D_x^{-1}(\varphi) v + v_0^0 \psi)
\end{aligned}$$

These expressions immediately follow from definitions of the variables u_α, v_β, w_β^α.

Direct computations [2] show that in the above defined setting, Equation (5.23) possesses five symmetries with the shadow

$$\begin{aligned}
Y_{\frac{1}{2}} &= v_1 \frac{\partial}{\partial u} + u \frac{\partial}{\partial v},\\
\overline{Y}_{\frac{1}{2}} &= (2w_0^1 u_1 - w_0^0 v_1 + uv - v_2)\frac{\partial}{\partial u} + (2w_0^1 v_1 - w_0^0 u + u_1)\frac{\partial}{\partial v},\\
X_1 &= u_1 \frac{\partial}{\partial u} + V_1 \frac{\partial}{\partial v},\\
\overline{X}_1 &= w_0^1 v_1 \frac{\partial}{\partial u} + (w_0^1 u - v_1)\frac{\partial}{\partial v},\\
v_0 &= (-3u_3 + (x + 18tu)u_1 + 2u - 9tvv_2)\frac{\partial}{\partial u} +\\
&\quad + \left(-3tv_3 + (x + 9tu)v_1 + \frac{3}{2}v + 9tvu_1\right)\frac{\partial}{\partial v}.
\end{aligned} \tag{5.26}$$

Applying recursion operator (5.24) to (5.26), one gets five infinite series of symmetries

$$Y_{n+\frac{1}{2}},\ \overline{Y}_{n+\frac{3}{2}},\ X_{n+1},\ \overline{X}_{n+1},\ V_n, \quad n = 0, 1, \dots,$$

for the super KdV equation.

Acknowledgements

The first author (I. S. K.) expresses his gratitude to the Faculty of Applied Mathematics, University of Twente, and to NWO (the Netherlands) for the financial support during his stay at the University of Twente in February–July 1993.

References

1. Gerstenhaber, M. and Schack, S.D.: Algebraic cohomology and deformation theory, in M. Hazewinkel and M. Gerstenhaber (eds), *Deformation Theory of Algebras and Structures and Applications*, Kluwer, Dordrecht, 1988, pp. 11–264.

2. Kersten, P. H. M.: Higher order symmetries and fermionic conservation laws of the supersymmetric extension of the KdV equation, *Phys. Lett. A* **134**(1) (1988), 25–30.
3. Khor'kova, N. G.: Conservation laws and nonlocal symmetries, *Mat. Zametki* **44** (1988), 134–144.
4. Krasil'shchik, I. S.: Some new cohomological invariants for nonlinear differential equations, *Differential Geom. Appl.* **2** (1992), 307–350.
5. Krasil'shchik, I. S.: Supercanonical algebras and Schouten brackets, *Mat. Zametki* **49** (1991), 70–76 (in Russian).
6. Krasil'shchik, I. S., Lychagin, V. V., and Vinogradov, A. M.: *Geometry of Jet Spaces and Nonlinear Partial Differential Equations*, Gordon and Breach, New York, 1986.
7. Krasil'shchik, I. S. and Kersten, P. H. M.: Deformations of differential equations and recursion operators, in A. Prastaro and Th. M. Rassias (eds), *Geometry in Partial Differential Equations*, World Scientific, Singapore, 1993.
8. Krasil'shchik, I. S. and Vinogradov, A. M.: Nonlocal trends in the geometry of differential equations: Symmetries, conservation laws, and Bäcklund transformations, *Acta Appl. Math.* **15** (1989), 161–209.
9. Lecomte, P. A. B., Michor, P. W., and Schiketanz, H.: The multigraded Nijenhuis–Richardson algebra, its universal property and applications, Preprint, April 1990.
10. Magri, F.: A simple model of the integrable Hamiltonian equation, *J. Math. Phys.* **19** (1978), 1156–1163.
11. Mathieu, P.: Supersymmetric extension of the Korteweg–de Vries equation, *J. Math. Phys.* **29** (1988), 2499.
12. Olver, P. J.: *Applications of Lie Groups to Differential Equations*, Graduate Texts in Mathematics 107, Springer-Verlag, New York, 1986.
13. Tsujishita, T.: Homological method of computing invariants of differential equations, *Differential Geom. Appl.* **1** (1991), 3–34.
14. Vinogradov, A. M.: The $\mathcal{C}$-spectral sequence, Lagrangian formalism, and conservation laws, *J. Math. Anal. Appl.* **100** (1984), 2–129.
15. Vinogradov, A. M.: *Category of Nonlinear Differential Equations*, Lecture Notes Math., Vol. 1108, Springer-Verlag, Berlin, 1984, pp. 77–102.
16. Vinogradov, A. M.: The logic algebra of linear differential operators, *Soviet Math. Dokl.* **13**(4) (1972), 1058–1062.
17. Vinogradov, A. M.: Unification of Schouten–Nijenhuis and Frölicher–Nijenhuis brackets, cohomology, and superdifferential operators, *Mat. Zametki* **47** (1990), 138–140 (in Russian).
18. Vinogradov, M. M.: The main functors of differential calculus in graded algebras, *Uspekhi Mat. Nauk* **44**(3) (1989), 151–152 (in Russian).

Acta Applicandae Mathematicae **41:** 193–226, 1995.

Colour Calculus and Colour Quantizations

V. LYCHAGIN
Sophus Lie Center, PB 546, 119618 Moscow, Russia and Center for Advanced Study at the Norwegian Academy of Science and Letters, PO Box 7606, Skillebekk 0205 Oslo, Norway

(Received: 28 February 1994)

Abstract. A colour calculus linked with an any discrete group G is developed. Colour differential operators and colour jets are introduced. Algebras colour differential forms and de Rham complexes are constructed. For colour differential equations, Spencer complexes are constructed. Relations between colour commutative algebras and quantizations of usual algebras are considered.

Mathematics Subject Classifications (1991): 18D20, 35A99, 58G99, 81R50, 58F07, 58A50, 35Q53.

Key words: monoidal categories, symmetries, colours, colour commutative algebras, colour differential forms and de Rham complexes, colour differential operators and equations, colour Spencer complexes and quantizations.

0. Introduction

Presently there exist a number of different approaches to the construction of a calculus: the universal construction for associative algebras [3, 6, 9], fermionic and colour calculus [2, 8, 10, 20] the calculus for quadratic algebras [17, 21], etc.

This paper is intended as an attempt to illustrate the general scheme [12, 13] of braided calculus on the example of colour calculus. The last one is a quantization of usual calculus, determined by some discrete group G of inner symmetries. We restrict ourself to colour calculus over groups only, but it is clear that the constructions may be carried over to the context of Hopf algebras also.

For any Hopf algebra which is not necessarily quasitriangular, Drinfeld [5] suggests to consider the quantum double – a new quasitriangular Hopf algebra. Our discussion in Section 1 shows a naturality of this construction as a basis for the colour calculus.

Briefly speaking, colour structures related to braidings on the Drinfeld quantum double of a group algebra and a colour is a 'diagonalizable' solution of the Yang–Baxter equation. A. B. Sletsjøe [19] recently found another description of colours in terms of twisted Hochschild complexes.

We start with an analysis of the notion of derivations and yield a structure which we will call a *colour*, arising from a natural requirement to get a reasonable analog of our usual calculus.

In Section 1, we introduce the important notion of colour commutative algebra and show that generalized (or colour) derivations of such algebras yield a colour Lie algebra structure. The notion of colour is linked with some discrete group G. For the case $G = \{e\}$ we get the usual commutative algebras and for the case $G = \mathbb{Z}_2$ we get the notion of super commutative algebra.

More important examples of colour commutative algebras appear from projective (or twisting) group algebras and crossed products. The crossed products are a natural generalization of group algebras. For example, all finite-dimensional division algebras are crossed products. It means that a calculus over these algebras is also a colour calculus.

Other examples are given by the Galois theory of rings where crossed products and skew group algebras may be considered as base objects. Therefore, a calculus in the Galois theory is based on a colour calculus also.

In Section 2, we introduce colour differential operators. We show that such objects as modules of colour differential forms and colour jets may be considered as representative objects for functors of colour derivations and colour differential operators. Our construction of colour de Rham complexes over colour commutative algebra A is based on two assumptions:

(1) An algebra $\Omega^*(A)$ of colour differential forms should be an extension of A to a new colour commutative algebra over the group $\bar{G} = \mathbb{Z} \times G$ generated by A and $\Omega^1(A)$.
(2) A colour de Rham differential d should be a colour derivation of the colour commutative algebra $\Omega^*(A)$, such that $\mathrm{d}^2 = 0$.

These conditions determine the pair $(\Omega^*(A), \mathrm{d})$ up to some group homomorphism of G, where G is the base group.

Note that in the particular case of our usual calculus $G = \{e\}$ the algebra $(\Omega^*(A), \mathrm{d})$ is a unique, but in the super case $G = \mathbb{Z}_2$, we get two equally possible constructions of super de Rham complexes.

One can continue the procedure and produce a chain of colour commutative algebras: $(\Omega^*(A), \Omega^*(\Omega^*(A))$, and so on. Their colour cohomologies determine a new chain of colour commutative algebras. Remark that our usual calculus actually is the second one and based on calculus over $\Omega^*(A)$ (cf. Lie derivations, Schouten and Nijenhuis brackets, etc.).

In Section 3, we put colour calculus within the framework of category theory. We show that colours are actually symmetries in the special monoidal category. We describe all quantizations in the monoidal category in the sense [12–15] and show that colour commutative algebras are quantizations of usual ones. In particular, it yields isomorphisms of algebras of differential operators. This result was noted by Ogievetsky [18] for algebras of quantum hyperplanes.

1. Derivations

In this section, we analyze some structures linked with the notion of derivations.

All algebras and rings are assumed to be associative with unit $1 \neq 0$ and all modules are unital.

1.1. Associativity

Let A be an associative algebra over a commutative ring k and let λ: $A \to A$ be a k-linear map.

We start with the following preliminary naive definition of derivations.

A k-linear map X_λ: $A \to A$ will be called a λ-*derivation* if the following version of *Leibniz rule* holds:

$$X_\lambda(ab) = X_\lambda(a)b + \lambda(a)X_\lambda(b),$$

where $a, b \in A$.

The associtivity law for A produces the following condition on λ. We compare

$$X_\lambda(a(bc)) = X_\lambda(a)(bc) + \lambda(a)(X_\lambda(b)c + \lambda(b)X_\lambda(c)),$$

and

$$X_\lambda((ab)c) = (X_\lambda(a)b + \lambda(a)X_\lambda(b))c + \lambda(ab)X_\lambda(c).$$

This gives

$$\lambda(ab)X_\lambda(c) = \lambda(a)\lambda(b)X_\lambda(c) \quad \text{for all } a, b, c \in A.$$

We will suggest that $X_\lambda(c)$ are arbitrary elements of arbitrary algebras or, equivalently, that our future definition of derivations should work in arbitrary algebras. Therefore, we should assume that λ is an algebra endomorphism.

Moreover, we will *require* λ to be an automorphism of A. In this case the λ-Leibniz rule is compatible with the associativity law.

EXAMPLES

(1) Let $A = C^\infty(M)$ be an algebra of smooth functions on manifold M. It is well known that derivations of A with $\lambda = \mathrm{id}$, are vector fields on M.

(2) Let $A = C^\infty(M)$ and α: $M \to M$ be a diffeomorphism of M. Denote by X_α: $A \to A$ the following 'difference' operator: $X_\alpha(f) = \alpha^*(f) - f$.

Then,

$$X_\alpha(fg) = \alpha^*(fg) - fg = (\alpha^*(f) - f)g + \alpha^*(f)(\alpha^*(g) - g)$$

and, therefore, X_α is an α^*-derivation of A.

1.2. Brackets

Here we amplify our definition of derivations by the requirement of a Lie structure analog.

Let X_α and X_β be two derivations of the algebra corresponding to automorphisms $\alpha, \beta \in \mathrm{Aut}(A)$.

Define a commutator $[X_\alpha, X_\beta]$ in the following way:

$$[X_\alpha, X_\beta] := X_\alpha \circ X_\beta - S_{\alpha,\beta} \circ X_\beta \circ X_\alpha,$$

for some k-linear map $S_{\alpha,\beta} : A \longrightarrow A$.

If one requires the commutator to be a derivation corresponding to an automorphism $\sigma \in \mathrm{Aut}(A)$, then

$$\begin{aligned}
&[X_\alpha, X_\beta](ab) \\
&\quad = X_\alpha X_\beta(a)b + \alpha(X_\beta(\overset{(4)}{a}))X_\alpha(b) + \\
&\qquad + X_\alpha(\beta(\overset{(5)}{a}))X_\beta(b) + \alpha\beta(a)\overset{(1)}{X_\alpha}X_\beta(b) - \\
&\qquad - S_{\alpha,\beta}(\overset{(3)}{X_\beta}X_\alpha(a)b) - S_{\alpha,\beta}(\beta(\overset{(5)}{X_\alpha}(a))X_\beta(b)) - \\
&\qquad - S_{\alpha,\beta}(X_\beta(\overset{(4)}{\alpha}(a))X_\alpha(b)) - S_{\alpha,\beta}(\beta\alpha(\overset{(2)}{a})X_\beta X_\alpha(b)),
\end{aligned}$$

and, on the other hand, by the definition of derivations, we get

$$\begin{aligned}
&[X_\alpha, X_\beta](ab) \\
&\quad = [X_\alpha, X_\beta](a)b + \sigma(a)[X_\alpha, X_\beta](b) \\
&\quad = X_\alpha X_\beta(a)b - S_{\alpha,\beta}(\overset{(3)}{X_\alpha}X_\beta(a))b + \\
&\qquad + \sigma(a)\overset{(1)}{X_\alpha}X_\beta(b) - \sigma(a)S_{\alpha,\beta}\overset{(2)}{(}X_\beta X_\alpha(b)) .
\end{aligned}$$

Comparing terms with mark (1), we get $\sigma = \alpha \circ \beta$. From terms with marks (2) and (3) one has:

$$S_{\alpha,\beta}(a) = s_{\alpha,\beta} \cdot a, \tag{1}$$

where $s_{\alpha,\beta} = S_{\alpha,\beta}(1) \in A$ and

$$s_{\alpha,\beta} \cdot \beta\alpha(a) = \alpha\beta(a) \cdot s_{\alpha,\beta}. \tag{2}$$

for all $a \in A$.

The terms with mark (4) and (5) give

$$\alpha \circ X_\beta = S_{\alpha,\beta} \circ X_\beta \circ \alpha, \tag{3}$$

and

$$X_\alpha \circ \beta = S_{\alpha,\beta} \circ \beta \circ X_\alpha. \tag{4}$$

Therefore, we should assume that

$$s_{\alpha,\beta} \cdot s_{\beta,\alpha} = 1.$$

Then relation (3) is a consequence of (4).

Now we take a derivation X_γ for some $\gamma \in \operatorname{Aut}(A)$ and consider the composition $\alpha \circ X_\beta \circ \gamma$. By formulae (3) and (4), we get the following action on the elements $s_{\beta,\gamma} \in A$:

$$\alpha(s_{\beta,\gamma}) = s_{\alpha,\beta} \cdot s_{\beta,\alpha\gamma}. \tag{5}$$

In particular, it follows that

$$s_{\alpha\beta,\gamma} = s_{\alpha,\beta} \cdot s_{\beta,\alpha\gamma} \cdot s_{\alpha,\gamma}. \tag{6}$$

EXAMPLE. Let X_α and X_β be the derivations corresponding to diffeomorphisms α: $M \to M$ and β: $M \to M$ (see, Example 1.1(2)). To introduce a commutator $[X_\alpha, X_\beta]$, one needs to enlarge the algebra $A = C^\infty(M)$ by means of an element ϑ with new commutative relation: $\vartheta \cdot f = [\alpha^*, \beta^*](f) \cdot \vartheta$, where $f \in A$ and $[\alpha^*, \beta^*]$ is a commutator of automorphisms α^*, β^*.

Actually it means that we should extend our initial algebra A to colour commutative algebra $A[G]$ (see 1.6. below).

1.3. JACOBI IDENTITY

Here we yield new properties of $S_{\alpha,\beta}$ by assumption of the following form of Jacobi identity:

$$[X_\alpha, [X_\beta, X_\gamma]] = [[X_\alpha, X_\beta], X_\gamma] + Z_{\alpha,\beta,\gamma} \circ [X_\beta, [X_\alpha, X\gamma]],$$

for some k-linear map $Z_{\alpha,\beta,\gamma}$: $A \to A$.

It follows by the same method as above that we should assume that

$$Z_{\alpha,\beta,\gamma} = S_{\alpha,\beta}, \tag{1}$$

$$X_\alpha(s_{\beta,\gamma}) = 0, \tag{2}$$

and the 'multiplicative' Jacobi identity

$$s_{\alpha\beta,\gamma} \cdot s_{\gamma\alpha,\beta} \cdot s_{\beta\gamma,\alpha} = 1 \tag{3}$$

hold.

Then the bracket $[X_\alpha, X_\beta]$ is an $\alpha\beta$-derivation and we get Jacobi and skew-symmetry properties in the following form:

(1) Jacobi identity:

$$[X_\alpha,[X_\beta,X_\gamma]] = [[X_\alpha,X_\beta],X_\gamma] + s_{\alpha,\beta}\cdot[X_\beta,[X_\alpha,X_\gamma]],$$

(2) skew-symmetry:

$$[X_\alpha,X_\beta] = -s_{\alpha,\beta}\cdot[X_\beta,X_\alpha].$$

Remark. Denote by $\mathrm{Ad}\,_\alpha(\beta) = {}^\alpha\beta = \alpha\cdot\beta\cdot\alpha^{-1}$ – the adjoint action of G. Then applying the map $(\ ) \mapsto \gamma\circ(\ )\circ\gamma^{-1}$ to the Jacobi identity, we get the following property of s: $\gamma(s_{\alpha,\beta}) = s_{^\gamma\alpha,^\gamma\beta}$.

1.4. A MODULE STRUCTURE

In this section we get an additional structure in the algebra by assuming that the set of all derivations has an A-module structure.

Let X_λ be a λ-derivation and let $a\in A$ be an element such that $a\cdot X_\lambda$ is a ν-derivation for an automorphism $\nu=\nu(a,\lambda)$.

In this case, we have

$$\begin{aligned}(a\cdot X_\lambda)(xy) &= a\cdot(X_\lambda(x)y+\lambda(x)X_\lambda(y))\\ &= (aX_\lambda)(x)y+\nu(x)(aX_\lambda)(y),\end{aligned}$$

for all $x,y\in A$.

Therefore, we should require that $a\cdot\lambda(x)=\nu(x)\cdot a$ for all elements $x\in A$.

To do this, we introduce a new version of the commutative law. We say that an element $a\in A$ is a '*simple*' if there exist an automorphism $\sigma=\sigma_a\in\mathrm{Aut}\,(A)$ such that

$$a\cdot x = \sigma_a(x)\cdot a, \tag{1}$$

for all elements $x\in A$.

Denote by $A_\alpha\subset A$ the k-module of all 'simple' elements corresponding to the given automorphism α:

$$A_\alpha = \{a\in A \mid ax=\alpha(x)a,\ \forall x\in A\}.$$

Then $A_* = \sum_\alpha A_\alpha$ is a graded algebra: $A_\alpha A_\beta\subset A_{\alpha\beta}$ and a subalgebra of A.

Let's assume that $G=\{\alpha\in\mathrm{Aut}\,(A)\mid A_\alpha\neq 0\}$ is a group and that $A = A_*$.

Then A becomes a G-graded algebra with the given action of group G such that $\alpha(A_\beta)\subset A_{\alpha\beta\alpha^{-1}}$.

DEFINITION. Let G be a multiplicative group with identity element $e\in G$. We say that G-graded G-algebra $A=\sum_{\alpha\in G}A_\alpha$ is a *graded commutative G-algebra* if

(1) A is a G-graded algebra: $A_\alpha A_\beta\subset A_{\alpha\beta}$,

(2) there is a G-action on A such that: $\alpha(A_\beta) \subset A_{\alpha\beta\alpha^{-1}}$,
(3) A is a graded commutative algebra: $a_\alpha \cdot x = \alpha(x) \cdot a_\alpha$,

for all $x \in A,\ a_\alpha \in A_\alpha;\ \alpha, \beta \in G$.

1.5. Consider derivations of a graded commutative algebra A. It is natural to assume that derivations are compatible with the graded structure: $X_\lambda(A_\alpha) \subset A_{\Lambda(\alpha)}$ for some map $\Lambda\colon G \to G$.

Comparing the degrees of terms incoming in the Leibniz rule

$$X_\lambda(a_\alpha \cdot a_\beta) = X_\lambda(a_\alpha)a_\beta + \lambda(a_\alpha)X_\lambda(a_\beta),$$

we get $\Lambda(\alpha\beta) = \Lambda(\alpha)\beta$ and, therefore, $\Lambda(\alpha) = \hat{\lambda} \cdot \alpha$, for some automorphism $\hat{\lambda} \in G$.

Throughout this paper, we will identify Λ and $\hat{\lambda}$.

In this case, the compatibility conditions take the form $X_\lambda(A_\alpha) \subset A_{\lambda\alpha}$.

Remark. Let H be a (discrete) subgroup of Aut(A). We restrict ourself to derivations X_λ such that $\lambda \in H$.

Then

(1) The Leibniz rule determines a group homomorphism $\varkappa\colon H \to G$, where $\varkappa(\lambda) = \hat{\lambda}$.
(2) The module structure $a_\alpha \cdot X_\lambda = X_\nu,\ \nu = \alpha \cdot \lambda$ determines an embedding $G \subset H$ such that $\varkappa|_G = \mathrm{id}$.
(3) Bracket conditions (1.2) yield $[H, H] \subset G$, where $[H, H]$ is the commutator group generated by the commutators $[x, y] = xyx^{-1}y^{-1}$.

Moreover, it follows that

(1) $K = \ker \varkappa \subset H$ is an Abelian group,
(2) $G \cap K = 1,\ [G, K] = 1$,

and, hence, H is a direct product of groups: $H = K \times G$.

1.6. Let X_λ be a derivation of a graded commutative G-algebra A.

Applying the Leibniz rule to products $a_\alpha x = \alpha(x)a_\alpha$, where $\alpha \in G$, $a_\alpha \in A_\alpha$ and $x \in A$, we get two terms:

$$X_\lambda(a_\alpha)x + \lambda(a_\alpha)X_\lambda(x)$$

and

$$X_\lambda(\alpha(x))a_\alpha + \lambda(\alpha(x))X_\lambda(a_\alpha).$$

But $X_\lambda(a_\alpha)x = \lambda(\alpha(x))X_\lambda(a_\alpha)$, because A is a graded commutative G-algebra and $X_\lambda(a_\alpha) \in A_{\lambda\alpha}$.

Hence, $\lambda(a_\alpha)X_\lambda(x) = X_\lambda(\alpha(x))a_\alpha$, and by using formulae 1.2(4) and 1.2(1) we get

$$\lambda(a_\alpha) = s_{\lambda,\alpha} \cdot a_\alpha, \tag{1}$$

for all $\lambda\alpha \in G, a_\alpha \in A_\alpha$.

We should remark that formula 1.2(5) means that (1) determines a G-action on A and $s_{\lambda,\alpha} \in A_{[\lambda,\alpha]}$.

Remark. The graded commutative property

$$a_\alpha \cdot a_\beta = s_{\alpha,\beta} \cdot a_\beta \cdot a_\alpha$$

gives the *hexagon equation* on s.

Indeed, if we consider the product $a_\alpha \cdot a_\beta \cdot a_\gamma$ in two different ways we get:

$$\begin{array}{ccccc} (a_\alpha \cdot a_\beta)a_\gamma & = & s_{\alpha\beta,\gamma}a_\gamma(a_\alpha a_\beta) & = & s_{\alpha\beta,\gamma}a_\gamma a_\alpha a_\beta \\ \| & & & & \| \\ a_\alpha(a_\beta a_\gamma) & = & (a_\alpha(s_{\beta,\gamma}a_\gamma))a_\beta & = & s_{\alpha,\gamma\beta}s_\beta(\gamma)a_\gamma a_\alpha a_\beta. \end{array}$$

Hence, we should require that the following form of hexagon equation holds:

$$s_{\alpha\beta,\gamma} = s_{\alpha,\beta\gamma} \cdot s_{\beta,\gamma}. \tag{2}$$

1.7. Starting from this point we will formalize the above constructions.

Let G be a discrete multiplicative group and $A = \sum_{g\in G} A_g$ be a G-graded algebra. Denote by $U(A)$ the group of units of algebra A.

DEFINITION. A map $s\colon\ G \times G \to U(A)$, where $s\colon\ \alpha \times \beta \mapsto s_{\alpha,\beta} \in A_{[\alpha,\beta]}$, and $s_{e,\alpha} = 1$, for all $\alpha \in G$, is called a *colour on the group* G if the following properties hold:

(1) *multiplicative skew symmetry:* $s_{\alpha,\beta} \cdot s_{\beta,\alpha} = 1$,
(2) the *hexagon* equation: $s_{\alpha\beta,\gamma} = s_{\alpha,\beta\gamma} \cdot s_{\beta,\gamma}$,
(3) the *compatibility* of colour with G-action: $s_{\alpha\beta,\gamma} = s_{\alpha,\beta} \cdot s_{\beta,\alpha\gamma} \cdot s_{\alpha,\gamma}$. In this case, the colour s determines an G-action:

$$\alpha(a_\beta) = s_{\alpha,\beta} \cdot a_\beta, \tag{1}$$

such that

$$\gamma(s_{\alpha,\beta}) = s_{\gamma,\alpha} \cdot s_{\alpha,\gamma\beta}, \tag{2}$$

for all $\alpha, \beta, \gamma \in G; a_\alpha \in A_\alpha$.

Remark. Condition (2) is equivalent to the multiplicative Jacobi identity.

EXAMPLES OF COLOURS

(1) Let G be a commutative group. Then $s\colon G\times G \to U(k) \subset A_e$, and conditions (1) and (2) mean that $s_{\alpha,\beta} \cdot s_{\beta,\alpha} = 1$ and s is a group bihomomorphism.
(2) Let $G = \mathbb{Z}$. We have two possible colours: $s_{\alpha,\beta} \equiv 1$ – *a trivial colour*, and $s_{\alpha,\beta} = (-1)^{\alpha\beta}$ – *a standard colour.*
Note that the trivial colour is the base of our usual calculus and the standard one is a base for super-calculus.
(3) Let $G = \mathbb{Z}^n$. All colours take the following forms:

$$s_{\alpha,\beta} = \prod_{1\leqslant i,j\leqslant n} a_{ij}^{(\alpha_i\beta_j-\alpha_j\beta_i)} \cdot \prod_{1\leqslant k\leqslant n} b_k^{\alpha_k\beta_k}, \tag{1}$$

for all $\alpha = (\alpha_1, \ldots, \alpha_n)$, $\beta = (\beta_1, \ldots, \beta_n) \in \mathbb{Z}^n$, and for some elements $a_{ij} \in k$, and $b_i = \pm 1$.
(4) Let G be an Abelian finite generated group, $G = \mathbb{Z}^n/K$, where $K \subset \mathbb{Z}^n$ is a subgroup with generators $\lambda_1 = (\lambda_{11}, \ldots, \lambda_{1n}), \ldots, \lambda_r = (\lambda_{r1}, \ldots, \lambda_{rn})$. Then formula (1) gives a colour on G if $s_{\alpha,\beta} = 1$, for all $\alpha \in K$, $\beta \in \mathbb{Z}^n$, or equivalently, if $\prod_i a_{ij}^{\lambda_{ki}} = 1$, for all $i, j = 1, \ldots, n$, $k = 1, \ldots, r$.

1.8. DEFINITION. Let A be a G-graded algebra and $s\colon G \times G \to U(A)$ be a colour. We say that A is a *colour commutative algebra* if A is a commutative graded G-algebra with respect to the action (1.7(1)).

EXAMPLES

(1) *Group algebras.* Let $k[G]$ be a group algebra of a multiplicative group G with the standard basis $\{\varepsilon_\alpha,\ \alpha \in G\}$. Then $A = k[G]$ is a colour commutative algebra with respect to the grading: $A_\alpha = k\cdot\varepsilon_\alpha$ and the colour: $s_{\alpha,\beta} = \varepsilon_{[\alpha,\beta]}$.
(2) *Crossed products.* Let G be a finite multiplicative group and k be a commutative G-ring with given action $\nu\colon G \to \mathrm{Aut}\,(k)$.
In some sense, the crossed product of k and G is a generalized group algebra. To define the algebra we consider a free k-module with a basis $\{\varepsilon_\alpha,\ \alpha \in G\}$, and determine a multiplication as follows:

$$\varepsilon_\alpha \cdot \varepsilon_\beta = \omega(\alpha, \beta)\varepsilon_{\alpha\beta}, \tag{1}$$

and

$$\varepsilon_\alpha \cdot x = \nu_\alpha(x)\varepsilon_\alpha, \tag{2}$$

where $\omega\colon G \times G \to U(k)$ is a *twisting cocycle*, and $\alpha, \beta \in G$, $x \in k$.
The cocycle conditions on ω we can get from the associativity law for the multiplication:

$$\nu_\alpha(\omega(\beta,\gamma)) \cdot \omega(\alpha, \beta\gamma) \cdot \omega(\alpha\beta, \gamma)^{-1} \cdot \omega(\alpha, \beta)^{-1} = 1. \tag{3}$$

Therefore,

$$\omega \in Z^2_{\text{mult}}(G, k).$$

We should also remark that the scale transformation:

$$\varepsilon_\alpha \mapsto \varepsilon^t_\alpha = t(\alpha)\varepsilon_\alpha,$$

for some function $t\colon G \to U(k)$ determines a new twisting cocycle:

$$\omega^t(\alpha, \beta) = t(\alpha)\nu_\alpha(t(\beta))t^{-1}(\alpha\beta)\omega(\alpha, \beta),$$

or

$$\omega^t = \delta t \cdot \omega,$$

where $\delta t(\alpha, \beta) = t(\alpha)\nu_\alpha(t(\beta))t^{-1}(\alpha\beta)$ is the coboundary of t. Therefore, up to scale automorphisms, properties (1) and (2) together with the cohomology class $[\omega] \in H^2_{\text{mult}}(G, k)$, determine the crossed product algebra $A = k_\omega[G]$. If we consider A as a k^G-algebra and define a grading by $A_\alpha = k \cdot \varepsilon_\alpha$, $\forall \alpha \in G$, and a G-action by

$$\alpha(\varepsilon_\beta) = \frac{\omega(\alpha, \beta)}{\omega({}^\alpha\beta, \alpha)}\varepsilon_{{}^\alpha\beta},$$

we get a colour algebra with colour $s_{\alpha,\beta} = \chi(\alpha, \beta)\varepsilon_{[\alpha,\beta]}$, where the function $\chi\colon G \times G \to k$ is given by the formula

$$\chi(\alpha, \beta) = \frac{\omega(\alpha, \beta)}{\omega({}^\alpha\beta, \alpha)\omega([\alpha, \beta], \alpha)}. \tag{4}$$

(3) We get *projective* or *twisting* group algebras if the action ν is trivial. If the twisting is trivial we get a *skew* group algebra.

(4) Let $A = \sum_{\alpha \in G} A_\alpha$ be a G-graded algebra such that $A_e = k$.
Assume that any A_α contains an invertible element and denote this element by g_α and define a G-action on k as $\alpha(x) = g_\alpha \cdot x \cdot g_\alpha^{-1}$, for any $x \in k = A_e$. Remark that any element $a_\alpha \in A_\alpha$ can be represented as $a_\alpha = g_\alpha \cdot x_\alpha$, with $x_\alpha \in k$. It is show that action (1) is independent of the choice of g_α. If we define a twisting 2-cocycle ω by the formula

$$\omega(\alpha, \beta) = g_\alpha \cdot g_\beta \cdot g_\alpha^{-1} \cdot g_\beta^{-1} \in A_{[\alpha,\beta]},$$

then

$$g_\alpha \cdot g_\beta = \omega(\alpha, \beta)g_\beta \cdot g_\alpha,$$

and we get an algebra isomorphism $\psi\colon k_\omega[G] \to A$, such that $\psi(\varepsilon_\alpha) = g_\alpha$.

(5) Let $A = \sum_{\alpha \in G} A_\alpha$ be a *strongly* G-graded algebra [4] with $A_e = k$. Then, by definition, $A_\alpha \cdot A_\beta = A_{\alpha\beta}$ and we can use the above procedure.

(6) Let $k = \mathbb{R}$ and $G = \mathbb{Z}_2 \oplus \mathbb{Z}_2$.
Denote by $a = (\bar{1}, \bar{0})$ and $b = (\bar{0}, \bar{1})$ the generators in the group and by

$$\omega(\alpha, \beta) = (-1)^{\alpha_1\beta_1 + \alpha_2\beta_2 + \alpha_2\beta_1},$$

a twisting cocycle.
As $\mathbb{R}$-vector space the projective group algebra $\mathbb{R}_\omega[G]$ is generated by $\varepsilon_0 = 1$, ε_a, ε_b, ε_{a+b} with relations

$$\varepsilon_a^2 = \varepsilon_b^2 = \varepsilon_{a+b}^2 = -1 \quad \text{and} \quad \varepsilon_a\varepsilon_b = \varepsilon_b\varepsilon_a = \varepsilon_{a+b}.$$

Hence, if we define the map ψ by $\varepsilon_a \mapsto \imath$, $\varepsilon_b \mapsto \jmath$, $\varepsilon_{a+b} \mapsto k$, we get an isomorphism between $\mathbb{R}_\omega[G]$ and the quaternion algebra $\mathbb{H}$. Therefore, $\mathbb{H}$ is a colour commutative algebra.

(7) Let $G = (\mathbb{Z}_2)^n$. Identify elements $\alpha = (\alpha_1, \ldots, \alpha_n) \in G$ with sequences $\hat{\alpha}_1 < \cdots < \hat{\alpha}_k$, where $1 \leqslant \alpha_i \leqslant n$ and $\alpha_{\hat{\alpha}_i} = 1$.
We define a twisting 2-cocycle on G as $\omega(\alpha, \beta) = (-1)^{\pi(\alpha,\beta)}$. Here $\pi(\alpha, \beta)$ is a number of inversions in the sequence $(\hat{\alpha}_1, \ldots, \hat{\alpha}_k, \hat{\beta}_1, \ldots, \hat{\beta}_l)$. Then $\mathbb{R}_\omega[G]$ is a Clifford algebra.

(8) Let $k = \mathbb{R}$, $G = \mathbb{Z}_2 \oplus \mathbb{Z}_2$ and $A = \mathrm{Mat}_2(\mathbb{R})$. Similarly to the construction of Example (6), we define a map ψ: $\mathbb{R}_\omega[G] \to \mathrm{Mat}_2(\mathbb{R})$, for some cocycle ω as follows:

$$\varepsilon_a \longmapsto \begin{bmatrix} 0 & 1 \\ 1 & 0 \end{bmatrix}, \qquad \varepsilon_b \longmapsto \begin{bmatrix} 1 & 0 \\ 0 & -1 \end{bmatrix},$$

$$\varepsilon_{a+b} \longmapsto \begin{bmatrix} x0 & -1 \\ 1 & 0 \end{bmatrix}, \qquad \varepsilon_0 \longmapsto \begin{bmatrix} 1 & 0 \\ 0 & 1 \end{bmatrix}.$$

Now we define a cocycle ω in such a way that ψ becomes an algebra morphism. Finally, we get the following conditions on ω:

$$\omega(a, a) = \omega(b, b) = \omega(a, b) = \omega(a, a + b) = -\omega(a + b, a + b) = 1,$$

and $\omega(x, y) = -\omega(y, x)$ if $x \neq y$.
Tensoring of the algebras $\mathrm{Mat}_2(\mathbb{R})$ we get a colour commutative structure on the algebras $\mathrm{Mat}_{2^n}(\mathbb{R})$, for all $n = 1, 2, \ldots$.

(9) Any matrix algebra $A = \mathrm{Mat}_n(\mathbb{C})$ is a colour commutative algebra.
To see this consider matrixes

$$p = \begin{bmatrix} 1 & 0 & \cdots & 0 \\ 0 & \eta & \cdots & 0 \\ \vdots & \vdots & \ddots & \vdots \\ 0 & 0 & \cdots & \eta^{n-1} \end{bmatrix}, \qquad q = \begin{bmatrix} 0 & 0 & \cdots & 0 & 1 \\ \eta & 0 & \cdots & 0 & 0 \\ 0 & \eta^2 & \cdots & 0 & 0 \\ \vdots & \vdots & \ddots & \vdots & \vdots \\ 0 & 0 & \cdots & \eta^{n-1} & 0 \end{bmatrix},$$

where $\eta = \exp(2\pi\imath/n)$. Then, $pq = \eta \cdot qp$, and the set $\{p^a q^b;\ a, b = 0, 1, \dots, n-1\}$ is an $\mathbb{R}$-vector space basis of the algebra $\mathrm{Mat}_n(\mathbb{C})$.
Consider now the group $G = \mathbb{Z}_n \oplus \mathbb{Z}_n$ and the twisting cocycle

$$\omega(\alpha, \beta) = \begin{cases} \eta^{\bar{\alpha}_1 \bar{\beta}_2} & \text{if} \quad \alpha_2 + \beta_2 < n, \\ (-1)^{n-1} \eta^{\bar{\alpha}_1 \bar{\beta}_2} & \text{if} \quad \alpha_2 + \beta_2 \geqslant n, \end{cases}$$

where

$$\alpha = (\bar{\alpha}_1, \bar{\alpha}_2), \qquad \beta = (\bar{\beta}_1, \bar{\beta}_2) \in G, \quad \text{and} \quad \alpha_i, \beta_j = 0, 1, \dots, n-1.$$

An isomorphism ψ between $\mathbb{C}_\omega[\mathbb{Z}_n \oplus \mathbb{Z}_n]$ and $\mathrm{Mat}_n(\mathbb{C})$ is given by the formula: $\psi\colon\ \varepsilon_\alpha \mapsto q^{\alpha_2} \cdot p^{\alpha_1}$ for all $\alpha \in G$.

(10) Let $G = \mathbb{Z}^n$, $k = \mathbb{C}$ and $\Theta = \|\theta_{ij}\| \in \mathrm{Mat}_n(\mathbb{R})$ be a matrix such that $\theta_{ij} = 0$, if $i \leqslant j$. Taking the twisting cocycle $\omega(x, y) = \exp(\pi\imath\langle Ax, y\rangle)$, where

$$x = (x_1, \dots, x_n), \quad y = (y_1, \dots, y_n) \in \mathbb{Z}^n, \quad \langle x, y\rangle = \sum_i x_i y_i,$$

we get the algebra $\mathbb{C}_\omega[\mathbb{Z}^n]$ called a *quantum torus.*

1.9. Now we suggest our main notion.

DEFINITION. A k-linear map $X_\lambda\colon\ A \to A$ is said to be a *colour λ-derivation,* $\lambda \in G$, if

(1) X_λ is a graded derivation: $X_\lambda(A_\alpha) \subset A_{\lambda\alpha}$,
(2) λ-Leibniz rule holds:

$$X_\lambda(a_\alpha \cdot a_\beta) = X_\lambda(a_\alpha)a_\beta + \lambda(a_\alpha)X_\lambda(a_\beta),$$

(3) values $s_{\alpha,\beta}$ of the colour map are constants: $X_\lambda(s_{\alpha,\beta}) = 0$,

where $a_\alpha \in A_\alpha$, $a_\beta \in A_\beta$; $\alpha, \beta \in G$.

Remark. Property (3) of colour derivations is equivalent to the following interaction between the G-action and derivations: $X_\lambda \circ \beta = s_{\lambda,\beta} \cdot \beta \circ X_\lambda$.

Summarizing results 1.1–1.6, we have the following

THEOREM. *Let A be a colour commutative algebra and let* $\mathrm{Der}_\lambda(A)$ *be a k-module of all colour λ-derivations $\lambda \in G$. Then*

(1) *A G-graded k-module* $\mathrm{Der}_*(A) = \sum_{\lambda\in G} \mathrm{Der}_\lambda(A)$ *is a left G-graded A-module*:

$$A_\alpha \cdot \mathrm{Der}_\lambda(A) \subset \mathrm{Der}_{\alpha\lambda}(A), \quad \alpha, \lambda \in G.$$

(2) $\mathrm{Der}_*(A)$ *is a G-module with respect to the action*:

$$\alpha(X_\lambda) = \alpha \circ X_\lambda \circ \alpha^{-1} = s_{\alpha,\lambda} \cdot X_\lambda,$$

and $\alpha(\mathrm{Der}_\lambda(A) \subset \mathrm{Der}_{\alpha\lambda}(A)$.

(3) *The bracket*

$$[X_\alpha, X_\beta] = X_\alpha \circ X_\beta - s_{\alpha,\beta} \cdot X_\beta \circ X_\alpha$$

determines a colour Lie algebra *structure on* $\mathrm{Der}_*(A)$:

$$[\mathrm{Der}_\alpha(A), \mathrm{Der}_\beta(A)] \subset \mathrm{Der}_{\alpha\beta}(A), \tag{1}$$
$$[X_\alpha, X_\beta] = -s_{\alpha,\beta} \cdot [X_\beta, X_\alpha], \tag{2}$$
$$[X_\alpha, [X_\beta, X_\gamma]] = [[X_\alpha, X_\beta], X_\gamma] + s_{\alpha,\beta} \cdot [X_\beta, [X_\alpha, X_\gamma]]. \tag{3}$$

Remark. Note that defining in an obvious way from the theorem colour Lie algebra structures are particular cases of generalized Lie algebras introduced by D. Gurevich [7].

1.10. Now we analyze a notion of derivation with values in A-modules. Let P be an A-module. Before we consider derivations with values in P we should remark that the definition of maps satisfying Leibniz rule requires an $A - A$ bimodule structure on P. Moreover, working with graded algebras we should consider graded modules too.

So we have to start with a notion of graded bimodule over a graded algebra.

Let $A = \sum_{\alpha \in G} A_\alpha$ be a G-graded G-algebra and let $P = \sum_{m \in M} P_m$ be a M-graded k-module where M is an arbitrary grading set.

To define an $A - A$ bimodule structure on P we must require that:

$$A_\alpha \cdot P_m \subset P_{\alpha \cdot m}, \qquad P_m \cdot A_\alpha \subset P_{m \cdot \alpha}, \tag{1}$$

and

$$(A_\alpha \cdot P_m) \cdot A_\beta = A_\alpha \cdot (P_m \cdot A_\beta). \tag{2}$$

Therefore, we need a left and a right G-action on the set M.

DEFINITIONS

(1) A set M with left $G \times M \to M$, $\alpha \times m \mapsto \alpha \cdot m$, and right $M \times G \to M$, $m \times \beta \mapsto m \cdot \beta$, G-actions is called a *G-bispace* if $(\alpha \cdot m) \cdot \beta = \alpha \cdot (m \cdot \beta)$, for all $\alpha, \beta \in G$, $m \in M$.

(2) Let M be a bispace and P be an $A - A$ bimodule. We say that P is a *M-graded* bimodule if conditions (1) and (2) hold.

1.11. It is natural to consider derivations compatible with the grading only. From the beginning, we define derivations X: $A \to P$ with values in M-graded $A - A$ bimodule P as k-linear maps of degree $m \in M$; $X = X_m$: $A_\alpha \to P_{m \cdot \alpha}$ satisfying the following form of Leibniz rule:

$$X_m(a_\alpha \cdot a_\beta) = X_m(a_\alpha)a_\beta + \tilde{\sigma}_m(a_\alpha)X_m(a_\beta).$$

Denote by $\mathrm{Aut}_{\Phi}(A)$ the set of automorphisms of the algebra A over the group automorphism $\Phi\colon G \to G$, that is, the set of automorphisms $\tilde{\sigma}$ of A such that $\tilde{\sigma}(A_\alpha) \subset A_{\Phi(\alpha)}$.

Then, for the automorphism $\tilde{\sigma}_m$ from the Leibniz rule, one has $\tilde{\sigma}_m \in \mathrm{Aut}_{\varphi_m}(A)$, for some automorphism $\varphi_m \in \mathrm{Aut}\, G$.

Comparing the degrees of terms interred in the Leibniz rule, one gets the following interaction between the left and the right G-structures:

$$m \cdot \alpha = \varphi_m(\alpha) \cdot m. \tag{1}$$

Moreover, if, analogously to (1.4), we require an A-module structure in the set of all derivations, we must assume that A is a commutative G-graded G-algebra and

$$\tilde{\sigma}_{\alpha m} = \alpha \circ \tilde{\sigma}_m, \qquad \varphi_{\alpha m} = \mathrm{Ad}_{\alpha} \circ \varphi_m. \tag{2}$$

1.12. Assume now that A is a colour commutative algebra and consider the action of X_m on the products $a_\alpha \cdot a_\beta = \alpha(a_\beta) \cdot a_\alpha$:

$$X_m(a_\alpha \cdot a_\beta) = X_m(a_\alpha) \cdot a_\beta + \tilde{\sigma}_m(a_\alpha) \cdot X_m(a_\beta),$$

and

$$X_m(\alpha(a_\beta) \cdot a_\alpha) = X_m(\alpha(a_\beta)) \cdot a_\alpha + \tilde{\sigma}_m(\alpha(a_\beta)) \cdot X_m(a_\alpha).$$

Comparing the terms with $X_m(a_\alpha)$, one gets the following symmetric properties for the bimodule P: $z_m \cdot a_\beta = \tilde{\sigma}_m(a_\beta) \cdot z_m$, and

$$\tilde{\sigma}_{m\alpha} = \tilde{\sigma}_m \circ \alpha, \qquad \varphi_{m\alpha} = \varphi_m \circ \mathrm{Ad}_{\alpha}. \tag{1}$$

To compare the terms with $X_m(a_\beta)$, we will require the presence of some G-action $G \times P \to P$, such that $\alpha\colon P_m \to P_{\alpha m \alpha^{-1}}$, and the following compatibility of the derivations with the group action: $X_m \circ \alpha = \sigma_m(\alpha) \cdot \alpha \circ X_m$, for some map $\sigma\colon M \times G \to U(A)$, where $\sigma\colon m \times \alpha \mapsto \sigma_m(\alpha) \in A_{\varphi_m(\alpha)\alpha^{-1}}$.

In this case we get

$$\tilde{\sigma}_m(a_\alpha) = \sigma_m(\alpha) \cdot a_\alpha. \tag{2}$$

DEFINITIONS

(1) Let M be a G-bispace and $\varphi\colon M \to \mathrm{Aut}(G)$ a map. We say that M is a *φ-commutative* G-space if relations 1.11(1) and (2), on automorphisms φ_m hold.

(2) Let M be a commutative G-bispace. A map

$$\sigma\colon M \times G \longrightarrow U(A), \qquad \sigma\colon m \times \alpha \longrightarrow \sigma_m(\alpha) \in A_{\varphi_m(\alpha)\alpha^{-1}},$$

we will call *a colour* on the bispace M over a colour $s\colon G \times G \to U(A)$ on the group G if

$$\sigma_m(\alpha\beta) = \sigma_m(\alpha)\alpha(\sigma_m(\beta)), \quad \sigma_m(e) = 1,$$
$$\sigma_{\alpha m} = \alpha(\sigma_m(\beta))s_{\alpha,\beta}$$

for all $\alpha, \beta \in G$, $m \in M$.

In other words, M is a commutative bispace if M is a left G-space considered together with a map φ. In this case, 1.11(1) may be considered as the definition of the right G-action.

1.13. Let M be a commutative G-bispace with a colour σ and let A be a colour commutative G-algebra.

DEFINITION. We say that an M-graded $A - A$ bimodule $P = \sum_{m \in M} P_m$ is a *colour symmetric bimodule* if there is a G-action on P, such that

$$\alpha(x_m) = \sigma_m(\alpha)^{-1} \cdot x_m$$

and

$$x_m \cdot a_\alpha = \sigma_m(\alpha) \cdot a_\alpha \cdot x_m,$$

for all $\alpha \in G$, $a_\alpha \in A_\alpha$, $x_m \in P_m$.

We should remark that any M-graded A-module, where M is a commutative G-bispace with a colour σ, may be considered as a colour symmetric bimodule if the properties from the definition one looks at as the definition of G-action and right multiplication.

1.14. Let E be a colour symmetric $A - A$ bimodule.

DEFINITION. A graded k-linear morphism $X\colon A \to P$ a degree $m \in M$, $X(A_\alpha) \subset P_{m\alpha}$ will be called *a colour derivation* with values in the bimodule P if the following properties hold:

(1) colour Leibniz rule:

$$X(a_\alpha \cdot a_\beta) = X(a_\alpha)a_\beta + \sigma_m(\alpha)a_\alpha X(a_\beta),$$

(2) constancy of colours:

$$X(\sigma_m(\alpha)) = X(s_{\alpha,\beta}) = 0,$$

or

$$X_m \circ \alpha = \sigma_m(\alpha) \cdot \alpha \circ X_m$$

for all $m \in M$; $\alpha, \beta \in G$; $a_\alpha \in A_\alpha$, $a_\beta \in A_\beta$.

1.15. Denote by $\mathrm{Der}_m(A, P)$ the module of all m-degree colour derivations with values in P and by

$$\mathrm{Der}_*(A, P) = \sum_{m \in M} \mathrm{Der}_m(A, P)$$

a M-graded k-module of all colour derivations.

THEOREM. $\mathrm{Der}_*(A, P)$ *is a* M*-graded* A*-module*:

$$A_\alpha \cdot \mathrm{Der}_m(A, P) \subset \mathrm{Der}_{\alpha m}(A, P)$$

for all $\alpha \in G$, $m \in M$.

2. Colour Differential Operators

In this section we define colour differential operators in a category of colour symmetric bimodules and introduce the main functors of colour calculus. We show that such objects as modules of colour differential forms and colour jets may be considered as representative objects of functors differential operators and derivations. We should stress also that the group G in the colour differential calculus are considered in the triple role:

(1) G is a group controlling the commutative laws,
(2) G is a grading group of the base algebra A,
(3) G is an action group.

2.1. We start with a reformulation of the notion of derivation. To do this we consider the Leibniz rule

$$X_m(a_\alpha a_\beta) = X_m(a_\alpha)a_\beta + \sigma_m(\alpha)a_\alpha X_m(a_\beta)$$

in the following form:

$$(X_m \circ l_{a_\alpha} - \sigma_m(\alpha)l_{a_\alpha} \circ X_m)\,(a_\beta) = l_{X_m(a_\alpha)}(a_\beta), \tag{1}$$

where l_a: $A \to A$ is an operator of the left multiplication $l_a(b) = ab$, $\forall a, b \in A$.

Denote by $[X_m, l_{a_\alpha}]$ the operator on the left-hand side of (1). Then the Leibniz rule takes the form: $[X_m, l_{a_\alpha}] = l_{X_m(a_\alpha)}$. We should note that the operator $\tilde{l} = l_{X_m(a_\alpha)}$: $A_\beta \to E_{(m\alpha)\beta}$ has a degree $m\alpha$ and satisfies the relation:

$$[\tilde{l}, l_{a_\beta}] = \tilde{l} \circ l_{a_\beta} - \sigma_{m\alpha}(\beta)l_{a_\beta} \circ \tilde{l} = 0. \tag{2}$$

2.2. To define differential operators, one can use the usual inductive definition of differential operators (see, for example, [11]) with an evident modification of the commutator notion.

To do this, we start with a notion of 0-order differential operators or 'colour' homomorphisms.

Let

$$P = P_M = \sum_{m \in M} P_m, \qquad Q = Q_N = \sum_{n \in N} Q_n$$

be graded modules over the colour algebra A.

Here M and N are G-bispaces.

Let σ: $A \to A$ be a graded automorphism. We say that a k-linear graded morphism f: $P \to Q$ is a σ-morphism if $f(ax) = \sigma(a)f(x)$, for all $a \in A$, $x \in P$.

Below, we will consider automorphisms σ which are generated by a colour $\sigma(a_\beta) = s_{\alpha,\beta}a_\beta$, for some elements $\alpha = \alpha(\sigma) \in G$.

Any σ-morphism f determines a map $\hat{f}$: $M \to N$, such that $f(P_m) \subset Q_{\hat{f}(m)}$. Compare degrees of $f(a_\beta x_m)$ and $\sigma(a_\beta)f(x_m)$ one gets $\hat{f}(\beta m) = {}^\alpha\beta\hat{f}(m)$. Therefore, the colour dictates a special type of G-morphisms between grading sets.

Returning to the definition of colour derivations we see that one needs special types of commutative bispaces and colours.

More precisely, we will require that in the definition of commutative bispace morphism φ has the following form $\varphi_m(\beta) = {}^{\hat{m}}\beta$, for some map $\hat{}$: $M \longrightarrow G$, satisfying the condition analogous 1.11: $\hat{\alpha m} = \alpha\hat{m}$ and $\sigma_m(\alpha) = s_{\hat{m},\alpha}$, for all $\alpha, \beta \in G, m \in M$.

DEFINITIONS

(1) Let M be a G-bispace. We say that M is a *symmetric bispace* if there is a G-bispace map $\hat{}$: $M \to G$ such that $m \cdot \alpha = {}^{\hat{m}}\alpha \cdot m$ for all $m \in M, \alpha \in G$.
(2) Let $\alpha \in G$. A map φ: $M \to N$ of symmetric bispaces will be called an *α-map* if $\varphi(\beta m) = {}^\alpha\beta\varphi(m)$.
(3) A k-morphism f: $P_M \to Q_N$ of colour symmetric bimodules over symmetric bispaces will be called *φ-morphism* if $f(P_m) \subset Q_{\varphi(m)}$, where φ is an α-map, and the following properties hold:

$$f(a_\beta \cdot x_m) = s_{\alpha,\beta} \cdot a_\beta \cdot f(x_m), \tag{1}$$

$$f \circ \beta = s_{\alpha,\beta} \cdot \beta \circ f \tag{2}$$

for all $\beta \in G$, $x_m \in P_m$, $m \in M$.

2.3. Let $\operatorname{Map}_\alpha(M, N)$ be a set of all α-maps and

$$\operatorname{Map}_*(M, N) = \bigcup_{\alpha \in G} \operatorname{Map}_\alpha(M, N).$$

Denote by $\mathrm{Hom}_{\varphi}(P_M, Q_N)$ a set of all φ-morphisms, where $\varphi \in \mathrm{Map}_{\alpha}(M, N)$, and

$$\mathrm{Hom}_{*}(P_M, Q_N) = \sum_{\varphi \in \mathrm{Map}_{*}(M,N)} \mathrm{Hom}_{\varphi}(P_M, Q_N).$$

The set $\mathrm{Map}_{*}(M, N)$ is a symmetric bispace with respect to G-actions

$$(\beta \cdot \varphi)(m) = \beta(\varphi(m)), \qquad (\varphi \cdot \beta)(m) = \varphi(\beta m)$$

and a map $\hat{}$: $\mathrm{Map}_{*}(M, N) \to G$, where $\hat{\varphi} = \alpha$, if $\varphi \in \mathrm{Map}_{\alpha}(M, N)$.

The k-module $\mathrm{Hom}_{*}(P_M, Q_N)$ is a $\mathrm{Map}_{*}(M, N)$-graded module by definition.

Moreover, if one defines the left and right A-structures by

$$(a_\beta \cdot f)(x_m) = a_\beta \cdot (f(x_m)), \qquad (f \cdot a_\beta)(x_m) = f(a_\beta \cdot x_m),$$

where

$$a_\beta \in A_\beta, \quad x_m \in P_m, \quad f \in \mathrm{Hom}_{\varphi}(P_M, Q_N),$$

one gets a symmetric A-module with

$$A_\beta \cdot \mathrm{Hom}_{\varphi}(P_M, Q_N) \subset \mathrm{Hom}_{\beta\varphi}(P_M, Q_N),$$

$$\mathrm{Hom}_{\varphi}(P_M, Q_N) \cdot A_\beta \subset \mathrm{Hom}_{\varphi\beta}(P_M, Q_N),$$

and

$$f \cdot a_\beta = s_{\alpha,\beta} \cdot a_\beta \cdot f,$$

if $\varphi \in \mathrm{Map}_{\alpha}(M, N)$.

Summarizing, we have the following result.

THEOREM. *Let M and N be symmetric bispaces and P_M, Q_N be M and N-graded A-modules, respectively. Then*:

(1) $\mathrm{Map}_{*}(M, N)$ *is a symmetric bispace,*
(2) $\mathrm{Hom}_{*}(P_M, Q_N)$ *is a symmetric $A - A$ bimodule, and*
(3) $f \in \mathrm{Hom}_{\varphi}(P_M, Q_N)$, $\varphi \in \mathrm{Map}_{\alpha}(M, N)$, *and* $g \in \mathrm{Hom}_{\psi}(Q_N, R_K)$, $\psi \in \mathrm{Map}_{\beta}(M, N)$, *then* $\varphi \circ \psi \in \mathrm{Map}_{\alpha\beta}(M, K)$ *and* $g \circ f \in \mathrm{Hom}_{\varphi\psi}(P_M, R_K)$.

2.4. Denote by $\mathrm{Hom}_k^{\varphi}(P_M, Q_N)$ the set of all k-linear morphisms Δ: $P_M \to Q_N$ of degree $\varphi \in \mathrm{Map}_{\alpha}(M, N)$, i.e. $\Delta(P_m) \subset Q_{\varphi(m)}$, and such that $\Delta \circ \beta = s_{\alpha,\beta}\beta \circ \Delta$ for all $\beta \in G$.

For any homogeneous element $a_\beta \in A_\beta$ we denote by l_{a_β}, r_{a_β} and δ_{a_β} the following operators:

$$l_{a_\beta}(\Delta) = a_\beta \cdot \Delta, \qquad r_{a_\beta}(\Delta) = \Delta \cdot a_\beta, \qquad \delta_{a_\beta} = r_{a_\beta}(\Delta) - l_{\alpha(a_\beta)}(\Delta).$$

DEFINITION. The element $\Delta \in \mathrm{Hom}_k^{\varphi}(P_M, Q_N)$ is a *colour homomorphism* if $\delta_{a_\beta}(\Delta) = 0$, for all homogeneous elements $a_\beta \in A_\beta$ and $\beta \in G$.

Denote by $\mathrm{Hom}_A^{\varphi}(P_M, Q_N)$ the set of all colour homomorphisms and define colour differential operators by induction on the order.

DEFINITIONS

(1) Colour differential operators of order 0 are colour homomorphisms.
(2) The element $\Delta \in \mathrm{Hom}_k^{\varphi}(P_M, Q_N)$ is a *colour differential operator* over colour commutative algebra A of order $\leqslant l+1$, $l = 0, 1, \ldots$, and degree φ acting from the colour symmetric module P to the colour symmetric module Q if $\delta_{a_\beta}(\Delta)$ is a colour differential operator of order $\leqslant l$ and degree $\beta \cdot \varphi$ for any homogeneous element $a_\beta \in A_\beta$, $\beta \in G$.

Denote by $\mathrm{Diff}_l^{\varphi}(P_M, Q_N)$ the set of all colour differential operators of order $\leqslant l$ and degree φ acting from P_M to Q_N.

The set of colour differential operators of order $\leqslant l$,

$$\mathrm{Diff}_l(P_M, Q_N) = \sum_{\varphi \in \mathrm{Map}\,(M,N)} \mathrm{Diff}_l^{\varphi}(P_M, Q_N)$$

is stable with respect to left l_{a_β} and right r_{a_β} multiplications in

$$\mathrm{Hom}_k(P_M, Q_N) = \sum_{\varphi \in \mathrm{Map}\,(M,N)} \mathrm{Hom}_k^{\varphi}(P_M, Q_N)$$

and therefore possesses an $A - A$ bimodule structure.

The definition of colour differential operators implies the existence of imbeddings:

$$\mathrm{Diff}_l^{\varphi}(P_M, Q_N) \subset \mathrm{Diff}_{l+1}^{\varphi}(P_M, Q_N).$$

Denote by

$$\mathrm{Diff}_*^{\varphi}(P_M, Q_N) = \bigcup_{l \geqslant 0} \mathrm{Diff}_l^{\varphi}(P_M, Q_N)$$

the k-module of all colour differential operators of degree φ and by

$$\mathrm{Diff}_*(P_M, Q_N) = \sum_{\varphi \in \mathrm{Map}\,(M,N)} \mathrm{Diff}_*^{\varphi}(P_M, Q_N)$$

the filtered $A - A$ bimodule of all colour differential operators.

2.5. Let

$$\Delta_1 \in \mathrm{Diff}_l^{\varphi}(P_M, Q_N), \qquad \Delta_2 \in \mathrm{Diff}_t^{\psi}(Q_N, R_K)$$

be colour differential operators of degrees $\varphi \in \mathrm{Map}_\alpha(M,N)$, $\psi \in \mathrm{Map}_\beta(N,K)$, respectively.

An easy computation shows that for any element $a = a_\gamma \in A_\gamma$ we have the following identity:

$$\delta_a(\Delta_2 \circ \Delta_1) = \Delta_2 \circ \delta_a(\Delta_1) + \delta_{\alpha(a)}(\Delta_2) \circ \Delta_1.$$

This implies the following proposition

PROPOSITION. *The composition $\Delta_2 \circ \Delta_1$ of colour differential operators: $\Delta_1 \in \mathrm{Diff}_l^\varphi(P_M, Q_N)$ and $\Delta_2 \in \mathrm{Diff}_t^\psi(Q_N, R_K)$ is a colour differential operator $\Delta_2 \circ \Delta_1 \in \mathrm{Diff}_{l+t}^{\psi\circ\varphi}(P_M, R_K)$.*

Any morphism $\varphi \in \mathrm{Map}(G, M)$ can be determined by the element $m = \varphi(e) \in M$. We denote by $\mathrm{Diff}_l^m(P_M) = \mathrm{Diff}_l^\varphi(A, P_M)$ the corresponding module of colour differential operators, and

$$\mathrm{Diff}_l(P_M) = \sum_{m\in M} \mathrm{Diff}_l^m(P_M), \qquad \mathrm{Diff}_*(P_M) = \bigcup_{l\geqslant 0} \mathrm{Diff}_l(P_M).$$

The composition of colour differential operators defines an associative algebra structure in $\mathrm{Diff}_*(A)$ and right $\mathrm{Diff}_*(A)$-module structure in the module $\mathrm{Diff}_*(P_M)$.

Moreover, the rule $\alpha(\Delta_m) = s_{\alpha,\hat{m}} \cdot \Delta_m$, where $\alpha \in G$, $\Delta_m \in \mathrm{Diff}_*^m(P_M)$, defines a G-module structure in $\mathrm{Diff}_*(P_M)$.

Suppose $\Delta\colon P_M \to Q_N$ is a colour differential operator of order $\leqslant l$. Then the composition $f_\Delta\colon \nabla \in \mathrm{Diff}_t^m(P_M) \mapsto \Delta \circ \nabla \in \mathrm{Diff}_{l+t}^{\varphi(m)}(Q_N)$ defines a homomorphism of filtered right $\mathrm{Diff}_*(A)$-modules:

$$f_\Delta\colon \mathrm{Diff}_*(P_M) \longrightarrow \mathrm{Diff}_*(Q_N).$$

EXAMPLE. Let A be the algebra of the quantum hyperplane generated by elements $x_1, \dots x_n$ and relations $x_i x_j = \omega_{ij} x_j x_i$, for some 2-cocycle ω (see Examples 1.8, 2.8.). Then the module

$$\mathrm{Der}_*(A,A) = \sum_{\bar{n}\in\mathbb{Z}^n} \mathrm{Der}_{\bar{n}}(A,A)$$

is generated by the partial derivations $\partial/\partial x_i \in \mathrm{Der}_{-1_i}(A,A)$, where $1_i = (0,\dots,0,\underset{i}{1},0,\dots,0) \in \mathbb{Z}^n$, and $\partial/\partial x_i(x_j) = \delta_{ij}$.

The colour Leibniz rule gives the relations:

$$\partial/\partial x_i \cdot x_j - \omega_{ij}^{-1} x_j \cdot \partial/\partial x_i = \delta_{ij},$$

and

$$\partial/\partial x_i \cdot \partial/\partial x_j - \omega_{ij} \partial/\partial x_j \cdot \partial/\partial x_i = 0.$$

The algebra of colour differential operators $\mathrm{Diff}_*(A, A)$ is a $\mathbb{Z}^n$- graded algebra generated by elements x_i of degree 1_i, $i = 1, \dots, n$, and $\partial/\partial x_j$ of degree -1_j, $j = 1, \dots, n$, and the relations above.

2.6. The graded module associated with the filtered module $\mathrm{Diff}_*(P_M)$,

$$\mathrm{Smbl}_*(P_M) = \sum_{t \geqslant 0, m \in M} \mathrm{Smbl}_t^m(P_M),$$

where

$$\mathrm{Smbl}_t^m(P_M) = \frac{\mathrm{Diff}_t^m(P_M)}{\mathrm{Diff}_{t-1}^m(P_M)},$$

is the *module of colour symbols* of P_M.

Since the map f_Δ is compatible with the filtration, it defines a morphism

$$\mathrm{smbl}_*(\Delta)\colon\ \mathrm{Smbl}_*(P_M) \longrightarrow \mathrm{Smbl}_*(Q_N)$$

and k-morphisms

$$\mathrm{smbl}_t(\Delta)\colon\ \mathrm{Smbl}_t^m(P_M) \longrightarrow \mathrm{Smbl}_{l+t}^{\varphi(m)}(Q_N)$$

called *t-symbols* of the operator Δ.

PROPOSITION. *Let $\Delta_\alpha \in \mathrm{Diff}_t^\alpha(A)$, $\Delta_\beta \in \mathrm{Diff}_s^\beta(A)$. Then the colour commutator*

$$[\Delta_\alpha, \Delta_\beta] = \Delta_\alpha \circ \Delta_\beta - s_{\alpha,\beta} \cdot \Delta_\beta \circ \Delta_\alpha$$

is a colour differential operator of order $\leqslant s + t - 1$ and degree $\alpha\beta$.

Proof. We have

$$\delta_a(\Delta_\alpha) = [\Delta_\alpha, l_a]$$

if $a = a_\gamma \in A_\gamma$.

Therefore, by using the colour Jacobi identity, we get

$$\begin{aligned} \delta_a[\Delta_\alpha, \Delta_\beta] &= [[\Delta_\alpha, \Delta_\beta], l_a] \\ &= [\Delta_\alpha, [\Delta_\beta, l_a]] + s_{\alpha,\beta} \cdot [\Delta_\beta, [\Delta_\alpha, l_a]] \\ &= [\Delta_\alpha, \delta_a(\Delta_\beta)] + s_{\alpha,\beta} \cdot [\Delta_\beta, \delta_a(\Delta_\alpha)]. \end{aligned}$$ □

Denote by $\mathrm{smbl}(\Delta) = \Delta \bmod \mathrm{Diff}_{t-1}^\alpha(A)$, if $\Delta \in \mathrm{Diff}_t^\alpha(A)$, the *symbol* of the colour differential operator.

Then the following theorem is a direct consequence of the above proposition.

THEOREM.

(1) *The symbol algebra* $\mathrm{Smbl}_*(A)$ *is a colour commutative algebra*:

$$\mathrm{smbl}\,(\Delta_\alpha)\cdot\mathrm{smbl}\,(\Delta_\beta)=s_{\alpha,\beta}\cdot\mathrm{smbl}\,(\Delta_\beta)\cdot\mathrm{smbl}\,(\Delta_\alpha).$$

(2) *The bracket*

$$\{\mathrm{smbl}\,(\Delta_a),\mathrm{smbl}\,(\Delta_\beta)\}=[\Delta_\alpha,\Delta_\beta]\ \mathrm{mod}\ \mathrm{Diff}\,^{\alpha\beta}_{s+t-1}(A),$$

where $\Delta_\alpha\in\mathrm{Diff}\,^{\alpha}_{t}(A)$ *and* $\Delta_\beta\in\mathrm{Diff}\,^{\beta}_{s}(A)$, *determines a* colour Poisson structure.

It means that the bracket

$$\{\ ,\ \}\colon\ \mathrm{Smbl}\,^{\alpha}_{t}(A)\times\mathrm{Smbl}\,^{\beta}_{s}(A)\longrightarrow\mathrm{Smbl}\,^{\alpha\beta}_{s+t-1}(A)$$

satisfies the following conditions:

(2.1) *the* colour skew symmetry:

$$\{\mathrm{smbl}\,(\Delta_\alpha),\mathrm{smbl}\,(\Delta_\beta)\}=-s_{\alpha,\beta}\cdot\{\mathrm{smbl}\,(\Delta_\beta),\mathrm{smbl}\,(\Delta_\alpha)\},$$

(2.2) *the* colour Jacobi identity:

$$\begin{aligned}&\{\mathrm{smbl}\,(\Delta_\alpha),\{\mathrm{smbl}\,(\Delta_\beta),\mathrm{smbl}\,(\Delta_\gamma)\}\}\\&\quad=\{\{\mathrm{smbl}\,(\Delta_\alpha),\mathrm{smbl}\,(\Delta_\beta)\},\mathrm{smbl}\,(\Delta_\gamma)\}+\\&\qquad+s_{\alpha,\beta}\cdot\{\mathrm{smbl}\,(\Delta_\beta),\{\mathrm{smbl}\,(\Delta_\alpha),\mathrm{smbl}\,(\Delta_\gamma)\}\},\end{aligned}$$

(2.3) *the* action condition:

$$\mathrm{smbl}\,(\alpha(\Delta_\beta))=s_{\alpha,\beta}\cdot\mathrm{smbl}\,(\Delta_\beta),$$

(2.4) *the* colour derivation condition:

$$\begin{aligned}&\{\mathrm{smbl}\,(\Delta_\alpha),\mathrm{smbl}\,(\Delta_\beta)\cdot\mathrm{smbl}\,(\Delta_\gamma)\}\\&\quad=\{\mathrm{smbl}\,(\Delta_\alpha),\mathrm{smbl}\,(\Delta_\beta)\}\cdot\mathrm{smbl}\,(\Delta_\gamma)+\\&\qquad+s_{\alpha,\beta}\cdot(\mathrm{smbl}\,(\Delta_\beta))\cdot\{\mathrm{smbl}\,(\Delta_\alpha),\mathrm{smbl}\,(\Delta_\gamma)\}.\end{aligned}$$

2.7. In this section, we build up the representative object for the functor of colour derivations.

Denote by

$$\Omega^1(A)=\sum_{\alpha\in G}\Omega^1_\alpha(A)$$

the A-module generated by elements $a_\beta da_\gamma$ of degree $\beta\gamma$, with

(1) the usual relations:

$$d(a_\alpha + a_\beta) = da_\alpha + da_\beta, \qquad d(a_\alpha \cdot a_\beta) = da_\alpha \cdot a_\beta + a_\alpha \cdot da_\beta,$$

and

(2) the relations of colour constancy, $ds_{\alpha,\beta} = 0$.

Note that relation (1) can be considered as the definition of a right A-module structure on $\Omega^1(A)$.

Let d: $A \to \Omega^1(A)$ be the operator d: $a_\alpha \mapsto da_\alpha$.

THEOREM. *For any colour derivation* X_m: $A \to P_M$ *of degree* $m \in M$ *there is a homomorphism* $f_m \in \mathrm{Hom}_m(\Omega^1(A), P_M)$ *of degree* $m \in M$ *such that* $X_m = f_m \circ d$. *The homomorphism* f_m *is uniquely determined, and the correspondence* $X_m \mapsto f_m$ *establishes an isomorphism between* $\mathrm{Der}_m(A, P_M)$ *and* $\mathrm{Hom}_m(\Omega^1(A), P_M)$.

Proof. If we set

$$f_m(a_\alpha da_\beta) = (s_{\hat{m},\alpha} \cdot a_\alpha) \cdot X_m(a_\beta),$$

we transform the usual Leibniz rule for the operator d into the colour Leibniz rule for the derivation X_m. □

2.8. Starting from $A - A$ bimodule $\Omega^1(A)$ and a colour commutative algebra $\Omega^0(A) = A$ we build up an algebra of colour differential forms over A.

This algebra will be a new colour commutative algebra

$$\Omega^*(A) = \sum_{i \in \mathbb{N}, \alpha \in G} \Omega^i_\alpha(A),$$

graded by the group $\bar{G} = \mathbb{Z} \times G$ and generated by elements $a_\alpha \in A_\alpha = \Omega^0_\alpha$ of degree $(0, \alpha)$ and their differentials $da_\alpha \in \Omega^1_\alpha$ of degree $(1, \alpha)$.

We will require also that the universal derivation d: $A \to \Omega^1(A)$ can be extended to a colour derivation of the algebra $\Omega^*(A)$ with degree $(1, e)$ in such a way that $d^2 = 0$.

The last condition on d dictates some restrictions on a colour in the algebra $\Omega^*(A)$. To get the restrictions we consider a some colour $\bar{s}$: $\bar{G} \times \bar{G} \to U(A)$ on $\Omega^*(A)$ such that $\bar{s}|_{G \times G} = s$.

Hexagon condition 1.6(2) for $\bar{s}$ may be written in the form

$$\bar{s}_{(n+m,\alpha\beta),(k,\gamma)} = \bar{s}_{(n,\alpha),(k,^{\beta}\gamma)} \cdot \bar{s}_{(m,\beta),(k,\gamma)}, \tag{1}$$

where $k, m, n \in \mathbb{N}$, $\alpha, \beta, \gamma \in G$. Denote by $\omega \wedge \theta \in \Omega^{n+m}_{\alpha\beta}(A)$ the product of forms $\omega \in \Omega^n_\alpha(A)$, $\theta \in \Omega^m_\beta$ in the algebra $\Omega^*(A)$. Then

$$d(\omega \wedge \theta) = d\omega \wedge \theta + \bar{s}_{(1,e),(n,\alpha)} \cdot \omega \wedge d\theta,$$

and

$$d^2(\omega \wedge \theta) = \bar{s}_{(1,e),(n+1,\alpha)} \cdot d\omega \wedge d\theta + \bar{s}_{(1,e),(n,\alpha)} \cdot d\omega \wedge d\theta = 0.$$

Hence, we should require that

$$\bar{s}_{(1,e),(n+1,\alpha)} + \bar{s}_{(1,e),(n,\alpha)} = 0. \quad (2)$$

From these relations, it follows that

$$\bar{s}_{(1,e),(k,\alpha)} = (-1)^k \varphi(\alpha),$$

where φ: $G \to U(A_e)$, $\varphi(\alpha) = \bar{s}_{(1,e),(0,\alpha)}$.

Taking $n = m = 0$, $k = 1$, $\gamma = e$ in hexagon equation (1), one gets that φ is a group homomorphism. So, if we fix a colour s: $G \times G \longrightarrow U(A)$ and a group homomorphism φ: $G \to U(A_e)$, then from Equation (1) we get

$$\bar{s}_{(n,\alpha),(m,\beta)} = (-1)^{nm} \varphi(\alpha)^{-m} \cdot s_{\alpha,\beta} \cdot \varphi(\beta)^n, \quad (3)$$

and compatibility conditions 1.7(2) are fulfilled.

PROPOSITION. *Let A be a colour commutative algebra with a colour s. Then any colour $\bar{s}$ on the group $\bar{G} = \mathbb{Z} \times G$ with conditions $\bar{s}|_{G\times G} = s$ and* 2.7(2) *are given by formula* (3) *for some group homomorphism φ: $G \to U(A_e)$.*

Below we will denote by $\Omega^*(A, \varphi)$, or simply by $\Omega^*(A)$, the colour commutative $\bar{G}$-graded algebra with the colour derivation $d = d_\varphi$ of degree $(1, e)$.

Therefore, for any colour commutative algebra A, a group homomorphism φ: $G \to U(A_e)$, and an element $\alpha \in G$, we have the complex:

$$\begin{aligned} 0 \longrightarrow A \xrightarrow{d_\varphi} \Omega^1_\alpha(A,\varphi) \xrightarrow{d_\varphi} \Omega^2_\alpha(A,\varphi) \xrightarrow{d_\varphi} \cdots \\ \xrightarrow{d_\varphi} \Omega^i_\alpha(A,\varphi) \xrightarrow{d_\varphi} \Omega^{i+1}_\alpha(A,\varphi) \xrightarrow{d_\varphi} \cdots. \end{aligned}$$

The cohomology of the complex at the term $\Omega^i_\alpha(A,\varphi)$ we will denote by $H^i_\alpha(A,\varphi)$, and will be called as the *colour de Rham cohomology* of the algebra A.

Because of d_φ is a colour derivation of the colour commutative algebra $\Omega^*(A,\varphi)$, then

$$H^*(A,\varphi) = \sum_{i\in\mathbb{N},\alpha\in G} H^i_\alpha(A,\varphi)$$

is a $\mathbb{Z} \times G$-graded colour commutative algebra with respect to the colour $\bar{s}$.

EXAMPLES

(1) Let $G = \{e\}$. Then homomorphism φ is a also trivial, and we get the standard colour on the algebra of differential forms: $\bar{s}_{m,n} = (-1)^{mn}$.

(2) Let $G = \mathbb{Z}_2$. There are two possibilities for φ: (1) φ is a trivial homomorphism, and (2) $\varphi(\bar{1}) = -1$. Thus, there are two algebras of differential forms linked with a commutative $\mathbb{Z}_2$-algebras.
(3) Consider $A = \mathrm{Mat}_n(\mathbb{C})$ as $G = \mathbb{Z}_n \oplus \mathbb{Z}_n$-graded colour commutative algebra. Then $\varphi \in \bar{G} = \mathrm{Hom}\,(G, \mathbb{C}^*)$ is a character of the group. Therefore, any character on G gives us the algebra of colour differential forms on $\mathrm{Mat}_n(\mathbb{C})$.
(4) The quantum hyperplane (cf. [1, 21]) is given by the following data $k = \mathbb{C}$, $G = \mathbb{Z}^n$, and the twisted 2-cocycle: $\omega(\bar{a}, \bar{b}) = q^{\langle \vartheta \bar{a}, \bar{b} \rangle}$, where ϑ is a skew-symmetric $n \times n$ matrix, $q \in \mathbb{C}^*$, $\bar{a}, \bar{b} \in \mathbb{Z}^n$.
Let A be the algebra generated by elements $x_1, \ldots, x_n$ and the relations: $x_i x_j = \omega_{ij} x_j x_i$, where ω_{ij} are matrix elements of ω. A is obviously a colour commutative algebra and we can build up the algebra of colour differential forms on A. For this end we note that any group homomorphism φ: $G \to \mathbb{C}^*$ has the form $\varphi(\bar{a}) = z^{\bar{a}}$, for some complex vector $z = (z_1, \ldots, z_n) \in (\mathbb{C}^*)^n$. The algebra $\Omega^*(A, \varphi)$ is now a colour commutative algebra generated by elements $x_i, y_j = \mathrm{d}x_j$, $1 \leqslant i, j \leqslant n$, and the relations:

$$x_i x_j = \omega_{ij} x_j x_i, \qquad x_i y_j = z_i \omega_{ij} y_j x_i, \qquad y_i y_j = -z_i z_j \omega_{ij} y_j y_i.$$

2.9. Here we outline the construction of Lie derivations and Nijenhuis brackets over colour commutative algebras.

Let us describe derivations of the algebra $\Omega^*(A)$. Denote by $\mathrm{Der}_*^{\mathrm{alg}}\Omega^*(A)$ submodule of all algebraic derivations, i.e. colour derivations X of $\Omega^*(A)$ such that $X|_{\Omega^0(A)} = 0$.

Since any algebraic derivation is determined by values on $\Omega^1(A)$ we get the isomorphism:

$$\mathrm{Der}_{(k,\alpha)}^{\mathrm{alg}}\Omega^*(A) \simeq \mathrm{Hom}_\alpha(\Omega^1(A), \Omega^{k+1}(A)) \simeq \mathrm{Der}_\alpha(\Omega^{k+1}(A)). \tag{1}$$

For any derivation $X \in \mathrm{Der}_\alpha(\Omega^{k+1}(A))$ we will denote by $\imath_X$ the corresponding algebraic (*inner*) derivation of $\Omega^*(A)$.

In other words, the operator may be defined as the colour derivation in $\Omega^*(A)$ such that:

(1) $\imath_X$: $\Omega^j_\beta(A) \longrightarrow \Omega^{j+k}_{\alpha\beta}(A)$,
(2) $\imath_X(\omega_1 \wedge \omega_2) = \imath_X(\omega_1) \wedge \omega_2 + (-1)^{jk}\varphi(\alpha)^{-j} \cdot s_{\alpha,\beta} \cdot \varphi(\beta)^k \cdot \omega_1 \wedge \imath_X(\omega_2)$,
(3) $\imath_X(a_\beta) = 0$, $\quad \imath_X(\mathrm{d}a_\beta) = X(a_\beta)$,

where $j, k \in \mathbb{N}$, $\alpha, \beta \in G$, $\omega_1 \in \Omega^j_\beta(A)$, $a_\beta \in A_\beta$.

The module of colour algebraic derivations is obviously closed with respect to the colour commutator of derivations.

Therefore, we get a colour Lie algebra structure on

$$\mathrm{Nij}(A) = \sum_{k \in \mathbb{Z}, \alpha \in G} \mathrm{Hom}_\alpha(\Omega^1(A), \Omega^{k+1}(A)).$$

The colour Lie algebra $\mathrm{Nij}(A)$ will be called a *Nijenhuis algebra* of the colour commutative algebra A and the bracket will be called a *colour algebraic Nijenhuis bracket.*

By definition the Nijenhuis bracket of elements $X \in \mathrm{Hom}_\alpha(\Omega^1(A), \Omega^{k+1}(A))$ and $Y \in \mathrm{Hom}_\beta(\Omega^1(A), \Omega^{l+1}(A))$ is given by the formula

$$[X,Y](\omega) = \imath_X(Y(\omega)) - (-1)^{kl} \cdot \varphi(\alpha)^{-l} \cdot s_{\alpha,\beta} \cdot \varphi(\beta)^k \cdot \imath_Y(X(\omega)),$$

for all $\omega \in \Omega^1(A)$.

Any derivation $X \in \mathrm{Der}_\alpha(A)$ determines an inner derivation

$$\imath_X \in \mathrm{Der}_{(-1,\alpha)}(\Omega^*(A))$$

and a *Lie derivation*: $\mathcal{L}_X = [\imath_X, d]$.

For Lie derivations we have the same properties as for the usual ones.

THEOREM. *$\mathcal{L}_X$ is a colour (k,α)-derivation of the algebra of colour differential forms $\Omega^*(A)$:*

$$\mathcal{L}_X(\omega_1 \wedge \omega_2) = \mathcal{L}_X(\omega_1) \wedge \omega_2 + \varphi(\alpha)^{-k} \cdot s_{\alpha,\beta} \cdot \omega_1 \wedge \mathcal{L}_X(\omega_2).$$

(1) *The bracket $[\mathcal{L}_X, \mathcal{L}_Y]$ is a Lie derivation $\mathcal{L}_Z$ for some element $Z = [X,Y]$, and is called the* Frölicher–Nijenhuis bracket.

(2) *The Frölicher–Nijenhuis bracket*

$$\mathrm{Hom}_\alpha(\Omega^1(A), \Omega^{k+1}(A)) \times \mathrm{Hom}_\beta(\Omega^1(A), \Omega^{l+1}(A)) \longrightarrow \mathrm{Hom}_{\alpha\beta}(\Omega^1(A), \Omega^{k+l+2}(A))$$

determines a $\bar{G}$-graded $\bar{s}$-colour Lie algebra structure in the Nijenhuis algebra. Here $\omega_1 \in \Omega^k_\beta(A), \omega_2 \in \Omega^l_\gamma(A)$.

2.10. In this section, we outline the construction of the modules of colour jets as representative objects for the functors of colour differential operators (see [11] for the usual case).

Let P_M be a symmetric colour module. Consider the tensor product $A \otimes_k P_M$ as a colour M-graded module too. To do this we assume that elements $a_\alpha \otimes x_m$, $a_\alpha \in A_\alpha$, $x_m \in P_m$, have the degree αm and the G-action is given by the formula $\beta(a_\alpha \otimes x_m) = \beta(\alpha) \otimes \beta(x_m)$. The left A-module structure in $A \otimes_k P_M$ is induced by multiplication in the first factor.

For any element $a_\alpha \in A_\alpha$ we define the endomorphism $\delta^{a_\alpha}\colon A \otimes_k P_M \to A \otimes_k P_M$ as follows:

$$\delta^{a_\alpha}(a_\beta \otimes x_m) = a_\beta \cdot a_\alpha \otimes x_m - a_\beta \otimes a_\alpha x_m.$$

Then δ^{a_α} are endomorphisms such that $\delta^{a_\alpha}(A_\beta \otimes_k P_m) \subset (A \otimes_k P_M)_{\beta\alpha m}$, and $\delta^{a_\alpha}(a_\beta \otimes z) = a_\beta \otimes \delta^{a_\alpha}(z)$.

Let μ_{l+1} be the M-graded submodule of $A \otimes_k P_M$ generated by all the elements of the form

$$\delta^{a_{\alpha_0}} \circ \delta^{a_{\alpha_1}} \circ \cdots \circ \delta^{a_{\alpha_l}}(a_\beta \otimes x_m).$$

Denote by $\mathcal{J}^l(P_M)$ the quotient module $A \otimes_k P_M/\mu_{l+1}$ and by j_l: $P_M \to \mathcal{J}^l(P_M)$ the map: $j_l(x_m) = (1 \otimes x_m) \bmod \mu_{l+1}$.

We will call the modules $\mathcal{J}^l(P_M)$, $l = 0, 1, \ldots$, as *modules of colour jets* and j_l as *colour l-jets operators.*

One has

$$\begin{aligned} \delta_{a_\alpha}(j_l)(x_m) &= j_l(a \cdot x_m) - a_\alpha \cdot j_l(x_m) \\ &= (1 \otimes a_\alpha \cdot x_m - a_\alpha \otimes x_m) \bmod \mu_{l+1} \\ &= -\delta^{a_\alpha}(1 \otimes x_m) \bmod \mu_{l+1} \end{aligned}$$

and $j_l \circ \beta = \beta \circ j_l$.

Therefore, $\delta_{a_{\alpha_0}} \circ \cdots \circ \delta_{a_{\alpha_l}}(j_l) = 0$ and j_l: $P_M \to \mathcal{J}^l(P_M)$ is a colour differential operator of degree id: $M \to M$ and order l.

THEOREM. *For any colour differential operator Δ: $P_M \to Q_N$ of degree $\varphi \in$* Map$_\alpha(M, N)$ *and order $l \geqslant 0$ there is a colour homomorphism f^Δ: $\mathcal{J}^l(P_M) \to Q_N$ of degree φ such that $\Delta = f^\Delta \circ j_l$. The homomorphism f^Δ is uniquely determined, and the correspondence $\Delta \mapsto f^\Delta$ establishes an isomorphism between* Diff$_l^\varphi(P_M, Q_N)$ *and* Hom$_\varphi(\mathcal{J}^l(P_M), Q_N)$.

Proof. The uniqueness of the morphism f^Δ is obvious. In order to show existence we define $\bar{f}^\Delta$: $A \otimes_k P_M \to Q_N$ by putting

$$\bar{f}^\Delta(a_\beta \otimes x_m) = \alpha(a_\beta) \cdot \Delta(x_m).$$

Then

$$\begin{aligned} &\bar{f}^\Delta(\delta^{a_\beta}(a_\gamma \otimes x_m)) \\ &\quad = \bar{f}^\Delta(a_\gamma \otimes a_\beta x_m - a_\gamma a_\beta \otimes x_m) \\ &\quad = \alpha(a_\gamma)\Delta(a_\beta x_m) - \alpha(a_\gamma a_\beta)\Delta(x_m) = \alpha(a_\gamma)\delta_{a_b e}(\Delta)(x_m). \end{aligned}$$

Therefore, $\bar{f}^\Delta|_{\mu_{l+1}} = 0$ and we get the morphism f^Δ: $\mathcal{J}^l(P_M) \to Q_N$, such that $\Delta = f^\Delta \circ j_l$. □

2.11. Let Δ: $P_M \to Q_N$ be a colour differential operator of order l and degree $\varphi \in$ Map$_\alpha(M, N)$. We define the *t-jet prolongation* of Δ in the usual way as the composition $\Delta^{(t)} = j_t \circ \Delta$: $P_M \to \mathcal{J}^t(Q_N)$.

Denote by $f^{\Delta^{(t)}}$ the corresponding homomorphisms: $\Delta^t = f^{\Delta^t} \circ j_{l+t}$.

The embeddings Diff$_t^m(P_M) \subset$ Diff$_u^m(P_M)$, $t \leqslant u$, generate epimorphisms $\pi_{u,t}$: $\mathcal{J}_m^u(P_M) \to \mathcal{J}_m^t(P_M)$, $m \in M$, such that $\pi_{u,t} \circ j_u = j_t$.

Moreover, $\pi_{u,t} \circ \pi_{v,u} = \pi_{v,t}$, for all $t \leqslant u \leqslant v$.

Denote by $\mathrm{Cosmbl}^{\,t}_{m}(P_M)$ the kernel of the projection $\pi_{t,t-1}\colon \mathcal{J}^t_m(P_M) \to \mathcal{J}^{t-1}_m(P_M)$ and by

$$\mathrm{Cosmbl}_{\,*}(P_M) = \sum_{t\geqslant 0, m\in M} \mathrm{Cosmbl}^{\,t}_{m}(P_M)$$

the *cosymbol module* of P_M.

One has $\pi_{t,t-1}\circ f^{\Delta^{(t)}} = f^{\Delta^{t-1}}\circ\pi_{t+l,t+l-1}$ and therefore any colour differential operator Δ determines a homomorphism

$$\mathrm{Cosmbl}^{\,l+t}_{m}(\Delta)\colon \mathrm{Cosmbl}^{\,t+l}_{m}(P_M) \longrightarrow \mathrm{Cosmbl}^{\,t}_{\varphi(m)}(Q_N)$$

of degree φ.

DEFINITIONS

(1) Let $\Delta\colon P_M \to Q_N$ be a colour differential operator of order l and degree $\varphi \in \mathrm{Mor}_{\varphi}(M,N)$. Then M-graded submodule $\mathcal{R}_l = \mathrm{Ker}\, f^{\Delta} \subset \mathcal{J}^l(P_M)$ is called a *colour differential equation.* $\mathcal{R}_{l+t} = \mathrm{Ker}\, f^{\Delta^{(t)}} \subset \mathcal{J}^{l+t}(P_M)$ is called *t-prolongations* of $\mathcal{R}_l$.

(2) The M-graded module $g_l = \sum_{m\in M} g^m_l$, where $g^m_l = \mathrm{Ker}\,\mathrm{Cosmbl}^{\,l}_{m}(\Delta)$, is called a *symbol of the equation* $\mathcal{R}_l$, and $g_{l+t} = \sum_{m\in M} g^m_{l+t}$, where $g^m_{l+t} = \mathrm{Ker}\,\mathrm{Cosmbl}^{\,m}_{l+t}(\Delta)$, is called *t-prolongations* of the symbol.

Similarly ([11]) by using of colour de Rham operator $\mathrm{d} = \mathrm{d}_{\varphi}$, we can define *Spencer colour operators*

$$S = S_{\varphi}\colon \Omega^i(A,\varphi)\otimes_A \mathcal{J}^k(P_M) \longrightarrow \Omega^{i+1}(A,\varphi)\otimes_A \mathcal{J}^{k-1}(P_M),$$

such that $S(\omega\otimes j_k(p)) = \mathrm{d}\omega\otimes j_{k-1}(p)$, for all $\omega\in\Omega^i(A), p\in P_M$.

For any colour differential equation we have $S(\mathcal{R}_{l+t}) \subset \Omega^1(A)\otimes_A \mathcal{R}_{l+t-1}$.

Therefore, Spencer operators produce the complexes:

$$0 \longrightarrow \mathcal{R}_{l+t} \longrightarrow \Omega^1(A,\varphi)\otimes_A \mathcal{R}_{l+t-1} \longrightarrow \Omega^2(A,\varphi)\otimes_A \mathcal{R}_{l+t-2} \longrightarrow \cdots,$$

which cohmologies are *Spencer colour cohomologies* of colour differential equation $\mathcal{R}_l$.

3. Graded Monoidal Categories and Colour Quantizations

In this section we compare the colour calculus developed above with the calculus in braided tensor categories suggested in [12]. To do this, we describe two monoidal categories linked with colour calculus. The first one is an underlying category of grading spaces and the second one is an corresponding category of graded k-modules.

3.1. Denote by Sym_G the category of symmetric G-bispaces. Thus, an object X in the category is a symmetric G-bispace or, in other words, a left G-space

X endowed with G-map $\hat{}: X \to G$. A morphism in the category is a G-map $f: X \to Y$ of left G-spaces such that $\widehat{f(x)} = \hat{x}$ for all $x \in X$.

One can convert Sym_G into a monoidal category by introducing a 'tensor product' of G-spaces.

At first we define the tensor product to be the pushout $X \times_G Y$ of X and Y, where X is considered as right G-space and Y as left G-space, respectively. Therefore, elements of $X \times_G Y$ can be identified with equivalence classes $x \otimes y$ of $(x, y) \in X \times Y$ with respect to the following equivalence relation: $(x\alpha, y) \sim (x, \alpha y)$, $\forall \alpha \in G$.

EXAMPLES

(1) Let $X = G$ be a standard symmetric G-bispace. Then $G \times_G Y = Y$, for all $Y \in \mathrm{Ob}(\mathrm{Sym}_G)$, and the isomorphism above is given by the formula $(\alpha, y) \mapsto \alpha y$, $\forall \alpha \in G$, $y \in Y$.

(2) Let $Y = *$ be a trivial G-bispace. Then, $X \times_G * = X/G$ is the orbit space.

We will consider $X \times_G Y$ as a G-bispace with left and right multiplications $\alpha \cdot (x \otimes y) = \alpha x \otimes y$ and $(x \otimes y) \cdot \alpha = x \otimes y\alpha$, and with the symmetry function: $\widehat{x \otimes y} = \hat{x} \cdot \hat{y}$.

3.2. Let $\mathcal{C}$ be an arbitrary monoidal category with a tensor product $\otimes$. Recall the following

DEFINITION [16]. A *symmetry* for monoidal category $\mathcal{C}$ is a natural isomorphism $\sigma_{X,Y} : X \otimes Y \to Y \otimes X$ for all $X, Y \in \mathrm{Ob}(\mathcal{C})$, such that

(1) the *hexagon condition:* $\sigma_{X \otimes Y, Z} = (\sigma_{X,Z} \otimes \mathrm{id}_Y) \circ (\mathrm{id}_X \otimes \sigma_{Y,Z})$ and

(2) the *unitary condition:* $\sigma_{X,Y} \circ \sigma_{Y,X} = \mathrm{id}_{X \otimes Y}$

hold.

THEOREM. *Any symmetry in the monoidal category* Sym_G *has the form*

$$\sigma_{X,Y}(x \otimes y) = [\hat{x}, \hat{y}] \cdot y \otimes x, \tag{1}$$

for all $X, Y \in \mathrm{Ob}(\mathrm{Sym}_G)$, $x \in X$, $y \in Y$.

Proof. We will identufy elements $x \in X$ with morphisms $t_x: G \to X$ in the category such that: $t_x(\alpha) = (\alpha \hat{x}^{-1}) \cdot x$.

From the naturality of the symmetry σ it follows that the diagram

$$\begin{array}{ccc}
X \times_G Y & \xrightarrow{\sigma_{X,Y}} & Y \times_G X \\
\uparrow {\scriptstyle t_X \times_G t_Y} & & \uparrow {\scriptstyle t_Y \times_G t_X} \\
G \times_G G & \xrightarrow{\sigma_{G,G}} & G \times_G G \\
\| & & \| \\
G & \xrightarrow{\lambda} & G
\end{array}$$

commutes.

But $\lambda = \sigma_{G,G}$ is a morphism in the category. Hence, $\lambda = \mathrm{id}$, and we get formula (1). An easy computation shows that the hexagon and the unitary conditions are valid for the given σ. □

Remark. The same result we get if we consider the new tensor product $X \times_{[G,G]} Y$, where $[G,G]$ is the commutator group.

3.3. Denote by $\mathcal{G}\mathrm{Mod}_k$ the category of graded k-modules equipped with a special G-action.

Thus objects in $\mathcal{G}\mathrm{Mod}_k$ are graded k-modules $P_M = \sum_{m \in M} P_m$, graded by G-bispaces $M \in \mathrm{Ob}(\mathrm{Sym}_G)$, and endowed with a left $k[G,G]$-module structure such that $\alpha(P_m) \subset P_{\alpha m}$ for all $\alpha \in [G,G]$.

The $k[G,G]$-module structure determine a special G-module structure on P_M: $\alpha(x_m) = [\alpha, \hat{m}]\cdot x_m$, and a right $k[G,G]$-module structure $x_m \cdot \beta = {}^{\hat{m}}\alpha \cdot x_m$, where $\alpha \in [G,G]$, $\beta \in G$, $x_m \in P_m$.

Morphisms in the category are graded $k[G,G]$-morphisms $f\colon P_M \to Q_N$ covering morphisms in the category Sym_G.

We convert $\mathcal{G}\mathrm{Mod}_k$ into a monoidal category by letting $P_M \otimes Q_N$ be the usual $k[G,G]$-tensor product graded by the space $M \times_G N$.

Any object $M \in \mathrm{Ob}(\mathrm{Sym}_G)$ determines an object $k(M) \in \mathrm{Ob}(\mathcal{G}\mathrm{Mod}_k)$, where $k(M)$ is the algebra of k-valued functions on M with basis $\{\delta_m\}$, $m \in M$, such that $\delta_m(m) = 1$, and $\delta_m(m') = 0$, otherwise. We will consider $k(M)$ as M-graded k-module, where functions δ_m have degree m, the G-action is given by $\alpha(\delta_m) = \delta_{\alpha m \alpha^{-1}}$, and the left $k[G,G]$-module structure is given by $\alpha \cdot \delta_m = \delta_{\alpha m}$.

Any element $x_m \in P_m$ we will identify with morphism: $t(x_m)\colon k(M) \to P_M$, such that $t(x_m)(\delta_m) = x_m$, $t(x_m)(\delta_{\alpha m \alpha^{-1}}) = \alpha(x_m)$, $\forall \alpha \in G$, and $t(x_m)(\delta_{m'}) = 0$ otherwise.

Let σ be a symmetry in the monoidal category. By using of naturality of symmetries we get the following commutative diagram:

$$
\begin{array}{ccc}
P_M \otimes Q_N & \xrightarrow{\ \sigma_{P,Q}\ } & Q_N \otimes P_M \\
\uparrow {\scriptstyle t(x_m)\otimes t(y_n)} & & \uparrow {\scriptstyle t(y_n)\otimes t(x_m)} \\
k(M) \otimes k(N) & \xrightarrow{\ \sigma_{k(M),\ k(N)}\ } & k(N) \otimes k(M)
\end{array}
$$

Any symmetry σ covers the symmetry in the category Sym_G. Therefore,

$$\sigma_{k(M),k(N)}\colon\ \delta_m \otimes \delta_n \longmapsto \chi(\hat{m},\hat{n})[\hat{m},\hat{n}]\delta_n \otimes \delta_m,$$

for some function $\chi\colon\ G \times G \to U(k)$, and

$$\sigma_{P,Q}(x_m \otimes y_n) = \chi(\hat{m},\hat{n})[\hat{m},\hat{n}]y_n \otimes x_m. \tag{1}$$

The condition that σ is a morphism in the category yields the following condition on χ:

$$\chi(\alpha\alpha_0, \beta) = \chi(\alpha, \alpha_0\beta) \tag{2}$$

for all $\alpha, \beta \in G$, $\alpha_0 \in [G, G]$.

The unitary equation produces the multiplicative skew-symmetry property on χ

$$\chi(\alpha, \beta) \cdot \chi(\beta, \alpha) = 1,$$

and the hexagon equation yields

$$\chi(\alpha\beta, \gamma) = \chi(\alpha, \gamma) \cdot \chi(\beta, \gamma)$$

for all $\alpha, \beta \in G$.

Summarizing we get the following

THEOREM. *Any symmetry σ in the monoidal category $\mathcal{G}\mathrm{Mod}_k$ is given by formula* (1), *where $\chi\colon G \times G \to U(k)$ is a skew-symmetric bihomomorphism such that property* (2) *holds.*

3.4. Let $A = \sum_{\alpha \in G} A_\alpha$ be an algebra in the category $\mathcal{G}\mathrm{Mod}_k$ with multiplication $\mu\colon A \otimes A \to A$, $\mu(a \otimes b) = a \cdot b$. Therefore $A_\alpha \cdot A_\beta \subset A_{\alpha\beta}$, and there is a left $[G, G]$-action such that $\alpha(A_\beta) \subset A_{\alpha\beta}$, $\forall \alpha \in [G, G]$.

By using the action, we can identify elements α of the group with the elements $\tilde{\alpha} = \alpha(1) \in A_\alpha$ of the algebra.

Let A be a σ-commutative algebra in the category with respect to some symmetry σ. Then from 3.3(1) we get the following law of commutativity:

$$a_\alpha \cdot a_\beta = \chi(\alpha, \beta)[\tilde{\alpha}, \tilde{\beta}] \cdot a_\beta \cdot a_\alpha, \tag{1}$$

for all $a_\alpha \in A_\alpha, a_\beta \in A_\beta$.

THEOREM.

(1) *Any G-graded σ-commutative algebra $A = \sum_{\alpha \in G} A_\alpha$ in $\mathcal{G}\mathrm{Mod}_k$ with given symmetry σ, is a colour commutative algebra with respect to the colour $s_{\alpha,\beta} = \chi(\alpha, \beta)[\tilde{\alpha}, \tilde{\beta}]$.*
(2) *Let A be a σ-commutative algebra in the category. Then any σ-symmetric $A - A$ bimodule in the category is colour symmetric.*

3.5. We start with recalling the definition [12, 13] of quantizations in monoidal categories.

Let $\mathcal{C}$ be a monoidal category with

(1) a bifunctor of tensor product: $\otimes\colon \mathcal{C} \times \mathcal{C} \to \mathcal{C}$, $\otimes\colon X \times Y \to X \otimes Y$, where $X, Y \in \mathrm{Ob}(\mathcal{C})$,

(2) a trivial associativity constraint: $X \otimes (Y \otimes Z) = (X \otimes Y) \otimes Z$,
(3) a unit object k, with natural isomorphisms: $\eta_X^l\colon k \otimes X \to X$, $\eta_X^r\colon X \otimes k \to X$, such that

$$\mathrm{id}_X \otimes \eta_Y^l = \eta_X^r \otimes \mathrm{id}_Y\colon X \otimes k \otimes Y \longrightarrow X \otimes k \otimes Y.$$

DEFINITION. A quantization $\mathcal{Q}$ of a monoidal category $\mathcal{C}$ is a natural isomorphism of tensor product bifunctor $\mathcal{Q}_{X,Y}\colon X \otimes Y \to X \otimes Y$, such that the coherence conditions

(1) $\mathcal{Q}_{X,Y\otimes Z} \circ (\mathrm{id}_X \otimes \mathcal{Q}_{Y,Z}) = \mathcal{Q}_{X\otimes Y,Z} \circ (\mathcal{Q}_{X,Y} \otimes \mathrm{id}_Z)$,
(2) $\eta_X^r \circ \mathcal{Q}_{X,k} = \eta_X^r$, $\eta_X^l \circ \mathcal{Q}_{k,X} = \eta_X^l$,

hold.

A quantization in monoidal category $\mathcal{C}$ preserves all natural algebraic structures linked with the tensor product in $\mathcal{C}$ [12, 13]. Thus, if $\mu\colon A \otimes A \to A$ is an algebra in the category we can define a *quantization* (A_q, μ^q) of the algebra as the object $A_q = A$ with new multiplication $\mu^q = \mu \circ \mathcal{Q}_{A,A}$. In the same way, for any left A-module X with multiplication $\mu_X\colon A \otimes X \to X$ we can define the A_q-module (X_q, μ_X^q), a *quantization* of X, as the object $X_q = X$ with new multiplication $\mu_X^q = \mu \circ \mathcal{Q}_{A,X}$. It is important that (A_q, μ^q) is an algebra and that (X_q, μ_X^q) is an A_q-module in the category too.

Moreover, if σ is a symmetry (or braiding) in the monoidal category then $\sigma_{X,Y}^q = \mathcal{Q}_{Y,X}^{-1} \circ \sigma_{X,Y} \circ \mathcal{Q}_{X,Y}$ is a symmetry too [12]. One can define σ-differential operators in braided tensor categories in such a way that quantizations generate transformations of modules of σ-differential operators [12, 13].

3.6. Here we describe quantizations in the category $\mathcal{G}\mathrm{Mod}_k$.

Let $\mathcal{Q}$ be a quantization. Then by using the naturality of $\mathcal{Q}$ we get the following commutative diagram (see 3.3):

$$\begin{array}{ccc}
X_M \otimes_G Y_N & \xrightarrow{\ \mathcal{Q}_{X,Y}\ } & X_M \otimes_G Y_N \\
\uparrow{\scriptstyle t(x_m)\otimes t(y_n)} & & \uparrow{\scriptstyle t(x_m)\otimes t(y_n)} \\
k[G] \otimes_G k[G] & \xrightarrow{\ \mathcal{Q}_{k[G],k[G]}\ } & k[G] \otimes_G k[G] \\
\Vert & & \Vert \\
k[G] & \xrightarrow{\ \ q\ \ } & k[G]
\end{array}$$

where $X_M, Y_N \in \mathrm{Ob}(\mathcal{G}\mathrm{Mod}_k)$, $x_m \in X_M$, $y_n \in Y_N$.

Because of $\hat{q} = \mathrm{id}$, we have $q(\varepsilon_\alpha) = q(\alpha)\varepsilon_\alpha$, for some function $q\colon G \to k^*$, and $\mathcal{Q}_{X,Y}(x_m \otimes y_n) = q(\hat{m}\hat{n}) x_m \otimes y_n$. From condition 3.5(1) we get $q \equiv 1$ and, therefore, $\mathcal{Q} = \mathrm{id}$.

THEOREM. *Any quantization in the monoidal category* $\mathcal{G}\mathrm{Mod}_k$ *is trivial.*

3.7. Similarly to Theorem 3.3, we get the following theorem

THEOREM. *Any quantization* $\mathcal{Q}$ *in the monoidal category* $\mathcal{G}\mathrm{Mod}_k$ *is given by the formula*

$$\mathcal{Q}_{P,Q}(x_m \otimes y_n) = q(\hat{m}, \hat{n}) x_m \otimes y_n,$$

where $q\colon G \times G \to U(k)$ *is a* quantizer [15], *i.e. an* $[G, G]$*-invariant multiplicator on the group*:

(1) $q(\alpha\alpha_0, \beta) = q(\alpha, \alpha_0\beta)$,
(2) $q(e, \alpha) = q(\alpha, e) = 1$,
(3) $q(\alpha, \beta\gamma)q(\beta, \gamma) = q(\alpha\beta, \gamma)q(\alpha, \beta)$,

where $\alpha_0 \in [G, G]$, $\alpha, \beta, \gamma \in G$.

EXAMPLE. Let $G = \mathbb{Z}^n$. Then the theorems above show that quantizations act on the set of all symmetries in a transitive way. Therefore all modules of differential operators on quantum hyperplanes are isomorphic. The isomorphisms can be given by the quantizations (cf. [18]).

Acknowledgements

I would like to thank the Department of Mathematics, the University of Oslo and the Centre for Advanced Study at the Norwegian Academy of Science and Letters for support during 1993 while this work was being developed.

References

1. Bongaarts, P. J. M. and Pijls, H. G. J.: Almost commutative algebra and differential calculus on the quantum hyperplane, Preprint Inst. Lorentz, 1992.
2. Borowiec, A., Marcinek, W., and Oziewicz, Z.: On multigraded differential calculus, in R. Gielerak (ed.), *Quantum Groups and Related Topics*, Kluwer Acad. Publ., Dordrecht, 1992, pp. 103–114.
3. Connes, A.: *Géométrie non commutative*, Inter. Editions, Paris, 1990.
4. Dade, E. C.: Group graded rings and modules, *Math. Z.* **174** (1980), 241–262.
5. Drinfeld, V. G.: Quantum groups, in *Proc. ICM*, Amer. Math. Soc., Berkeley, 1986, pp. 798–820.
6. Dubois-Violette, M.: Dérivations et calcul différentiel non commutatif, *C.R. Acad. Sci.* **307** (1988), 403–408.
7. Gurevich, D.: Hecke symmetries and braided Lie algebras, in Z. Oziewicz (ed.), *Spinors, Twistors, Clifford Algebras and Quantum Deformations*, Kluwer Acad. Publ., Dordrecht, 1993, pp. 317–326.
8. Jadczyk, A. and Kastler, D.: The fermionic differential calculus, *Ann. Physics* **179** (1987), 169–200.
9. Karoubi, M.: Homologie cyclique des groupes et des algebres, *C.R. Acad. Sci.* **297** (1983), 381–384.
10. Kersten, P. H. M. and Krasil'shchik, I. S.: Graded Frölicher–Nijenhuis brackets and the theory of recursion operators for super differential equations, Preprint Univ. of Twente, 1993.
11. Krasil'shchik, I. S., Lychagin, V. V., and Vinogradov, A. M.: *Geometry of Jet Spaces and Nonlinear Partial Differential Equations*, Gordon and Breach, New York, 1986.

12. Lychagin, V.: Differential operators and quantizations, I, Preprint Math. Inst. Univ. Oslo 44, 1993.
13. Lychagin, V.: Quantizations of braided differential operators, Preprint ESI 51, 1993.
14. Lychagin, V.: Braided differential operators and quantizations in ABC-categories, *C. R. Acad. Sci. Serie I* **318** (1994), 857–862.
15. Lychagin, V.: Braidings and quantizations over bialgebras, Preprint ESI 61, 1993.
16. MacLane, S.: *Categories for the Working Mathematician*, Springer-Verlag, Berlin, 1971.
17. Manin, Y.: Notes on quantum groups and quantum de Rham complexes, Preprint MPI 60, 1991.
18. Ogievetsky, O.: Differential operators on quantum spaces for $\mathrm{Gl}_q(n)$ and $\mathrm{SO}_q(n)$, Preprint MPI 91–103, 1991.
19. Sletsjøe, A. B.: Twisted Hochschild cohomologies (to appear).
20. Vinogradov, M. M.: The main functors of differential calculus in graded algebras, *Usp. Mat. Nauk* **44**(3) (1989), 151–152 (in Russian).
21. Wess, J. and Zumino, B.: Covariant differential calculus on the quantum hyperplane, *Nuclear Phys.* **18B** (1990), 303–312.

Acta Applicandae Mathematicae **41:** 227–245, 1995.

Spencer Cohomologies and Symmetry Groups

V. LYCHAGIN and L. ZIL'BERGLEIT
"Sophus Lie" Centre (Moscow Branch), PB 546, 119618 Moscow, Russia

(Received: 28 February 1994)

Abstract. We propose the construction of a spectral sequence converging to Spencer cohomologies. By using symmetry groups of differential equations systems, we manage to unify computations by reduction to the invariant systems over a homogeneous space. The conditions of coincidence of Spencer cohomologies with the cohomologies of an invariant Spencer complex we obtain from the arithmetic of a $\mathbb{C}$-characteristic manifold with respect to fundamental weights of the homogeneous space.

Mathematics Subject Classifications (1991): 58G05, 58G37, 58H10, 35Q53.

Key words: differential equations, Spencer cohomologies, spectral sequences, symmetries, characters, homogeneous space, characteristic manifold, degenerated bicomplexes.

Introduction

We propose the construction of a spectral sequence determined by natural filtrations in a Spencer complex. This sequence for the de Rham operator leads to the classical Leray–Serre theorem. Our result links the stable Spencer cohomologies of a SDE on the space of a noncharacteristic bundle with the de Rham cohomologies of the base with the values in stable Spencer cohomologies of the fiberwise image equations.

By using symmetry groups of SDE, we, in the majority of cases, manage to unify computations by reduction to the invariant SDE over a homogeneous space. The coincidence of stable cohomologies with the cohomologies of an invariant Spencer complex is the crucial point in the computation of Spencer cohomologies. We obtain the conditions of coincidence from the arithmetic of a (complex) characteristic manifold of a SDE with respect to fundamental weights of a homogeneous space. Hence, the computation problem for stable Spencer cohomologies of invariant SDE over a homogeneous manifold acquires a purely algebraic nature.

1. Spencer Complexes

This section contains some basic definitions and properties relating to Spencer complexes.

1.1. JETS, DIFFERENTIAL EQUATIONS AND CHARACTERISTICS

For a smooth manifold M, $\dim M = n$, and a smooth vector bundle α: $E(\alpha) \to M$ by $C^\infty(\alpha)$ we denote the module of smooth sections of bundle α. Let α_k: $J^k(\alpha) \to M$ be the corresponding bundle of k-jets of sections. By $\mathcal{J}^k(\alpha)$, we denote the module of smooth sections of α_k. For any section $h \in C^\infty(\alpha)$, a section $j_k(h) \in C^\infty(\alpha_k)$ is defined as

$$j_k(h)\colon\ x \longmapsto [h]_x^k,$$

where $[h]_x^k$ is the k-jet of h at point $x \in M$.

Let

$$\alpha_{k,l}\colon\ \mathcal{J}^k(\alpha) \longrightarrow \mathcal{J}^l(\alpha),$$

be module epimorphisms, such that $\alpha_{k,l} \circ j_k(h) = j_l(h)$ for all $h \in C^\infty(\alpha)$. We denote by $\mathrm{Diff}_k(\alpha, \beta)$ the module of linear differential operators of the order k acting from the sections of α into the sections of β. Any operator $\Delta \in \mathrm{Diff}_k(\alpha, \beta)$ defines a homomorphism $\varphi_\Delta \in \mathrm{Hom}(\alpha_k, \beta)$ by the following rule:

$$\varphi_\Delta([h]_x^k) = \Delta(h)(x).$$

Any operator $\Delta \in \mathrm{Diff}_k(\alpha, \beta)$ satisfying natural regularity conditions defines a system of differential equations $R_\Delta = \varphi_\Delta^{-1}(0)$. We shall consider this system as a subbundle r of the bundle α_k:

$$r\colon\ E(r) = R \longrightarrow M$$

A section $h \in C^\infty(\alpha)$, such that $j_k(h) \subset C^\infty(r)$, is called a *smooth solution* of R. The module of smooth sections of the bundle r is denoted by $\mathcal{R}$. We also denote by $\Delta^{(l)}$ and $r^{(l)}$ the lth prolongations of the operator $\Delta \in \mathrm{Diff}_k(\alpha, \beta)$ and of the SDE $r \in \alpha_k$, respectively:

$$\Delta^{(l)} \in \mathrm{Diff}_{k+l}(\alpha, \mathcal{J}^l\beta), \qquad r^{(l)} \subset \alpha_{k+l}.$$

The symbol of l-prolongation of $\Delta \in \mathrm{Diff}_k(\alpha, \beta)$ is the mapping

$$\sigma(\Delta^{(l)})(x)\colon\ S^{k+l}\tau_x^* \otimes \alpha_x \longrightarrow S^l\tau_x^* \otimes \beta_x,$$

$$\mathrm{d}_x f_1 \circ \mathrm{d}_x f_2 \circ \cdots \circ \mathrm{d}_x f_{k+l} \otimes h_x \longmapsto [\Delta(f_1 \cdots f_{k+l} h)]_x^l,$$

where $f_i \in C^\infty(M)$, $f_i(x) = 0$, $i = 1, 2, \ldots, k+1$, $h \in C^\infty(\alpha)$, and τ^* is the cotangent bundle.

EXAMPLE. Consider the exterior differential d: $\Lambda^k(M) \to \Lambda^{k+1}(M)$. Symbols of its l-prolongation

$$\sigma(\mathrm{d}^{(l)})(x)\colon\ S^{l+1}\tau_x^* \otimes \Lambda^k\tau_x^* \longrightarrow S^l\tau_x^* \otimes \Lambda^{k+1}\tau_x^*,$$

$$\mathrm{d}_x f_1 \circ \cdots \circ \mathrm{d}_x f_{l+1} \otimes \omega_k$$
$$\longmapsto \sum_{i=1}^{l+1} \mathrm{d}_x f_1 \circ \cdots \circ \widehat{\mathrm{d}_x f_i} \circ \cdots \circ \mathrm{d}_x f_{l+1} \otimes \mathrm{d}_x f_i \wedge \omega_k, \quad \omega_k \in \Lambda^k \tau_x^*,$$

are called Spencer δ-operators and are denoted by δ.

A vector bundle $g = \operatorname{Ker} \sigma(\Delta) \subset S^k\tau^* \otimes \alpha$ is called *a symbol bundle of* SDE $r_\Delta \subset \alpha_k$ associated with the operator $\Delta \in \mathrm{Diff}_k(\alpha, \beta)$. We use the notations

$$g^l = \operatorname{Ker} \sigma(\Delta^{(l)}) \subset S^{k+l}\tau^* \otimes \alpha$$

Using the fact that $\delta(C^\infty(g^{l-1})) \subset C^\infty(g^{l-1}) \otimes \Lambda^1(M)$, we define a complex

$$0 \longrightarrow C^\infty(g^l) \xrightarrow{\delta} C^\infty(g^{l-1}) \otimes \Lambda^1(M) \xrightarrow{\delta} \cdots$$
$$\xrightarrow{\delta} C^\infty(g^{l-n}) \otimes \Lambda^n(M) \longrightarrow 0$$

which is called the *Spencer δ-complex.*

We denote by $H^{l,k}(g)$ the cohomologies of this complex (i.e. *the Spencer δ-cohomologies* of an SDE r) at the term $C^\infty(g^l) \otimes \Lambda^k(M)$.

POINCARE'S δ-LEMMA [1]. *There exist a number l_0 such that $H^{l,i}(g) = 0$ for any $l \geqslant l_0$, $i \geqslant 0$.*

A subspace U will be called *noncharacteristic*, if

$$U \otimes g_x^{(l)} \cap g_x^{(l+1)} = 0$$

for $l \geqslant l_0$.

Consider a bundle κ: $M \to B$ with a fiber F. The bundle κ is called *noncharacteristic* with respect to a system of differential equations $r \subset \alpha_k$, if all subspaces $\operatorname{Ann} F_{b,y}$ are noncharacteristic, $b \in B$, $y \in F_b$.

1.2. The Spencer operator

The Spencer operator D: $\mathcal{J}^l(\alpha) \otimes \Lambda^k(M) \to \mathcal{J}^{l-1}(\alpha) \otimes \Lambda^{k+1}(M)$ is a uniquely determined natural operator [1]. More exactly, let Ψ be a fiberwise map from the vector bundle α^1 over M into the vector bundle α^2 over M_2:

$$\begin{array}{ccc} E(\alpha^1) & \xrightarrow{\Psi} & E(\alpha^2) \\ \downarrow & & \downarrow \\ M_1 & \xrightarrow{\bar{\Psi}} & M_2, \end{array}$$

i.e. Ψ: $\alpha_x^1 \to \alpha^2_{\Psi(x)}$ are isomorphisms for all $x \in M$. Then the diagrams

$$\begin{array}{ccc} \mathcal{J}^l(\alpha^2) \otimes \Lambda^k(M_2) & \xrightarrow{\Psi_l^* \otimes \bar{\Psi}^*} & \mathcal{J}^l(\alpha^1) \otimes \Lambda^k(M_1) \\ D \downarrow & & \downarrow D \\ \mathcal{J}^{l-1}(\alpha^2) \otimes \Lambda^{k+1}(M_2) & \xrightarrow{\Psi_{l-1}^* \otimes \bar{\Psi}^*} & \mathcal{J}^{l-1}(\alpha^1) \otimes \Lambda^{k+1}(M_1) \end{array}$$

are commutative. Here $\bar{\Psi}^*\colon C^\infty(M_2) \to C^\infty(M_1)$, $\Psi_k^*\colon \mathcal{J}^k(\alpha^2) \to \mathcal{J}^k(\alpha^1)$ are induced mappings. In particular, if Ψ is a restriction on a submanifold $M_0 \subset M$, $\alpha|_{M_0} = \alpha^0$, the diagrams have the following form

$$\begin{array}{ccc} \mathcal{J}^l(\alpha) \otimes \Lambda^k(M) & \longrightarrow & \mathcal{J}^l(\alpha^0) \otimes \Lambda^k(M_0) \\ {\scriptstyle D}\downarrow & & \downarrow{\scriptstyle D} \\ \mathcal{J}^{l-1}(\alpha) \otimes \Lambda^{k+1}(M) & \longrightarrow & \mathcal{J}^{l-1}(\alpha^0) \otimes \Lambda^{k+1}(M_0). \end{array}$$

Let us clarify the geometrical sense of the Spencer operator. Let $\theta \in \mathcal{J}^{k+1}(\alpha)$. At any point $x \in M$ the $k+1$-jet $x_{k+1} = \theta(x)$ determines a decomposition

$$T_{x_k} J^k\alpha = L(x_{k+1}) \oplus J_x^k(\alpha),$$

where $x_k = \alpha_{k+1,k}(x_{k+1})$, and, due to linearity of α, the fibre $J_x^k\alpha$ is identified with the tangent space to the fibre $\alpha_k^{-1}(x)$,

$$L(x_{k+1}) = T_{x_k}[j_k(h)], \qquad x_{k+1} = [h]_x^{k+1}, \quad h \in C^\infty(\alpha).$$

Let $\widehat{\theta} = \alpha_{k+1,k}(\theta) \in \mathcal{J}^k(\alpha)$. Then the submanifold $\widehat{\theta}(M)$ passes through the point x_k and projects diffeomorphically on M. Hence,

$$(\alpha_k)_*\colon\ T_{x_k}(\widehat{\theta}(M)) \longrightarrow T_x(M)$$

is an isomorphism. Using this isomorphism, one can lift any vector $v \in T_x(M)$ into $T_{x_k}(J^k(\alpha))$ and then project along $L(x_{k+1})$ using the decomposition above. The element constructed in such way we consider as a value of some 1-form $D\theta \in \mathcal{J}^k(\alpha) \otimes \Lambda^1(M)$ on the vector $v \in T_xM$.

1.3. Spencer complexes

Let r be a formal integrable SDE. Restricting the Spencer operator on $\mathcal{R}^{(l)}$ we obtain a differential operator of order 1:

$$D\colon\ \mathcal{R}^{(l)} \otimes \Lambda^k(M) \longrightarrow \mathcal{R}^{(l-1)} \otimes \Lambda^{k+1}(M).$$

It defines a complex called the naive Spencer complex associated with the system r:

$$0 \longrightarrow \mathcal{R}^{(l)} \xrightarrow{D} \mathcal{R}^{l-1} \otimes \Lambda^1(M) \longrightarrow \cdots \longrightarrow \mathcal{R}^{k-n} \otimes \Lambda^n(M) \longrightarrow 0. \tag{1}$$

From Poincare's δ-lemma, it follows that for $l \geqslant l_0 + n$ the cohomologies of a naive Spencer complex become stable. This stable cohomologies are called *Spencer cohomologies* of the formal integrable system r; we denote them by $\mathcal{H}^i(r)$, $i = 0, 1, \ldots$.

The second (sophisticated) Spencer complex is more simple. It arises as follows. For the operator $\Delta_1\colon C^\infty(\alpha^1) \to C^\infty(\alpha^2)$, we consider the compatibility operator Δ_2: $C^\infty(\alpha^2) \to C^\infty(\alpha^3)$ which determines a complex $C^\infty(\alpha^1) \xrightarrow{\Delta_1} C^\infty(\alpha^2) \xrightarrow{\Delta_2} C^\infty(\alpha^3)$. For the operator Δ_2 we consider the compatibility operator Δ_3, etc. In general, the complex $C^\infty(\alpha^1) \xrightarrow{\Delta_1} C^\infty(\alpha^2) \xrightarrow{\Delta_2} C^\infty(\alpha^3) \xrightarrow{\Delta_3} \cdots$ does not exist. But for the formally integrable system, such a complex always exists and is unique in the class of formally exact ones (the Kuranishi theorem [3]). The second Spencer complex has the form

$$0 \longrightarrow C_l^0 \xrightarrow{D} C_l^1 \xrightarrow{D} \cdots \xrightarrow{D} C_l^n \longrightarrow 0, \tag{2}$$

where $C_l^s = \mathcal{R}^{(l)} \otimes \Lambda^s(M) \;/\; \delta(C^\infty(g^{(l+1)}) \otimes \Lambda^{s-1}(M))$. Cohomologies of this complex stabilize for large l's and coincide with the Spencer cohomologies $\mathcal{H}^\bullet(r)$.

The exterior product on differential forms defines the structure of $\Lambda^\bullet(M) = \bigoplus_{i\geqslant 0} \Lambda^i(M)$-module on $C_l^\bullet = \bigoplus_{i\geqslant 0} C_l^i$. Passing to the cohomologies, this product generates a module structure in Spencer cohomologies $\mathcal{H}^\bullet(r)$ over the de Rham cohomologies $H^\bullet(M)$. Below, we shall regard large (stable) values of l and omit the subscript l in C_l^k: $C^k = C_l^k$.

EXAMPLES

1. The zero group of Spencer cohomologies of a formally integrable system r_Δ is isomorphic to the space of solutions: $\mathcal{H}^0(r_\Delta) = \operatorname{Ker}\Delta$.

2. For any determined system $r_\Delta \subset \alpha_k$, one has $\mathcal{H}^0(r_\Delta) = \operatorname{Ker}\Delta$, $\mathcal{H}^1(r_\Delta) = \operatorname{Coker}\Delta$, and $\mathcal{H}^i(r_\Delta) = 0$, for all $i > 1$.

3. Spencer cohomologies of a system determined by the operator d: $C^\infty(M) \to \Lambda^1(M)$ coincide with de Rham's cohomologies.

4. A natural generalization of the previous example is a SDE determined by a flat connection

$$\nabla\colon\ C^\infty(\alpha) \longrightarrow C^\infty(\alpha) \otimes \Lambda^1(M).$$

Spencer cohomologies of this system coincide with de Rham's cohomologies of the manifold M with the coefficients in sections of bundle α.

5. Let $r \subset \alpha_k$ be a formally integrable SDE of a finite type and let l be such number that $r^{(l+1)} \simeq r^{(l)}$. Then the Spencer operator

$$\mathcal{R}^{(l)} \simeq \mathcal{R}^{(l+1)} \xrightarrow{D} \mathcal{R}^{(l)} \otimes \Lambda^1(M)$$

defines a flat connection in the bundle r^l. Thus, the stable Spencer cohomologies are isomorphic to the cohomologies considered in the previous example.

6. Let M be a complex manifold and let α be a holomorphic bundle over M. Denote by $\Lambda^{p,q}(\alpha)$ differential (p,q)-forms on the manifold M with the values in the bundle α and by $\Omega^p(\alpha)$ holomorphic p-forms with the value in the bundle α. The stable Spencer cohomologies determined by the Cauchy–Riemann differential equation operator

$$\bar{\partial}\colon\ \Lambda^{p,0}(\alpha) \longrightarrow \Lambda^{p,1}(\alpha)$$

are Dolbeault cohomologies $H^j(M, \Omega^p(\alpha))$.

7. Let us consider a determined system $r_\Delta \subset \alpha_k$ and define a mapping

$$H^1(M) \otimes \mathcal{H}^0(r_\Delta) \longrightarrow \mathcal{H}^1(r_\Delta).$$

Any element of $H^1(M) \otimes \mathcal{H}^0(r_\Delta)$ can be locally represented in the form $\sum_i \mathrm{d} f_i \otimes s_i$, where $f_i \in C^\infty(M)$, $s_i \in \operatorname{Ker} \Delta$. Consider elements $\Delta(\sum_i f_i \otimes s_i)$ glue them together over all the local charts and factorise with respect to ImΔ. As the result we obtain an element of Coker$\Delta \simeq \mathcal{H}^1(r_\Delta)$.

2. Spectral Sequences

In this section, we expose the construction of Spencer cohomology spectral sequences analogous to the classical Leray–Serre sequence for a fibre bundle.

2.1. D- AND δ-SPENCER COHOMOLOGIES

Let $r \subset \alpha_k$ be a formally integrable system over M. We use the notations $R^\infty = \lim_{l \to \infty} R^{(l)}$ and $\mathcal{K}^{(l)} = \operatorname{Ker} \alpha_{\infty, l}|_{R^\infty}$. Introduce a filtration in the first Spencer complex:

$$F_p(\mathcal{R}^{(\infty)} \otimes \Lambda^{p-q}(M)) = \mathcal{K}^{(k+q)} \otimes \Lambda^{p-q}(M), \quad q \geqslant -1. \tag{1}$$

Spencer operator D preserves this filtration.

THEOREM [6]. *Let $r \subset \alpha_k$ be a formally integrable normal system over a smooth manifold M. Then filtration* 2.1(1) *determines a spectral sequence converging to Spencer cohomologies of system r. The first term of this sequence has the form*:

$$E_1^{p,q} = H^{q,p-q}(g), \quad q \geqslant 0$$

and $\{E_1^{p,-1}, d_1^{p,-1}\}$ is the following complex:

$$0 \longrightarrow \mathcal{J}^{k-1}(\alpha) \xrightarrow{d_1} \frac{\mathcal{J}^{k-1}(\alpha) \otimes \Lambda^1(M)}{\delta(C^\infty(g))} \longrightarrow \cdots$$
$$\xrightarrow{d_1} \frac{\mathcal{J}^{k-1}(\alpha) \otimes \Lambda^n(M)}{\delta(C^\infty(g) \otimes \Lambda^{n-1}(M))} \longrightarrow 0 \tag{2}$$

If r is involutive, then the term $(E_1^{p,q}, d_1^{p,q})$ coincides with complex 2.1(2) *and the spectral sequence degenerates on the second step.*

2.2. SYSTEMS OVER FIBRE BUNDLES

Let $r \subset \alpha_k$ be a formally integrable normal system over a smooth manifold M. We suppose that M is a total space of a fibre bundle κ: $M \to B$, over a smooth manifold B. Keeping in mind the Spencer complex filtration, we call a vector field on M vertical if $\kappa_*(X) = 0$. The module $\mathcal{D}_\kappa(M)$ of vertical fields is a Lie algebra. A differential i-form $\omega \in \Lambda^i(M)$ is called *q-horizontal* if $X_1 \wedge \cdots \wedge X_{q+1} \lrcorner \omega = 0$ for any $X_1, \ldots, X_{q+1} \in \mathcal{D}_\kappa(M)$. A submodule of q-horizontal forms is denoted by $\Lambda_q^\bullet(M)$. Let

$$F_{p,q} = F_p(\mathcal{R}^{(l)} \otimes \Lambda^{p+q}(M)) = \mathcal{R}^{(l)} \otimes \Lambda_q^{p+q}(M). \tag{1}$$

PROPOSITION 1. *Modules* 2.2(1) *define a filtration in Spencer complexes* 1.7(1).

Describe the initial terms of the Spencer cohomology spectral sequence associated with the filtration.

PROPOSITION 2. *There exists an isomorphism of $C^\infty(B)$-modules*

$$\Lambda_q^i(M)/\Lambda_{q-1}^i(M) = \Lambda^q(\kappa) \bigotimes_{C^\infty(B)} \Lambda^{i-q}(B),$$

$$E_0^{p\,q} = F_{p,q}/F_{p+1,q+1} \simeq \mathcal{R}^l \bigotimes_{C^\infty(M)} \Lambda^q(\kappa) \bigotimes_{C^\infty(B)} \Lambda^p(B),$$

where $\Lambda^p(\kappa) = \Lambda^p(M)/\kappa^*\Lambda^p(B)$ are differential forms on fibers of bundle κ.

PROPOSITION 3. *The differentials*

$$d_0^{p\,q} \colon E_0^{p\,q} \longrightarrow E_0^{p\,q+1}$$

are first-order fiberwise differential operators preserving the $C^\infty(B)$-module structure:

$$d_0^{p\,q}(fv) = f d_0^{p\,q}(v),$$

$f \in C^\infty(B),\ v \in E_0^{p\,q}$.

Proof. By the definition of the Spencer operator D, we have $D(fv) = fDv + \mathrm{d}f \wedge \alpha_{l+1,l}v$. For $\mathrm{d}f \in \Lambda_0^1(M)$, it follows that $\mathrm{d}f \wedge \alpha_{l+1,l}v \in \mathcal{R}^{l-1} \otimes \Lambda_q^{p+q+1} \subset F_{p+1,q}$. □

Let bundle κ be noncharacteristic and α_b: $E(\alpha_b) \to F_b$ be the restriction of α on to the fiber F_b. Consider a restriction $h_l(b)$: $R_b^l \to (\alpha_b)_{l+k}$. The image of $h_l(b)$ is denoted by r_b^l.

PROPOSITION 4. *If κ is a noncharacteristic bundle with respect to r, then $h_l(b)$ is a monomorphism.*

Proof. By formal integrability and normality r, one obtains the result from the following commutative diagram

$$\begin{array}{ccccccccc} 0 & \longrightarrow & g^l & \longrightarrow & r^l & \longrightarrow & r^{l-1} & \longrightarrow & 0 \\ & & \big\downarrow h_l(b) & & \big\downarrow h_l(b) & & \big\downarrow h_{l-1}(b) & & \\ 0 & \longrightarrow & \alpha_b \otimes S^{l+k}\tau^*(B) & \longrightarrow & (\alpha_b)_{l+k} & \longrightarrow & (\alpha_b)_{l+k-1} & \longrightarrow & 0. \end{array}$$

PROPOSITION 5. $(r_b^l)^{(1)} \supset r_b^{l+1}$.

Proof. Let $a \in C^\infty(\alpha)$, $a_b = a|_{F_b}$ and

$$[a_b]_y^{l+k} \in R_b^l, \qquad [a]_y^{l+k} \in R^l.$$

Then the section $j_{l+k}(a)$ is tangent to R^l at the point $[a]_y^{l+k}$ and, consequently, the section $j_{l+k}(a_b)$ is tangent to R_b^l at the point $[a_b]_y^{l+k}$. Therefore, $[a_b]_y^{l+k+1} \in (R_b^l)^{(1)}$. □

THEOREM 1. *Let κ be a noncharacteristic bundle with respect to a formally integrable normal system $r \subset \alpha_k$. Then there exists an integer $l_0 \geqslant 0$, such that*

$$(r_b^l)^{(1)} = r_b^{l+1}$$

and the image system

$$r_b^{l_0}\colon\ R_b^{l_0} \longrightarrow F_b$$

is formally integrable.

Proof. It follows from the Cartan–Kuranishi theorem [3] and from Proposition 5. □

Consider a fiber bundle $r_\kappa^{l_0}\colon R_\kappa^{l_0} \to M$ with the fiber $R_{b,y}^{l_0}$ over $y \in F_b$. There exists a 'fiberwise' Spencer complex

$$0 \longrightarrow \mathcal{R}_\kappa^l \xrightarrow{D_\kappa} \mathcal{R}_\kappa^{l-1} \otimes \Lambda^1(\kappa) \xrightarrow{D_\kappa} \cdots \longrightarrow 0. \tag{2}$$

Stable cohomologies of this complex we denote by $\mathcal{H}_\kappa^\bullet(r)$.

From the naturality of D, we obtain the following commutative diagram

$$\begin{array}{ccc} \mathcal{R}^l \otimes \Lambda^q(\kappa) \otimes \Lambda^p(B) & \xrightarrow{d_0^{p\,q}} & \mathcal{R}^{l-1} \otimes \Lambda^{q+1}(\kappa) \otimes \Lambda^p(B) \\ \big\downarrow h_l(\kappa)\otimes 1 & & \big\downarrow h_l(\kappa)\otimes 1 \\ \mathcal{R}_\kappa^l \otimes \Lambda^q(\kappa) \otimes \Lambda^p(B) & \xrightarrow{D_\kappa \otimes 1} & \mathcal{R}_\kappa^{l+1} \otimes \Lambda^{q+1}(\kappa) \otimes \Lambda^p(B) \end{array}$$

Therefore, $E_1^{p\,q}$ can be represented as

$$E_1^{p\,q} = \mathcal{H}_\kappa^q(r) \otimes \Lambda^p(B).$$

PROPOSITION 6. *The differential*

$$d_1^{p\,q}\colon\ E_1^{p\,q} \longrightarrow E_1^{p+1\,q}$$

is defined by a flat connection in the module $\mathcal{H}_\kappa^\bullet(r)$*:*

$$d_1^{0\,q}\colon\ \mathcal{H}_\kappa^q(r) \longrightarrow \mathcal{H}_\kappa^q(r)\otimes\Lambda^1(B).$$

Putting all this together, we get the following result.

THEOREM 2. *Let* $\kappa\colon M \to B$ *be a smooth vector bundle, noncharacteristic with respect to a normal formally integrable system* $r \subset \alpha_k$. *Then filtration* 2.2(1) *determines spectral sequence* $(E_r^{p\,q}, d_r^{p\,q})$ *converging to the Spencer cohomologies of* r *and*

1^0. *The terms* $E_1^{p\,q}$ *are isomorphic to* $\mathcal{H}_\kappa^q(r)\otimes\Lambda^p(B)$.

2^0. *Differentials* $d_1^{0\,q}\colon\ \mathcal{H}_\kappa^q(r)\to\mathcal{H}_\kappa^q(r)\otimes\Lambda^1(B)$ *determine a flat connection in the module* $\mathcal{H}_\kappa^\bullet(r)$.

3^0. *The terms* $E_2^{p\,q}$ *are isomorphic to the de Rham cohomologies of manifold* B *with coefficients in the module* $\mathcal{H}_\kappa^\bullet(r)$ *equipped with the natural flat connection* $d_1^{p\,q}\colon\ E_2^{p\,q} = H^p(B, \mathcal{H}_\kappa^q(r))$.

Remarks

1. Denote by $(\widehat{E}_r^{p\,q}, \widehat{d}_r^{p\,q})$ the Leray–Serre spectral sequence for the de Rham cohomologies of the bundle κ. Then the sequence $(E_r^{p\,q}, d_r^{p\,q})$ is a graded module over $(\widehat{E}_r^{p\,q}, \widehat{d}_r^{p\,q})$.

2. If κ is an elliptic noncharacteristic bundle [7], then the Spencer cohomologies are finite-dimensional.

3. If r is an elliptic SDE satisfying the conditions of the theorem, then its index $\operatorname{ind} R$, understood as the Euler characteristic of the stable Spencer complex, is $\operatorname{ind} R = X(B)\,\operatorname{ind} R_\kappa$ when the SDE R_κ is elliptic.

3. Symmetries

We use spectral sequence 2.2 for the computation of Spencer cohomologies of a SDE with a compact connected symmetry group. The crucial point of our computations lies in the term $E_2^{p\,q}$ which, in the case under consideration, reduces to Spencer cohomologies of an invariant SDE over a homogeneous space. Therefore, below we accentuate on the systems of such a type.

3.1. Definitions

Consider a Lie group G and its representation ρ by automorphisms of a fiber bundle α: $E(\alpha) \to M$, ρ: $G \to \mathrm{Aut}(\alpha)$. This representation covers the action of G on M in such a way that the following diagram

$$\begin{array}{ccc} E(\alpha) & \xrightarrow{\rho(g)} & E(\alpha) \\ {\scriptstyle\alpha}\downarrow & & {\scriptstyle\alpha}\downarrow \\ M & \xrightarrow{g} & M \end{array}$$

is commutative for all $g \in G$.

Automorphisms $\rho(g)$ induce automorphisms $\rho^*(G)$ of the module $C^\infty(\alpha)$ over algebra automorphisms

$$(g^{-1})^*\colon\ C^\infty(M) \longrightarrow C^\infty(M).$$

Namely,

$$\rho^*(g)(h) = \rho(h) \circ h \circ g^{-1}, \qquad \rho^*(g)(fh) = (g^{-1})^*(f)\rho^*(g)h$$

for any function $f \in C^\infty(M)$ and any section $h \in C^\infty(\alpha)$. The representation ρ: $G \to \mathrm{Aut}(\alpha)$ is lifted up to a representation $\rho^{(k)}$: $G \to \mathrm{Aut}(\alpha_k)$ in such a way that

$$\rho^{(k)}(g)([h]_x^k) = [\rho^*(g)(h)]_{g(x)}^k, \quad h \in C^\infty(\alpha),\ g \in G,\ x \in M.$$

Representations $\rho^{(k)}$, $k = 0, 1, \dots$, commute with the projections $\alpha_{k,l}$:

$$\alpha_{k,l} \circ \rho^{(k)}(g) = \rho^{(l)}(g) \circ \alpha_{k,l}$$

for all $g \in G$, $k \geqslant l$. We call a representation ρ: $G \to \mathrm{Aut}(\alpha)$ a *symmetry of a linear system* $r \subset \alpha_k$, if $\rho^{(k)}(g)(r) = r$ for all elements $g \in G$.

3.2. Symmetries and Spencer cohomologies

Let G be a compact connected symmetry group of a system $r \subset \alpha_k$. Then the representations $\rho^{(k)}$: $G \to \mathrm{Aut}(\alpha_k)$ induce representations $\rho^{(k)}$: $G \to \mathrm{Aut}(C^i)$ for all $i = 0, 1, \dots, n$. In what follows, we consider all the modules and fiber bundles to be complexified. By this reason, any $C^\infty(M)$-module C^k is decomposed in isotopic components of the representation $\rho^{(k)}$: $C^k = \bigoplus_{\mu \in \mathrm{Ch}(G)} C_\mu^k$, where $\mathrm{Ch}(G)$ is the group of characters for group G.

PROPOSITION 1. *The Spencer operator* D: $C^k \to C^{k+1}$ *is decomposed into the direct sum of its restrictions* $D_\mu = D|_{C_\mu^k}$: $C_\mu^k \to C_\mu^{k+1}$, $\mu \in \mathrm{Ch}(G)$.

Proof. Let $v \in C_\mu^k$ be the dominant vector associated with the irreducible representation of character μ, T be maximal torus of group G, and B be Borel's subgroup of group G. To prove the proposition, it is sufficient to check that Dv is the dominant vector in module C_μ^{k+1}.

One can determine v by the conditions

$$\rho^{(k)}(t)v = \mu(t)v, \qquad \rho^{(k)}(b)v = v,$$

for all $t \in T$, $b \in B$. Since D is a natural operator (see 1.6), we have

$$\rho^{(k+1)}(t)Dv = \mu(t)Dv, \qquad \rho^{(k+1)}(b)Dv = Dv$$

for all $t \in T$, $b \in B$. □

Hence, for every μ, one has a complex

$$0 \longrightarrow C_\mu^0 \xrightarrow{D_\mu} C_\mu^1 \xrightarrow{D_\mu} \cdots \xrightarrow{D_\mu} C_\mu^n \longrightarrow 0 \qquad (1_\mu)$$

Denote by $\mathcal{H}_\mu^q$, $q = 0, 1, \ldots, n$ the corresponding cohomologies. From Proposition 3.2(1) one has

PROPOSITION 2. *The second Spencer complex is the direct sum of complexes* 3.2(1_μ) *and Spencer cohomology* $\mathcal{H}^q(R)$ *is the direct sum of cohomologies* $\mathcal{H}_\mu^q$ $\mathcal{H}^\bullet(r) = \bigoplus_{\mu \in \mathrm{Ch}(G)} \mathcal{H}_\mu^\bullet$.

We call complex 3.2(1_e) associated with the trivial character e invariant Spencer complex; its cohomologies are called invariant cohomologies and are denoted by $\mathcal{H}_\mu^\bullet$.

COROLLARY. *Let* $\mathcal{H}_\mu^q = 0$ *for all nontrivial characters* $\mu \in \mathrm{Ch}(G)$. *Then* $\mathcal{H}^q(r) \simeq \mathcal{H}_e^q$, *where* $q = 0, 1, \ldots, n$.

3.3. Differential equations over homogeneous space

Let α be a vector bundle over a homogeneous space G/G_0, where G is a compact connected Lie group, G_0 is a closed connected subgroup of G, and $R \subset E(\alpha_k)$ be a G-invariant formally integrable system.

One has a bundle $\widehat{\alpha} = \kappa^*(\alpha)$: $E(\widehat{\alpha}) \to G$ induced by the natural projection κ: $G \to G/G_0$. Then the module $C^\infty(\alpha)$ is identified with a submodule of module $C^\infty(\widehat{\alpha})$ with the trivial G_0-action. Conversely, G-fibre bundle β: $E(\beta) \to G$ is induced from some fibre bundle α: $E(\alpha) \to G/G_0$ if the action of the stability subgroup G_0 on fibre bundle β is trivial.

The following result is a consequence of the isotypic decomposition.

PROPOSITION 1. *For any G-fibre bundle* α: $E(\alpha) \to G/G_0$ *G-module* $C^\infty(\alpha)$ *is decomposed into the direct sum of isotopic components* $C^\infty(\alpha)_\mu$, *where* μ *are elements of character group* $\mathrm{Ch}(G, G_0)$ *consisting of G-characters trivial on Cartan's subgroup of* G_0.

Using this proposition for differential module $C^\bullet$, we get the decomposition

$$C^k = \bigoplus_{\mu \in \mathrm{Ch}(G,G_0)} C^k_\mu.$$

Remark. The following result was pointed out by D. P. Jelobenko. Let $\mathcal{G}$ be a reductive Lie algebra over $\mathbb{C}$, $\dim \mathcal{G} < \infty$, $\mathcal{G}_0$ be a regular imbedded subalgebra $\mathcal{G}$, $\mathcal{G} = \mathcal{G}_0 \oplus \mathcal{G}_1$, where $\mathcal{G}_1$ is the $\mathcal{G}_0$-module, and K be a lattice generated by the weights (roots) of module $\mathcal{G}_1$.

THEOREM. *A representation π_λ of $\mathcal{G}$ with the dominant weight λ* ($\dim \pi_\lambda < \infty$) *contains the representation π_μ of a subalgebra $\mathcal{G}_0$ if and only if*

$$\lambda|_{\mathcal{H}_0} - \mu \in K|_{\mathcal{H}_0}, \tag{1}$$

where $\mathcal{H}_0$ is Cartan's subalgebra of the algebra $\mathcal{G}_0$.

COROLLARY. *Condition* 3.3(1) *for the representation π_0 of subalgebra $\mathcal{G}_0$* ($\mu =$ 0) *is $\lambda|_{\mathcal{H}_0} \in K|_{\mathcal{H}_0}$.*

Returning to complexes 3.2(1_μ), $\mu \in \mathrm{Ch}(G, G_0)$ we consider operators ∂_μ: $C^k_e \to C^{k+1}_e$ which are defined by the following commutative diagram:

$$\begin{array}{ccc} C^k_\mu & \xrightarrow{D_\mu} & C^{k-1}_\mu \\ {\scriptstyle \mu^{-1}}\downarrow & & \downarrow{\scriptstyle \mu^{-1}} \\ C^k_e & \xrightarrow{\partial_\mu} & C^{k-1}_e \end{array}$$

PROPOSITION 2. *Operator ∂_μ has the form $\partial_\mu = D_e + \sigma_\mu$, where σ_μ is the symbol of D on the infinitesimal character $\mu^* = \mathrm{d}(\ln \mu) = \mu^{-1}\mathrm{d}\mu$.*

Proof. Let $\mu \in \mathrm{Ch}(G, G_0)$, $f \in C^\infty(G)_\mu$, and $\omega \in C^k_e$. Then, obviously,

$$D_\mu(f\omega) = d_\mu f \cdot \omega + f \cdot D_e \omega,$$

where d_μ: $\Lambda^k_\mu(\tau^*_G) \to \Lambda^{k+1}_\mu(\tau^*_G)$ is the isotopic μ-component of de Rham's operator.

Choosing $f = \mu \in \mathrm{Ch}(G, G_0)$ we get

$$[D_\mu(\mu\omega)] = d_\mu \mu \cdot \omega + \mu D_e \omega \;=\; \mu[D_e + \sigma_\mu]\omega$$

for all $\omega \in C^\bullet_e$. □

PROPOSITION 3. *The cochain map μ^{-1}*

$$\begin{array}{ccccccccccc} 0 & \longrightarrow & C^0_\mu & \xrightarrow{D_\mu} & C^1_\mu & \xrightarrow{D_\mu} & \cdots & \xrightarrow{D_\mu} & C^n_\mu & \longrightarrow & 0 \\ & & {\scriptstyle \mu^{-1}}\downarrow & & \downarrow{\scriptstyle \mu^{-1}} & & & & \downarrow{\scriptstyle \mu^{-1}} & & \\ 0 & \longrightarrow & C^0_e & \xrightarrow{\partial_\mu} & C^1_e & \xrightarrow{\partial_\mu} & \cdots & \xrightarrow{\partial_\mu} & C^n_e & \longrightarrow & 0 \end{array}$$

defines isomorphism of complexes.

The proof is similar to the previous one.

It should be noted that

$$D_e^2 = 0, \qquad \sigma_\mu^2 = 0, \qquad \partial_\mu^2 = 0,$$

and, therefore, operators σ_μ and D_e anticommute:

$$D_e \circ \sigma_\mu + \sigma_\mu \circ D_e = 0.$$

3.4. THEOREM OF COINCIDENCE

As was proved above, Spencer cohomologies coincide with invariant ones if differential modules $(C_e^\bullet, \partial_\mu)$, where $\partial_\mu = \sigma_0 + \sigma_{\mu^*}$ are acyclic for any nontrivial infinitesimal character $\mu^* \in \mathrm{Ch}(\mathcal{G}, \mathcal{G}_0)$. Here

$$\mathcal{G} = \mathrm{Lie}(G), \qquad \mathcal{G}_0 = \mathrm{Lie}(G_0), \qquad \sigma_0 = D_e, \qquad \mu^* = \frac{\mathrm{d}\mu}{\mu}.$$

Fundamental infinitesimal characters $\nu_1, \ldots, \nu_r$ generate the space of all infinitesimal characters over $\mathbb{C}$.

Use Remark 3.3 and represent any symbol in the form

$$\sigma_{\mu^*} = \sum_{i=1}^{r} k_i \sigma_i, \quad k_i \in \mathbb{Z}, \ \sigma_i = \sigma_{\nu_i}.$$

Therefore, by the results from the Appendix combined with Quillen's theorem [2], we obtain

THEOREM. *Let r be a G-invariant formally integrable* SDE *over a homogeneous space G/G_0, where G is a compact connected Lie group and G_0 is a regular imbedded, connected subgroup of group G. Then the Spencer cohomologies of R coincide with invariant ones up to the order m $\mathcal{H}^q(r) = \mathcal{H}_e^q$, $q \leqslant m$, if nontrivial characters $\mu^* \in \mathrm{Ch}(\mathcal{G}, \mathcal{G}_0)$ are noncharacteristic over $\mathbb{C}$ and the following conditions hold*:

A. *Operators*

$$\Phi_i \circ \sigma_k \colon \mathcal{H}_k^q(C_e/\sigma_i C_e) \longrightarrow \mathcal{H}_k^q(C_e/\sigma_i C_e),$$

where $k = (1, k_1, \ldots, k_{i-1}, 0, k_{i+1}, \ldots, k_r)$, $k_j \in \mathbb{Z}$, $i = 1, 2, \ldots, r$, $q \leqslant m$ have no nonzero integer eigenvalues.

B. *Differential modules*

$$(\mathcal{H}_k^\bullet, \sigma_i), \quad k = (1, k_1, \ldots, k_{i-1}, 0, k_{i+1}, \ldots, k_r), \ k_j \in \mathbb{Z}, \ i = 1, 2, \ldots, r,$$

are $(m+1)$-acyclic.

C. *Differential modules*

$$(\mathcal{H}_e^\bullet, \sigma_t), \quad t = (0, t_1, \dots, t_r),\ t_j \in \mathbb{Z},\ t \neq 0$$

are $(m+1)$*-acyclic.*

D. *Differential module (see Theorem 4 in the Appendix below)* $(E \otimes \mathcal{H}_e^\bullet, \sigma)$ *is* $(m+1)$*-acyclic.*

Appendix. Degenerated Bicomplexes

We discuss here methods of cohomology computations for complexes equipped with a set of anticommuting differentials. Let

$$P^\bullet = \bigoplus_{s=0}^{n} P^s$$

be a graded module. We suppose that module $P^\bullet$ is equipped with the set of anticommuting operators $\sigma_0, \sigma_1, \dots, \sigma_r$ of the degree $+1$. Therefore the operators $\sigma_k = \sum_{i=0}^{r} k_i \sigma_i$, $k = (k_0, k_1, \dots, k_r)$, $k_i \in \mathbb{R}$, $i = 0, 1, \dots,$ define cohomological complexes. We denote the corresponding cohomologies by $H_k^\bullet(P)$. We find out the triviality conditions for the cohomologies $H_k^\bullet(P)$, $k = (1, k_1, \dots, k_r)$, $k_i \in \mathbb{R}$, $k \neq 1_0$, where $1_0 = (1, 0, \dots, 0)$. For this purpose, we consider a submodule $\sigma_i P^\bullet$ of $P^\bullet$:

$$\sigma_i P^\bullet = \bigoplus_{s=1}^{n} \sigma_i P^{s-1},$$

where $i = 1, \dots, r$.

We suppose cohomologies $H_k^q(P)$ to be trivial for $q \leqslant m$ and $k = (0, k_1, \dots, k_r)$, $k \neq 0$. Then the natural mappings

$$\begin{aligned} &\Phi_i\colon\ \sigma_i P^q \longrightarrow P^q/\sigma_i P^{q-1}, \\ &\Phi_i\colon\ \sigma_i(x) \longmapsto x \bmod \operatorname{im} \sigma_i, \quad x \in P^q \end{aligned} \tag{1}$$

are isomorphisms for $i = 1, \dots, r$ and $q \leqslant m$.

From the exact differential modules sequence

$$0 \longrightarrow (\sigma_i P^\bullet, \sigma_k) \longrightarrow (P^\bullet, \sigma_k) \longrightarrow (P^\bullet/\sigma_i P^{\bullet-1}, \sigma_k) \longrightarrow 0, \tag{2}$$

where $k = (k_0, k_1, \dots, k_{i-1}, 1, k_{i+1}, \dots, k_r)$, $k_j \in \mathbb{R}$, $j = 0, 1, \dots, r$, $k_0 \neq 0$, and from isomorphisms (1), we obtain an exact cohomology sequence

$$\begin{aligned} &0 \longrightarrow H_k^0(P) \longrightarrow H_{k-1_i}^0(P/\sigma_i P) \xrightarrow{\delta_i} H_{k-1_i}^0(P/\sigma_i P) \longrightarrow \cdots \\ &\longrightarrow H_k^1(P) \longrightarrow \cdots \xrightarrow{\delta_i} H_{k-1_i}^{m-1}(P/\sigma_i P) \longrightarrow H_k^m(P) \longrightarrow \\ &\longrightarrow H_{k-1_i}^m(P/\sigma_i P) \xrightarrow{\delta_i} H_{k-1_i}^m(P/\sigma_i P) \longrightarrow \cdots, \end{aligned} \tag{3}$$

where $1_i = (0, 0, \ldots, 0, \underset{i}{1}, 0, \ldots, 0)$.

The boundary operator δ_i in this sequence is $1 + \Phi_i \circ \sigma_{k-1_i}$. It should be noted that

$$H^0_{k-1_i}(P/\sigma_i P) = \operatorname{Ker}(\sigma_i \circ \sigma_{k-1_i}),$$

when $n > 1$.

If $m = n$, then

$$H^{n-1}_{k-1_i}(P/\sigma_i P) = P^{n-1}/\sigma_i P^{n-2}, \qquad H^n_{k-1_i}(P/\sigma_i P) = 0.$$

The corollary of the exactness for (3) is

PROPOSITION 1. *Cohomologies $H^q_k(P)$, where $k = (k_0, k_1, \ldots, k_{i-1}, 1, k_{i+1}, \ldots, k_r), k_j \in \mathbb{R}$, $j = 0, 1, \ldots, r$, $k_0 \neq 0$, and $q \leqslant m$, are trivial, if the operators*

$$\delta_i\colon\ H^q_{k-1_i}(P/\sigma_i P) \longrightarrow H^q_{k-1_i}(P/\sigma_i P)$$

are isomorphisms.

Remark. The condition of Proposition 1 means that the operators

$$\Phi_i \circ \sigma_{k-1_i}\colon\ H^q_{k-1_i}(P/\sigma_i P) \longrightarrow H^q_{k-1_i}(P/\sigma_i P), \quad q \leqslant m,$$

have no eigenvalues equal to -1.

THEOREM 1. *Let $P^\bullet = \bigoplus_{s=0}^n P^s$ be a graded module equipped with a set of $r+1$ anticommuting operators $\sigma_0, \sigma_1, \ldots, \sigma_r$ of the degree $+1$. Cohomologies $H^q_k(P)$ of P with the differential $\sigma_k = \sum_{j=1}^r k_j \sigma_j + \sigma_0$, $k_j \in \mathbb{Z}$, $k = (1, k_1, \ldots, k_r)$ are trivial for $k \neq 1_0$ and $q \leqslant m$, if:*

1⁰. *Cohomologies $H^q_t(P)$ are trivial for $q \leqslant m$ and $t = (0, t_1, \ldots, t_r)$, $t_j \in \mathbb{Z}$, $t \neq 0$.*

2⁰. *Operators*

$$\Phi_i \circ \delta_k\colon\ H^q_k(P/\sigma_i P) \longrightarrow H^q_k(P/\sigma_i P)$$

for $q \leqslant m$ and $k = (1, k_1, \ldots, k_{i-1}, 0, k_{i+1}, \ldots, k_r)$, $k_j \in \mathbb{Z}$ for $i = 1, \ldots, r$ have no nonzero integer eigenvalues.

Proof. Consider module P with the differential σ_k, $k = (1, k_1, \ldots, k_r)$, $k \neq 1_0$, and choose a number i, $1 \leqslant i \leqslant r$ such that $k_i \neq 0$. Denote by k/k_i a vector $(1/k_i, k_1/k_i, \ldots, k_r/k_i)$.

The next equality is obvious: $H^\bullet_k(P) = H^\bullet_{k/k_i}(P)$.

From Proposition 1 applied to cohomologies $H^q_{k/k_i}(P)$, we obtain that they are trivial for $q \leqslant m$, if the operators

$$\delta_i\colon\ H^q_{k/k_i - 1_i}(P/\sigma_i P) \longrightarrow H^q_{k/k_i - 1_i}(P/\sigma_i P)$$

are isomorphisms. In this case the boundary operators δ_i are equal to $1 + \Phi_i \circ \sigma_{k/k_i - 1_i}$. Using the equality $\sigma_{k/k_i-1_i} = (1/k_i) \cdot \sigma_{k-1_i k_i}$, this operators may be expressed in the form $1 + (1/k_i)\Phi \circ \sigma_{k-1_i k_i}$. Therefore, the boundary operators δ_i, $i = 1, \ldots, r$, are isomorphisms iff the mappings

$$\Phi_i \circ \sigma_{k-1_i k_i} \colon H^q_{k-1_i k_i}(P/\sigma_i P) \longrightarrow H^q_{k-1_i k_i}(P/\sigma_i P), \quad i = 1, 2, \ldots, r,$$

have no nonzero integer eigenvalues. □

Keeping in view the pure homological conditions for triviality of $H^q_k(P)$, $q \leqslant m$, $k = (1, k_1, \ldots, k_r)$, $k_i \in \mathbb{Z}$, $k \neq 1_0$, we consider a differential module $(H^\bullet_k(P), \sigma_i)$, $k = (1, k_1, \ldots, k_{i-1}, 0, k_{i+1}, \ldots, k_r)$, $k_j \in \mathbb{Z}$, $i \neq 0$.

PROPOSITION 2. *If the modules $H^q_t(P)$ are trivial for $q \leqslant m$ and $t = (0, t_1, \ldots, t_r)$, $t_j \in \mathbb{Z}$, $t \neq 0$, and the modules $(H_k(P), \sigma_i)$, $k = (1, k_1, \ldots, k_{i-1}, 0, k_{i+1}, \ldots, k_r)$, $k_j \in \mathbb{Z}$ are $(m+1)$-acyclic, then*

$$H^q_k(P/\sigma_i P) \simeq H^q_k(P)/\sigma_i H^{q-1}_k(P)$$

for $q \leqslant m$, $i \neq 0$ and $k = (1, k_1, \ldots, k_{i-1}, 0, k_{i+1}, \ldots, k_r)$, $k_j \in \mathbb{Z}$.

Since differentials σ_i and σ_k anticommute, the operator σ_i, $i \neq 0$, defines a mapping from $H^q_k(P/\sigma_i P)$ into $H^{q+1}_k(P)$. Obviously, the image of this mapping lies in the kernel of σ_i and, moreover, these images coincide with the kernels of σ_i. From the conditions of $(m+1)$-acyclicity for the complexes $(H^\bullet_k(P), \sigma_i)$, we obtain that these images are contained in $\sigma_i H^q_k(P)$.

We shall prove that the mappings

$$\bar{\sigma}_i \colon H^q_k(P/\sigma_i P) \longrightarrow \sigma_i H^q_k(P)$$

are isomorphisms. For this purpose, consider an element $[x] \in H^q_k(P/\sigma_i P)$, which belongs to the kernel of $\bar{\sigma}_i$. It means that there exists an element $y \in P^{q-1}$ such that $\sigma_i x = \sigma_i \sigma_k y$. Hence, there exist $z \in P^q$, such that $x = \sigma_k y + \sigma_i z$.

The last equality means that the class $[x] \in H^q_k(P/\sigma_i P)$ is trivial and, therefore, $\bar{\sigma}_i$ is an isomorphism.

The conditions of $(m+1)$-acyclicity for the complexes $(H^\bullet_k(P), \sigma_i)$ give isomorphisms

$$\bar{\Phi}_i \colon \sigma_i H^q_k(P) \longrightarrow H^q_k(P)/\sigma_i H^{q-1}_k(P).$$

The needed isomorphisms are given by the composition $\bar{\Phi}_i \circ \bar{\sigma}_i$

$$\bar{\Phi}_i \circ \bar{\sigma}_i \colon H^q_k(P/\sigma_i P) \longrightarrow H^q_k(P)/\sigma_i H^{q-1}_k(P).$$ □

From the proposition above, it follows

THEOREM 2. *Let* $P^\bullet = \bigoplus_{s=0}^{n} P^s$ *be a graded module equipped with* $r+1$ *anticommuting operators* $\sigma_0, \sigma_1, \ldots, \sigma_r$ *of the degree* $+1$. *Then cohomologies* $H_k^q(P)$ *of* P *with the differential*

$$\sigma_k = \sum_{j=1}^{r} k_j \sigma_j + \sigma_0, \quad k_j \in \mathbb{Z},\ k = (1, k_1, \ldots, k_r)$$

are trivial for $k \neq 1_0$ *and* $q \leqslant m$, *if*

1^0. *Cohomologies* $H_t^q(P)$ *are trivial for* $q \leqslant m$ *and* $t = (0, t_1, \ldots, t_r)$, $t_j \in \mathbb{Z}$, $t \neq 0$.

2^0. *Differential modules* $(H_k^\bullet(P), \sigma_i)$, $i = 1, 2, \ldots, r$, *are* $(m+1)$*-acyclic for* $k = (1, k_1, \ldots, k_{i-1}, 0, k_{i+1}, \ldots, k_r)$, $k_j \in \mathbb{Z}$.

Proof. Choose a number i, $1 \leqslant i \leqslant r$ such that $k_i \neq 0$. Just like in the proof of theorem triviality of $H_k^q(P)$ for $q \leqslant m$ follows from the fact that the operators

$$\delta_i\colon\ H^q_{k-k_i 1_i}(P/\sigma_i P) \longrightarrow H^q_{k-k_i 1_i}(P/\sigma_i P)$$

are isomorphisms. In our case, operators δ_i are of the form $1 + (1/k_i) \cdot \Phi_i \circ \sigma_{k-k_i 1_i}$.

From Proposition 2, one obtains that the differentials $\sigma_{k-k_i 1_i}$ act trivially on the modules $H^q_{k-k_i 1_i}(P/\sigma_i P)$ and, therefore, the boundary operators δ_i in exact sequence Equation (3) are identical $\delta_i = \mathrm{id}$ for $q \leqslant m$. □

The following result is also valid:

THEOREM 3. *Let* $P^\bullet = \bigoplus_{s=0}^{n} P^s$ *be a graded module equipped with the set of* $r+1$ *anticommuting operators* $\sigma_0, \sigma_1, \ldots, \sigma_r$ *of the degree* $+1$. *Then cohomologies* $H_k^q(P)$ *of* P *with the differential*

$$\sigma_k = \sum_{j=1}^{r} k_j \sigma_j + \sigma_0, \quad k_j \in \mathbb{Z},\ k = (1, k_1, \ldots, k_r)$$

are trivial for $k \neq 1_0$ *and* $q \leqslant m$, *if*

1^0. *Cohomologies* $H_l^q(P)$ *are trivial for* $q \leqslant m$ *and* $l = (0, l_1, \ldots, l_r)$, $l_j \in \mathbb{Z}$, $l \neq 0$.

2^0. *Differential modules* $(H_{1_0}^\bullet(P), \sigma_k)$ *are* $(m+1)$*-acyclic for all* $k = (0, k_1, \ldots, k_r)$, $k_j \in \mathbb{Z}$, $k \neq 0$.

Proof. Change operator σ_i to operator σ_k in Definition 1, exact sequence (3) and Proposition 2. Similar to the proof of Theorem 2, we obtain that the boundary operator

$$\delta_k\colon\ H^q_{1_0}(P/\sigma_k P) \longrightarrow H^q_{1_0}(P/\sigma_k P)$$

is identical. □

Remark. The triviality conditions for $H^q_k(P)$, $q \leqslant m$, and $k = (1, k_1, \ldots, k_r)$, $k \neq 1_0$, are similar to the conditions of 'complete intersection' [9].

Let $N = \bigoplus_{s=0}^n N^s$ be a graded module equipped with the set of r anticommuting operators $\sigma_1, \ldots, \sigma_r$ of the degree $+1$. Operators

$$\sigma_k = \sum_{j=1}^r k_j \sigma_j, \quad k_j \in \mathbb{Z}, \ k = (0, k_1, \ldots, k_r),$$

determine the cohomological complexes in N. We denote its cohomologies by $H^\bullet_k(N)$.

Introduce the notations

$$E = \mathbb{Z}[y_1 \ldots y_r], \qquad E_k = \mathbb{Z}(y_1^{k_1} \ldots y_r^{k_r}),$$

$$k = (0, k_1, \ldots, k_r), \qquad \sigma = \sum_{j=1}^r y_j \frac{\partial}{\partial y_j} \otimes \sigma_j.$$

Consider differential modules $(E \otimes N^\bullet, \sigma)$ and $(E_k \otimes N^\bullet, \sigma)$ and denote by $H^\bullet(E, N)$ and $H^\bullet(E_k, N)$ the corresponding cohomologies.

PROPOSITION 3. *Differential modules $(E \otimes N^\bullet, \sigma)$ are decomposed into the direct sum of modules $(E_k \otimes N^\bullet, \sigma)$. Cohomologies $H^\bullet(E, N)$ are decomposed into the direct sum of the form*

$$H^\bullet(E, N) = \bigoplus_{k \neq 0} H^\bullet(E_k, N)$$

and

$$H^\bullet(E_k, N) = E_k \otimes H^\bullet_k(N), \quad k \neq 0.$$

Apply this proposition to the module $N^\bullet = H^\bullet_{1_0}(P)$. Using Theorem 2, we obtain

THEOREM 4. *Let $P^\bullet = \bigoplus_{s=0}^n P^s$ be a graded module equipped with the set of $r+1$ anticommuting operators $\sigma_0, \sigma_1, \ldots, \sigma_r$ of the degree $+1$. Then cohomologies $H^q_k(P)$ of P with the differential*

$$\sigma_k = \sum_{j=1}^r k_j \sigma_j + \sigma_0, \quad k_j \in \mathbb{Z}, \ k = (1, k_1, \ldots, k_r)$$

are trivial for $k \neq 1_0$ and $q \leqslant m$, if:

1^0. *Cohomologies $H^q_l(P)$ are trivial for $q \leqslant m$ and $l = (0, l_1, \ldots, l_r)$, $l_j \in \mathbb{Z}$, $l \neq 0$.*

2^0. *Differential module $(E \otimes H_{1_0}(P), \sigma)$ is acyclic.*

Proof. From condition $\mathbf{2}^0$ and Proposition 3, one obtains $H^q(E_k, H_{1_0}(P)) = 0$ for all $k = (0, k_1, \ldots, k_r)$, $k \neq 0$, $q \leqslant m+1$.

Therefore, differential module $(H^{\bullet}_{1_0}(P), \sigma_k)$ is $m+1$ acyclic and the statement of theorem coincides with the statement of Theorem 3. □

References

1. Spencer, D. C.: Overdetermined systems of linear partial differential equations, *Bull. Amer. Math. Soc.* **75**(2) (1969), 179–239.
2. Quillen, D. G.: Formal properties of overdetermined systems of linear partial differential equations, Thesis, Harvard University, Cambridge, Mass., 1964, p. 96.
3. Kuranishi, M.: On E. Cartan's prolongations theorem of exterior differential systems, *Amer. J. Math.* **79**(1) (1957), 1–47.
4. Goldshmidt, H.: Existence theorems for analytic linear partial differential equations, *Ann. Math.* **86**(2) (1967), 246–270.
5. Guillemin, V.: Some algebraic results concerning the characteristics of overdetermined partial differential equations, *Amer. J. Math.* **90** (1968), 270–284.
6. Lychagin, V. V. and Zil'bergleit, L. V.: Spencer cohomology of differential equations, in *Lecture Notes in Math.* 1453, Springer-Verlag, Berlin, 1990, pp. 121–136.
7. Lychagin, V. V.: Differential operators on fiber bundles, *Uspekhi Mat. Nauk* **40**(2) (1985), 187–188 (in Russian); *Russian Math. Surveys* **40** (1985).
8. Lychagin, V. V. and Zil'bergleit, L. V.: Spencer cohomologies of invariant systems of differential equations above homogeneous spaces, *Russian Math. Dokl.* **328**(5) (1992), 544–546 (in Russian).
9. Bourbaki, N.: *Algèbre. Chap. X, Algèbre homologique*, Masson, Paris, 1980, p. 192.

Acta Applicandae Mathematicae **41:** 247–270, 1995.

On the Geometry of Soliton Equations*

FRANCO MAGRI
Dipartimento di Matematica, Università di Milano, Via C. Saldini 50, I-20133 Milan, Italy

(Received: 28 February 1994)

Abstract. The paper aims to suggest a geometric point of view in the theory of soliton equations. The belief is that a deeper understanding of the origin of these equations may provide a better understanding of their remarkable properties. According to the geometric point of view, soliton equations are the outcome of a specific reduction process of a bi-Hamiltonian manifold. The suggestion of the paper is to pay attention also to the 'unreduced form' of soliton equations.

Mathematics Subject Classifications (1991): 58F07, 35Q53.

Key words: soliton equations, integrable systems, bi-Hamiltonian manifolds.

Introduction

This is an expository paper on the bi-Hamiltonian approach to soliton equations. Soliton equations are, by now, a well-known class of partial differential equations displaying several interesting properties [3]. The most famous example is the so-called Korteweg–de Vries equation. However, no previous knowledge of soliton equations is required to read this paper. The present approach is self-contained. It rests on two ideas. The first concerns the structure of integrable Hamiltonian systems. According to the point of view followed in this paper, the characteristic property of these systems is to be Hamiltonian with respect to more than one Poisson bracket. This means that, given an integrable Hamiltonian system which we write in the usual Poisson form $\dot{f} = \{f, h\}$, we have to be able to find a second Poisson bracket $\{\cdot, \cdot\}'$ and a second Hamiltonian function h' such that the *same* equation can also be written in the form $\dot{f} = \{f, h'\}'$. These systems are called bi-Hamiltonian, and the manifolds endowed with special pairs of Poisson brackets are called bi-Hamiltonian manifolds. According to this point of view, the integrability of the system is the outcome of a suitable interplay of the two Poisson brackets. The second idea is that the dual of Lie algebras are bi-Hamiltonian manifolds. In particular, this is true for the special class of Lie algebras called '*loop algebras*'. Combining the two ideas, the point of view of this paper is that the 'soliton theory' is the study of bi-Hamiltonian systems on loop algebras.

* This work has been supported by the Italian MURST and by the GNFM of the Italian CNR.

Of course, to make precise this point of view is not short work. In this paper we concentrate on the interplay of ideas and the algorithms to be used in the applications, with sacrifice of the proves. Instead of giving proofs, we have chosen to work out in detail an explicit example at the end of the paper.

The structure of the paper is bipartited according with the two ideas on which it rests. The first part (Sections 1–3) is an Introduction to the '*technique of Casimir's functions of a Poisson pencil*' on a bi-Hamiltonian manifold. This is the main technique used to produce integrable systems on such kind of manifolds. The second part (Section 4) deals with the application of this technique on loop algebras. The results presented in this part are the outcome of a collaboration with P. Casati, G. Falqui, and M. Pedroni. The proofs missing in the paper, and the study of some of the consequences which have not found a place in this paper, will be the object of a joint work with them.

1. Poisson Pencil

The objects we are interested in are manifolds endowed with a pair of Poisson brackets. A Poisson bracket is a $\mathbb{R}$-bilinear composition law on C^∞-functions on a manifold $\mathcal{M}$ verifying the usual conditions

$$\{f,g\} = -\{g,f\} \quad \text{(antisymmetry)}, \tag{1.1}$$

$$\{fg,h\} = f\{g,h\} + \{f,h\}g \quad \text{(Leibnitz rule)}, \tag{1.2}$$

$$\{\{f,g\},h\} + \{\{h,f\},g\} + \{\{g,h\},f\} = 0 \quad \text{(Jacobi identity)}. \tag{1.3}$$

Given a pair of Poisson brackets $\{f,g\}_0$ and $\{f,g\}_1$, we form the linear pencil

$$\{f,g\}_\lambda = \{f,g\}_0 - \lambda\{f,g\}_1. \tag{1.4}$$

The manifold $\mathcal{M}$ is said to be a *bi-Hamiltonian manifold* if $\{f,g\}_\lambda$ is a Poisson bracket for any value of the real or complex parameter λ. In this case, $\{f,g\}_\lambda$ is called the *Poisson pencil* defined on $\mathcal{M}$.

The condition to be imposed on $\{f,g\}_0$ and $\{f,g\}_1$ in order to define a Poisson pencil is

$$\begin{aligned}&\{f,\{g,h\}_0\}_1 + \{h,\{f,g\}_0\}_1 + \{g,\{h,f\}_0\}_1 + \\ &\quad + \{f,\{g,h\}_1\}_0 + \{h,\{f,g\}_1\}_0 + \{g,\{h,f\}_1\}_0 = 0\end{aligned} \tag{1.5}$$

for any triple of functions (f,g,h) on $\mathcal{M}$. A simple way to fulfill this condition is to consider a single bracket $\{f,g\}_0$ and a vector field X, and to define the second bracket

$$\{f,g\}_1 = \{L_X(f),g\}_0 + \{f,L_X(g)\}_0 - L_X\{f,g\}_0. \tag{1.6}$$

It is easily seen that this bracket is compatible with $\{f, g\}_0$, fulfilling condition (1.5). So, if $\{f, g\}_1$ is a Poisson bracket, $\mathcal{M}$ is a bi-Hamiltonian manifold (for more details see [6]).

A simple instance of this situation is provided by the *Lie–Poisson pencil* defined on the dual of any Lie algebra. For loop algebras, the construction goes as follows. Let $\mathfrak{g}$ be a finite-dimensional simple Lie algebra, and let $\mathcal{M}$ be the space of C^∞-maps from S^1 into $\mathfrak{g}$. We denote by x the coordinate on S^1, and either by $S(x)$ or simply by S a map from S^1 into $\mathfrak{g}$. The main example dealt with in this paper is $\mathfrak{g} = \mathfrak{sl}(2, \mathbb{C})$. Accordingly, we write

$$S = \begin{pmatrix} p & r \\ q & -p \end{pmatrix}, \tag{1.7}$$

where p, q, r are three arbitrary periodic functions playing the role of 'global coordinates' on our infinite-dimensional manifold. A curve $S(t)$ on $\mathcal{M}$ is a one-parameter family of maps from S^1 into $\mathfrak{g}$. The tangent vector at S is denoted by

$$\dot{S} = \begin{pmatrix} \dot{p} & \dot{r} \\ \dot{q} & -\dot{p} \end{pmatrix}, \tag{1.8}$$

where the dot means differentiation with respect to the parameter t. A covector is a second map

$$V = \begin{pmatrix} v_1 & v_2 \\ v_3 & -v_1 \end{pmatrix} \tag{1.9}$$

from S^1 into $\mathfrak{g}$, whose value on the tangent vector $\dot{S}$ is given by

$$(V, \dot{S}) = \int_{S^1} \left(V(x), \dot{S}(x)\right)_{\mathfrak{g}} \mathrm{d}x = \int_{S^1} (2\dot{p}v_1 + \dot{q}v_2 + \dot{r}v_3)\, \mathrm{d}x. \tag{1.10}$$

In this formula, $(V(x), \dot{S}(x))_{\mathfrak{g}}$ is the Killing form on the algebra $\mathfrak{g}$.

The space $\mathcal{M}$ is an infinite-dimensional Lie algebra with the pointwise commutator

$$[S_1, S_2]_M(x) = [S_1(x), S_2(x)]_{\mathfrak{g}}. \tag{1.11}$$

It is also a Poisson manifold endowed with the Poisson bracket

$$\{f, g\}_0 = \omega(\mathrm{d}f(S), \mathrm{d}g(S)) + (S, [\mathrm{d}f(S), \mathrm{d}g(S)]), \tag{1.12}$$

where ω is the cocycle

$$\omega(\mathrm{d}f(S), \mathrm{d}g(S)) = \int_{S^1} \left(\mathrm{d}f(S), \frac{\mathrm{d}}{\mathrm{d}x}\mathrm{d}g(S)\right)_{\mathfrak{g}} \mathrm{d}x, \tag{1.13}$$

and where $\mathrm{d}f(S)$ and $\mathrm{d}g(S)$ are the differentials of $f\colon \mathcal{M} \to \mathbb{R}$ and $g\colon \mathcal{M} \to \mathbb{R}$ regarded as elements of the loop algebra itself. If we denote these differentials by V and W, respectively, for the example $\mathfrak{g} = \mathfrak{sl}(2, \mathbb{C})$ we explicitly get

$$\begin{aligned}\{f,g\}_0(S) &= \int_{S^1} \mathrm{Tr}\, V W_x \,\mathrm{d}x + \int_{S^1} \mathrm{Tr}\,(S\,[V,W])\,\mathrm{d}x \\ &= \int_{S^1} \mathrm{Tr}\, V\,(W_x + [W,S])\,\mathrm{d}x \\ &= \int_{S^1} 2v_1\,(w_{1x} + q w_2 - r w_3)\,\mathrm{d}x + \\ &\quad + \int_{S^1} v_2\,(w_{3x} + 2p w_3 - 2q w_1) + \\ &\quad + \int_{S^1} v_3\,(w_{2x} + 2r w_1 - 2p w_2)\,\mathrm{d}x.\end{aligned} \tag{1.14}$$

Let now A be a fixed element in $\mathfrak{g}$ (a 'constant loop') to be chosen below, and let X be the vector field on $\mathcal{M}$ defined by

$$\dot{S} = A. \tag{1.15}$$

According to Equation (1.6), we get

$$\{f,g\}_1 = (A, [V,W]). \tag{1.16}$$

This bracket verifies the Jacobi identity (1.3) as a consequence of the Jacobi identity on the commutator (1.11). Therefore, $\mathcal{M}$ is a bi-Hamiltonian manifold endowed with the Poisson pencil

$$\{f,g\}_\lambda = \omega(V,W) + (S + \lambda A, [V,W]). \tag{1.17}$$

If, in our example,

$$A = \begin{pmatrix} a_1 & a_2 \\ a_3 & -a_1 \end{pmatrix}, \tag{1.18}$$

we get

$$\begin{aligned}\{f,g\}_1 &= \int_{S^1} 2a_1\,(v_2 w_3 - v_3 w_2)\,\mathrm{d}x + \int_{S^1} 2a_2\,(v_1 w_3 - v_3 w_1)\,\mathrm{d}x + \\ &\quad + \int_{S^1} 2a_3\,(v_1 w_2 - v_2 w_1)\,\mathrm{d}x.\end{aligned} \tag{1.19}$$

The brackets (1.14) and (1.19) are central in the theory of KdV equation.

The Poisson brackets of the pencil are used to associate vector fields with functions. The vector field X_h defined by

$$X_h(f) = \{f,h\} \tag{1.20}$$

is called the *Hamiltonian vector field* associated with the function h with respect to the given Poisson bracket. This correspondence depends on the choice of the Poisson bracket. On a bi-Hamiltonian manifold, we have several choices. Consequently, we can either associate different vector fields with the same function or different functions with the same vector field. Vector fields for which this is possible are called bi-Hamiltonian. The Hamiltonian vector fields

$$X_h(f) = \{f, h\}_0, \qquad Y_h(f) = \{f, h\}_1 \tag{1.21}$$

associated with a pair of compatible brackets verify the commutation relations

$$\begin{aligned} &[X_f, X_g] = X_{\{f,g\}_0}, \\ &[X_f, Y_g] + [Y_f, X_g] = X_{\{f,g\}_1} + Y_{\{f,g\}_0}, \\ &[Y_f, Y_g] = Y_{\{f,g\}_1}. \end{aligned} \tag{1.22}$$

They can also be written in the form

$$X_f = P_0\,\mathrm{d}f, \qquad Y_f = P_1\,\mathrm{d}f \tag{1.23}$$

by introducing the *Poisson tensors* P_0 and P_1 associated with the brackets $\{\cdot,\cdot\}_0$ and $\{\cdot,\cdot\}_1$ through the relations

$$\{f, g\}_0 = \langle \mathrm{d}f, P_0\,\mathrm{d}g\rangle, \qquad \{f, g\}_1 = \langle \mathrm{d}f, P_1\,\mathrm{d}g\rangle. \tag{1.24}$$

These tensors are linear maps from the cotangent to the tangent bundle of $\mathcal{M}$, uniquely associated with the Poisson brackets. Equations (1.14) and (1.16) show that the Poisson tensors associated to the Lie–Poisson pencil on a loop algebra are

$$P_0(V) = V_x + [V, S], \tag{1.25}$$
$$P_1(V) = [A, V]. \tag{1.26}$$

The Hamiltonian vector fields of the Lie–Poisson pencil are therefore defined by

$$\dot{S} = V_x + [V, S + \lambda A], \tag{1.27}$$

where $V = \mathrm{d}f(S)$.

2. Casimir's Functions

The next objects we are interested in are Casimir's functions of the Poisson pencil. We recall that a Casimir function of a Poisson bracket is a function which commutes with any other function f:

$$\{f, h\} = 0, \quad \forall f \in C^\infty(\mathcal{M}). \tag{2.1}$$

Casimir's functions of the Poisson pencil are therefore functions $h(\lambda)$, usually depending on the parameter λ of the pencil, which verify the equation

$$\{f, h(\lambda)\}_\lambda = \{f, h(\lambda)\}_0 - \lambda\{f, h(\lambda)\}_1 = 0. \tag{2.2}$$

We assume that the second bracket $\{f, g\}_1$ of the pencil has a Casimir function h_0:

$$\{f, h_0\}_1 = 0, \quad \forall f \in C^\infty(\mathcal{M}) \tag{2.3}$$

and we concentrate our attention to the subclass of Casimir functions $h(\lambda)$ which admit a development in a Laurent series

$$h(\lambda) = h_0 + h_1\lambda^{-1} + h_2\lambda^{-2} + \cdots \tag{2.4}$$

with leading term h_0. To find such Casimir functions is a nontrivial problem. A possible technique is discussed in the next section. However, it is a worthwhile problem. Indeed, it is easily proved [1] that the coefficients h_j fulfill the '*Lenard recursion relations*'

$$\{f, h_j\}_0 = \{f, h_{j+1}\}_1 \tag{2.5}$$

and that they are in involution with respect to the whole pencil of Poisson brackets:

$$\{h_j, h_k\}_0 = \{h_j, h_k\}_1 = \{h_j, h_k\}_\lambda = 0. \tag{2.6}$$

Such Casimir functions provide, therefore, hierarchies of functions which are in involution with respect to the brackets of the pencil.

Consider now the *pencil of vector fields*

$$Y_\lambda(f) = \{f, h(\lambda)\}_1 \tag{2.7}$$

canonically associated with the Casimir's function $h(\lambda)$. It will be referred to as the *canonical hierarchy* associated with the Casimir's function. Assume that there exists a vector field X such that

$$\begin{aligned}\{L_X(f), g\}_0 + \{f, L_X(g)\}_0 - L_X\{f, g\}_0 &= \{f, g\}_1,\\ \{L_X(f), g\}_1 + \{f, L_X(g)\}_1 - L_X\{f, g\}_1 &= 0.\end{aligned} \tag{2.8}$$

This assumption is fulfilled by the Lie–Poisson pencil (1.17). Then

$$\{f, g\}_1 = \{L_X(f), g\}_\lambda + \{f, L_X(g)\}_\lambda - L_X\{f, g\}_\lambda \tag{2.9}$$

and, therefore, for any Casimir's function $h(\lambda)$ of $\{\cdot, \cdot\}_\lambda$ we get

$$Y_\lambda(f) = \{f, h(\lambda)\}_1 = \{f, L_X(h)\}_\lambda. \tag{2.10}$$

This shows that the canonical hierarchy is bi-Hamiltonian. The first Hamiltonian is the Casimir function; the second Hamiltonian is its Lie derivative

$$k(\lambda) = L_X(h) \tag{2.11}$$

along the vector field X used to define the pencil.

The aim of this paper is to study the Casimir functions of the Lie–Poisson pencil and to show that the soliton equations are the 'canonical hierarchies' defined by them.

3. Two Reduction Theorems

To construct Casimir's functions of a Poisson pencil is a difficult problem; too difficult to discuss it globally on $\mathcal{M}$. Furthermore, we are not even sure of the global existence of such functions on $\mathcal{M}$. So, a possible strategy is to reduce the manifold $\mathcal{M}$ and the corresponding Poisson pencil, and to look for the Casimir functions of the reduced pencil. This can be done according to the general reduction scheme for bi-Hamiltonian manifolds presented in this section.

We consider two generalized distributions on $\mathcal{M}$. The first is spanned by the Hamiltonian vectors fields

$$Y_h(f) = \{f, h\}_1, \quad h \in C^\infty(\mathcal{M}). \tag{3.1}$$

This distribution is integrable and its maximal integral submanifold are endowed with a canonical symplectic structure. They are called the symplectic leaves of the bracket $\{\cdot,\cdot\}_1$ [5]. The second distribution (let us call it D) is spanned by the Hamiltonian vector fields

$$X_k(f) = \{f, k\}_0 \tag{3.2}$$

associated with the Casimir functions k of $\{\cdot,\cdot\}_1$:

$$\{f, k\}_1 = 0, \quad \forall f \in C^\infty(\mathcal{M}). \tag{3.3}$$

This distribution too is integrable. Let $\mathcal{S}$ be a specific symplectic leaf of $\{\cdot,\cdot\}_1$, and let us denote by E the foliation induced on $\mathcal{S}$ by D. The leaves of E are the intersections of $\mathcal{S}$ with the leaves of D. We assume that E is sufficiently regular for the quotient space $\mathcal{N} = \mathcal{S}/E$ to be a manifold, and we denote the canonical immersion of $\mathcal{S}$ in $\mathcal{M}$ and the canonical projection of $\mathcal{S}$ onto $\mathcal{N}$ by i: $\mathcal{S} \to \mathcal{M}$ and π: $\mathcal{S} \to \mathcal{N}$, respectively.

PROPOSITION 3.1. *The quotient space $\mathcal{N} = \mathcal{S}/E$ is a bi-Hamiltonian manifold. On $\mathcal{N}$ there exists a unique Poisson pencil $\{\cdot,\cdot\}^\lambda_\mathcal{N}$ such that*

$$\{f, g\}^\lambda_\mathcal{N} \circ \pi = \{F, G\}^\lambda_\mathcal{M} \circ i \tag{3.4}$$

for any pair of functions, F and G, which extend the functions f and g of $\mathcal{N}$ into $\mathcal{M}$, and are constant on D. Technically, this means that F satisfies the conditions,

$$F \circ i = f \circ \pi, \tag{3.5}$$

$$\{F, k\}_0 = 0, \tag{3.6}$$

for any Casimir function k of $\{\cdot,\cdot\}_1$.

The proof (see [1]) is based on a reduction theorem for Poisson manifolds, first pointed out by Marsden and Ratiu in [7].

To show how the reduction may simplify the problem of finding Casimir's functions of the Poisson pencil, let us suppose that, for a suitable choice of $\mathcal{S}$, we are able to find a 1-form V which solves the equation

$$(P_0 - \lambda P_1)V = 0 \tag{3.7}$$

at the points of $\mathcal{S}$. We stress that we are not requiring that V is defined outside $\mathcal{S}$. We are only requiring that V belongs to the kernel of the Poisson pencil at the points of $\mathcal{S}$. Furthermore, let us assume that V, restricted to $T\mathcal{S}$, is exact. This means that we assume the existence of a function $H \in C^\infty(\mathcal{S})$ such that

$$\langle V, X\rangle = L_X(H) \tag{3.8}$$

for any vector field X on $\mathcal{S}$. By using the symplectic 2-form ω induced by $\{\cdot,\cdot\}_1$ on $\mathcal{S}$, we associate the Hamiltonian vector field X_H with H:

$$\omega(X_H, \cdot) = -\mathrm{d}H. \tag{3.9}$$

PROPOSITION 3.2.

(1) *The function H is constant along E, and therefore there exists a function h on $\mathcal{N}$ such that $H = h \circ \pi$.*
(2) *The function h is a Casimir's function of the reduced Poisson pencil on $\mathcal{N}$.*
(3) *The Hamiltonian vector field X_H on $\mathcal{S}$ projects onto the bi-Hamiltonian vector field X_h defined on $\mathcal{N}$ by*

$$X_h(f) = \{f, h\}_1. \tag{3.10}$$

Proof. We limit ourselves to prove the first point. To show that H is invariant along E, we show that V annihilates the distribution D. Indeed, for any element $K \in \operatorname{Ker} P_1$, we have:

$$\begin{aligned}\langle V, X_D\rangle &= \langle V, P_0(K)\rangle = -\langle K, P_0(V)\rangle \\ &= -\lambda\langle K, P_1(V)\rangle = \lambda\langle V, P_1(K)\rangle = 0.\end{aligned}$$

The rest of the theorem follows immediately. □

This theorem clearly fixes the steps of our approach to soliton equations:

(1) we choose a simple Lie algebra $\mathfrak{g}$, and we compute the corresponding Lie–Poisson pencil (1.17);
(2) we select a suitable symplectic leaf $\mathcal{S}$, and we determine the distributions D, E, and the quotient space $\mathcal{N}$;
(3) we solve Equation (3.4) at the points of $\mathcal{S}$, and we pick up among its solutions the 1-forms which are exact on $\mathcal{S}$; for the Lie–Poisson pencil this means that we have to solve the equation

$$V_x + [V, S + \lambda A] = 0 \tag{3.11}$$

and to find the solutions for which there exists a function H on $\mathcal{S}$ such that

$$\langle V, \dot{S} \rangle = \frac{\mathrm{d}}{\mathrm{d}t} H(S) \tag{3.12}$$

for every curve $S(t)$ on $\mathcal{S}$;
(4) we compute the Hamiltonian H, its projection h, and the Hamiltonian vector fields X_H and X_h on $\mathcal{S}$ and $\mathcal{N}$, respectively.

To detail these steps is the aim of the rest of the paper.

4. Lie–Poisson Pencil

In this section, we give the general algorithm for computing Casimir's functions of a special Lie–Poisson pencil on a loop algebra. For the sake of clarity, we divide the discussion into four parts. First, we make explicit the choice of the element A which enters the definition of the Lie–Poisson pencil (1.17), and we select a specific symplectic leaf $\mathcal{S}$. Next, we study the kernel of the Poisson pencil at a specific point B of the symplectic leaf. Then, we extend the analysis to a generic point S of the symplectic leaf, showing that the properties which hold in B are true everywhere on the symplectic leaf. This technique may be considered a geometric version of the 'method of dressing transformations'. Finally, we give the explicit formula for computing the Hamiltonian function H for the algebras of type A_n. No proofs are given in this section. The aim is simply to emphasize the logical coherence of the theory, and to provide a clear algorithm to be used in the example of the next section.

4.1. First Step: The Canonical Symplectic Leaf

What is needed to associate a soliton hierarchy with a simple Lie algebra $\mathfrak{g}$, of rank n, is the explicit form of the generators $(E_1, \ldots, E_n; F_1, \ldots, F_n)$ of a Chevalley basis, in a faithful representation of minimal dimension of $\mathfrak{g}$. By means of $(F_1, \ldots, F_n)$, one then constructs a basis of the nilpotent subalgebra $\mathfrak{n}^-$, in the Cartan decomposition $\mathfrak{g} = \mathfrak{n}^- \oplus \mathfrak{h} \oplus \mathfrak{n}^+$ of $\mathfrak{g}$. The element A of minimal weight, and the covector $\langle e|$ of minimal weight of the representation are then uniquely defined (up to a multiplicative constant) by the equations

$$[A, F_j] = 0, \qquad \langle e| F_j = 0 \tag{4.1}$$

for every $F_j \in \mathfrak{n}^-$. For instance, for the algebra $B_2 = so(5, \mathbb{C})$ the representation has dimension five, and a possible Chevalley basis is:

$$E_1 = \begin{bmatrix} 0 & 0 & 1 & 0 & 0 \\ 0 & 0 & 0 & 0 & 0 \\ 0 & 0 & 0 & 0 & 0 \\ 0 & 0 & 0 & 0 & 0 \\ -1 & 0 & 0 & 0 & 0 \end{bmatrix}, \qquad E_2 = \begin{bmatrix} 0 & 0 & 0 & 0 & 0 \\ 0 & 0 & 0 & 0 & 0 \\ 0 & 1 & 0 & 0 & 0 \\ 0 & 0 & 0 & 0 & -1 \\ 0 & 0 & 0 & 0 & 0 \end{bmatrix}, \tag{4.2}$$

$$F_1 = \begin{bmatrix} 0 & 0 & 0 & 0 & -1 \\ 0 & 0 & 0 & 0 & 0 \\ 1 & 0 & 0 & 0 & 0 \\ 0 & 0 & 0 & 0 & 0 \\ 0 & 0 & 0 & 0 & 0 \end{bmatrix}, \qquad F_2 = \begin{bmatrix} 0 & 0 & 0 & 0 & 0 \\ 0 & 0 & 1 & 0 & 0 \\ 0 & 0 & 0 & 0 & 0 \\ 0 & 0 & 0 & 0 & 0 \\ 0 & 0 & 0 & -1 & 0 \end{bmatrix}. \tag{4.3}$$

Therefore, $\mathfrak{n}^-$ is generated by F_1, F_2 and by

$$F_3 = \begin{bmatrix} 0 & 1 & 0 & 0 & 0 \\ 0 & 0 & 0 & 0 & 0 \\ 0 & 0 & 0 & 0 & 0 \\ -1 & 0 & 0 & 0 & 0 \\ 0 & 0 & 0 & 0 & 0 \end{bmatrix}, \qquad F_4 = \begin{bmatrix} 0 & 0 & 0 & 0 & 0 \\ 0 & 0 & 0 & 0 & 1 \\ 0 & 0 & 0 & -1 & 0 \\ 0 & 0 & 0 & 0 & 0 \\ 0 & 0 & 0 & 0 & 0 \end{bmatrix} \tag{4.4}$$

and the elements of minimal weight are given by

$$A = \begin{bmatrix} 0 & 0 & 0 & 0 & 0 \\ 0 & 0 & 0 & 0 & 1 \\ 0 & 0 & 0 & 1 & 0 \\ 0 & 0 & 0 & 0 & 0 \\ 0 & 0 & 0 & 0 & 0 \end{bmatrix}, \tag{4.5}$$

$$\langle e| = [0, 1, 0, 0, 0] .$$

Following an idea of Drinfeld and Sokolov [4, p. 2010], we choose A to define the Lie–Poisson pencil (1.17) on the loop algebra $\mathcal{M}$ associated with $\mathfrak{g}$. It is easily shown that the bracket

$$\{F, G\}_1 = \langle A, [\mathrm{d}F(S), \mathrm{d}G(S)]\rangle \tag{4.6}$$

has constant rank, and that its symplectic leaves are affine hyperplanes modelled on the vector space of loops with value in $\mathfrak{g}_A^\perp$, where $\mathfrak{g}_A^\perp$ is the orthogonal space, with respect to the Killing form, of the isotropy algebra $\mathfrak{g}_A$ of A. Following a second idea of Drinfeld and Sokolov [4, p. 2010], we pick up among these leaves the one passing through the point

$$B = \sum_{j=1}^{n} E_j, \tag{4.7}$$

where the sum is over the element of the Chevalley basis associated with the simple positive roots. This choice of A and B completely specifies the submanifold $\mathcal{S}$ in Proposition 3.1. It will be referred to as the *canonical symplectic leaf* of the algebra $\mathcal{M}$.

4.2. Second Step: The Kernel of the Poisson Pencil at the Point B

At the point B, the equation

$$V_x + [V, S + \lambda A] = 0, \tag{4.8}$$

defining the kernel of the Poisson pencil, becomes

$$V_x + [V, B + \lambda A] = 0. \tag{4.9}$$

Let us set

$$\Lambda = B + \lambda A. \tag{4.10}$$

For the algebras of type A_n, B_n, and C_n the spectrum of Λ is *simple*. (The case of the algebra D_n requires a specific study, which will not be considered in this paper.) If we introduce new parameters z and ω according to

$$\begin{array}{lll} A_n\colon & z^{n+1} = \lambda, & \omega^{n+1} = 1, \\ B_n\colon & z^{2n} = \lambda, & \omega^{2n} = 1, \\ C_n\colon & z^{2n} = \lambda, & \omega^{2n} = 1. \end{array} \tag{4.11}$$

we can set [4]

$$\begin{array}{lll} \text{spectrum}(\Lambda) & = \{z, \omega z, \ldots, \omega^n z\} & \text{for } A_n, \\ \text{spectrum}(\Lambda) & = \{0, z, \omega z, \ldots, \omega^{2n-1} z\} & \text{for } B_n, \\ \text{spectrum}(\Lambda) & = \{z, \omega z, \ldots, \omega^{2n-1} z\} & \text{for } C_n. \end{array} \tag{4.12}$$

Therefore, Λ possesses a basis of eigenvectors. To construct this basis, we use the technique of the cyclic covector. Consider the iterated covectors

$$\langle \mathrm{e}^{i+1}| = \langle \mathrm{e}^i|\Lambda \tag{4.13}$$

starting from

$$\langle \mathrm{e}^0| = \langle \mathrm{e}|. \tag{4.14}$$

Let d be the dimension of the representation, and let us denote by $|l_a\rangle$ the eigenvector of Λ corresponding to the eigenvalue $\omega_a z$, where ω_a is either zero or some root of unity. Then

PROPERTY 4.1.

(1) *the covectors* $\{\langle \mathrm{e}^i|\}_{i=0,\dots,d-1}$ *form a basis of the representation space, called the (left)-Frobenius basis associated with* Λ;
(2) *the covectors* $\{\langle \mathrm{e}^i|\}_{i=0,\dots,d}$ *obey the equation*

$$\langle \mathrm{e}^d| + \sum_{k=0}^{d-1} \langle \mathrm{e}^k|a_k(\lambda) = 0, \tag{4.15}$$

where $a_k(\lambda)$ *are the coefficients of the characteristic polynomial of* Λ;
(3) *the components of* $|l_a\rangle$ *on the (left)-Frobenius basis are the powers of the eigenvalues* $\omega_a z$:

$$\langle \mathrm{e}^j|l_a\rangle = (\omega_a z)^j\,. \tag{4.16}$$

This means that the matrix of the components of the (right)-eigenvectors of Λ *on the (left)-Frobenius basis is the* Vandermonde matrix *of the eigenvalues of* Λ.

Let us now return to Equation (4.9). Our aim is to study the solutions of this equation admitting a Laurent expansion of the type

$$V(z) = \sum_{k=-j}^{\infty} V_k z^{-k}. \tag{4.17}$$

By using the properties of the eigenvectors of Λ, one can show the following results.

PROPOSITION 4.2.

(1) *the solution* $V(z)$ *of Equation* (4.9) *form an abelian subalgebra of the algebra of linear endomorphisms of the representation space;*
(2) *they have the vectors* $|l_a\rangle$ *as common eigenvectors:*

$$V|l_a\rangle = v_a|l_a\rangle; \tag{4.18}$$

(3) *the eigenvalues* v_a *are independent of* x:

$$v_{ax} = 0; \tag{4.19}$$

(4) *every solution* $V(z)$ *can be expressed as a linear combination of powers of* Λ, *with coefficients which are scalar-valued Laurent series in* z *independent of* x.

Not all the powers of Λ, however, belong to the algebra $\mathfrak{g}$. For the algebras A_n, B_n, and C_n the allowed powers are:

$$\begin{array}{ll} A_n: & (\Lambda, \Lambda^2, \dots, \Lambda^n)\,, \\ B_n: & (\Lambda, \Lambda^3, \dots, \Lambda^{2n-1})\,, \\ C_n: & (\Lambda, \Lambda^3, \dots, \Lambda^{2n-1})\,. \end{array} \tag{4.20}$$

So, to stay in the algebra, only linear combination of these powers have to be considered. The following are especially important:

$$\begin{aligned} A_n\colon &\quad V(B;z) = z^n\left(\tfrac{\Lambda}{z} + \tfrac{\Lambda^2}{z^2} + \cdots + \tfrac{\Lambda^n}{z^n}\right), \\ B_n\colon &\quad V(B;z) = z^{2n-1}\left(\tfrac{\Lambda}{z} + \tfrac{\Lambda^3}{z^3} + \cdots + \tfrac{\Lambda^{2n-1}}{z^{2n-1}}\right), \\ C_n\colon &\quad V(B;z) = z^{2n-1}\left(\tfrac{\Lambda}{z} + \tfrac{\Lambda^3}{z^3} + \cdots + \tfrac{\Lambda^{2n-1}}{z^{2n-1}}\right). \end{aligned} \tag{4.21}$$

They will be referred to as the *fundamental solutions* of Equation (4.8) at the point B, for the algebras of type A_n, B_n, C_n, respectively.

4.3. Third Step: The Kernel of the Poisson Pencil at the Point S

Let us now pass to the generic point S. Since the canonical symplectic leaf is an affine hyperplane, we can introduce the 'displacement vector' Q from B to S, and write

$$S = B + Q, \tag{4.22}$$

where Q is any element of $\mathfrak{g}_A^\perp$ (the vector space of the translations on $\mathcal{S}$). Therefore

$$S + \lambda A = Q + (B + \lambda A) = Q + \Lambda. \tag{4.23}$$

PROPERTY 4.3. *The matrix representing Q on the (left)-Frobenius basis associated with Λ is lower triangular, that is we have*

$$\langle \mathrm{e}^i|Q = \sum_{k=0}^{j} \langle \mathrm{e}^k| q_k^j \tag{4.24}$$

for $j = 0, 1, \ldots, d-1$.

This property allows us to introduce a new basis in the representation space associated with the point S. Consider the sequence of covectors $\langle g^j|$ iteratively defined by

$$\langle g^{i+1}| = \langle g^i|_x + \langle g^i|(S + \lambda A), \tag{4.25}$$

where, as usual,

$$\langle g^0| = \langle \mathrm{e}|. \tag{4.26}$$

PROPERTY 4.4.

(1) *the covectors $\{\langle g^i|\}_{i=0,\ldots,d-1}$ form a basis of the representation space called the (left)-Frobenius basis at the point S;*

(2) *for* $S = B$ *this basis coincides with the previous Frobenius basis at the point* B*:*

$$\langle g^j(B)| = \langle e^j|. \tag{4.27}$$

(3) *the covectors* $\{\langle g^i|\}_{i=0,\dots,d}$ *obey an equation of the kind*

$$\langle g^d| + \sum_{i=0}^{d-1} \langle g^i|c_i(S,\lambda) = 0, \tag{4.28}$$

called the characteristic equation *of the canonical symplectic leaf;*

(4) *the coefficients* $c_i(S,\lambda)$ *of this equations are invariant along the fibration* E *on* S*, and therefore they are functions on the quotient space* $\mathcal{N}$*;*

(5) *the coefficients* $c_i(S,\lambda)$ *are perturbations of the coefficients* $a_i(\lambda)$ *of the characteristic equation of* Λ*, that is we can write*

$$c_i(S,\lambda) = a_i(\lambda) + b_i(Q) \tag{4.29}$$

where the coefficients $b_i(Q)$ *depend on* Q*, but not on* λ*, and vanish for* $Q = 0$.

For instance, for the algebras of type A_n, the characteristic equation at the point B is

$$\langle e^{n+1}| - \langle e^0|z^{n+1} = 0, \tag{4.30}$$

while the characteristic equation at the point S is

$$\langle g^{n+1}| - \langle g^0|z^{n+1} = \sum_{j=0}^{n-1} \langle g^j|b_j(Q), \tag{4.31}$$

where the functions $b_j(Q)$ are the coordinates on the quotient space $\mathcal{N}$.

We now introduce what may be considered as the perturbation of the eigenvalues of Λ. Given any function μ, we define the sequence of functions $\mu^{(i)}$ according to

$$\mu^{(j+1)} = \mu^{(j)}_x + \mu\mu^{(j)} \tag{4.32}$$

starting from

$$\mu^{(0)} = 1. \tag{4.33}$$

These functions are called *Faà di Bruno's polynomials* of μ [3]. We say that μ is a *characteristic function* on the leaf S if it verifies the differential equation

$$\mu^{(d)} + \sum_{j=0}^{d-1} c_j(S,z)\mu^{(j)} = 0, \tag{4.34}$$

where the coefficients $c_j(S, z)$ are the same coefficients appearing in Equation (4.28). We look for the solutions of this equation which can be expanded in Laurent series in z. A detailed study shows that the leading term of μ necessarily is an eigenvalue of Λ:

$$\mu(z)_a = \omega_a z + \sum_{j=0}^{\infty} \mu_j z^{-j}, \tag{4.35}$$

while the following coefficients μ_j can be determined iteratively by recurrence. Thus, on S, we have as many characteristic functions as eigenvalues of Λ.

With every characteristic function μ_a we associate the vector $|\psi_a\rangle$ defined by

$$\langle g^i|\psi_a\rangle = \mu_a^{(j)}. \tag{4.36}$$

The components of $|\psi_a\rangle$ on the (left)-Frobenius basis associated with the point S are the Faà di Bruno's polynomials of the characteristic function μ_a. We remark that, at the point B, the functions μ_a and the vectors $|\psi_a\rangle$ coincide with the eigenvalues and the eigenvectors of Λ respectively. At the points S, they are the eigenvalues and the eigenvectors of the operator $\partial_x - (S + \lambda A)$. Indeed one can prove that the vectors $|\psi_a\rangle$ satisfy the equation

$$|\psi_a\rangle_x = (S + \lambda A - \mu_a \mathrm{I})|\psi_a\rangle, \tag{4.37}$$

where I is the identity matrix on the representation space. This fact is the key for discussing the solutions of the equation

$$V_x + [V, S + \lambda A] = V_x + [V, Q + \Lambda] = 0 \tag{4.38}$$

defining the kernel of the Poisson pencil at the point S.

PROPOSITION 4.5.

(1) *the solutions $V(z)$ of Equation* (4.38) *form an Abelian subalgebra of the algebra of linear endomorphisms on the representation space;*
(2) *they have the vectors $|\psi_a\rangle$ as common eigenvectors:*

$$V|\psi_a\rangle = v_a|\psi_a\rangle; \tag{4.39}$$

(3) *the eigenvalues v_a are independent of x:*

$$v_{ax} = 0; \tag{4.40}$$

(4) *there exists a unique solution which assume a given value at the point B; in particular, there exists a unique fundamental solution which obey the condition* (4.21) *at the point B.*

We give the explicit form of this solution for the algebra A_n. Let us denote by $|\psi\rangle = |\psi_1\rangle$ the first eigenvector of (4.37), corresponding to the characteristic function μ_1 whose leading term is z (the first eigenvalue of Λ). Moreover, let us denote by $\langle\psi^*|$ the dual eigenvector, defined by the equations

$$\langle\psi^*|\psi_a\rangle = \delta_a^1. \tag{4.41}$$

PROPOSITION 4.6. *The unique solution of Equation* (4.28) *in the algebra* A_n *which, at the point* B, *assume the value* (4.21) *is*

$$V = (n+1)z^n(|\psi\rangle\langle\psi^*| - \mathrm{I}). \tag{4.42}$$

4.4. Fourth Step: Hamiltonian Function

In [2], it has been proved that a sufficient condition for the existence of a function H such that

$$\langle V, \dot{S}\rangle = \int_{S^1} (V, \dot{S})_{\mathfrak{g}}\, \mathrm{d}x = \frac{\mathrm{d}H}{\mathrm{d}t} \tag{4.43}$$

for every vector field $\dot{S}$ is that the spectrum of V be constant on S. This is the case for the solution (4.42). To compute its Hamiltonian function, we remark that from Equation (4.31) one gets

$$|\dot{\psi}\rangle_x = (S + \lambda A - \mu_1\mathrm{I})|\dot{\psi}\rangle + (\dot{S} - \dot{\mu}_1\mathrm{I})|\psi\rangle. \tag{4.44}$$

Therefore

$$\begin{aligned}\langle V, \dot{S}\rangle & \qquad\qquad (4.45)\\ &= (n+1)z^n \int_{S^1} \langle\psi^*|\dot{S}|\psi\rangle\, \mathrm{d}x = (n+1)z^n \int_{S^1} \dot{\mu}_1\, \mathrm{d}x + \\ &\quad +(n+1)z^n \int_{S^1} \langle\psi^*|\dot{\psi}_x\rangle\, \mathrm{d}x - (n+1)z^n \int_{S^1} \langle\psi^*|S + \lambda A - \mu_1\mathrm{I}|\dot{\psi}\rangle\, \mathrm{d}x \\ &= (n+1)z^n \int_{S^1} \dot{\mu}_1\, \mathrm{d}x - (n+1)z^n \int_{S^1} \left(\langle\psi^*|\dot{\psi}\rangle + \langle\psi^*|S + \lambda A - \mu_1\mathrm{I}|\dot{\psi}\rangle\right) \mathrm{d}x.\end{aligned}$$

Since the eigenvector $\langle\psi^*|$ obeys the equation

$$\langle\psi_x^*| + \langle\psi^*|(S + \lambda A - \mu_1\mathrm{I}) = 0 \tag{4.46}$$

we finally get

$$H = (n+1)z^n \int_{S^1} \mu_1\, \mathrm{d}x. \tag{4.47}$$

This is the Hamiltonian function for the fundamental solution of the algebras A_n. Similar formulas can be obtained also for algebras of type B_n and C_n.

4.5. FIFTH STEP: THE ALGORITHM

Let us collect the main points discussed so far in a kind of algorithm to be used in the examples. As a starting point we assume to know the explicit form of the generators $(E_1, \ldots, E_n; F_1, \ldots, F_n)$ of a Chevalley basis of a simple Lie algebra $\mathfrak{g}$, in a faithful representation of minimal dimension of $\mathfrak{g}$. Then we proceed as follows.

(1) First, we compute the elements of minimal weight A and $\langle e|$ of the representation. Given A we compute its isotropy algebra $\mathfrak{g}_A$ and its orthogonal space $\mathfrak{g}_A^{\perp}$, and we determine the canonical symplectic leaf $\mathcal{S}$.
(2) Next, we reduce this leaf. By using $\mathfrak{g}_A$ we compute the distributions D and E of Proposition 3.1 on $\mathcal{S}$, and we determine the quotient manifold $\mathcal{N} = \mathcal{S}/E$. Finally, we compute the reduced Poisson pencil on $\mathcal{N}$ by using Equation (3.4).
(3) Then we compute the (left)-Frobenius basis (4.25) at the generic point of $\mathcal{S}$. This allows to compute the coefficients of the characteristic Equation (4.28), and then to determine the characteristic functions μ_a associated with the canonical symplectic leaf $\mathcal{S}$, by solving Equation (4.44).
(4) We use the Faà di Bruno polynomials of the characteristic functions to compute the eigenvectors $|\psi_a\rangle$ of the matrices $V(z)$ which belong to the kernel of the Poisson pencil at the points of $\mathcal{S}$. The eigenvalues of these matrices are determined by imposing the condition (4.21) at the point B.
(5) Finally, we compute the Casimir's function of the reduced Poisson pencil on $\mathcal{N}$ by using Equation (4.43). This function is used to construct the soliton hierarchy according to Equation (3.10).

For algebras of type A_n, the last two steps are summarized by Equations (4.42) and (4.47), respectively.

5. An Explicit Example: The KdV Hierarchy

This section is devoted to perform the previous program for the simplest algebra $\mathfrak{g} = A_1 = \mathfrak{sl}(2, \mathbb{C})$. The resulting hierarchy is the well-known KdV hierarchy. A peculiarity of the present approach, compared with the usual one, is to provide two different formulations of this hierarchy: on the canonical symplectic leaf $\mathcal{S}$ and on the quotient space $\mathcal{N}$. The usual formulation coincides with the one on the quotient space. Our point of view is that the formulation on the symplectic leaf too is worthwhile of interest. It shows features of the flows which are caught with difficulty after the reduction on the quotient space. We shall give an example in the next section.

5.1. THE SYMPLECTIC LEAF

For $\mathfrak{g} = A_1 = \mathfrak{sl}(2, \mathbb{C})$ we have

$$E_1 = \begin{pmatrix} 0 & 1 \\ 0 & 0 \end{pmatrix}, \qquad F_1 = \begin{pmatrix} 0 & 0 \\ 1 & 0 \end{pmatrix}, \quad \lambda = z^2. \tag{5.1}$$

Consequently

$$A = \begin{pmatrix} 0 & 0 \\ 1 & 0 \end{pmatrix}, \qquad B = \begin{pmatrix} 0 & 1 \\ 0 & 0 \end{pmatrix}, \tag{5.2}$$

and the eigenvalues of

$$\Lambda = B + \lambda A = \begin{pmatrix} 0 & 1 \\ \lambda & 0 \end{pmatrix} \tag{5.3}$$

are $\{z, -z\}$. The covector of minimal weight is

$$\langle e| = [1, 0], \tag{5.4}$$

while the isotropy algebra $\mathfrak{g}_A$ and its orthogonal subspace $\mathfrak{g}_A^\perp$ are:

$$\mathfrak{g}_A \colon \begin{pmatrix} 0 & 0 \\ w & 0 \end{pmatrix}, \qquad \mathfrak{g}_A^\perp \colon \begin{pmatrix} p & 0 \\ q & -p \end{pmatrix}. \tag{5.5}$$

Therefore the canonical symplectic leaf is

$$S = \begin{pmatrix} p & 1 \\ q & -p \end{pmatrix}. \tag{5.6}$$

5.2. THE REDUCTION

The distribution D of Proposition 3.1 is given by

$$\dot{S} = \begin{pmatrix} -w & 0 \\ w_x + 2pw & w \end{pmatrix} \tag{5.7}$$

or, in components, by

$$\dot{p} = -w, \qquad \dot{q} = w_x + 2pw, \qquad \dot{r} = 0. \tag{5.8}$$

In this example, D is completely contained in the leaf S, so that $E = D$. By eliminating the arbitrary parameter w, we get

$$\dot{p}_x + 2p\dot{p} + \dot{q} = \frac{\mathrm{d}}{\mathrm{d}t}(p_x + p^2 + q) = 0, \tag{5.9}$$

showing that the projection π is given by

$$u = \pi(S) = p_x + p^2 + q = u. \tag{5.10}$$

The loop $u\colon S^1 \to \mathbb{R}$ plays the role of coordinate on the quotient manifold $\mathcal{N} = \mathcal{S}/E$.

It is interesting to note that the fibers of the projection π are orbits of a 'gauge' group acting on $\mathcal{S}$. Indeed let us consider

$$\mathfrak{g}_{AB} = \{T \in \mathfrak{g}_A \mid T_x + [T, B] \in \mathfrak{g}_A^{\perp}\}. \tag{5.11}$$

It is a nilpotent subalgebra of the isotropy algebra of A. Let G_{AB} be the subgroup of G whose Lie algebra is $\mathfrak{g}_{AB}$, and $\Phi\colon G_{AB} \times \mathcal{S} \to \mathcal{S}$ the action given by

$$\Phi(g, S) = gSg^{-1} + g_x g^{-1}. \tag{5.12}$$

Due to condition (5.11), this action preserve the submanifold $\mathcal{S}$ and its orbits coincide with the fibers of π. In the example, $\mathfrak{g} = \mathfrak{sl}(2, \mathbb{C})$

$$\mathfrak{g}_{AB} = \mathfrak{g}_A : \begin{pmatrix} 0 & 0 \\ w & 0 \end{pmatrix}, \tag{5.13}$$

so that the generic element g of G_{AB} is

$$g = \begin{pmatrix} 1 & 0 \\ t & 1 \end{pmatrix}. \tag{5.14}$$

Hence, a simple calculation shows that $S' = \Phi_g(S)$ means

$$p' = p - t, \qquad q' = t_x + 2pt + p'_x + qt^2, \tag{5.15}$$

which implies $q' + p'^2 + p'_x = q + p^2 + p_x$, and the check is complete.

To compute the Poisson pencil on $\mathcal{N}$, we have to compute the bracket

$$\{F_1, F_2\}_\lambda = \omega(V_1, V_2) + (S + \lambda A, [V_1, V_2]) \tag{5.16}$$

on functions which are constant along D. The differentials $V = \mathrm{d}F$ of these functions belong to the annihilator D^0 of D. Consequently, they have the form

$$V = \mathrm{d}F(S) = \begin{pmatrix} -\frac{1}{2}v_{2x} + pv_2 & v_2 \\ v_3 & \frac{1}{2}v_{2x} - pv_2 \end{pmatrix}. \tag{5.17}$$

A covector on $\mathcal{N}$ will be denoted by v, and its value on the tangent vector $\dot{u}$ by

$$(v, \dot{u}) = \int_{S^1} v(x)\dot{u}(x)\,\mathrm{d}x. \tag{5.18}$$

The differential $v = \mathrm{d}f$ of the function f on $\mathcal{N}$ corresponding to F according to $f \circ \pi = F \circ i$ is related to V by

$$(V, \dot{S}) = (v, \dot{u}). \tag{5.19}$$

Explicitly

$$\int_{S^1} [2\dot{p}(-\tfrac{1}{2}v_{2x} + pv_2) + \dot{q}v_2]\,\mathrm{d}x = \int_{S^1} \dot{u}v\,\mathrm{d}x. \tag{5.20}$$

Since $\dot{u} = \mathrm{d}\pi(\dot{S}) = \dot{p}_x + 2p\dot{p} + \dot{q}$, we get $v_2 = v$ and, therefore, we can write the differential (5.17) as

$$V = \mathrm{d}F = \begin{pmatrix} -\frac{1}{2}v_x + pv & v \\ v_3 & \frac{1}{2}v_x - pv \end{pmatrix}, \tag{5.21}$$

where v_3 is a remnant arbitrary parameter which will disappear after the reduction. By inserting this expression into Equation (5.16), we obtain

$$\{f_1, f_2\}_\lambda = \int_{S^1} \tfrac{1}{2}v_{1x}v_{2xx} + (u+\lambda)(v_1v_{2x} - v_2v_{1x}) \tag{5.22}$$

or

$$\{f_1, f_2\}_\lambda = \omega(v_1, v_2) + (u, [v_1, v_2]_N) + \lambda(1, [v_1, v_2]_N), \tag{5.23}$$

by setting

$$[v_1, v_2]_N = v_1v_{2x} - v_2v_{1x}, \tag{5.24}$$

$$\omega(v_1, v_2) = \frac{1}{2}\int_{S^1} v_{1x}v_{2xx}\,\mathrm{d}x. \tag{5.25}$$

Therefore, the reduced bracket has the same form of the initial bracket (5.16), up to the replacement of the commutator on the loop algebra with the commutator (5.24) of the Virasoro algebra of the vector fields on the circle. We then can conclude that the quotient space $\mathcal{N}$, in this example, is the *Virasoro algebra* endowed with the canonical Lie–Poisson bracket (5.23), which is the well-known Poisson pencil associated with the KdV equation.

5.3. THE CHARACTERISTIC FUNCTIONS

The Frobenius basis at the point S is given by

$$\langle g^0| = [1, 0], \qquad \langle g^1| = [p, 1]. \tag{5.26}$$

Since

$$\langle g^2| = [p_x + p^2 + q + z^2, 0], \tag{5.27}$$

the characteristic equation is

$$\langle g^2| = \langle g^0|(u + z^2). \tag{5.28}$$

The characteristic functions μ_1 and μ_2 are the solutions of the equation

$$\mu_x + \mu^2 = u + z^2 \tag{5.29}$$

whose expansions in Laurent series start from z and $-z$, respectively:

$$\mu_1(z) = z + \sum_{j\geqslant 0} \frac{\mu_{1j}}{z^j}, \tag{5.30}$$

$$\mu_2(z) = -z + \sum_{j\geqslant 0} \frac{\mu_{2j}}{z^j}. \tag{5.31}$$

Obviously $\mu_2(z) = \mu_1(-z)$. By inserting expansion (5.30) in Equation (5.29) we iteratively get:

$$\begin{aligned}\mu_1(z) &= z + \frac{1}{2z}u - \frac{1}{4z^2}u_x + \frac{1}{8z^3}(u_{xx} - u^2) + \\ &+\frac{1}{16z^4}(-u_{xxx} + 4uu_x) + \\ &+\frac{1}{32z^5}(u_{xxxx} - 6uu_{xx} + 5u_x^2 + 2u^3) + \cdots .\end{aligned} \tag{5.32}$$

The other coefficients may be determined recursively according to

$$\mu_{1,k} = -\frac{1}{2}(\mu_{1,k-1})_x - \frac{1}{2}\sum_{l=0}^{k-1} \mu_{1,l}\,\mu_{1,k-1-l} \tag{5.33}$$

for $k \geqslant 1$. The eigenvector $|\psi_1\rangle$ defined by

$$\langle g^0|\psi_1\rangle = 1, \qquad \langle g^1|\psi_1\rangle = \mu_1 \tag{5.34}$$

is explicitly given by

$$|\psi\rangle = |\psi_1\rangle = \begin{bmatrix} 1 \\ \mu_1 - p \end{bmatrix}. \tag{5.35}$$

Therefore

$$|\psi_2\rangle = |\psi(-z)\rangle = \begin{bmatrix} 1 \\ \mu_2 - p \end{bmatrix} \tag{5.36}$$

and the dual covector $\langle\psi^*|$ is

$$\langle\psi^*| = (\mu_1 - \mu_2)^{-1}[p - \mu_2, 1], \tag{5.37}$$

as is easily checked by using the equations

$$\langle\psi^*|\psi\rangle = 1, \qquad \langle\psi^*|\psi_2\rangle = 0. \tag{5.38}$$

Finally, the fundamental solution

$$V = z\,(2|\psi\rangle\langle\psi^*| - \mathrm{I}) \tag{5.39}$$

is given by

$$V = \frac{1}{\mu_1 - \mu_2}\begin{bmatrix} 2z(p-\mu_2) & 2z \\ 2z(p-\mu_2)(\mu_1 - p) & 2z(\mu_1 - p)\end{bmatrix} - z\mathrm{I}. \tag{5.40}$$

The first coefficients of its expansion $V(z) = V_{-2}z^2 + V_{-1}z + V_0 + V_1 z^{-1} + \cdots$ are

$$V_{-2} = A = \begin{pmatrix} 0 & 0 \\ 1 & 0 \end{pmatrix}, \qquad V_0 = \begin{pmatrix} p & 1 \\ \frac{1}{2}(p_x + q - p^2) & -p \end{pmatrix},$$

$$V_2 = \begin{pmatrix} \frac{1}{4}p_{xx} + \frac{1}{4}q_x - \frac{1}{2}p^3 - \frac{1}{2}pq & -\frac{1}{2}(p_x + q + p^2) \\ \frac{1}{8}p_{xxx} - \frac{1}{4}pp_{xx} + \frac{1}{8}q_{xx} - \frac{3}{4}p^2p_x - \frac{1}{2}pq_x + \\ -\frac{1}{4}p_xq + \frac{1}{8}{p_x}^2 + \frac{3}{8}p^4 + \frac{1}{4}p^2q - \frac{1}{8}q^2 & * \end{pmatrix} \tag{5.41}$$

while the odd coefficients vanish: $V_{2k+1} = 0\ \forall k$.

5.4. The KdV hierarchy

The pencil of vector fields $X_z = [A, V(z)]$ corresponding to V is given by $X_z = \sum_{k\geqslant -1} X_{2k}z^{-2k}$, where

$$X_{-2} = 0, \qquad X_0 = \begin{bmatrix} -1 & 0 \\ 2p & 1 \end{bmatrix},$$

$$X_2 = \begin{bmatrix} \frac{1}{2}(p_x + q + p^2) & 0 \\ \frac{1}{2}p_{xx} + \frac{1}{2}q_x + \\ -qp - p^3 & -\frac{1}{2}(p_x + q + p^2) \end{bmatrix},$$

$$X_4 = \begin{bmatrix} \frac{1}{8}p_{xxx} + \frac{1}{8}q_{xx} - \frac{3}{4}p^2p_x - \frac{3}{4}p_xq + \frac{1}{4}pp_{xx} + \\ -\frac{1}{8}{p_x}^2 - \frac{3}{8}p^4 - \frac{3}{4}p^2q - \frac{3}{8}q^2 & 0 \\ \frac{1}{8}p_{xxxx} + \frac{1}{8}q_{xxx} - \frac{5}{4}p^2p_{xx} - \frac{5}{4}p{p_x}^2 - \frac{1}{4}pq_{xx} - \frac{3}{4}p_xq_x + \\ -\frac{3}{4}p_{xx}q - \frac{3}{4}qq_x - \frac{3}{4}p^2q_x + \frac{3}{4}p^5 + \frac{3}{2}p^3q + \frac{3}{4}pq^2 & * \end{bmatrix}. \tag{5.42}$$

It defines the KdV hierarchy on the canonical symplectic leaf. According to the results of the previous sections, the vector fields X_{2k} fulfill the following properties:

(1) they are bi-Hamiltonian: $X_k = [A, V_k] = V_{k-2_x} + [V_{k-2}, S]$;
(2) they can be projected on the quotient space $\mathcal{N}$;
(3) their projections $\dot{u}_k$ are the vector fields:

$$\dot{u}(z) = P_1^{\mathcal{N}}\,\mathrm{d}h(z) = -2(\mathrm{d}h(z))_x \tag{5.43}$$

associated with the Hamiltonian function

$$h(z) = 2z \int_{S^1} \mu_1 \, \mathrm{d}x. \tag{5.44}$$

Let us check these statements for the first vector fields of the hierarchy. It is easy to see that X_0 projects on the null vector field:

$$\dot{u}_0 = (\dot{p}_0)_x + 2p\dot{p}_0 + \dot{q}_0 = 0. \tag{5.45}$$

Similarly, it can be shown that X_2 projects on $\dot{u}_2 = u_x$, while X_4 projects on the KdV vector field

$$\dot{u}_4 = \tfrac{1}{4}(u_{xxx} - 6uu_x). \tag{5.46}$$

On the other hand, the first terms of the Laurent expansion $h(z) = 2z^2 + \sum_{j\geqslant 0} h_j z^{-j}$ are

$$\begin{aligned} h_0 &= \int u \, \mathrm{d}x, \qquad h_2 = -\frac{1}{4}\int u^2 \, \mathrm{d}x, \\ h_4 &= \frac{1}{16}\int (-6uu_{xx} - 5u_x^2 + 2u^3) \, \mathrm{d}x, \ldots \end{aligned} \tag{5.47}$$

while the odd coefficients vanish: $h_{2j+1} = 0, \forall j$. Hence Equation (5.42) implies

$$\begin{aligned} \dot{u}_2 &= -2(\mathrm{d}h_2)_x = u_x, \\ \dot{u}_4 &= -2(\mathrm{d}h_4)_x = \frac{1}{4}u_{xxx} - \frac{3}{2}uu_x, \end{aligned} \tag{5.48}$$

proving our statement. Equation (5.43) defines the KdV hierarchy on the quotient space $\mathcal{N}$.

6. Concluding Remarks

We have spent a lot of work in defining the equations from several points of views, guided by the conviction that this will allow to display features of the flows which are not easily seen in the conventional approach. We shall now show one of these features.

As explained in [2], besides the usual Hamilton equations on functions, one may consider Hamilton equations on 1-forms. In the present context, this means that instead of studying the equation

$$\frac{\partial S}{\partial t} = [A, V] \tag{6.1}$$

on the matrix S, we study how the matrix V, depending on S, evolves along the flow. Since the eigenvalues of V are constant, this amounts to study how the eigenvectors of V evolve in time. The fundamental solution (4.42) for the algebras A_n is characterized by a single eigenvector $|\psi\rangle$. So, it is enough to see how this eigenvector evolves. Our claim is that the equation of motion for $|\psi\rangle$ is the *linear* equation

$$\frac{\partial}{\partial t}|\psi\rangle = \frac{\partial}{\partial z}|\psi\rangle. \tag{6.2}$$

To recover this property on the quotient space $\mathcal{N}$ is more difficult. Indeed, one has to introduce the tau function, the Baker–Akhiezer function, and the vertex operators. It is possible to show that Equation (6.2) is equivalent to the well-known equation [3]

$$w(z;t_1,t_2,\ldots) = \frac{\tau(t_1 - \frac{1}{z}, t_2 - \frac{1}{2z^2}, \ldots)}{\tau(t_1,t_2,\ldots)} \tag{6.3}$$

connecting these three objects. However, a discussion of these topics is completely outside the limits of this paper. Its aim was simply to display how the bi-Hamiltonian point of view may systematically guide us in the construction of the soliton hierarchies, by starting merely from a Chevalley basis of a simple Lie algebra.

References

1. Casati, P., Magri, F., and Pedroni, M.: Bi-Hamiltonian manifolds and τ-function, in M. J. Gotay *et al.* (eds), *Mathematical Aspects of Classical Field Theory 1991*, Contemporary Mathematics 132, Amer. Math. Soc., Providence, 1992, pp. 213–234.
2. Casati, P., Magri, F., and Pedroni, M.: Bi-Hamiltonian manifolds and Sato's equations, in O. Babelon *et al.* (eds), *The Verdier Memorial Conference on Integrable Systems, Actes du Colloque International de Luminy, 1991*, Progress in Mathematics, Birkhäusèr, Basel, pp. 251–272.
3. Dickey, L. A.: *Soliton Equations and Hamiltonian Systems*, Adv. Series in Math. Phys., 12, World Scientific, Singapore, 1991.
4. Drinfeld, V. G. and Sokolov, V. V.: Lie algebras and equations of Korteweg–de Vries type, *J. Soviet Math.* **30** (1985), 1975–2036.
5. Libermann, P. and Marle, C. M.: *Symplectic Geometry and Analytical Mechanics*, Reidel, Dordrecht, 1987.
6. Magnano, G. and Magri, F.: Poisson–Nijenhuis structures and Sato hierarchy, *Rev. Math. Phys.* **3**(4) (1991), 403–466.
7. Marsden, J. E. and Ratiu, T.: Reduction of Poisson manifolds, *Lett. Math. Phys.* **11** (1986), 161–169.

Acta Applicandae Mathematicae **41:** 271–284, 1995.

Differential Invariants

PETER J. OLVER*
School of Mathematics, University of Minnesota, Minneapolis, MN 55455, U.S.A.
e-mail: olver@ima.umn.edu

(Received: 27 October 1993)

Abstract. This paper summarizes recent results on the number and characterization of differential invariants of transformation groups. Generalizations of theorems due to Ovsiannikov and to M. Green are presented, as well as a new approach to finding bounds on the number of independent differential invariants.

Mathematics Subject Classifications (1991): 53A55, 22E70, 58G35.

Key words: differential invariant, transformation group, point transformation, contact transformation, jet space.

Consider a group of transformations acting on a jet space coordinatized by the independent variables, the dependent variables, and their derivatives. Scalar functions which are not affected by the group transformations are known as differential invariants. Their importance was emphasized by Sophus Lie [9], who showed that every invariant system of differential equations [10], and every invariant variational problem [11], could be directly expressed in terms of the differential invariants. As such they form the basic building blocks of many physical theories, where one begins by postulating the invariance of the equations or the variational principle under a prescribed symmetry group. Lie also demonstrated [10], how differential invariants could be used to integrate invariant ordinary differential equations, and succeeded in completely classifying all the differential invariants for all possible finite-dimensional Lie groups of point transformations in the case of one independent and one dependent variable. Lie's results were pursued by Tresse [18], and, much later, Ovsiannikov [17]. In this paper, I will summarize some new results extending these earlier classification theorems, which were discovered in the course of writing the recent book [16]. Space considerations preclude the inclusion of proofs and significant examples here. It is worth remarking that, surprisingly, the complete classification of differential invariants for many of the groups of physical importance, including the affine, conformal, and Poincaré groups, does not yet seem to be known!

* Supported in part by NSF Grants DMS 91–16672 and DMS 92–04192.

Consider the space* $M = X \times U \simeq \mathbb{R}^p \times \mathbb{R}^q$ whose coordinates represent our independent variables $x = (x^1, \ldots, x^p) \in X$ and dependent variables $u = (u^1, \ldots, u^q) \in U$. Let J^n denote the associated jet bundle of order n, whose coordinates $(x, u^{(n)})$ represent the independent variables and the derivatives $u_I^\alpha = \partial^k u^\alpha / \partial x^{i_1} \cdots \partial x^{i_k}$, $\alpha = 1, \ldots, q$, $1 \leqslant i_\nu \leqslant p$, of the dependent variables of orders $0 \leqslant k = \#I \leqslant n$. Thus, $\dim \mathrm{J}^n = p + q^{(n)}$, where $q^{(n)} = q\binom{p+n}{n}$. The number of derivative coordinates of order exactly n is denoted by

$$q_n = \dim \mathrm{J}^n - \dim \mathrm{J}^{n-1} = q^{(n)} - q^{(n-1)} = qp_n = q\binom{p+n-1}{n}. \tag{1}$$

A smooth function (or section) $u = f(x)$ from X to U has nth *prolongation* (or n-jet) $u^{(n)} = \mathrm{pr}^{(n)} f(x)$, which is the section of J^n given by $u_I^\alpha = \partial_I f^\alpha(x)$.

A differential one-form θ on the jet space J^n is called a *contact form* if it is annihilated by all prolonged functions. It is easy to prove that every contact form on J^n is a linear combination of the basic contact forms

$$\theta_I^\alpha = \mathrm{d}u_I^\alpha - \sum_{k=1}^{p} u_{I,k}^\alpha \mathrm{d}x^k, \quad \alpha = 1, \ldots, q, \ 0 \leqslant \#I < n. \tag{2}$$

We call $\#I$ the *order* of the contact form θ_I^α, so the contact forms on J^n have orders at most $n - 1$. A one-form on J^n is called *horizontal* if it annihilates all vertical tangent directions, i.e., just involves the $\mathrm{d}x^i$'s.

A smooth, real-valued function F: $\mathrm{J}^n \to \mathbb{R}$ is called a *differential function* of order n. Note that any differential function $F(x, u^{(n)})$ of order n automatically defines a differential function on any higher order jet space merely by treating the coordinates $(x, u^{(n)})$ of J^n as a subset of the coordinates $(x, u^{(n+k)})$ of J^{n+k} – this is the same as composing F with the natural projection π_n^{n+k}: $\mathrm{J}^{n+k} \to \mathrm{J}^n$. In the sequel, we will not distinguish between F and $F \circ \pi_n^{n+k}$. Given a differential function F: $\mathrm{J}^n \to \mathbb{R}$, its differential is the one-form

$$\mathrm{d}F = \sum_{i=1}^{p} \frac{\partial F}{\partial x^i}\, \mathrm{d}x^i + \sum_{\alpha=1}^{q} \sum_{\#I \leqslant n} \frac{\partial F}{\partial u_I^\alpha}\, \mathrm{d}u_I^\alpha.$$

On the next higher order jet space J^{n+1}, we can uniquely decompose $\mathrm{d}F$ into a horizontal one-form plus a contact form. The horizontal component is called the *total differential* of F, and given by $DF = \sum_{i=1}^p D_i F\, \mathrm{d}x^i$, where D_i denotes the total derivative with respect to x^i.

DEFINITION 1. A local diffeomorphism Ψ: $\mathrm{J}^n \to \mathrm{J}^n$ defines a *contact transformation* of order n if it preserves the space of contact forms, meaning that if θ is any contact form on J^n, then $\Psi^*\theta$ is also a contact form.

* More generally, M can be a vector or fiber bundle, or even an arbitrary smooth manifold [14]. However, as all our considerations are local, there is no loss in generality in restricting our attention to open subsets of Euclidean space.

In particular, any point transformation $\Phi\colon M \to M$ defines a 0th-order contact transformation. Contact transformations act on functions by point-wise transforming their prolonged graphs. If $\Psi\colon \mathrm{J}^n \to \mathrm{J}^n$ defines an nth-order contact transformation, then, for any $k \geqslant 0$, there is a uniquely defined $(n+k)$th-order contact transformation $\mathrm{pr}^{(k)}\Psi\colon \mathrm{J}^{n+k} \to \mathrm{J}^{n+k}$, the kth *prolongation* of Ψ, which projects back down to Ψ. Bäcklund's Theorem [3], demonstrates that, except in the case of a single dependent variable, all contact transformations are merely prolonged point transformations.

THEOREM 2. *If the number of dependent variables is greater than one, $q > 1$, then every contact transformation is the prolongation of a point transformation $\Phi\colon M \to M$. If $q = 1$, then there are first order contact transformations which do not come from point transformations, but every* nth-*order contact transformation is the* $(n-1)$st *prolongation of a first order contact transformation* $\Psi\colon \mathrm{J}^1 \to \mathrm{J}^1$.

By a *transformation group G*, then, we mean either a local Lie group of point transformations acting on (an open subset of) the space M of independent and dependent variables, or, in the single dependent variable case, a local Lie group of contact transformations acting on (an open subset of) the first jet space J^1. (In this paper, a transformation group is always a finite-dimensional Lie group.) To keep the notation uniform, we let $G^{(n)}$ denote the associated prolonged group action (by contact transformations) on the jet space J^n, so that $G^{(n)} = \mathrm{pr}^{(n)}G$ in the case of point transformations, whereas $G^{(n)} = \mathrm{pr}^{(n-1)}G$ in the case of first order contact transformations. (In the latter case we assume that $n \geqslant 1$.) The (prolonged) infinitesimal generators of $G^{(n)}$ form a Lie algebra $\mathfrak{g}^{(n)}$ of vector fields $\mathbf{v}^{(n)}$ on J^n satisfying the same commutation relations as the Lie algebra $\mathfrak{g}$ of G, and determined by the standard prolongation formula, cf. [15].

DEFINITION 3. Let G be a group of point or contact transformations. A *differential invariant* is a real-valued function $I\colon \mathrm{J}^n \to \mathbb{R}$ which satisfies

$$I\big(g^{(n)} \cdot (x, u^{(n)})\big) = I(x, u^{(n)})$$

for all $g^{(n)} \in G^{(n)}$, and all $(x, u^{(n)}) \in \mathrm{J}^n$, where the prolonged transformation $g^{(n)} \cdot (x, u^{(n)})$ is defined.

As usual, the differential invariant I may only be defined on an open subset of the jet space, although we shall still write $I\colon \mathrm{J}^n \to \mathbb{R}$ to indicate its jet space of definition. Differential invariants (of connected groups) are most easily determined using infinitesimal methods.

PROPOSITION 4. *A function $I\colon \mathrm{J}^n \to \mathbb{R}$ is a differential invariant for a connected transformation group G if and only if $\mathbf{v}^{(n)}(I) = 0$ for every prolonged infinitesimal generator $\mathbf{v}^{(n)} \in \mathfrak{g}^{(n)}$.*

In applications, the determination of a complete set of functionally independent differential invariants of a given group action is of significant importance. Functional independence is guaranteed if their differentials are linearly independent: $\mathrm{d}F_1 \wedge \cdots \wedge \mathrm{d}F_k \neq 0$. Since, as noted above, any lower order differential invariant $I(x, u^{(k)})$, $k < n$, is automatically an nth-order differential invariant, it will be important to distinguish differential invariants of order exactly n from lower order differential invariants. We will call a set of differential functions on J^n *strictly independent* if, as functions of the nth-order derivative coordinates alone, they are functionally independent. The functions $F_1, \ldots, F_k$ are *strictly independent* if their nth-order differentials, given (intrinsically) by

$$\mathrm{d}_n F = \sum_{\alpha=1}^{q} \sum_{\#I=n} \frac{\partial F}{\partial u_I^\alpha}\, \mathrm{d}u_I^\alpha, \tag{3}$$

are linearly independent at each point: $\mathrm{d}_n F_1 \wedge \mathrm{d}_n F_2 \wedge \cdots \wedge \mathrm{d}_n F_k \neq 0$. Note that, in particular, this implies that none of the F's, or any function thereof, can be of order strictly less than n.

In order to study the differential invariants of a transformation group, a more detailed knowledge of the structure of the prolonged group actions is required. The following remarks all follow directly from Frobenius' Theorem, cf. [15]. Let G be an r-dimensional Lie group of transformations. Let s_n denote the maximal (generic) orbit dimension of the prolonged action $G^{(n)}$, so that $G^{(n)}$ acts semi-regularly on the open subset $V^{(n)} \subset \mathrm{J}^n$ consisting of all points contained in the orbits of maximal dimension. (If G acts analytically, then the subset $V^{(n)}$ is dense in J^n.) In the remainder of this paper, we shall restrict our attention to the subset $V^{(n)}$, thereby avoiding more delicate questions concerning singularities of the prolonged group actions. Let h_n denote the dimension of any isotropy subgroup $H_z^{(n)} = \{g \mid g^{(n)} \cdot z = z\}$ for $z \in V^{(n)}$. (These isotropy subgroups are, in general, different, but all have the same dimension.) Then the orbit dimensions satisfy $s_n = r - h_n$. Locally, near any point $z \in V^{(n)}$, there are

$$i_n = p + q^{(n)} - s_n = p + q^{(n)} - r + h_n \tag{4}$$

functionally independent differential invariants of order at most n. Since each differential invariant of order less than n is included in this count, the integers i_n form a nondecreasing sequence: $i_0 \leqslant i_1 \leqslant i_2 \leqslant \cdots$. The difference

$$j_n = i_n - i_{n-1} = q_n - s_n + s_{n-1} = q_n + h_n - h_{n-1}, \tag{5}$$

cf. (1), will count the number of strictly independent nth-order differential invariants. For groups of point transformations, we set $j_0 = i_0$ to be the number of ordinary invariants; for contact transformation groups, where i_0 and s_0 are not defined, we set $j_1 = i_1$. Note that j_n cannot exceed q_n, the number of independent derivative coordinates of order n; this implies the elementary inequalities

$$i_{n-1} \leqslant i_n \leqslant i_{n-1} + q_n. \tag{6}$$

If $\mathcal{O}^{(n)} \subset \mathrm{J}^n$ is any orbit of $G^{(n)}$, then, for any $k < n$, its projection $\pi_k^n(\mathcal{O}^{(n)}) \subset \mathrm{J}^k$ is an orbit of the kth prolongation $G^{(k)}$. Therefore, the maximal orbit dimension s_n of $G^{(n)}$ is a *nondecreasing* function of n, bounded by $r = \dim G$:

$$s_0 \leqslant s_1 \leqslant s_2 \leqslant \cdots \leqslant r. \tag{7}$$

On the other hand, since the orbits cannot increase in dimension any more than the increase in dimension of the jet spaces themselves, we have the the elementary inequalities

$$s_{n-1} \leqslant s_n \leqslant s_{n-1} + q_n, \tag{8}$$

governing the orbit dimensions. Note that, in view of Equations (1) and (4), the inequalities (8) are equivalent to those in (6). Condition (7) implies that the maximal orbit dimension eventually stabilizes, so that there exists an integer s such that $s_m = s$ for all m sufficiently large. In particular, if the orbit dimension is ever the same as that of G, meaning $s_n = r$ for some n, then $s_m = r$ for all $m \geqslant n$. We will call s the stable orbit dimension, and the minimal order n for which $s_n = s$ the *order of stabilization* of the group.

EXAMPLE 5. Consider the three-parameter group action $(x, u) \mapsto (\lambda x + a, \lambda u + b)$, $(x, u) \in M \simeq \mathbb{R}^2$, generated by the vector fields ∂_x, ∂_u, $x\partial_x + u\partial_u$. There are no ordinary invariants since the group is transitive on $M = \mathbb{R}^2$. Furthermore, all three vector fields happen to coincide with their first prolongations, and hence there is one independent first order differential invariant, namely u_x. The second prolongations are $\partial_x, \partial_u, x\partial_x + u\partial_u - u_{xx}\partial_{u_{xx}}$, and hence there are no differential invariants of (strictly) second order. There is a single third order differential invariant, namely $u_{xx}^{-2}u_{xxx}$, a single fourth order invariant, $u_{xx}^{-3}u_{xxxx}$, and, in general, a single nth order differential invariant $u_{xx}^{-n-1}D_x^n u$. Therefore, the number of strictly independent differential invariants is given by

$$j_0 = j_2 = 0, \qquad j_1 = j_3 = \cdots = j_n = 1, \quad n \geqslant 3.$$

Therefore,

$$i_0 = 0, \qquad i_1 = i_2 = 1, \quad i_3 = 2, \ldots, i_n = n - 1, \quad n \geqslant 3,$$

which implies that the maximal orbit dimensions are

$$s_0 = s_1 = 2, \qquad s_2 = s_3 = \cdots = 3 = \dim G.$$

We observe that the orbit dimensions 'pseudo-stabilized' at order 0 since $s_0 = s_1$, but the correct order of stabilization is $n = 2$. More generally, the r-dimensional group generated by

$$\partial_x,\ \partial_u,\ x\partial_u, \ldots, x^{r-3}\partial_u,\ x\partial_x + (r-2)u\partial_u$$

has orbit dimensions

$$s_0 = 1, \quad s_1 = 2, \quad \ldots, \quad s_{r-3} = s_{r-2} = r - 1, \quad s_{r-1} = s_r = \cdots = r,$$

so the orbit dimensions pseudo-stabilize at order $r - 3$.

A transformation group acts *effectively* if different group elements have different actions, so that $g \cdot x = h \cdot x$ for all $x \in M$ if and only if $g = h$. The *global isotropy subgroup* $G_M = \{g \mid g \cdot x = x \text{ for all } x \in M\}$, which is a closed normal subgroup of G, measures the 'effectiveness' of the action of G in the sense that G acts effectively if and only if $G_M = \{e\}$ is trivial. If G does not act effectively, we can replace it by the quotient group G/G_M, which does act effectively on M in essentially the same way as G itself. Thus, there is no loss in generality in assuming that all our group actions are (locally) effective. A Lie group G is said to act *locally effectively* if the global isotropy group G_M is a discrete subgroup of G, in which case G/G_M has the same dimension (and the same Lie algebra) as G. Remarkably, the local effectiveness of a group action is characterized by its stable orbit dimension [17].

THEOREM 6. *A transformation group G acts locally effectively if and only if its dimension is the same as its stable orbit dimension, so that $s_m = r = \dim G$ for all m sufficiently large.*

The basic method for constructing a complete system of differential invariants of a given transformation group is to use invariant differential operators. A differential operator is said to be G-invariant if it maps differential invariants to higher-order differential invariants, and thus, by iteration, produces hierarchies of differential invariants of arbitrarily large order. For n sufficiently large, we can guarantee the existence of sufficiently many such differential operators so as to completely generate all the higher-order independent differential invariants of the group by successively differentiating lower order differential invariants. Thus, a complete description of all the differential invariants is provided by a collection of low order 'fundamental' differential invariants along with the requisite invariant differential operators.

DEFINITION 7. A differential one-form ω on J^n is called *contact-invariant* under a transformation group G if and only if, for every $g \in G$, the pull-back $(g^{(n)})^*\omega = \omega + \theta$ for some contact form $\theta = \theta_g$.

The infinitesimal criterion for contact-invariance is that the Lie derivative of the form with respect to each prolonged infinitesimal generator be a contact form. Contact forms are trivially contact-invariant, so only the horizontal contact-invariant forms are of interest. If $I(x, u^{(n)})$ is any nth-order differential invariant, its total differential $DI = \sum D_j I \, dx^j$ is a contact-invariant one-form on J^{n+1}. The construction of high order differential invariants is facilitated by the existence of enough horizontal contact-invariant forms.

DEFINITION 8. Let G be a transformation group acting on a space having p independent variables. An nth-*order contact-invariant coframe* is a collection of $p = \dim X$ independent contact-invariant horizontal one-forms $\omega^1, \ldots, \omega^p$, defined locally on the jet space J^n.

Contact-invariant coframes are the jet space counterparts of the coframes from differential geometry that form the foundation of the Cartan equivalence method [4, 5]. If $F(x, u^{(m)})$ is any differential function, we can rewrite its total differential in terms of the coframe,

$$DF = \sum_{k=1}^{p} \mathcal{D}_k F \, \omega^k. \tag{9}$$

The resulting 'coframe differential operators' $\mathcal{D}_k$ are G-invariant differential operators:

PROPOSITION 9. *If $I(x, u^{(m)})$ is any differential invariant of order m, then $\mathcal{D}_k I$ is a differential invariant of order* $\leqslant \max\{n, m+1\}$.

In local coordinates, if $\omega^i = \sum_k P_k^i(x, u^{(n)}) \, \mathrm{d}x^k$, then

$$\mathcal{D}_k = \sum_i Q_k^i(x, u^{(n)}) D_i,$$

where $\mathbf{Q} = (Q_j^i(x, u^{(n)})) = \mathbf{P}^{-T}$. In particular, if the $\omega^i = DI_i$ are obtained from functionally independent differential invariants $I_1, \ldots, I_p$, then

$$\mathbf{P} = (D_j I_i(x, u^{(n)}))$$

is their total Jacobian matrix.

Let us first consider the case $p = 1$, so there is a single independent variable x. A contact-invariant coframe is just a nonvanishing contact-invariant one-form $\omega = P(x, u^{(n)}) \, \mathrm{d}x$. The associated invariant differential operator $\mathcal{D} = (1/P) D_x$ maps a differential invariant J to the differential invariant $\mathcal{D}J = D_x J / P$. In particular, if $I(x, u^{(n)})$ is any (non-constant) differential invariant, the corresponding invariant differential operator is $\mathcal{D} = (D_x I)^{-1} D_x$, which maps J to $\mathrm{d}J/\mathrm{d}I = D_x J / D_x I$. Therefore, starting from a pair of differential invariants (or, more generally, a single differential invariant and a contact-invariant horizontal one-form) we construct an infinite sequence of higher and higher order differential invariants $\mathcal{D}^k J$, $k = 0, 1, 2, \ldots$. The functional independence of the resulting differential invariants is guaranteed by the following lemma.

LEMMA 10. *Suppose $J_1, \ldots, J_r$ are strictly independent nth-order differential invariants, and I is either a differential invariant of order strictly less than n, or an nth-order differential invariant which is strictly independent of the J_ν's. Let $\mathcal{D} = (D_x I)^{-1} D_x$ denote the invariant differential operator associated with DI. Then the differential functions $\mathcal{D}J_1, \ldots, \mathcal{D}J_r$ are strictly independent $(n+1)$st-order differential invariants.*

THEOREM 11. *Suppose that G is a group of point or contact transformations acting on a space M having one independent variable and q dependent variables. Then, locally, there exist $q+1$ fundamental, independent differential invariants $I, J_1, \ldots, J_q$, such that every differential invariant can be written as a function of these differential invariants and their derivatives $\mathcal{D}^m J_\nu$, where $\mathcal{D} = (D_x I)^{-1} D_x$ is the invariant differential operator associated with the first differential invariant I.*

Both results can be readily generalized by using a contact-invariant one-form instead of DI.

Theorem 11 has some important consequences governing the order of stabilization n of a transformation group. As we saw in Example 5, it is possible for the orbit dimension to pseudo-stabilize at some lower order, meaning that $s_k = s_{k+1} < s_{k+2}$ for some $k < n$. First, we note that a pseudo-stabilization of the orbit dimensions can only occur if the orbit dimension is rather high.

THEOREM 12. *Suppose that, for some $n \geqslant 0$, the maximal orbit dimensions of the prolonged group actions satisfy $s_{n-1} < s_n = s_{n+1} \leqslant q^{(n)}$. Then n is the order of stabilization of G.*

The next result, which follows directly from Theorem 12, demonstrates that there can be at most one such pseudo-stabilization.

THEOREM 13. *Suppose that the maximal orbit dimensions of the prolonged group actions satisfy $s_k = s_{k+1}$ and, also, $s_n = s_{n+1}$ for some $n > k$. Then $s_m = s_n$ for all $m \geqslant n$.*

Both of these results are valid as stated in the general case of several independent variables and several dependent variables – see below. Theorem 13 provides a significant strengthening of Ovsiannikov's stabilization theorem, [17, p. 313], which states that if $s_{n-1} < s_n = s_{n+1} = s_{n+2}$, then the orbit dimension stabilizes at order n. Indeed, even in this case, the proof of Theorem 13 in [16] is new and much more direct than that of Ovsiannikov.

As a consequence of Theorems 12 and 13, there are essentially only two possibilities for the orders of the fundamental differential invariants, as described in Theorem 11, of a group of transformations acting on a space with just one independent variable. Assume G is an r-dimensional group acting locally effectively, and let n denote the order of stabilization. Then either

(a) The fundamental differential invariants have order at most $n+1$. In this case, there exist $q+1$ differential invariants, I, of order $\leqslant n$, $J_1, \ldots, J_{q-1}$ of order $\leqslant n+1$, and J_q of order $= n+1$. In this case $\dim G = r \leqslant 1 + (n+1)q$.

(b) The fundamental differential invariants have order at most $n+2$. In this case, there exist $q+1$ differential invariants $I, J_1, \ldots, J_{q-1}$, all of order $= n+1$, and J_q of order $= n+2$. This case can *only* occur if the dimension of G equals $r = 1 + (n+1)q$.

Consequently, in the case of a single independent variable, the order of stabilization n of an r-dimensional locally effective group action obeys the inequalities

$$\frac{r-1}{q} - 1 \leqslant n \leqslant r - 1. \tag{10}$$

In his study of the differential invariants of curves in a homogeneous space [6], M. Green discovered a striking formula relating the number of fundamental differential invariants to the dimensions of the isotropy subgroups of the prolonged group action. Our Theorem 11 implies that Green's results are, in fact, valid for completely general transformation groups on spaces with one independent variable! Let k_n denote the number of strictly independent fundamental differential invariants of order n, i.e., those differential invariants which are not expressed in terms of differentiated invariants of any lower order. Since, according to Lemma 10, the differentiated invariants coming from strictly independent invariants are themselves strictly independent, these numbers satisfy $k_n = j_n - j_{n-1}$ provided either $i_{n-2} \geqslant 1$, so there is at least one lower order invariant to provide the require invariant differential operator, or, more generally, there exists a contact-invariant one-form of order at most $n-1$.

THEOREM 14. *Let G be a transformation group. Then the number k_n of fundamental differential invariants of order n is given in terms of the minimal dimension h_n of the isotropy subgroups of $G^{(n)}$ according to*

$$k_n = \begin{cases} h_n - 2h_{n-1} + h_{n-2} + 1, & \text{if } i_{n-2} = 0,\ i_{n-1} > 0, \\ h_n - 2h_{n-1} + h_{n-2}, & \text{otherwise.} \end{cases} \tag{11}$$

Equation (11) *is valid for all $n \geqslant 0$ provided we set $i_n = 0$ and $h_n = r - 1 - (n+1)q$ whenever the action $G^{(n)}$ is not defined, i.e., for $n = -2, -1$, and, in the case of a contact transformation group, $n = 0$.*

EXAMPLE 15. Let G be an r-dimensional Lie group and $H \subset G$ a closed subgroup of dimension s. Let $M = \mathbb{R} \times (G/H)$, so that the functions $u = f(x)$ are described by curves in the homogeneous space G/H. The group G acts on M by the Cartesian product of the trivial action on the independent variable $x \in \mathbb{R}$ and its usual action via left multiplication on G/H. In this case, $h_0 = s = \dim H$, and $i_0 = j_0 = k_0 = 1$, since there is a single ordinary invariant x, with consequential invariant differential operator D_x. Formula (11) implies M. Green's main result that there are $k_n = h_n - 2h_{n-1} + h_{n-2}$ fundamental differential invariants of order $n \geqslant 2$. For $n = 1$, we have $k_1 = h_1 - 2h_0 + h_{-1} + 1 = h_1 - 2s + r$ since $i_{-1} = 0$, $i_0 = 1$, while $h_{-1} = r - 1$ according to our convention. (Green sets $h_{-1} = r$, but this does not conform with our general formula.) See [6] for a wide variety of applications and explicit examples, including affine, projective, and conformal geometry.

Let us now specialize even further, to the case $p = q = 1$, so there is just one independent and one dependent variable. Here $q_n = 1$ for all n, so (6) implies that, for each $n \geqslant 1$, there is at most one independent differential invariant of order $= n$, i.e., $j_n \leqslant 1$. Theorem 11 implies that there are precisely two fundamental differential invariants $I(x, u^{(s)})$ and $J(x, u^{(t)})$, having orders $0 \leqslant s < t$, respectively. (Here we are leaving aside the trivial case when $G = \{e\}$ acts trivially on M, where there are two independent 0th-order invariants, namely x and u, and every differential function is a differential invariant.) Every other differential invariant can be written in terms of I, J, and the differentiated invariants $\mathcal{D}^m J = \mathrm{d}^m J/\mathrm{d}I^m$, so that $j_s = 1$, and $j_m = 1$ for every $m \geqslant t$. Equation (5) implies that the orbit dimensions and number of invariants are given by

$$\begin{aligned} &s_k = k+2, && i_k = 0, && \text{when} \quad k \leqslant s-1, \\ &s_k = k+1, && i_k = 1, && \text{when} \quad s \leqslant k \leqslant t-1, \\ &s_k = t, && i_k = k-t+2, && \text{when} \quad k \geqslant t. \end{aligned} \tag{12}$$

Assuming G acts locally effectively, we conclude that the second fundamental differential invariant J necessarily has order $t = r = \dim G$. There are three subcases: If the first fundamental invariant I has order $s = 0$, the group acts intransitively on M, the orbit dimension stabilizes at order $r - 1$, and there is one ordinary invariant (of order 0) and one rth order differential invariant. At the other extreme, if $s = r - 1$, then the group has fundamental differential invariants of orders $r - 1$ and r, and the orbit dimension stabilizes at order $r - 2$. The intermediate cases $0 < s < r - 1$ are when the orbit dimension pseudo-stabilizes at order s, and finally stabilizes at order $r - 1$. Thus, our methods provide rather detailed information on the possible orbit dimensions of prolonged group actions in the single variable case. However, even this is not as detailed as possible. Lie [8] (see also [12]) completely classified the Lie groups of both point and contact transformations acting on a two-dimensional complex manifold. Moreover, in [10] he determined the differential invariants for each of the point transformation groups. (See [16] for the differential invariants of the contact transformation groups.) Inspecting Lie's tables, we are led to the following remarkable result.

THEOREM 16. *Let G be a locally effective r-dimensional Lie group of point or contact transformations acting on $M \simeq \mathbb{R}^2$. Then G has fundamental differential invariants $I(x, u^{(s)})$ and $J(x, u^{(r)})$ having orders $s < r$. Moreover, $s = r - 1$ unless either* (a) *G acts intransitively, in which case $s = 0$, or* (b) *the prolonged orbit dimensions pseudo-stabilize, in which case $s = r - 2$, and the pseudo-stabilization occurs at order $r - 3$. In fact, the orbit dimensions pseudo-stabilize if and only if the group action is equivalent, under a change of variables, to the r-dimensional Lie group action described in Example* 5!

We now discuss generalizations of these results in the case of several independent variables. We have already seen how to constuct a contact-invariant coframe and the consequent invariant differential operators $\mathcal{D}_1, \ldots, \mathcal{D}_p$. To proceed further, we must find an independence result for the differentiated invariants similar to that in Lemma 10. The multi-variable case, though, is more complicated since the strict independence of nth-order differential invariants $J_1, \ldots, J_r$ does not necessarily imply the independence of the differentiated invariants $\mathcal{D}_i J_\nu$. A new approach to this problem relies on a combinatorial theorem proved by Macaulay [13], in his study of the Hilbert function of an algebraic polynomial ideal.*

LEMMA 17. *Let $p \geqslant 1$ be an integer. Then any nonnegative integer $r \in \mathbb{N}$ can be uniquely written in the form*

$$r = \binom{k_1 + p - 1}{p} + \binom{k_2 + p - 2}{p - 1} + \cdots + \binom{k_s + p - s}{p - s + 1}, \tag{13}$$

where $k_1 \geqslant k_2 \geqslant \cdots \geqslant k_s \geqslant 1$ form a nonincreasing sequence of positive integers with $1 \leqslant s \leqslant p$.

Write $k_p(r) = (k_1, \ldots, k_s)$ for the integer sequence associated with $r \in \mathbb{N}$. Define the function μ_p: $\mathbb{N} \to \mathbb{N}$ which takes an integer r, represented by the sequence $k_p(r) = (k_1, \ldots, k_s)$, to the integer $\mu_p(r)$ which satisfies $k_p(\mu_p(r)) = (k_1 + 1, k_2 + 1, \ldots, k_s + 1)$. Macaulay's Theorem provides lower bounds for the dimensions of the homogeneous** components of polynomial ideals using the functions μ_p.

THEOREM 18. *Let $\mathcal{I} \subset \mathbb{R}[x_1, \ldots, x_p]$ be a homogeneous polynomial ideal in p variables. Let $\mathrm{d}_n = \dim \mathcal{I}^{(n)}$ be the dimension of the set*

$$\mathcal{I}^{(n)} = \{P \in \mathcal{I} \mid P(\lambda x) = \lambda^n P(x)\}$$

of polynomials of degree n in $\mathcal{I}$. (Note that, by homogeneity, $\mathcal{I} = \bigoplus_{n \geqslant 0} \mathcal{I}^{(n)}$.) Then $\mathrm{d}_{n+1} \geqslant \mu_p(\mathrm{d}_n)$.

Given $n \geqslant 1$, we define μ_p^n: $\mathbb{N} \to \mathbb{N}$ as follows: for $r \in \mathbb{N}$, we write $r = s p_n + t$, where s is the quotient, and t the remainder, when r is divided by $p_n = \binom{p+n-1}{n}$. Then $\mu_p^n(r) = s p_{n+1} + \mu_p(t)$. Note, in particular, $\mu_p^n(q_n) = q_{n+1}$. We can now state the basic (new) inequality for the number of differential invariants.

THEOREM 19. *Let G be a transformation group acting on a space with p independent variables and q dependent variables. Suppose $\mathcal{D}_1, \ldots, \mathcal{D}_p$ form a*

---

* In more recent years, Macaulay's theorem has been considerably generalized in the combinatorial theory of extremal multi-sets, and forms a special case of the Kruskal–Katona Theorem, cf. [7].

** Macaulay also extends this result to nonhomogeneous ideals.

complete set of invariant differential operators coming from a contact-invariant coframe of order n or less. Suppose $J_1, \ldots, J_r$ are strictly independent nth-order differential invariants. Then the set of differentiated invariants $\mathcal{D}_i J_\nu$, $i = 1, \ldots, p$, $\nu = 1, \ldots, r$, contains at least $\mu_p^n(r)$ strictly independent $(n+1)$st-order differential invariants. In particular, if there are a maximal number of strictly independent nth-order differential invariants, $J_1, \ldots, J_{q_n}$, then the set of differentiated invariants $\mathcal{D}_i J_\nu$, $i = 1, \ldots, p$, $\nu = 1, \ldots, q_n$, contains a complete set of q_{n+1} strictly independent $(n+1)$st-order differential invariants.

THEOREM 20. *Suppose that G is a group of point or contact transformations. Let n denote the order of stabilization of the group action. Then there exists a contact-invariant coframe $\omega^1, \ldots, \omega^p$ on J^{n+2}, with corresponding invariant differential operators $\mathcal{D}_1, \ldots, \mathcal{D}_p$, and differential invariants $J_1, \ldots, J_m$, of order at most $n+2$, such that, locally, every differential invariant can be written as a function of these differential invariants and their derivatives $\mathcal{D}_{j_1} \cdots \mathcal{D}_{j_\kappa} J_\nu$, $\kappa \geqslant 0$, $\nu = 1, \ldots, m$.*

Note that it is not asserted (in contrast to the single variable case in Theorem 11) that the differentiated invariants are necessarily functionally independent. Indeed, the classification of syzygies, or functional dependencies, among the differentiated invariants is an interesting problem that, as far as I know, has not been investigated in any degree of generality. Theorem 20 states that if the orbit dimension stabilizes at order n, then all the differential invariants can be obtained from those of order at most $n+2$ by applying the invariant differential operators. Moreover, if the stable orbit dimension satisfies $r = s_n \leqslant q^{(n)}$, and so there are at least p independent differential invariants $I_1, \ldots, I_p$ of order n, then the differential invariant coframe $DI_1, \ldots, DI_p$ lives on J^{n+1} and, moreover, the fundamental differential invariants have orders at most $n+1$. Finally, Theorem 20 also implies that our earlier stabilization and pseudo-stabilization Theorems 12 and 13 remain valid in the general multi-dimensional context. I do not know of any interesting examples of multi-dimensional transformation groups whose orbit dimensions pseudo-stabilize. Moreover, I do not know precisely how many fundamental differential invariants are required, although the dimension bounds of Theorem 19 should provide some useful estimates. Indeed, as remarked in [6], the generalization of Theorem 14 to the multi-dimensional case would be of great importance for studying, for example, the differential invariants of surfaces and higher dimensional submanifolds of homogeneous spaces.

Finally, for completeness, we recall how differential invariants are used to characterize systems of differential equations and variational problems which admit the transformation group as a symmetry group. The following results go back to Lie [9, 11]; see also [15, 16].

THEOREM 21. *Let G be a transformation group, and let $I_1, \ldots, I_k$, $k = i_n$, be a complete set of functionally independent nth-order differential invariants.*

A system of differential equations admits G as a symmetry group if and only if (restricted to the subset $V^{(n)}$) it can be rewritten in terms of the differential invariants:

$$\Delta_\nu(x, u^{(n)}) = F_\nu(I_1(x, u^{(n)}), \ldots, I_k(x, u^{(n)})) = 0, \quad \nu = 1, \ldots, l. \tag{14}$$

THEOREM 22. *Let G be a transformation group, and $\omega^1, \ldots, \omega^p$ a contact-invariant coframe. A variational problem admits G as a variational symmetry group if and only if it has the form*

$$\begin{aligned} \mathcal{L}[u] &= \int L(x, u^{(n)})\, \mathrm{d}x \\ &= \int F(I_1(x, u^{(n)}), \ldots, I_k(x, u^{(n)}))\, \omega^1 \wedge \cdots \wedge \omega^p, \end{aligned} \tag{15}$$

where $I_1, \ldots, I_k$ are functionally independent differential invariants of G.

Acknowledgement

It is a pleasure to thank Paul Edelman for pointing out the references [7, 13].

References

1. Ackerman, M. and Hermann, R.: *Sophus Lie's 1880 Transformation Group Paper*, Math. Sci. Press, Brookline, Mass., 1975.
2. Ackerman, M. and Hermann, R.: *Sophus Lie's 1884 Differential Invariant Paper*, Math. Sci. Press, Brookline, Mass., 1976.
3. Bäcklund, A.V.: Ueber Flachentransformationen, *Math. Ann.* **9** (1876), 297–320.
4. Cartan, É.: Les problèmes d'équivalence, in *Oeuvres Complètes*, Part II, Vol. 2, Gauthiers-Villars, Paris, 1952, pp. 1311–1334.
5. Gardner, R. B.: *The Method of Equivalence and Its Applications*, SIAM, Philadelphia, 1989.
6. Green, M. L.: The moving frame, differential invariants and rigidity theorems for curves in homogeneous spaces, *Duke Math. J.* **45** (1978), 735–779.
7. Greene, C. and Kleitman, D. J.: Proof techniques in the theory of finite sets, in G.-C. Rota (ed.), *Studies in Combinatorics*, Studies in Math. 17, Math. Assoc. Amer., Washington, D.C., 1978, pp. 22–79.
8. Lie, S.: Theorie der Transformationsgruppen I, *Math. Ann.* **16** (1880), 441–528; also *Gesammelte Abhandlungen*, vol. 6, B.G. Teubner, Leipzig, 1927, pp. 1–94; see [2] for an English translation.
9. Lie, S.: Über Differentialinvarianten, *Math. Ann.* **24** (1884), 537–578; also *Gesammelte Abhandlungen*, vol. 6, B.G. Teubner, Leipzig, 1927, pp. 95–138; see [2] for an English translation.
10. Lie, S.: Klassifikation und Integration von gewohnlichen Differentialgleichungen zwischen x, y, die eine Gruppe von Transformationen gestatten I, II, *Math. Ann.* **32** (1888), 213–281; also *Gesammelte Abhandlungen*, vol. 5, B.G. Teubner, Leipzig, 1924, pp. 240–310.
11. Lie, S.: Über Integralinvarianten und ihre Verwertung für die Theorie der Differentialgleichungen, *Leipz. Berichte* **4** (1897), 369–410; also *Gesammelte Abhandlungen*, vol. 6, B. G. Teubner, Leipzig, 1927, pp. 664–701.
12. Lie, S.: Gruppenregister, in B. G. Teubner (ed.), *Gesammelte Abhandlungen*, vol. 5, Leipzig, 1924, pp. 767–773.

13. Macaulay, F. S.: Some properties of enumeration in the theory of modular systems, *Proc. London Math. Soc.* **26**(2) (1927), 531–555.
14. Olver, P. J.: Symmetry groups and group invariant solutions of partial differential equations, *J. Differential Geom.* **14** (1979), 497–542.
15. Olver, P. J.: *Applications of Lie Groups to Differential Equations*, 2nd edn, Graduate Texts in Mathematics 107, Springer-Verlag, New York, 1993.
16. Olver, P. J.: *Equivalence, Invariants, and Symmetry*, Cambridge University Press, 1995.
17. Ovsiannikov, L. V.: *Group Analysis of Differential Equations*, Academic Press, New York, 1982.
18. Tresse, A.: Sur les invariant différentiels des groupes continus de transformations, *Acta Math.* **18** (1894), 1–88.

Acta Applicandae Mathematicae **41:** 285–296, 1995.

Spencer Sequence and Variational Sequence

J. F. POMMARET
Centre de Mathématiques Appliquées (CERMA), Ecole Nationale des Ponts et Chaussées (ENPC), La Courtine, 93 167 Noisy-le Grand Cedex, France

(Received: 28 February 1994)

Abstract. The purpose of this paper is to revisit the construction of the variational sequence existing within the formal calculus of variations, in order to stabilize the order of jets involved and to establish a link with the dual of the Spencer sequence existing within the formal theory of systems of partial differential equations.

Mathematics Subject Classifications (1991): 35Qxx, 53Bxx.

Key words: variational calculus, variational sequence, partial differential equations, jet theory, dual operator, Spencer sequence.

The importance of variational calculus is clear enough for engineering sciences, through the finite element approach to elasticity, heat, and electromagnetism, but also for mathematical physics, through the well-known foundations of gauge theory and general relativity [1, 2, 4, 6, 20]. During the last twenty years, advances have been done towards higher-order generalizations by exhibiting the so-called *variational sequence* in the bicomplex formalism [3, 16–18]. In spite of the powerfulness of the method, one last challenge was left, namely to avoid jet spaces of infinite order. A major step has been achieved very recently by D. Krupka for stabilizing the order of jets to a finite value in order to exhibit a '*variational sequence on finite order jet spaces*' [8, 9].

Meanwhile, but independently, we have been studying and applying the formal theory of systems of partial differential equations (PDE) and Lie pseudogroups initiated by D. C. Spencer in the U.S.A., during the period 1960–1975 [10, 14]. In particular, we have been able, for the first time, to unify the finite element approach to elasticity, heat, and electromagnetism, by means of these new group-theoretical tools. The reader will find in [11, 12] a modern exposition of the pioneering ideas of E. and F. Cosserat on elasticity theory [5] and of H. Weyl on electromagnetism [19] along the work of Spencer. The key concept is the *duality* existing between GEOMETRY (differential sequence) and PHYSICS (variational sequence) in any physical theory. The differential sequence will allow us to produce the *field* from the *potential* in such a way that the *field equations* are satisfied. At that time, the so-called *induction* and *induction equations* may be obtained by duality in a 'certain' variational calculus. The main discovery of the brothers Cosserat is that the various operators appearing *on both sides* in such an approach, only depend on group theory through the construction of the Spencer sequence. Roughly speaking, we may say that the passage from the potential to the field is expressed by the first Spencer operator, the field equations are expressed by the second Spencer operator, while the induction equations are expressed by the dual of the first Spencer operator. In this paper, for simplicity, we shall restrict to the '*linear*' or '*infinitesimal*' framework but the nonlinear situation could be treated as well, with more work, by means of the so-called '*vertical machinery*' in jet theory [11, 12].

The aim of this paper is to exhibit the close link existing between the Krupka variational sequence and the second Spencer sequence by showing that, with a slight abuse of language, they are dual to each other. At the same time, we shall discover that the induction equations in physics, even if they come indeed from a variational calculus, cannot be considered as Euler–Lagrange equations, contrary to a well-established belief. The brand-new results of this research paper will be presented in more details and extended to gravitation in a forthcoming book [13].

For this purpose, we start recalling the construction of the first and second Spencer sequences. Then we recall the bicomplex technique and exhibit the variational sequence in its standard form and in the form proposed by D. Krupka. Finally, we prove that the latter variational sequence is the dual of the second Spencer sequence in a certain sense, by using combinatoric arguments taken from the use of the Spencer δ-cohomology. The application of these results to Killing equations and elasticity theory will point out the lack of coherence with such a scheme, held by any moderm interpretation of the mathematical foundation of field theory.

Let X be a manifold of dimension n and local coordinates (x^i), with tangent and cotangent bundles, respectively, denoted by $T = T(X)$ and $T^* = T^*(X)$. We denote by $\otimes$, S and $\wedge$, respectively, the tensor, symmetric, and exterior products. If $\mu = (\mu_1, \dots, \mu_n)$ is a multi-index with length $|\mu| = \mu_1 + \cdots + \mu_n$, we introduce $\mu + 1_i = (\mu_1, \dots, \mu_{i-1}, \mu_i + 1, \mu_{i+1}, \dots, \mu_n)$ and we decompose any r-form on the basis $dx^I = dx^{i_1} \wedge \cdots \wedge dx^{i_r}$ with $I = (i_1 < \cdots < i_r)$. If E is a vector bundle over X with local coordinates (x^i, y^k) and (fiber) dimension m, we may introduce the map:

$$\delta\colon \bigwedge^r T^* \otimes S_{q+1}T^* \otimes E \longrightarrow \bigwedge^{r+1} T^* \otimes S_q T^* \otimes E$$

by considering the forms $\omega_\mu^k = v_{\mu,I}^k \, dx^I$ and using the formula: $(\delta\omega)_\mu^k = dx^i \wedge \omega_{\mu+1_i}^k$. We have $\delta \circ \delta = 0$ and the resulting δ-sequence is exact. We may now introduce the q-jet bundle $J_q(E)$ and define the Spencer operator:

$$D\colon J_{q+1}(E) \longrightarrow T^* \otimes J_q(E)\colon \xi_{q+1} \longrightarrow j_1(\xi_q) - \xi_{q+1}$$

by the local formulas

$$(D\xi_{q+1})^k_{\mu,I} = \partial_i \xi^k_\mu - \xi^k_{\mu+1_i} \quad \text{and} \quad j_q(\xi)\colon (x) \longrightarrow (x, \partial_\mu \xi^k(x))$$

on sections. We may extend D to

$$D\colon \bigwedge^r T^* \otimes J_{q+1}(E) \longrightarrow \bigwedge^{r+1} T^* \otimes J_q(E)$$

by setting

$$D(\omega \otimes \xi_{q+1}) = d\omega \otimes \xi_q + (-1)^r \omega \wedge D\xi_{q+1}$$

in such a way that $D \circ D = 0$. The resulting finite length differential sequence:

$$\begin{aligned} 0 \longrightarrow E \xrightarrow{j_{q+1}} J_{q+1}(E) \xrightarrow{D} T^* \otimes J_q(E) \xrightarrow{D} \cdots \\ \xrightarrow{D} \bigwedge^n T^* \otimes J_{q+1-n}(E) \longrightarrow 0 \end{aligned}$$

is called the *first Spencer sequence* and has the following two bad properties:

- The order of jets is not stabilized.
- Each operator D does not represent all the compatibility conditions for the preceding one.

One can 'factor out' these difficulties by introducing the *Spencer bundles*:

$$C_r(E) = C_{q,r}(E) = \bigwedge^r T^* \otimes J_q(E)/\delta(\bigwedge^{r-1} T^* \otimes S_{q+1}T^* \otimes E)$$

and inducing therefore the *second Spencer sequence*:

$$0 \longrightarrow E \xrightarrow{j_q} C_0(E) \xrightarrow{D_1} C_1(E) \xrightarrow{D_2} \cdots \xrightarrow{D_n} C_n(E) \longrightarrow 0$$

For a later purpose, we recall the short exact sequences:

$$0 \longrightarrow \delta\Big(\bigwedge^r T * \otimes S_q T * \otimes E\Big) \longrightarrow C_{q,r}(E) \longrightarrow \bigwedge^r T * \otimes J_{q-1}(E) \longrightarrow 0$$

that allow to split the Spencer bundles in actual practice.

If now $R_q \subset J_q(E)$ is a linear involutive system of PDE on E, we may define the restricted Spencer bundles:

$$C_r = \bigwedge^r T^* \otimes R_q/\delta\Big(\bigwedge^{r-1} T^* \otimes M_{q+1}\Big),$$

where

$$M_{q+1} = R_{q+1} \cap S_{q+1}T^* \otimes E \subset J_{q+1}(E),$$

whenever

$$R_{q+1} = J_1(R_q) \cap J_{q+1}(E) \subset J_1(J_q(E))$$

is the first prolongation of R_q (see [11] for more details). By restriction, we obtain the second Spencer sequence:

$$0 \longrightarrow \Theta \xrightarrow{j_q} C_0 \xrightarrow{D_1} C_1 \xrightarrow{D_2} \cdots \xrightarrow{D_n} C_n \longrightarrow 0,$$

where Θ is the solution sheaf of R_q. In actual practice, the determination of the dimensions of the $C_r \subset C_r(E)$ cannot be done without a computer. Of course, the first and the second Spencer sequences both admit same projective limit:

$$0 \longrightarrow E \xrightarrow{j_\infty} J_\infty(E) \xrightarrow{D} T^* \otimes J_\infty(E) \xrightarrow{D} \cdots$$
$$\xrightarrow{D} \bigwedge^n T^* \otimes J_\infty(E) \longrightarrow 0$$

and the corresponding restriction.

Let us turn to variational calculus. In analytical dynamics, using standard notations without any index for simplicity, with any Lagrangian function $L(t, q, \dot{q})$, we may associate the variational problem:

$$\delta \int L(t, q, \dot{q})\, \mathrm{d}t = 0$$

and obtain the well-known *Euler–Lagrange equations* (EL):

$$\frac{\mathrm{d}}{\mathrm{d}t}\left(\frac{\partial L}{\partial \dot{q}}\right) - \frac{\partial L}{\partial q} = 0.$$

On the one side, we notice that

$$L = \frac{\mathrm{d}\Phi}{\mathrm{d}t} = \frac{\partial \Phi}{\partial t} + \dot{q}\frac{\partial \Phi}{\partial q}$$

is a solution of (EL) for any function $\Phi(t, q)$, while the condition $\mathrm{d}\Phi/\mathrm{d}t = 0$ is called a *conservation law* (CL).

On the other side, introducing a second member $Q(t, q, \dot{q}, \ddot{q})$ in the EL equations, the formal existence of L depends on the *Helmholtz conditions* as a system of PDE that must be satisfied by Q. For example, in the single q situation just considered, we have:

$$\frac{\partial^2 L}{\partial t \partial \dot{q}} + \dot{q}\frac{\partial^2 L}{\partial q \partial \dot{q}} + \ddot{q}\frac{\partial^2 L}{\partial \dot{q} \partial \dot{q}} - \frac{\partial L}{\partial q} = Q(t, q, \dot{q}, \ddot{q})$$

and the Helmholtz conditions reduce to:

$$\frac{\partial Q}{\partial \dot{q}} - \frac{\mathrm{d}}{\mathrm{d}t}\left(\frac{\partial Q}{\partial \ddot{q}}\right) = 0$$

Collecting these results, we have a kind of differential sequence:

$$\Phi \xrightarrow{\mathrm{CL}} L \xrightarrow{\mathrm{EL}} Q \xrightarrow{\mathrm{HC}} H$$

called *variational sequence*.

We now recall, in a way convenient for our purpose, the construction of the variational bicomplex given by many authors [3, 16, 17].

If we consider a fibered manifold $\mathcal{E}$ over X instead of the vector bundle E, let us introduce the *fundamental contact forms*:

$$\theta_\mu^k = \mathrm{d}y_\mu^k - y_{\mu+1_i}^k \mathrm{d}x^i$$

and study exterior calculus on $J_\infty(\mathcal{E})$ as *it is the only jet space that can be stabilized* by the formal derivative d_i such that $\mathrm{d}_i y_\mu^k = y_{\mu+1_i}^k$. We may now decompose the exterior derivative d by setting $\mathrm{d} = \mathrm{d}_H + \mathrm{d}_V$, where H stands for '*horizontal*' while V stands for '*vertical*'. We may define d_H and d_V by the following rules:

$$\mathrm{d}_H\Phi = \mathrm{d}_i\Phi\,\mathrm{d}x^i, \qquad \mathrm{d}_H\,\mathrm{d}x^i = 0, \qquad \mathrm{d}_H\theta_\mu^k = \mathrm{d}x^i \wedge \theta_{\mu+1_i}^k,$$

$$\mathrm{d}_V\Phi = \frac{\partial\Phi}{\partial y_\mu^k}\theta_\mu^k, \qquad \mathrm{d}_V\,\mathrm{d}x^i = 0, \qquad \mathrm{d}_V\theta_\mu^k = 0$$

and we obtain $\mathrm{d}_H \circ \mathrm{d}_H = 0$, $\mathrm{d}_V \circ \mathrm{d}_V = 0$. If we denote the set of forms containing $\mathrm{d}x$ r-times and θ s-times by $\Omega^{r,s}$, then d_H increases r by 1 while d_V increases s by 1 in the resulting bicomplex and we obtain the so-called *variational sequence*:

$$0 \longrightarrow \mathbb{R} \longrightarrow \Omega^{0,0} \xrightarrow{\mathrm{d}_H} \cdots$$
$$\xrightarrow{\mathrm{d}_H} \Omega^{n-1,0} \xrightarrow{\mathrm{d}_H} \Omega^{n,0} \xrightarrow{\mathrm{d}_V} \Omega^{n,1}/\mathrm{d}_H\Omega^{n-1,1} \xrightarrow{\mathrm{d}_V} \Omega^{n,2}/\mathrm{d}_H\Omega^{n-1,2}$$

where the last three operators on the right are, respectively, CL, EL, and HC.

In the case $m = n = 1$ already considered, we may set $\theta = \mathrm{d}q - \dot{q}\mathrm{d}t$, $\dot{\theta} = \mathrm{d}\dot{q} - \ddot{q}\mathrm{d}t$ and the variational sequence simply becomes:

$$0 \longrightarrow \mathbb{R} \longrightarrow \Omega^{0,0} \xrightarrow{\mathrm{d}_H} \Omega^{1,0} \xrightarrow{\mathrm{d}_V} \Omega^{1,1}/\mathrm{d}_H\Omega^{0,1} \xrightarrow{\mathrm{d}_V} \Omega^{1,2}/\mathrm{d}_H\Omega^{0,2}$$

where the three operators are just CL, EL, and HC.

We check at once that

$$\Phi \in \Omega^{0,0} \Longrightarrow \mathrm{d}_H\Phi = \frac{\mathrm{d}\Phi}{\mathrm{d}t}\mathrm{d}t \in \Omega^{1,0}$$

and that, whenever $L(t,q,\dot{q})\,\mathrm{d}t \in \Omega^{1,0}$, we get:

$$\mathrm{d}_V(L\,\mathrm{d}t) = \frac{\partial L}{\partial q}\theta \wedge \mathrm{d}t + \frac{\partial L}{\partial \dot{q}}\dot{\theta} \wedge \mathrm{d}t \in \Omega^{1,1}.$$

Now any element in $\Omega^{0,1}$ can be written $A(t,q)\theta + B(t,q)\dot{\theta}$ and we obtain, by taking d_H:

$$\frac{\mathrm{d}A}{\mathrm{d}t}\,\mathrm{d}t \wedge \theta + \left(A + \frac{\mathrm{d}B}{\mathrm{d}t}\right)\mathrm{d}t \wedge \dot{\theta} + B\,\mathrm{d}t \wedge \ddot{\theta} \in \Omega^{1,1}$$

Identifying the terms in order to exhibit the residue modulo the contact forms, we may choose $A = -(\partial L/\partial \dot{q})$, $B = 0$ and a representative element of $\Omega^{1,1}/\mathrm{d}_H\Omega^{0,1}$ may be therefore written as:

$$\left(\frac{\mathrm{d}}{\mathrm{d}t}\left(\frac{\partial L}{\partial \dot{q}}\right) - \frac{\partial L}{\partial q}\right)\mathrm{d}t \wedge \theta$$

a result providing the Euler–Lagrange operator.

Finally, if $Q(t, q, \dot{q}, \ddot{q})\,\mathrm{d}t \wedge \theta \in \Omega^{1,1}$, the image by d_V is:

$$\frac{\partial Q}{\partial \dot{q}}\dot{\theta} \wedge \mathrm{d}t \wedge \theta + \frac{\partial Q}{\partial \ddot{q}}\ddot{\theta}\,\mathrm{d}t \wedge \theta \in \Omega^{1,2},$$

while the element $A(t, q, \dot{q})\theta \wedge \dot{\theta} \in \Omega^{0,2}$ admits under d_H the image:

$$\frac{\mathrm{d}A}{\mathrm{d}t}\,\mathrm{d}t \wedge \theta \wedge \dot{\theta} - A\theta \wedge \mathrm{d}t \wedge \ddot{\theta} \in \Omega^{1,2}$$

Identifying the terms similarly as before, we may choose $A = \partial Q/\partial \ddot{q}$ and obtain by residue the Helmholtz condition already given.

Despite the apparent usefulness of the latter construction, two problems can arise:

- One must work on $J_{\infty}(\mathcal{E})$.
- The variational sequence does not seem to have, *at first sight*, any relation with the Spencer sequences and their common projective limit.

The solution to the first problem has been given recently by D. Krupka in [8, 9] after tedious computations that we shall greatly simplify by exhibiting their link with the images and kernels of the δ-maps. Meanwhile, we notice that the higher order EL equations, namely [2, 7]:

$$\sum_{|\mu|=0}^{q} (-1)^{|\mu|}\mathrm{d}_\mu\left(\frac{\partial L}{\partial y^k_\mu}\right) = 0$$

comes from the *linear* operator of order q:

$$\lambda_q \longrightarrow \sum_{|\mu|=0}^{q} (-1)^{|\mu|}\partial_\mu \lambda^\mu_k$$

whenever we use $\lambda^\mu_k(x)$ instead of $\partial L/\partial y^k_\mu$ as a generalized momentum. As a byproduct, we just recognize the dual of j_q in the usual *adjoint* sense and we may understand that the variational sequence *must* have something to do with the dual of the second (and not first!) Spencer sequence. As the dimensions of the Spencer bundles and their dual are equal but cannot be computed by hand, this result explains why such an identification has been missed up to now or hidden behind the infinite-order jet framework. In fact, the history of PDE clearly proves that any progress in their formal study has always been done by avoiding, in a systematic and intrinsic way, the use of inifinite order jets.

Following Krupka, let us introduce the space Ω^r of r-forms on $J_q(\mathcal{E})$ and the subspace Θ^r generated as an ideal by the θ and $\mathrm{d}\theta$ whenever they can be defined. In fact, we notice that the Θ^r for $2 \leqslant r \leqslant n$ are generated by the θ only. Hence, in order to project Ω^r modulo Θ^r onto $A^r = \Omega^r/\Theta^r$ for $2 \leqslant r \leqslant n$, we just need to replace each $\mathrm{d}y^k_\mu$ by $y^k_{\mu+1_i}\mathrm{d}x^i$ when $0 \leqslant |\mu| \leqslant q-1$ but no simple interpretation can be given for the forms still involving $\mathrm{d}y^k_\mu$ with $|\mu| = q$ after reduction. The situation is much more complicate when $r \geqslant n+1$ and one must consider only the so-called '*strongly contact forms*' involving (at least) $r-n+1$ forms θ or $\mathrm{d}\theta$ after reduction. In any case, we shall explain later on why we do not have much trust in the usefulness, at least for applications to physics, of the variational sequence when $r \geqslant n+1$.

Using now the fact that the exterior derivative $\mathrm{d}\colon \Omega^r \to \Omega^{r+1}$ restricts to $\mathrm{d}\colon \Theta^r \to \Theta^{r+1}$, the main idea is to introduce the following induced stabilized variational sequence:

$$0 \longrightarrow \mathbb{R} \longrightarrow A^0 \longrightarrow A^1 \longrightarrow \cdots \longrightarrow A^{n-1} \longrightarrow A^n \longrightarrow A^{n+1}$$

which is locally exact according to the Poincaré lemma. Of course, the injective (CARE) limit of A^r for $q \to \infty$ is $\Omega^{r,0}$ when $0 \leqslant r \leqslant n$ if we set $A^0 = \Omega^0$ and we recover the previous results.

We first prove that the last operator in the above sequence just induces the Euler–Lagrange operator when $\mathcal{E}$ is a vector bundle over X, by considering only linear horizontal forms on $J_q(E)$. The general case could be treated as well by using the vertical bundle $V(\mathcal{E})$ of $\mathcal{E}$ instead of E, while taking into account the canonical isomorphism $J_q(V(\mathcal{E})) \simeq V(J_q(\mathcal{E}))$.

The *key trick* of the paper will be to introduce the vector bundles:

$$A^{n-r}(E) = A^{n-r}_q(E) = \bigwedge^n T^* \otimes C_{q,r}(E)^*$$

and to describe $A^n = A^n_q(E) = \wedge^n T^* \otimes J_q(E)^*$ by the n-form:

$$(a_k(x)y^k + a^i_k(x)y^k_i + a^{ij}_k(x)y^k_{ij} + \cdots)\mathrm{d}x^1 \wedge \cdots \wedge \mathrm{d}x^n$$

Taking the exterior derivative, we obtain the $(n+1)$-form:

$$a_k(x)\mathrm{d}y^k \wedge \mathrm{d}x^1 \wedge \cdots \wedge \mathrm{d}x^n + a^i_k(x)\mathrm{d}y^k_i \wedge \mathrm{d}x^1 \wedge \cdots \wedge \mathrm{d}x^n + \cdots$$

which cannot be reduced modulo contact forms ... as it is already the contact form:

$$a_k(x)\theta^k \wedge \mathrm{d}x^1 \wedge \cdots \wedge \mathrm{d}x^n + a^i_k(x)\theta^k_i \wedge \mathrm{d}x^1 \wedge \cdots \wedge \mathrm{d}x^n + \cdots$$

Now we have the relations:

$$\begin{aligned}&\mathrm{d}((-1)^i a^i_k \theta^k \wedge \mathrm{d}x^1 \wedge \cdots \wedge \widehat{\mathrm{d}x^i} \wedge \cdots \wedge \mathrm{d}x^n)\\ &\quad = \partial_i a^i_k \mathrm{d}y^k \wedge \mathrm{d}x^1 \wedge \cdots \wedge \mathrm{d}x^n + a^i_k \mathrm{d}y^k_i \wedge \mathrm{d}x^1 \wedge \cdots \wedge \mathrm{d}x^n\end{aligned}$$

and so on. Reducing inductively modulo these forms, we get:

$$(a_k(x) - \partial_i a_k^i(x) + \partial_{ij} a_k^{ij}(x) - \cdots)\mathrm{d}y^k \wedge \mathrm{d}x^1 \wedge \cdots \wedge \mathrm{d}x^n$$

that is to say:

$$\left(\sum_{|\mu|=0}^{q} (-1)^{|\mu|} \partial_\mu a_k^\mu(x) \right) \mathrm{d}y^k \wedge \mathrm{d}x^1 \wedge \cdots \wedge \mathrm{d}x^n$$

and we recognize the Euler–Lagrange operator of order q that dualizes j_q.

More generally, for $0 \leqslant r \leqslant n$ we have:

$$\bigwedge^r T^* \otimes J_{q-1}(E)^* \subset A_q^r(E) \subset \bigwedge^r T^* \otimes J_q(E)^*$$

and a representative may be:

$$\sum_{|\mu|=0}^{q} a_{k,I}^\mu y_\mu^k \mathrm{d}x^I.$$

The only difficulty left is to characterize the linear relations that must be satisfied by the $a_{k,I}^\mu$ with $|\mu| = q$ in order to describe the above inclusion. For this, we have the exact sequences:

$$\bigwedge^{r-1} T^* \otimes S_{q+1}T^* \otimes E \xrightarrow{\delta} \bigwedge^r T^* \otimes S_q T^* \otimes E \xrightarrow{\delta} \bigwedge^{r+1} T^* \otimes S_{q-1}T^* \otimes E$$

and the corresponding dual sequences (with r in place of $n - r$):.

$$\bigwedge^{r+1} T^* \otimes S_{q+1}T \otimes E^* \xleftarrow{\tilde{\delta}} \bigwedge^r T^* \otimes S_q T \otimes E^* \xleftarrow{\tilde{\delta}} \bigwedge^{r-1} T^* \otimes S_{q-1}T \otimes E^*$$

are exact because *duality preserves exactness*. Accordingly, we deduce that the conditions:

$$\sum_{\mu+1_i=\nu} a_{k,I}^\mu(x)\mathrm{d}x^i \wedge \mathrm{d}x^I = 0, \quad |\nu| = q+1$$

defining the kernel of the left $\tilde{\delta}$ are satisfied if and only if the q-order term in $A_q^r(E)$ is a linear combination of r-forms of the type $y_{\lambda+1_i}^k \mathrm{d}x^i \wedge \mathrm{d}x^J$ with $|\lambda| = q-1$ and $\mathrm{d}x^J \in \bigwedge^{r-1} T^*$. Indeed, such forms go to zero under the exterior derivative followed by reduction modulo contact forms, because $y_{\lambda+1_i+1_j}^k \mathrm{d}x^i \wedge \mathrm{d}x^j = 0$. Therefore, the induced dual (in the adjoint sense) operator $\widetilde{D}_{n-r}\colon A_q^r(E) \to A_q^{r+1}(E)$ is well defined.

Now, taking the exterior derivative of the general term, we get:

$$\partial_i a^{\mu}_{k,I}(x) y^k_{\mu} \mathrm{d}x^i \wedge \mathrm{d}x^I + a^{\mu}_{k,I}(x) \mathrm{d}y^k_{\mu} \wedge \mathrm{d}x^I$$

and, reducing modulo the contact forms, we finally get:

$$(\partial_i a^{\mu}_{k,I}(x) y^k_{\mu} + a^{\mu}_{k,I}(x) y^k_{\mu+1_i}) \mathrm{d}x^i \wedge \mathrm{d}x^I.$$

However, the dual of the Spencer operator is defined by the identity:

$$\begin{aligned} & a^{\mu}_k(x)(\partial_i \xi^k_{\mu}(x) - \xi^k_{\mu+1_i}(x)) \\ & \quad = \partial_i (a^{\mu}_k(x) \xi^k_{\mu}(x)) - (\partial_i a^{\mu}_k(x) \xi^k_{\mu}(x) + a^{\mu}_k(x) \xi^k_{\mu+1_i}(x)). \end{aligned}$$

It follows that the locally exact sequence:

$$\begin{aligned} & 0 \longrightarrow \mathbb{R} \longrightarrow A^0(E) \longrightarrow A^1(E) \longrightarrow \cdots \\ & \quad \longrightarrow A^n(E) \longrightarrow \textstyle\bigwedge^n T^* \otimes E^* \longrightarrow 0 \end{aligned}$$

is just the dual of the second Spencer sequence.

In order to help the reader, we describe with more details the first operator $A^0(E) \to A^1(E)$. As $C_{q,n}(E) = \bigwedge^n T^* \otimes J_{q-1}(E)$, we have the relations:

$$J_{q-1}(E)^* = A^0_q(E) \subset J_q(E)^*,$$

where the inclusion dualizes the canonical projection $J_q(E) \to J_{q-1}(E)$. Hence, we may consider a representative of the form: $\sum_{|\mu|=0}^{q-1} a^{\mu}_k(x) y^k_{\mu}$. Taking the exterior derivative, we get:

$$\sum_{|\mu|=0}^{q-1} (\partial_i a^{\mu}_k(x) y^k_{\mu} \mathrm{d}x^i + a^{\mu}_k(x) \mathrm{d}y^k_{\mu}).$$

Reducing modulo the contact forms, we get:

$$\sum_{|\mu|=0}^{q-1} (\partial_i a^{\mu}_k(x) y^k_{\mu} + a^{\mu}_k(x) y^k_{\mu+1_i}) \mathrm{d}x^i$$

and thc image in $A^1_q(E) \subset T^* \otimes J_q(E)^*$ is well defined, according to the above comment, because the only term of order q is of the form:

$$\sum_{|\mu|=q} a^{\mu}_k(x) y^k_{\mu+1_i} \mathrm{d}x^i.$$

In view of the powerfulness of the preceding arguments, the comparison with [8, 9] needs no comment, even if we discover that the first operator of the variational sequence dualizes the last operator of the Spencer sequence which is *absolutely never* used for applications as we shall see.

Considering now the restricted second Spencer sequence associated with a linear system $R_q \subset J_q(E)$, we notice that the dual of $\Theta \subset E$ cannot be defined any longer in term of vector bundle but only in term of sheaf. Hence, *we may have no hope for finding induction equations as Euler–Lagrange equations.*

We have proved in [11] that induction equations are indeed dualizing D_1 but it is not possible in such a short paper to recall the corresponding variational calculus on Lie groupoids [11, 12]. However, for convincing the reader about the novelty and striking usefulness of these methods, we shall just compute the dual of D_1 in the case $E = T$, $q = 1$ and $R_1 \subset J_1(T)$ is defined by the Killing equations for sections:

$$\omega_{rj}(x)\xi_i^r + \omega_{ir}(x)\xi_j^r + \xi^r \partial_r \omega_{ij}(x) = 0,$$

where ω is the standard Euclidean metric. We may construct at once the dual operator (in the adjoint sense) through integration by part along the variational condition:

$$\delta \int [\sigma_k^i(x)(\partial_i \xi^k - \xi_i^k) + \mu_k^{j,i}(x)(\partial_i \xi_j^k - \xi_{ij}^k)] \mathrm{d}x^1 \wedge \cdots \wedge \mathrm{d}x^n = 0.$$

As the Christoffel symbols defined from $j_1(\omega)$ are all zero, the second-order jets are all zero and we obtain, from an integration by part, the conditions:

$$\int (\partial_i \sigma_k^i \delta\xi^k + \sigma_k^i \delta\xi_i^k + \partial_i \mu_k^{j,i} \delta\xi_j^k) \mathrm{d}x^1 \wedge \cdots \wedge \mathrm{d}x^n = 0.$$

However, in the first summation of $\delta\xi_i^k$, the indices are arbitrary, while, in the second, we have $j < k$ because we consider only different first-order jets. Raising the indices by means of the metric, we finally recover the well-known induction equations for Cosserat media [5, 15]:

$$\partial_r \sigma^{ri} = 0, \qquad \partial_r \mu^{ij,r} + \sigma^{ij} - \sigma^{ji} = 0.$$

These equations *must* not be confused with the Euler–Lagrange equations in the variational sequence because the left members are sections of the vector bundle $\bigwedge^n T^* \otimes R_1^*$. Hence, *potentials* (moving frame) are sections of $C_0 = R_1$ while *fields* (Cosserat strain) are sections of $C_1 = T^* \otimes R_1$ in that case. *Field equations* (compatibility conditions for strain) are expressed by D_2, while *induction equations* (stress and couple-stress equations) are expressed by $\tilde{D}_1$ as above, in a purely group theoretic and geometric way.

More generally, whenever the symbol M_q of R_q is zero, like in the preceding situation, we have simply $C_r = \bigwedge^r T^* \otimes R_q$. In that case, the first and second Spencer sequences are isomorphic to the tensor product of the Poincaré sequence by a Lie algebra $\mathcal{G}$ as follows:

$$0 \longrightarrow \Theta \longrightarrow \bigwedge^0 T^* \otimes \mathcal{G} \longrightarrow \bigwedge^1 T^* \otimes \mathcal{G} \longrightarrow \cdots \longrightarrow \bigwedge^n T^* \otimes \mathcal{G} \longrightarrow 0$$

and this is *exactly* the usual differential geometric framework of gauge theory.

Let us finally prove briefly that electromagnetism comes from the conformal group of spacetime, as Weyl was claiming [19], and *not* from U(1), as gauge theory is claiming [6, 20].

Indeed, conformal Killing equations are obtained by setting the second members of the latter Killing equations no longer zero but equal to $A(x)\omega_{ij}(x)$, where $A(x)$ is an arbitrary function and ω is the Minkowski metric $(+++-)$ of spacetime with $n = 4$. Accordingly, second-order jets do appear and are of the form:

$$\xi^k_{ij} = \delta^k_i A_j + \delta^k_j A_i - \omega_{ij}\omega^{kr} A_r,$$

where now $\xi^r_{ri}(x) = nA_i(x)$ are arbitrary 1-forms, while third-order jets are zero because $n \geqslant 3$. Among the *new components* of the Spencer operator, we find $\partial_i \xi^r_{rj} - \xi^r_{rij} = \partial_i \xi^r_{rj}$ and we may therefore only pay attention to the linear combinations $F_{ij} = \partial_i A_j - \partial_j A_i$. Hence, we just discover that Weyl was looking fo the Spencer operator in 1919, though such an operator has been introduced by Spencer in 1960 for the linear framework [14] and in 1970 for the nonlinear framework [10].

The duality principle therefore furnishes at once *exactly* the second set of Maxwell equations along the same calculations as the ones done by Weyl in [19]. It follows that the 2-form describing the electromagnetic field comes from a section of C_1, a result that *cannot* be obtained from classical gauge theory where it should come from a section of C_2.

Collecting the preceding results, we arrive at the following dilemma produced by the use of the Spencer sequence (GEOMETRY) or of the variational sequence (PHYSICS):

$$\begin{array}{lll} \text{Classical theory} & \Longrightarrow & \text{Lagrangians are functions on } C_0, \\ \text{New theory} & \Longrightarrow & \text{Lagrangians are functions on } C_1, \\ \text{Gauge theory} & \Longrightarrow & \text{Lagrangians are functions on } C_2. \end{array}$$

In view of the agreement with engineering sciences (elasticity, heat, electromagnetism) and their couplings (thermoelasticity, photoelasticity, thermoelectricity) expressed by the preceding applications, we hope to have convinced the reader that there is no doubt about the interpretation that must be chosen! As a byproduct, many well-established theories must be revisited within this new framework.

Acknowledgement

The author thanks D. Krupka for the many fruitful discussions held in Opava, Czechoslovakia during June 1993.

References

1. Abers, E. S. and Lee, B. W.: Gauge theories, *Phys. Rep. C Phys. Lett.* **9**(1) (1973), 1–59.
2. Aldaya, V. and de Azcarraga, J. A.: Variational principle on rth order jets of fibre bundles in field theory, *J. Math. Phys.* **19** (1978), 1869–1975.
3. Anderson, I. and Thompson, G.: The inverse problem of the calculus of variations for ordinary differential equations, *Mem. Amer. Math. Soc.* **473**(98) (1992), 110.
4. Bleecker, D.: Gauge theory and variational principle, in *Global Analysis, Pure and Applied, 1*, Advanced Book Program, World Science Division XVIII, Addison-Wesley, Reading, Mass., 1981.
5. Cosserat, E. and Cosserat, F.: *Théorie des corps déformables*, Hermann, Paris, 1909.
6. Drechsler, W. and Mayer, M. E.: *Fiber Bundle Techniques in Gauge Theories*, Lecture Notes in Physics 67, Springer, Berlin, 1977.
7. Francaviglia, M. and Krupka, D.: The Hamilton formalism in higher order variational problems, *Ann. Inst. Henri Poincaré*, **37** (1982), 295–315.
8. Krupka, D.: Variational sequences on finite order jet spaces, *Proc. Conf. 'Differential Geometry and its Applications'*, Aug. 27–Sept. 2, 1989, Brno, Czchoslovakia, World Scientific, Singapore, 1990, pp. 236–254.
9. Krupka, D.: Topics in the calculus of variations: variational sequences, *Proc. Conf. 'Differential Geometry and its Applications'*, Aug. 1992, Opava, Czechoslovakia (to appear).
10. Kumpera, A. and Spencer, D. C.: *Lie Equations*, Ann. of Math. Stud. 73, Princeton University Press, Princeton, New Jersey, 1972.
11. Pommaret, J. F.: *Lie Pseudogroups and Mechanics*, Gordon and Breach, London, New York, 1983.
12. Pommaret, J. F. and Lazzarini, S.: Lie pseudogroups and differential sequences: new perspectives in two-dimensional conformal geometry, *J. Geom. Phys.* **10** (1993), 47–91.
13. Pommaret, J. F.: *Partial Differential Equations and Group Theory: New Perspectives for Applications*, Kluwer Acad. Publ., Dordrecht, 1994.
14. Spencer, D. C.: Overdetermined systems of partial differential equations, *Bull. Amer. Math. Soc.* **75** (1965), 1–114.
15. Teodorescu, P. P.: *Dynamics of Linear Elastic Bodies*, Editura Academiei, Bucuresti, 1972; Tunbridge Wells, Kent, 1975.
16. Tsujishita, T.: On variation bicomplexes associated to differential equations, *Osaka J. Math.* **19**(2) (1982), 311–363.
17. Tulczyjew, W. M.: The Euler–Lagrange resolution, *Int. Coll. on Differential Geom. Methods in Math. Physics*, Aix-en-Provence, Sept. 1979; Lecture Notes in Math. 836, Springer-Verlag, Berlin, 1980, pp. 22–48.
18. Vinogradov, A. M.: The C-spectral sequence: Lagrangian formalism and conservation laws, I, II, *J. Math. Anal. Appl.* **100**(1) (1984), 1–129.
19. Weyl, H.: *Space, Time, Matter*, Springer, Berlin, 1918, 1958.
20. Yang, C. N.: Geometry and physics, in Y. Ne'eman (ed.), *Jerusalem Einstein Centennial Symposium on Gauge Theories and Unification of Physical Forces*, Addison-Wesley, 1981.

Acta Applicandae Mathematicae **41:** 297–298, 1995.

Super Toda Lattices

E. D. VAN DER LENDE AND H. G. J. PIJLS
Department of Mathematics and Computer Science, University of Amsterdam, Amsterdam, The Netherlands

(Received: 28 February 1994)

The Lax formalism described by Adler [1] and Oevel and Ragnisco [2] can be generalized to the case where anticommuting variables are involved. In this contribution we apply this super Lax formalism to the Lie superalgebra $\mathfrak{g} = \mathrm{Mat}(m, n, \Lambda)$ to derive a superversion of the Toda lattice. Here Λ is a Grassmann algebra with some unspecified number of odd generators. We only give an outline of this construction and refer to [3] for all the details.

The super commutative algebra $\mathcal{A}$ of polynomial functions $f : \mathfrak{g}_0 \to \Lambda$, where $\mathfrak{g}_0$ denotes the even part of $\mathfrak{g}$, is a super Poisson algebra with bracket

$$\{f, g\} = \langle [L, \mathrm{d}f(L)], \mathrm{d}g(L) \rangle . \tag{1}$$

Now consider a decomposition

$$\mathfrak{g} = \mathfrak{g}_+ \oplus \mathfrak{g}_- \tag{2}$$

and let $\pi := \pi_+ - \pi_-$, where $\pi_\pm : \mathfrak{g} \to \mathfrak{g}_\pm$ are the natural projections. A second Poisson structure on $\mathcal{A}$ is then defined by the bracket

$$\begin{aligned} \{f, g\}_1 : &= \langle [L, \pi \mathrm{d}f] + \pi^* [L, \mathrm{d}f], \mathrm{d}g \rangle \\ &= \langle P_1(L) \mathrm{d}f(L), \mathrm{d}g(L) \rangle . \end{aligned}$$

Then a hierarchy of super Lax equations is defined by

$$\dot{L} = P_1(L) \mathrm{d}h_k(L),$$

where $h_k(L) = \frac{1}{k}\,\mathrm{str}(L^k)$ for $k \geq 1$ ($\mathrm{str}(L)$ denotes the supertrace of L). It turns out that this Poisson structure can be restricted to the subspace $\mathfrak{g}_{-1,1} := \{L \in \mathfrak{g} \mid L^{ij} = 0 \text{ for } |i - j| > 1\}$. This leads to the super extensions of nonrelativistic Toda lattices. In the simplest case, where only nearest neighbour interaction is involved, these equations can be solved explicitly using the concept of super determinant in a way analogeous to [4]. A description of the super extension of the relativistic Toda lattice can also be found in [3].

References

1. Adler, M.: On a trace functional for formal pseudo-differential operators and the symplectic structure of Korteweg–de Vries type equations, *Invent. Math.* **50** (1979), 219–248.

2. Oevel, W. and Ragnisco, O.: R-matrices and higher Poisson brackets for integrable systems, *Phys. A* **161**(1) (1989), 181–220.
3. van der Lende, E. D.: Super Toda Lattices, *J. Math. Phys.* (to appear in 1994).
4. Ruijsenaars, S. N. M.: Relativistic Toda systems, *Comm. Math. Phys.* **133** (1991), 217–247.

Acta Applicandae Mathematicae **41:** 299–309, 1995.

Decay of Conservation Laws and Their Generating Functions

ALEXEY V. SAMOKHIN
Department of Mathematics, Moscow Institute of Civil Aviation Engineers (MIIGA), 6a Pulkovskaya Street, 125838 Moscow, Russia

(Received: 30 May 1993)

Abstract. Quantities which are conserved in nondissipative media decay in the presence of dissipation. The velocity of such a decay (or the 'balance law') may be found explicitly using the generating function of the conservation law. The general approach is illustrated with a system of MHD equations for incompressible magnetofluids.

Mathematics Subject Classifications (1991): 35Q35, 58G35.

Key words: conservation law, balance law, dissipation, MHD equations, generating functions.

1. Introduction

This paper deals with decay velocity of the quantities which are conserved in absense of dissipation.

To be precise, let $\mathbf{E}(\mathbf{u}) = 0$ be a system of equations describing an ideal media state (i.e. without dissipation). A scalar H depending on $\mathbf{u}$ and its derivatives is a conservation law if an integral of H over some fixed spatial domain, denoted by $\langle H\rangle$, is independent of time: $\partial\langle H\rangle/\partial t|_{\mathbf{E}} = 0$. Here, the restriction to $\mathbf{E}$ means that $\partial\langle H\rangle/\partial t = 0$ on any solution $\mathbf{u}(x)$ of $\mathbf{E} = 0$.

With dissipation taken into account, the quantity H is constant no more and $\partial\langle H\rangle/\partial t \neq 0$ is called the decay velocity of H, cf. [3]. A dissipative media state usually satisfies the equation $\mathbf{E}(\mathbf{u}) + \eta\mathbf{F}(\mathbf{u}) = 0$, where η is some small parameter; for $\eta = 0$ we get the ideal state equation. The decay velocity depends naturally on the additional summand $\eta\,\mathbf{F}(\mathbf{u})$. The connection between decay velocity and $\eta\mathbf{F}(\mathbf{u})$ was called a 'balance law' in [1]. This connection was obtained there in case $\mathbf{E}(\mathbf{u}) = 0$ is an evolution equation.

Here, we obtain the balance laws for nonevolution equations (Section 2). These balance laws express $\partial\langle H\rangle/\partial t$ in terms of scalar product of $\eta\mathbf{F}(\mathbf{u})$ and the generating function $\mathbf{g}$ of the conserved quantity H. Namely

$$\frac{\partial\langle H\rangle}{\partial t} = -\eta\langle \mathbf{g}\cdot\mathbf{F}\rangle. \tag{1.1}$$

In more general case when dissipation enters the equation nonlinearly, $\mathbf{E} = \mathbf{E}(\mathbf{u}, \eta)$, we get

$$\frac{\partial\langle H\rangle}{\partial t} = -\eta\Big\langle \mathbf{g}\cdot\frac{\partial\mathbf{E}(\mathbf{u},\eta)}{\partial\eta}\Big\rangle\Big|_{\eta=0} + O(\eta^2). \tag{1.2}$$

The general approach is illustrated by a certain system of MHD equations studied previously, cf. [3], for comparative decay velocities of its conservation laws. A considerable difference in these velocities leads to a simple method, first discovered by Taylor [5], for finding quasistationary states of plasma which are of great practical importance. Some effects of self-organization also occur in this context, see [4].

The system itself is a system for incompressible two-dimensional magnetofluids

$$\begin{aligned} &\frac{\partial\omega}{\partial t} + \mathbf{v}\cdot\nabla\omega = \mathbf{B}\cdot\nabla j + \nu\nabla^2\omega,\\ &\frac{\partial a}{\partial t} + \mathbf{v}\cdot\nabla a = \eta\nabla^2 a. \end{aligned} \tag{1.3}$$

Here $\mathbf{B} = \nabla\times\mathbf{e}_z a$ is a magnetic field, $\mathbf{v} = \nabla\times\mathbf{e}_z\psi$ is a velocity field; $\omega = -\nabla^2\psi$ and $j = -\nabla^2 a$; ν and η are reciprocal of mechanical and magnetic Reynolds numbers, respectively.

In Section 3, we compute all low order conservation laws of the ideal variant of (1.3) and corresponding balance laws in presence of dissipation.

The symmetry algebra of Equation (1.3) is presented in the Appendix.

2. Decay Velocity

First, we introduce notions and notations which are necessary to formulate and prove results.

2.1. Preliminaries

Let $\mathbf{E} = (E_1, \ldots, E_l) = 0$ be a Nth-order system of l nonlinear differential equations on m-vector function $(f^1, \ldots, f^m)$ of $(n+1)$-independent variables $(x_0 = t, x_1, \ldots, x_n)$. We interpret the equation as a submanifold in a jet space $J^N(\pi)$, where $\pi\colon \mathbb{R}^m \times \mathbb{R}^{n+1} \to \mathbb{R}^{n+1}$ is a trivial bundle. If x_i and u^j are base and fiber coordinates of π, then $u^j = f^j(x_0, x_1, \ldots, x_n)$, $j = 1, \ldots, m$ are sections of π denoted by $j_0(\mathbf{f})$. The bundle $\pi_N\colon J^N(\pi) \to \mathbb{R}^{n+1}$ has $\mathbf{x}$ for base coordinates and u^j_σ, $j = 1, \ldots, m$, $\sigma = (i_0, \ldots, i_n)$, $|\sigma| \leqslant N$ for its fiber coordinates. The sections $j_N(\mathbf{f})$ of π_N are given by the formula

$$u^j_\sigma = \frac{\partial^{|\sigma|}}{\partial\mathbf{x}^\sigma} f^j(\mathbf{x}) = \frac{\partial^{|\sigma|}}{\partial x_0^{i_0}\cdots\partial x_n^{i_n}} f^j(\mathbf{x}).$$

Now $\{\mathbf{E}(\mathbf{x},\mathbf{u},\dots,\mathbf{u}_\sigma)=0\}\subset J^N(\pi)$ defines a submanifold in the jet space and we denote this submanifold by $\mathcal{E}$. Solutions of $\mathbf{E}=0$ are such $\mathbf{f}(x)$ that $j_N(\mathbf{f})(\mathbb{R}^{n+1})\subset\mathcal{E}$. Introduce the total differentiations D_i with respect to x_i:

$$D_i=\frac{\partial}{\partial x_i}+\sum_{j,\sigma}u^j_{\sigma+1_i}\frac{\partial}{\partial u^j_\sigma},\quad \text{where}\quad 1_k=(\underbrace{0,\dots,0}_{k},1,0,\dots,0).$$

All differential prolongations of $\mathbf{E}$, i.e. the differential ideal $\mathcal{J}$ generated by $\mathbf{E}$ and the total differentiations D_i, define the submanifold $\mathcal{E}^\infty\subset J^\infty(\pi)$.

An evolution differentiation on $J^\infty(\pi)$ is a vector field

$$\mathcal{X}_\phi=\sum_\sigma D_\sigma\frac{\partial}{\partial u^i_\sigma},$$

where $\phi=(\phi^1,\dots,\phi^m)$ is a function on $J^\infty(\pi)$. The *symmetry* is an evolution differentiation tangent to $\mathcal{E}^\infty$ or, equivalently, satisfying the condition $\mathcal{X}_\phi\mathcal{J}\subset\mathcal{J}$. The last condition is equivalent to

$$\mathcal{X}_\phi(\mathbf{E})=0\quad\text{on }\mathcal{E}^\infty.$$

This may be rewritten to the form

$$\ell_{\mathbf{E}}(\phi)=0\quad\text{on }\mathcal{E}^\infty,\tag{2.1}$$

where $\ell_{\mathbf{E}}$ is some $l\times m$ matrix (l is the codimension of $\mathbf{E}$ in $J^N(\pi)$),

$$(\ell_{\mathbf{E}})_{rs}=\sum_\sigma\frac{\partial E_s}{\partial u^r_\sigma}D_\sigma,$$

or

$$\ell_{\mathbf{E}}=\begin{pmatrix}\sum_\sigma\frac{\partial E_1}{\partial u^1_\sigma}D_\sigma & \dots & \sum_\sigma\frac{\partial E_1}{\partial u^r_\sigma}D_\sigma & \dots & \sum_\sigma\frac{\partial E_1}{\partial u^m_\sigma}D_\sigma\\ \vdots & & \vdots & & \vdots\\ \sum_\sigma\frac{\partial E_s}{\partial u^1_\sigma}D_\sigma & \dots & \sum_\sigma\frac{\partial E_s}{\partial u^r_\sigma}D_\sigma & \dots & \sum_\sigma\frac{\partial E_s}{\partial u^m_\sigma}D_\sigma\\ \vdots & & \vdots & & \vdots\\ \sum_\sigma\frac{\partial E_l}{\partial u^1_\sigma}D_\sigma & \dots & \sum_\sigma\frac{\partial E_l}{\partial u^r_\sigma}D_\sigma & \dots & \sum_\sigma\frac{\partial E_l}{\partial u^m_\sigma}D_\sigma\end{pmatrix}.\tag{2.2}$$

Equation (2.1) is a *symmetry equation*. To get the point symmetries one should take

$$\phi=\sum_{i=0}^n\xi_i(x,\mathbf{u})\mathbf{u}_{1_i}+\boldsymbol{\eta}(x,\mathbf{u}).\tag{2.3}$$

A *conservation law* for the equation $\mathbf{E}$ is a differential n-form $\omega=\sum_{i=0}^n\omega_i\hat{\mathrm{d}x_i}$, such that $\mathrm{d}\omega=0$ on $\mathcal{E}^\infty$; here $\hat{\mathrm{d}x_i}=\mathrm{d}x_0\wedge\dots\wedge\mathrm{d}x_{i-1}\wedge\mathrm{d}x_{i+1}\wedge\dots\wedge\mathrm{d}x_n$ and

ω_i's are some functions on $J^\infty(\pi)$. (In established terminology, ω_0 is called a conserved density, while $(-\omega_1, \omega_2, \dots, (-1)^n\omega_n)$ is called a flux.)

The method for finding of conservation laws is as follows. Let $\ell^*_{\mathbf{E}}$ be a formal conjugate of $\ell_{\mathbf{E}}$,

$$\ell^*_{\mathbf{E}} = \tag{2.4}$$
$$\begin{pmatrix} \sum_\sigma(-1)^{|\sigma|}D_\sigma \circ \frac{\partial E_1}{\partial u^1_\sigma} \dots \sum_\sigma(-1)^{|\sigma|}D_\sigma \circ \frac{\partial E_r}{\partial u^1_\sigma} \dots \sum_\sigma(-1)^{|\sigma|}D_\sigma \circ \frac{\partial E_l}{\partial u^1_\sigma} \\ \vdots \qquad\qquad \vdots \qquad\qquad \vdots \\ \sum_\sigma(-1)^{|\sigma|}D_\sigma \circ \frac{\partial E_1}{\partial u^s_\sigma} \dots \sum_\sigma(-1)^{|\sigma|}D_\sigma \circ \frac{\partial E_r}{\partial u^s_\sigma} \dots \sum_\sigma(-1)^{|\sigma|}D_\sigma \circ \frac{\partial E_l}{\partial u^s_\sigma} \\ \vdots \qquad\qquad \vdots \qquad\qquad \vdots \\ \sum_\sigma(-1)^{|\sigma|}D_\sigma \circ \frac{\partial E_1}{\partial u^m_\sigma} \dots \sum_\sigma(-1)^{|\sigma|}D_\sigma \circ \frac{\partial E_r}{\partial u^m_\sigma} \dots \sum_\sigma(-1)^{|\sigma|}D_\sigma \circ \frac{\partial E_l}{\partial u^m_\sigma} \end{pmatrix}.$$

Solutions of the equation

$$\ell^*_{\mathbf{E}}(\psi)|_{\mathcal{E}^\infty} = 0 \tag{2.5}$$

are so-called *generating functions* of conservation laws, cf. [2]. They are connected to conservation laws themselves in a following way. By the definition of a conservation law, the equation $\mathrm{d}\omega|_{\mathcal{E}^\infty} = 0$ holds, which is equivalent to $\mathrm{d}\omega = \mathcal{O}(\mathbf{E})\mathrm{d}x_0 \wedge \dots \wedge \mathrm{d}x_n$, where $\mathcal{O}(\mathbf{E}) \in \mathcal{J}$, $\mathcal{O}(\mathbf{E}) = \sum_{\sigma,r} \mathcal{O}_\sigma D_\sigma(E_r)$. Now $\mathcal{O}^*(1)$ is the generating function of this conservation law or a solution of (2.5) (*stand for formal conjugation). It remains, however, to find the conservation law itself and to check whether it is trivial (*trivial* by definition are are conservation laws ω which are exact, i.e. $\omega = \mathrm{d}w$ for some $(n-1)$-differential form w).

The general procedure for finding the conservation law, starting with its generating function, is connected to C-spectral sequence of the equation and is not used here: for a low order operator $\mathcal{O}$ (as in examples discussed here) it is usually not hard to discover the conservation law corresponding to any given $\mathcal{O}^*(1)$.

2.2. Main Theorem

THEOREM. *Consider an equation $\mathbf{E}(\eta)$ depending on a small parameter η in such a way that $\mathbf{E}_0 = \mathbf{E}(0)$ is a nondissipative system. Let $\omega = \sum_0^n \omega_i \hat{\mathrm{d}x_i}$ be the conservation law of $\mathbf{E}_0$. Then the decay velocity of the conserved quantity $\langle\omega_0\rangle$ in the presence of dissipation is given by*

$$\frac{\mathrm{d}}{\mathrm{d}t}\langle\omega_0\rangle\Big|_{\mathbf{E}(\eta)} = -\eta\left\langle \mathcal{O}^*(1)\cdot \frac{\partial \mathbf{E}}{\partial\eta}\Big|_{\eta=0}\right\rangle \quad \textit{up to } \mathrm{O}(\eta^2).$$

Proof. For any domain $V \subset \mathbb{R}^{n+1}$ we have

$$\int_{\partial V} \omega = \int_V \mathrm{d}\omega = 0 \quad \text{on } \mathcal{E}_0^\infty, \tag{2.6}$$

by the definition of a conservation law. If V is a cylinder over spatial domain S, $V = S \times [t_0, t_1]$, then

$$Q = \int_S \omega_0 \mathrm{d}x_1 \wedge \cdots \wedge \mathrm{d}x_n$$

is a function of variable t and the former integral equals $Q(t) - Q(t_0)$ plus the flow of a vector $(-\omega_1, \ldots, (-1)^n \omega_n)$ through the $\partial S \times [t_0, t_1]$.

In case this flow is trivial (such is a case, for instance, when $S = \mathbb{R}^n$ and $\omega_i|_{\mathbf{E}}$ are functions rapidly decreasing at infinity), we have $Q(t) - Q(t_0) = 0$. Hence, the function $Q(t)$ is constant, i.e. $Q(t)$ is a conserved quantity:

$$\frac{\mathrm{d}}{\mathrm{d}t} Q(t) = \frac{\mathrm{d}}{\mathrm{d}t} \int_S \omega_0 \mathrm{d}x_1 \wedge \cdots \wedge \mathrm{d}x_n \bigg|_{\mathcal{E}_0^\infty} = \frac{\mathrm{d}}{\mathrm{d}t} \langle \omega_0 \rangle = 0. \tag{2.7}$$

On the other hand

$$\begin{aligned} &\frac{\mathrm{d}}{\mathrm{d}t} \int_S \omega_0 \mathrm{d}x_1 \wedge \cdots \wedge \mathrm{d}x_n \\ &= \frac{\mathrm{d}}{\mathrm{d}t} \int_{\partial V} \omega = \frac{\mathrm{d}}{\mathrm{d}t} \int_V \mathrm{d}\omega \\ &= \frac{\mathrm{d}}{\mathrm{d}t} \int_V \left(\sum \mathcal{O}_\sigma D_\sigma(\mathbf{E}_0) \right) \mathrm{d}t \wedge \mathrm{d}x_1 \wedge \cdots \wedge \mathrm{d}x_n \\ &= \frac{\mathrm{d}}{\mathrm{d}t} \int_V \mathcal{O}^*(1) \mathbf{E}_0 \, \mathrm{d}t \wedge \mathrm{d}x_1 \wedge \cdots \wedge \mathrm{d}x_n = 0. \end{aligned} \tag{2.8}$$

The last equality is a result of integration by parts. Note that $\mathcal{O}^*(1)$ is l-vector and $\mathcal{O}^*(1)\mathbf{E}_0$ is a scalar product.

Since $V = S \times [t_0, t_1]$, we have

$$\begin{aligned} &\int_V \mathcal{O}^*(1) \mathbf{E}_0 \, \mathrm{d}t \wedge \mathrm{d}x_1 \wedge \cdots \wedge \mathrm{d}x_n \\ &= \int_{t_0}^{t_1} \left(\int_S \mathcal{O}^*(1) \mathbf{E}_0 \mathrm{d}x_1 \wedge \cdots \wedge \mathrm{d}x_n \right) \mathrm{d}t. \end{aligned} \tag{2.9}$$

Differentiation of the last integral by the upper limit t_1 imply

$$\frac{\mathrm{d}}{\mathrm{d}t} \int_S \omega_0 \mathrm{d}x_1 \wedge \cdots \wedge \mathrm{d}x_n = \int_S \mathcal{O}^*(1) \mathbf{E}_0 \, \mathrm{d}x_1 \wedge \cdots \wedge \mathrm{d}x_n. \tag{2.10}$$

The right-hand side of (2.10) is zero on $\mathbf{E}_0 = 0$, but when restricting (2.10) to $\mathbf{E}(\eta) = \mathbf{E}_0 + \eta \cdot \mathbf{F} = 0$, we get

$$\frac{\mathrm{d}}{\mathrm{d}t} \int_S \omega_0 \, \mathrm{d}x_1 \wedge \cdots \wedge \mathrm{d}x_n = \int_S \mathcal{O}^*(1)(-\eta \mathbf{F}) \, \mathrm{d}x_1 \wedge \cdots \wedge \mathrm{d}x_n \tag{2.11}$$

or

$$\frac{\mathrm{d}}{\mathrm{d}t} \langle \omega_0 \rangle \bigg|_{\mathbf{E}(\eta)} = -\eta \langle \mathcal{O}^*(1) \cdot \mathbf{F} \rangle. \tag{2.12}$$

In a more general case of

$$\mathbf{E}(\eta) = \mathbf{E}_0 + \eta \cdot \frac{\partial \mathbf{E}}{\partial \eta} + \mathrm{O}(\eta^2),$$

it follows from (2.10) that

$$\frac{\mathrm{d}}{\mathrm{d}t}\langle \omega_0 \rangle \bigg|_{\mathbf{E}(\eta)=0} = -\left\langle \eta \mathcal{O}^*(1) \cdot \frac{\partial \mathbf{E}}{\partial \eta}\bigg|_{\eta=0} \right\rangle \quad \text{up to } \mathrm{O}(\eta^2). \tag{2.13}$$

□

Remark 1. If there is more than one small parameter, formula (2.13) is readily generalized:

$$\frac{\mathrm{d}}{\mathrm{d}t}\langle \omega_0 \rangle \bigg|_{\mathbf{E}(\eta_1,\dots,\eta_s)=0} = -\sum_{r=1}^{s} \left\langle \eta_r \mathcal{O}^*(1) \cdot \frac{\partial \mathbf{E}}{\partial \eta_r}\bigg|_{\boldsymbol{\eta}=0} \right\rangle \quad \text{up to } \mathrm{O}(\eta^2). \tag{2.14}$$

Remark 2. In case of evolution equation $\partial \mathbf{u}/\partial t = \mathbf{E}_0(\mathbf{u}) + \eta \mathbf{F}(\mathbf{u})$ it is not hard to get another explicit form of the balance law, cf. [1]. Let ω_0 be an ideal conserved density, $\mathrm{d}\omega_0/\mathrm{d}t \equiv \ell_{\omega_0}(\mathbf{E}_0) = 0$ on $\mathbf{E}_0$ and $\mathrm{d}\omega_0/\mathrm{d}t \equiv \ell_{\omega_0}(\mathbf{E}_0) + \eta \ell_{\omega_0}(\mathbf{F})$ in the presence of dissipation. Therefore, the equality

$$\frac{\mathrm{d}\langle \omega_0 \rangle}{\mathrm{d}t} = -\eta \langle \ell_{\omega_0}(\mathbf{F}) \rangle$$

is the balance law for evolution equation.

Remark 3. It is noteworthly that in case of a system (i.e. $l > 1$) it is possible for any given conserved quantity to add dissipative-like summands in such a way that this quantity still remains conserved: for any $\mathcal{O}^*(1)$ one can choose such an $\mathbf{F}$ that right-hand side of (2.12) will be zero.

3. Example from Magnetohydrodynamics

3.1. Equation

The three-dimensional MHD equation, describing incompressible magnetofluids in dimensionless variables may be taken in the following form:

$$\begin{aligned} &\frac{\partial \mathbf{v}}{\partial t} + \mathbf{v} \cdot \boldsymbol{\nabla} \mathbf{v} = -\boldsymbol{\nabla} p^* + \mathbf{B} \cdot \boldsymbol{\nabla} \mathbf{B} + \nu \nabla^2 \mathbf{v}, \\ &\frac{\partial \mathbf{B}}{\partial t} + \mathbf{v} \cdot \boldsymbol{\nabla} \mathbf{B} = \mathbf{B} \cdot \boldsymbol{\nabla} \mathbf{v} + \eta \nabla^2 \mathbf{B}, \\ &\boldsymbol{\nabla} \cdot \mathbf{v} = 0 = \boldsymbol{\nabla} \cdot \mathbf{B}. \end{aligned} \tag{3.1}$$

Here ν and η are reciprocal of mechanical and magnetic Reynolds numbers respectively; $\mathbf{v}$ and $\mathbf{B}$ are velocity and magnetic fields and p^* is the total pressure. It is assumed that mass density is constant and uniform, and that $\mathbf{v}$ and $\mathbf{B}$ are in Alfven speed units.

Equation (3.1) may be simplified in case of two spatial variables (x, y) assuming $\partial/\partial z \equiv 0$.. In this case $\mathbf{B} = (B_x, B_y, 0)$, $\mathbf{v} = (v_x, v_y, 0)$. Moreover, $\mathbf{v} = \nabla \times \psi \mathbf{e}_z$ and $\mathbf{B} = \nabla \times \mathbf{a}$ for some stream function $\psi(x, y, t)$ and potential $\mathbf{a} = a(x, y, t)\mathbf{e}_z$. Introduce dimensionless vorticity and current by $\nabla \times \mathbf{v} = \omega \mathbf{e}_z$, $\nabla \mathbf{B} = j\mathbf{e}_z$, where $j = -\nabla^2 a$ and $\omega = -\nabla^2 \psi$. Then the last equation in the system (3.1) is automatically true, while the rest comes to

$$\begin{aligned} &\frac{\partial \omega}{\partial t} + \mathbf{v} \cdot \boldsymbol{\nabla} \omega = \mathbf{B} \cdot \boldsymbol{\nabla} j + \nu \nabla^2 \omega, \\ &\frac{\partial a}{\partial t} + \mathbf{v} \cdot \boldsymbol{\nabla} a = \eta \nabla^2 a. \end{aligned} \tag{3.2}$$

We rewrite it to coordinate form. Denote $u = -\psi$, $v = -a$ and $\Delta = \nabla^2$. Then

$$\begin{aligned} &\Delta u_t + u_x \Delta u_y - u_y \Delta u_x + v_y \Delta v_x - v_x \Delta v_y = \nu \Delta^2 u, \\ &v_t + u_x v_y - u_y v_x = \eta \Delta v. \end{aligned} \tag{3.3}$$

Here subscripts mean partial differentiation: $u_x = \partial u/\partial x$ and so on. Note that it is not an evolution equation in chosen variables. Equation (3.3) will be denoted $E(\nu, \eta)$ from now onwards. The ideal state is described by $E(0,\ 0)$ and denoted by E_0.

3.2. Conservation laws of ideal state

To obtain conservation laws one starts by solving

$$\ell^*_{E_0} \mathbf{f} |_{\mathcal{E}_0^\infty} = 0. \tag{3.4}$$

Here $\mathbf{f} = \binom{S}{T}$ is a possible generating function of a would be conservation law; components S and T are some functions on $J^\infty(\mathbb{R}^3, \mathbb{R}^2)$. We recall that * stands for formal conjugation. The universal linearization operator ℓ_{E_0} for (3.3) is given by the formula

$$\ell_{E_0} = \begin{pmatrix} \begin{matrix} D_t \Delta + u_x \Delta D_y + \Delta u_y \cdot D_x - \\ u_y \Delta D_x - \Delta u_x \cdot D_y \end{matrix} & \begin{matrix} v_y \Delta D_x - v_x \Delta D_y + \\ \Delta v_x \cdot D_y - \Delta v_y \cdot D_x \end{matrix} \\ & \\ v_y D_x - v_x D_y & D_t + u_x D_y - u_y D_x \end{pmatrix}. \tag{3.5}$$

Remind that solutions of the equation

$$\ell_{E_0}(\mathbf{f}) |_{\mathcal{E}_0^\infty} = 0 \tag{3.6}$$

are the (generating functions of) symmetries of Equation (3.3). The solutions of (3.6) which produce classical symmetries of (3.3) are given in Appendix.

Now $\ell^*_{E_0}$ is of the form

$$\ell^*_{E_0} = \begin{pmatrix} -D_t\Delta - u_x D_y\Delta + u_y D_x\Delta + & -v_y D_x + v_x D_y \\ 2(u_{yy} - u_{xx})D_x D_y + 2u_{xy}(D_x^2 - D_y^2) & \\ & \\ v_x D_y\Delta - v_y D_x\Delta - & -D_t - u_x D_y + u_y D_x \\ 2(v_{yy} - v_{xx})D_x D_y - 2v_{xy}(D_x^2 - D_y^2) & \end{pmatrix}. \quad (3.7)$$

We restrict ourselves to low-order conservation laws, that is to such a $\mathbf{f} = \binom{S}{T}$ in (3.4) that S and T are functions on $J^0(\mathbb{R}^3, \mathbb{R}^2)$ and $J^2(\mathbb{R}^3, \mathbb{R}^2)$, respectively. This choice may be understood by considering the structure of $\ell^*_{E_0}$ matrix: its second column is a first-order operator while the first column is of third-order. Solving Equation (3.4), which depends polynomially on higher derivatives u_σ, v_σ, is very tedious but a straightforward job. We simply produce the results.

The kernel of $\ell^*_{E_0}|_{\mathcal{E}_0^\infty}$ is linearly generated by

$$\begin{pmatrix} h(t) \\ 0 \end{pmatrix}, \quad \begin{pmatrix} x^2 + y^2 \\ 0 \end{pmatrix}, \quad \begin{pmatrix} p(t)x \\ 0 \end{pmatrix}, \quad \begin{pmatrix} q(t)y \\ 0 \end{pmatrix},$$

$$\begin{pmatrix} u \\ \Delta u \end{pmatrix}, \quad \begin{pmatrix} f(v) \\ f'(v)\Delta v \end{pmatrix}, \quad \begin{pmatrix} 0 \\ \Phi'(v) \end{pmatrix}, \quad (3.8)$$

where h, p, q, f and Φ are arbitrary functions. Most of them produce trivial conservation laws. There are three nontrivial conserved densities (two of them depending on arbitrary functions): the total energy E (magnetic plus kinetik energy), generalized 'cross helicity' H_c and mean magnetic potential A,

$$\begin{aligned} E &= \tfrac{1}{2}\langle u_x^2 + u_y^2 + v_x^2 + v_y^2\rangle, \\ H_c &= \langle f'(v) \cdot (u_x v_x + u_y v_y)\rangle, \qquad (3.9) \\ A &= \langle \Phi(v)\rangle. \end{aligned}$$

Their generating functions are placed on the second line of (3.8) in respective order. Recall that f and Φ are arbitrary functions of v.

3.3. Decay Rates

Once dissipaton coefficients ν or η are allowed to have small but finite values, quantities (3.8) are conserved no more. In accordance with general formulas of Section 2, their decay rates are

$$\begin{aligned}
\frac{\mathrm{d}E}{\mathrm{d}t} &= -\nu \int_S u\Delta^2 u \,\mathrm{d}x\,\mathrm{d}y - \eta \int_S (\Delta v)^2 \,\mathrm{d}x\,\mathrm{d}y \\
&= -\int_S \left[\nu(\Delta u)^2 + \eta(\Delta v)^2\right] \mathrm{d}x\,\mathrm{d}y; \\
\frac{\mathrm{d}H_c}{\mathrm{d}t} &= \frac{1}{2}\int_S \left[\nu f(v)\Delta^2 u + \eta f'(v)\Delta u \Delta v\right] \mathrm{d}x\,\mathrm{d}y \\
&= -\frac{1}{2}(\nu+\eta)\int_S f'(v)\Delta u\Delta v \,\mathrm{d}x\,\mathrm{d}y - \\
&\quad -\frac{1}{2}\nu \int_S f''(v)\Delta u(v_x^2+v_y^2)\,\mathrm{d}x\,\mathrm{d}y; \\
\frac{\mathrm{d}A}{\mathrm{d}t} &= -\eta\int_S \Phi'(v)\Delta v \,\mathrm{d}x\,\mathrm{d}y \\
&= \eta \int_S \Phi''(v)(v_x^2+v_y^2)\,\mathrm{d}x\,\mathrm{d}y.
\end{aligned} \tag{3.10}$$

These formulas in familiar physical variables become

$$\begin{aligned}
\frac{\mathrm{d}E}{\mathrm{d}t} &= -\nu \int_S [\nu\omega^2 + \eta j^2]\,\mathrm{d}x\,\mathrm{d}y; \\
\frac{\mathrm{d}H_c}{\mathrm{d}t} &= -\frac{1}{2}(\nu+\eta)\int_S f'(a)\,a\psi \,\mathrm{d}x\,\mathrm{d}y - \\
&\quad -\frac{1}{2}\nu\int_S f''(a)\,\omega(\nabla a)^2\,\mathrm{d}x\,\mathrm{d}y; \\
\frac{\mathrm{d}A}{\mathrm{d}t} &= \eta\int_S \Phi''(a)(\nabla a)^2\,\mathrm{d}x\,\mathrm{d}y.
\end{aligned} \tag{3.11}$$

One can see that the decay of E is monotonic but those of H_c and A are not necessarily so. Such an inequality in decay rates leads to a distinct physical phenomenon of 'self-organization' or quasi-stable states of plasma. Depending on initial conditions competing processes called 'selective decay' or 'dynamic alignment' occur: in selective decay energy decays relatively to mean potential, and in dynamic alignment energy decays relatively to cross-helicity (velocity and magnetic field being aligned). There are also some more delicate possibilities of self-organization.

There exist a very simple procedure for finding solutions of the above described behavior. It was suggested in [5], and is known as the 'Taylor trick'. The reason why or when it works is not so far entirely mathematically clear, but it allows us to predict and calculate quasistable states, cf. [3]. The procedure is as follows.

Taking into consideration their comparative decay rates, let us minimize E with H_c and A as constrains. We put $\delta(E+\lambda H_c+\mu A)=0$, A and H_c presumed constant, λ and μ being Lagrange multipliers. The Euler–Lagrange equations are

$$\Delta\left[u - F(v)\right] = 0,$$
$$\Delta v = f(v)\Delta u + g(v), \tag{3.12}$$

where $F' = f$ and $g = \pm\Phi'$.

The system (3.12) generally is not compatible with (3.3). But it is compatible if $\eta = \nu$ which is in particular true in the ideal case $\eta = \nu = 0$. In this case, combining (3.3) and (3.12) we get

$$\begin{aligned}
&\Delta\left[u - F(v)\right] = 0,\\
&\Delta v = \frac{ff'}{1-f^2}(v_x^2 + v_y^2) + \frac{g}{1-f^2},\\
&v_t = u_y v_x - u_x v_y,\\
&(u_{xy} - f v_{xy})(v_x^2 - v_y^2) + \left[(u_{yy} - f v_{yy}) - (u_{xx} - f v_{xx})\right]v_x v_y = 0.
\end{aligned} \tag{3.13}$$

Solutions of (3.13) describe the quasistationary states with remarkable accuracy as it was demonstrated numerically for special types of f and Φ in [3, 4].

Remark 1. The first and the last equations of (3.13) form the closed system

$$\Delta w = 0, \qquad z_t + w_x z_y - w_y z_x = 0,$$

where $w = u - F(v)$ and $z = v_x^2 + v_y^2$.

Remark 2. The second equaton in (3.13) may be written in a closed form $\Delta R = \Psi(R)$, where $R = R(v)$, $R' = \sqrt{1-f^2}$

Remark 3. The case of $u = F(v)$ in (3.13) is a generalization of dynamic alignment studied in [5] (aligned are gradients of u and v). It implies stationary solutions

$$u = F(v), \qquad v_t = 0, \qquad \Delta R = \Psi(R),$$

where $R'(v) = \sqrt{1 - f^2(v)}$ as in previous remark.

Appendix

Classical symmetries of Equation (3.3) correspond to generating functions of the form

$$\mathbf{f} = a\begin{pmatrix} u_x \\ v_x \end{pmatrix} + b\begin{pmatrix} u_y \\ v_y \end{pmatrix} + c\begin{pmatrix} u_t \\ v_t \end{pmatrix} + \begin{pmatrix} e_1 \\ e_2 \end{pmatrix},$$

where a, b, c and e_i are functions on $J^0(\mathbb{R}^3, \mathbb{R}^2)$, i.e. functions of x, y, t, u, v. On the other hand, a classical symmetry is a vector field on $J^\infty(\mathbb{R}^3, \mathbb{R}^2)$ which uniquely defined by its restriction to $J^0(\mathbb{R}^3, \mathbb{R}^2)$ [2]. In this way $\mathbf{f}$ corresponds to

$$X_{\mathbf{f}} = a\frac{\partial}{\partial x} + b\frac{\partial}{\partial y} + c\frac{\partial}{\partial t} + e_1\frac{\partial}{\partial u} + e_2\frac{\partial}{\partial v}$$

We list here all classical symmetries in a form of vector fields on $J^0(\mathbb{R}^3, \mathbb{R}^2)$:

(1) $u\frac{\partial}{\partial u} + v\frac{\partial}{\partial v} - t\frac{\partial}{\partial t}$ (scaling)
(2) $x\frac{\partial}{\partial x} + y\frac{\partial}{\partial y} + 2t\frac{\partial}{\partial t}$ (scaling)
(3) $\frac{\partial}{\partial v}$ (gauge)
(4) $\pi(t)\frac{\partial}{\partial u}$ (gauge)
(5) $x\rho'(t)\frac{\partial}{\partial u} + \rho(t)\frac{\partial}{\partial y}$ (generalized translation in y)
(6) $y\mu'(t)\frac{\partial}{\partial u} - \mu(t)\frac{\partial}{\partial x}$ (generalized translation in x)
(7) $y\frac{\partial}{\partial x} - x\frac{\partial}{\partial y}$ (rotation)
(8) $\frac{1}{2}(x^2 + y^2)\frac{\partial}{\partial u} - t(y\frac{\partial}{\partial x} - x\frac{\partial}{\partial y})$ (generalized rotation)
(9) $\frac{\partial}{\partial t}$ (translation in t),

where π, ρ, μ are arbitrary functions of t.

Remark. All but first of listed fields are symmetries of the dissipative equation, while the first one is valid only in ideal ($\eta = \nu = 0$) case.

References

1. van Groesen, E. and Mainardi, F.: Balance laws and centro velocity in dissipative systems, *J. Math. Phys.* **31** (1990), 2136–2140.
2. Vinogradov, A. M.: Symmetries and conservation laws of partial differential equations, *Acta Appl. Math.* **15** (1989), 3–21.
3. Ting, A. C., Matthaeus, M. H. and Montgomery, D.: Turbulent relaxation processes in magnetohydrodynamics, *Phys. Fluids* **29** (1986), 3261–3274.
4. Hasegawa, A.: Self organisation processes in continious media, *Adv. Phys.* **34** (1985), 1–42.
5. Taylor, J. B.: Relaxation of toroidal plasma and generation of reverse magnetic fields, *Phys. Rev. Lett.* **33** (1974), 1139–1141.

Acta Applicandae Mathematicae **41:** 311–322, 1995.

Arbitrariness of the General Solution and Symmetries

WERNER M. SEILER
Institut für Algorithmen und Kognitive Systeme, Universität Karlsruhe, D-76128 Karlsruhe, Germany, e-mail: Seilerw@ira.uka.de

(Received: 28 February 1994)

Abstract. The computation of a number of arbitrary functions in the general solution is briefly reviewed. The results are used to study normal systems and their symmetry reduction. We discuss the treatment of gauge systems, especially the analysis of gauge fixing conditions. As examples, the Yang–Mills equations with the Lorentz gauge and Einstein's vacuum field equations with harmonic coordinates are considered.

Mathematics Subject Classifications (1991): 35G20, 58G35, 70G50.

Key words: involution, symmetry reduction, gauge theory.

1. Introduction

If one cannot determine the general solution of an equation, as it is usually the case with differential equations, one wants to know at least the dimension of the solution space. This is, however, not trivial for systems, especially if they are overdetermined. It turns out, that involution [10] provides the key. We showed in a recent paper [13], how to compute the number of arbitrary functions and their arity for closed representations of the general solution of an involutive system.

The purpose of this paper is to provide more applications of these results. Normal systems are the natural generalization of single equations with respect to the arbitrariness. Symmetry reductions [1, 7] are most often performed for such systems. We will compute in Section 4 the loss of generality during such a reduction.

Gauge systems are a central theme in modern theoretical physics. We will present in Section 5 an improved treatment of them, compared with [13], through the introduction of gauge corrected Cartan characters. Einstein [2] pioneered the analysis of the arbitrariness of field theories. His philosophy was to choose by otherwise equal properties the system which restricts the fields stronger.

Special emphasis will be put on the analysis of gauge fixing conditions. As concrete examples, the last two sections treat the Yang–Mills equations and Einstein's vacuum field equations in arbitrary dimensions. For the former, we will analyse the Lorentz gauge; for the latter, harmonic coordinates.

2. Formal Theory

Formal theory [10] is based on jet bundle formalism. Let $x_1, \dots, x_n$ and $u^1, \dots,$ u^m be a local coordinate system on a bundle $\mathcal{E}$. We define a differential equation of order q as a fibred submanifold $\mathcal{R}_q$ in the jet bundle $J_q\mathcal{E}$. Locally, $\mathcal{R}_q$ is given by a system of equations $\Phi^\tau(x^i, u^\alpha, p^\alpha_\mu) = 0$, where

$$p^\alpha_\mu = \partial^{\mu_1+\cdots+\mu_n}u^\alpha/\partial(x^1)^{\mu_1}\cdots\partial(x^n)^{\mu_n}$$

for the multi-index $\mu = [\mu_1, \dots, \mu_n]$.

The *prolongation* $\mathcal{R}_{q+1} \subset J_{q+1}\mathcal{E}$ is obtained by formally differentiating all equations with respect to the independent variables x^i. It is well known that *integrability conditions* can arise during prolongation. They can occur at any prolongation order. A system that does not generate integrability conditions is called *formally integrable*, because it is possible to construct, order by order, a formal power series solution.

A formally integrable system is *involutive*, if it has an involutive *symbol*. The symbol $\mathcal{M}_q$ of $\mathcal{R}_q$ is a system of linear (algebraic, not differential!) equations in some unknowns v^α_μ defined by

$$\mathcal{M}_q\colon\ \sum_{\alpha,|\mu|=q} \frac{\partial\Phi^\tau}{\partial p^\alpha_\mu} v^\alpha_\mu = 0\,. \tag{1}$$

For a quasi-linear equation, the symbol is essentially obtained by substituting v^α_μ for p^α_μ in the highest-order part of the equation.

A jet variable p^α_μ is said to be of *class* k, if μ_k is the first nonvanishing entry of the multi-index. We order the columns of the symbol $\mathcal{M}_q$ by class (highest class first) and compute a row echelon form. If v^α_μ is the leading term of an equation in this solved form of the symbol, then p^α_μ is called a *principal derivative* [3]. All other derivatives of order q are *parametric*.

We define $\beta^{(k)}_q$ as the number of principal derivatives of class k. The symbol $\mathcal{M}_q$ is involutive, if

$$\operatorname{rank} \mathcal{M}_{q+1} = \sum_{k=1}^{n} k\beta^{(k)}_q\,. \tag{2}$$

Associating with each equation whose principal derivative is of class k its *multiplicative variables* $x^1, \dots, x^k$, we see that if we prolong each equation with respect to its multiplicative variables only, we get algebraically independent equations. Equation (2) tells us, that in the case of an involutive symbol no further independent equations exist.

The *Cartan characters*

$$\alpha^{(k)}_q = m\binom{q+n-k-1}{q-1} - \beta^{(k)}_q\,, \quad k = 1, \dots, n, \tag{3}$$

count the parametric derivatives of order q and class k. The *Cartan–Kähler theorem* states that for analytic involutive systems there exists a unique analytic solution for a certain initial value problem prescribing $\alpha_q^{(k)}$ analytic functions of $x^1, \ldots, x^k$ for $k = 1, \ldots, n$ as Cauchy data. This generalizes the *Cauchy–Kowalevsky theorem*.

Although these definitions appear to be coordinate dependent, one can show that, with the exception of certain singular systems, every coordinate system yields the same values for the $\beta_q^{(k)}$. For lack of space, we cannot discuss here the delicate question of δ-regularity of a coordinate system but refer to the literature [10].

Finally, one should note that any system can be algorithmically completed to an involutive one [10, 12]. This is ensured by the *Cartan–Kuranishi theorem*. Thus it poses no real restriction, if we assume from now on that we deal only with involutive systems.

3. Arbitrary Functions

In a recent paper [13], we have shown how one can derive the number of arbitrary functions in the general solution of an involutive system. We briefly repeat the results here. The Cartan characters are the key tool for this analysis.

The parametric derivatives represent the arbitrariness in the general solution allowed by the differential equation, whereas the principal derivatives can be computed through the differential equation or its prolongations. The number of parametric derivatives of order $q + r$ is given by the *Hilbert polynomial* of the equation

$$H_q(r) = \sum_{i=0}^{n-1} \left(\sum_{k=i}^{n-1} \frac{\alpha_q^{(k+1)}}{k!} s_{k-i}^{(k)}(0) \right) r^i . \tag{4}$$

Here we have used the *symmetric q-product*

$$s_k^{(n)}(q) = \sigma_k^{(n)}(q+1, q+2, \ldots, q+n) \quad \text{for } 0 \leqslant k \leqslant n , \tag{5}$$

where $\sigma_k^{(n)}$ ($\sigma_0^{(n)} = 1$) denotes the elementary symmetric polynomial of degree k in n unknowns.

We assume now that the general solution of $\mathcal{R}_q$ can be written in a closed form containing some arbitrary functions and that its Taylor series expansion can be constructed order by order, i.e. there exists a bijection between its coefficients of order $q+r$ and the Taylor coefficients of order $q+r+j$ of the arbitrary functions of differentiation order $j^\star$. Let there be $f_{k,j}$ such functions with k arguments.

---

* This means that the function appears for $j < 0$ as integrand of a j-dimensional integral and $j > 0$ as a derivative of jth order.

If we write the Hilbert polynomial as $H_q(r) = \sum h_i r^i$, we can derive the following linear diophantine system for the $f_{k,j}$ [13]

$$\sum_{j\in J}\sum_{k=i}^{n-1} \frac{n!}{k!} s_{k-i}^{(k)}(q-j) f_{k,j} = n! h_i \,, \quad i = 0, \ldots, n-1 \,, \tag{6}$$

where J denotes the set of considered values for j (usually $J \subset \{-q, \ldots, 0\}$). In general, this system has no unique solution, if $|J| > 1$. Setting $J = \{j_0\}$ we obtain the following theorem.

THEOREM 1. *If there exists a representation of the general solution whose power series can be constructed order by order and whose arbitrary functions are all of differentiation order $j_0 \geqslant -q$, then it contains f_k arbitrary functions with k arguments, where the f_k are determined by the recursion relation*

$$\begin{aligned} f_n &= \alpha_q^{(n)} \,, \\ f_k &= \alpha_q^{(k)} + \sum_{i=k+1}^{n} \frac{(k-1)!}{(i-1)!} \left(\alpha_q^{(i)} s_{i-k}^{(i-1)}(0) - f_i s_{i-k}^{(i-1)}(q+j_0) \right) . \end{aligned} \tag{7}$$

Such a representation can only exist if the solution of this recursion relation contains no negative integers.

Equation (7) has a simple solution for $j_0 = -q$, namely $f_k = \alpha_q^{(k)}$. This corresponds to the representation of the general solution given by the above-mentioned Cartan–Kähler theorem. This means, that the arbitrary functions in the Cauchy data will in general enter as q-dimensional integrals.

Einstein [2] introduced as a simpler measure for the arbitrariness the *strength* of a differential equation. If we expand the rational function

$$Z_q(r) = H_q(r) \Big/ \binom{n+q+r-1}{n-1} \tag{8}$$

in powers of $1/r$, the strength $Z_q^{(1)}$ is defined as the coefficient of $1/r$. It is related to the Cartan characters [13]

$$Z_q^{(1)} = (n-1)\left(\tfrac{1}{2} n \alpha_q^{(n)} + \alpha_q^{(n-1)}\right) . \tag{9}$$

4. Normal Systems

Normal systems satisfy the conditions of the Cauchy–Kowalevsky theorem [4]. With a coordinate transformation a normal system of order q can always be brought into the following solved form

$$p^{\beta}_{[0,\ldots,q]} = \Phi^{\beta}(x^i, u^\alpha, p^\alpha_\mu) \,, \quad \beta = 1, \ldots, m, \tag{10}$$

where the right-hand sides do not contain any derivative of order q, which is purely with respect to x^n.

In the terms introduced in Section 2, we see that (10) is solved for the derivatives of class n. Since we have as many equations as dependent functions, all such derivatives appear. Thus we find $\beta_q^{(n)} = m$, $\beta_q^{(n-1)} = \cdots = \beta_q^{(1)} = 0$ and

$$\alpha_N^{(n)} = 0, \qquad \alpha_N^{(n-k)} = m\binom{q+k-1}{q-1}. \tag{11}$$

Theorem 1 with $j_0 = 0$ yields $f_n = 0$, $f_{n-1} = mq$ and $f_{n-2} = mq(1-q)/2$. Thus we have proven the following perhaps surprising result:

THEOREM 2. *An algebraic representation, i.e. one without integrals or derivatives, of the general solution of a normal system, whose power series can be constructed order by order, can exist only if either there are only two independent variables or the system is of the first order.*

In the case of a first-order system, the recursion relation (7) can be solved in closed form for $j_0 = 0$ [13]: $f_n = \alpha_1^{(n)}$, $f_k = \alpha_1^{(k)} - \alpha_1^{(k+1)}$. For a normal system, this yields the well-known values of the Cauchy–Kowalevsky theorem: $f_{n-1} = m$, all other f_k vanish.

A standard technique for the construction of solutions is the reduction with respect to a symmetry group [1, 7]. It is well known that only for ordinary differential equations one can reconstruct the general solution of the original system from the general solution of the reduced system. In the case of partial differential equations, one obtains only special solutions.

Let us assume that the symmetry group operates with s-dimensional orbits and that after the reduction the system is still normal. Then we get a system with $\bar{n} = n - s$ independent and $\bar{m} = m$ dependent variables of order $\bar{q} = q - \Delta q$. The Cartan characters of the reduced system are

$$\bar{\alpha}_{\bar{q}}^{(\bar{n})} = 0, \qquad \bar{\alpha}_{\bar{q}}^{(\bar{n}-k)} = m\binom{\bar{q}+k-1}{\bar{q}-1}.$$

Thus if $\Delta q = 0$, then $\bar{\alpha}_q^{(k)} = \alpha_q^{(k+s)}$. If $\Delta q > 0$, we get an additional loss of generality as to expect.

Since the Cartan characters form a descending sequence [10]

$$\alpha_q^{(1)} \geqslant \alpha_q^{(2)} \geqslant \cdots \geqslant \alpha_q^{(n)}, \tag{12}$$

this result means that we do not only loose all arbitrary functions with more than $n - s$ arguments, but that even the number of arbitrary functions with less arguments decreases.

Even in more general situations, the loss of generality can be computed without explicit construction of the reduced system. One must simply compare the

Cartan characters of the original system alone with the characters of the system plus the invariant surface condition (after completion to an involutive system!). An extension of this method to reductions with respect to generalized or to weak symmetries [7, 8] is straightforward.

For instance, the analysis of Lie–Bäcklund symmetries of first-order, normal systems [6] leads to the addition of differential constraints of order r

$$\Psi^\tau(x^i, u^\alpha, p^\alpha_\mu) = 0 \qquad \begin{cases} \tau &= 1, \ldots, p\,, \\ |\mu| &\leqslant r\,, \\ \mu_n &= 0\,. \end{cases} \tag{13}$$

Note that x^n occurs here only as a parameter, because all derivatives with respect to x^n can be eliminated using (10).

We assume that (13) is in involution and that there are $\tilde{\beta}^{(k)}_r$ equations of class k for $k = 1, \ldots, n-1$. The combined system (10), (13) has an involutive symbol, if and only if there exist matrices $(M^i)^\tau_\rho$ such that

$$(M^i)^\tau_\rho \frac{\partial \Psi^\rho}{\partial p^\alpha_\mu} = -\frac{\partial \Psi^\tau}{\partial p^\beta_\mu}\frac{\partial \Phi^\beta}{\partial p^\alpha_i} \qquad \begin{cases} i = 1, \ldots, n-1\,, \\ |\mu| = r\,. \end{cases} \tag{14}$$

No integrability conditions arise, i.e. the system is involutive, if and only if

$$D_n\Psi^\tau - \sum_{\alpha, |\mu| = r} \frac{\partial \Psi^\tau}{\partial p^\alpha_\mu}(p^\alpha_{\mu+1_n} - D_\mu \Phi^\alpha) - \sum_{i=1}^{n-1} (M^i)^\tau_\rho D_i \Psi^\rho = 0, \tag{15}$$

where D_i denotes the formal derivative with respect to x^i.

Since (13) contains no derivatives with respect to x^n, we get for the combined system

$$\beta^{(n)}_r = m, \qquad \beta^{(n-k)}_r = m\binom{r+k-2}{r-2} + \tilde{\beta}^{(n-k)}_r.$$

This yields the Cartan characters

$$\alpha^{(n)}_r = 0, \qquad \alpha^{(n-k)}_r = m\binom{r+k-2}{r-1} - \tilde{\beta}^{(n-k)}_r\,. \tag{16}$$

This generalizes the results of [6] to nonlinear systems and to constraints which depend on x^n and which do not form a normal system themselves.

5. Gauge Systems

In gauge theories, determining the number of arbitrary functions in the general solution is not sufficient, because the arbitrariness stems partially from the symmetry. Really interesting here, is the number of physical degrees of freedom. We showed already in [13] how it can be computed. Here, we will give an improved version of these results and apply them to the analysis of gauge fixing conditions.

We consider gauge transformation of the form:

$$\begin{aligned} \bar{x}^i &= \Omega^i(x^j)\,, \\ \bar{u}^\alpha &= \Lambda(x^i, u^\beta, \lambda_a^{(0)}(x), \partial\lambda_a^{(1)}(x), \ldots, \partial^p\lambda_a^{(p)}(x)), \end{aligned} \tag{17}$$

where γ_0 gauge functions $\lambda_a^{(0)}$ are entering algebraically, γ_1 gauge functions $\lambda_a^{(1)}$ are entering through their first derivatives etc.

We assume that all derivatives of order l of the functions $\lambda^{(l)}$ are explicitly occurring in (17) and that all gauge functions depend on all independent variables. The only reason for these assumptions is simplicity. The formalism can handle more complicated situations, but the results become rather akward. Most gauge theories satisfy our assumptions anyway.

The symmetry (17) implies that

$$G_q(r) = \sum_{l=0}^{p} \gamma_l \begin{bmatrix} n \\ q+r+l \end{bmatrix} \tag{18}$$

Taylor coefficients of order $q+r$ can be arbitrarily set by a gauge fixing. Thus they must be subtracted from the Hilbert polynomial to obtain the physical relevant number of free coefficients. We introduce the *gauge corrected Hilbert polynomial*

$$\bar{H}_q(r) = H_q(r) - G_q(r)\,. \tag{19}$$

Its coefficient of r^k is given by

$$\bar{h}_k = h_k - \frac{1}{(n-1)!}\sum_{l=0}^{p} \gamma_l s_{n-k-1}^{(n-1)}(q+l)\,, \quad k = 0, \ldots, n-1\,. \tag{20}$$

A closer look at (4) reveals that the coefficients of the Hilbert polynomial and the Cartan characters are in a one-to-one correspondence. Thus we can associate *gauge corrected Cartan characters* with $\bar{H}_q(r)$. They can be expressed in a recursion relation

$$\bar{\alpha}_q^{(k)} = (k-1)!\,\bar{h}_{k-1} - \sum_{i=k+1}^{n} \frac{(k-1)!}{(i-1)!}\,\bar{\alpha}_q^{(k)} s_{i-k}^{(i-1)}(0)\,, \quad k = 1, \ldots, n\,. \tag{21}$$

To now obtain the the number of physical relevant free functions in the general solution of our systems, we simply apply the results of Section 3 to the gauge corrected Cartan characters and Hilbert polynomial, respectively. This yields for instance the following modification to Theorem 1.

THEOREM 3. *If the physical solution space has a representation satisfying the assumptions of Theorem* 1, *then it is spanned by* p_k *arbitrary functions with* k *arguments, where* $p_k = f_k - g_k$ *with*

$$g_n = \sum_{l=0}^{p} \gamma_l ,$$

$$g_k = \frac{(k-1)!}{(n-1)!}\sum_{l=0}^{p} \gamma_l\, s_{n-k}^{(n-1)}(q+l) - \sum_{i=k+1}^{n} \frac{(k-1)!}{(i-1)!}\, g_i\, s_{i-k}^{(i-1)}(q) . \qquad (22)$$

Similiarly, we obtain a correction for the strength of a gauge system

$$\bar{Z}_q^{(1)} = (n-1)\left(\frac{1}{2}n\alpha_q^{(n)} + \alpha_q^{(n-1)} - \sum_{l=0}^{p}\left(\frac{1}{2}n + q + l\right)\gamma_l\right). \qquad (23)$$

Finally, we can easily compute the remaining gauge freedom after some gauge conditions are imposed, if they take the form of differential equations. Then these equations are added to the original system. This yields in general an overdetermined system. It should be invariant only under a proper subgroup of the gauge group. (It is, however, possible that new symmetries arise!)

A necessary condition for a complete gauge fixing is that the Cartan characters of this system are the same as the gauge corrected Cartan characters of the original system (one can, of course, also use the Hilbert polynomial). Note, however, that this is only a necessary criterium, as there may still remain a finite-dimensional symmetry depending on some arbitrary *constants*. This cannot be checked with the Cartan characters, as they are only related with the number of arbitrary functions.

6. Yang–Mills Equations

As the first example, we consider the field equations of a Yang–Mills theory with a d-dimensional gauge group in an n-dimensional spacetime. The dependent variables are the $m = dn$ components A_μ^a of the vector potential. In terms of the field strengths $F_{\mu\nu}^a = \partial_\mu A_\nu^a - \partial_\nu A_\mu^a + C_{bc}^a A_\mu^b A_\nu^c$, where C_{bc}^a are the structure constants of the Lie algebra of the gauge group, the dn field equations are

$$\partial_\mu F_{\mu\nu}^a + C_{bc}^a A_\mu^b F_{\mu\nu}^c = 0 \qquad \begin{cases} a = 1,\dots,d ,\\ \nu = 1,\dots,n . \end{cases} \qquad (24)$$

One can easily check that they are involutive.

To compute the Cartan characters, we must analyse the symbol or the principal part of (24). It is given by $\partial_{\mu\mu}A_\nu^a - \partial_{\mu\nu}A_\mu^a$. Thus, for $\nu \neq n$, we can always take $\partial_{nn}A_\nu^a$ as principal derivative. If, however, $\nu = n$, then this term disappears and we get $\partial_{n,n-1}A_{n-1}^a$ as the principal derivative. This yields $\beta_{\mathrm{YM}}^{(n)} = d(n -$

1), $\beta_{\text{YM}}^{(n-1)} = d$. All other $\beta_{\text{YM}}^{(k)}$ vanish. Thus the Yang–Mills equations do not form a normal system. For the Cartan characters we get

$$\alpha_{\text{YM}}^{(n)} = d, \qquad \alpha_{\text{YM}}^{(n-1)} = d(2n-1), \qquad \alpha_{\text{YM}}^{(n-k)} = dn(k+1)\,. \tag{25}$$

The field equations (24) are invariant under gauge transformations

$$\bar{A}_\mu^a = A_\mu^a + \partial_\mu \lambda^a + C_{bc}^a A_\mu^b \lambda^c\,, \tag{26}$$

i.e. we have $\gamma_1 = d$ gauge functions λ^a. This yields for the gauge corrected strength

$$\bar{Z}_{\text{YM}}^{(1)} = 2(n-1)(n-2)d\,. \tag{27}$$

For simplicity, we restrict ourselves now to $n = 4$. Equation (4) leads to the Hilbert polynomial

$$H_{\text{YM}}(r) = 36d + \tfrac{73}{3}dr + \tfrac{9}{2}dr^2 + \tfrac{1}{6}dr^3\,. \tag{28}$$

Adjusting for the gauge symmetry (26) yields the corrected polynomial

$$\bar{H}_{\text{YM}}(r) = 16d + 12dr + 2dr^2 \tag{29}$$

and the corrected Cartan characters

$$\bar{\alpha}_{\text{YM}}^{(4)} = 0, \qquad \bar{\alpha}_{\text{YM}}^{(3)} = 4d, \quad \bar{\alpha}_{\text{YM}}^{(2)} = \bar{\alpha}_{\text{YM}}^{(1)} = 6d. \tag{30}$$

We see that the dimension d of the gauge group appears everywhere linearly. Thus we get just d-times the values of the Maxwell equations. For them, there also exists a first-order formulation using the field strengths as dependent variables. Since then, only gauge invariant objects are used, the equations no longer exhibit the gauge symmetry (26) and lead directly to the Cartan characters (30) [13].

Finally, we note that the results are independent of the structure of the gauge algebra, as the structure constants do not appear in the symbol. Especially, it makes no difference whether we treat an Abelian or a non-Abelian theory.

To conclude the analysis of the Yang–Mills equations, we consider as an example of a gauge fixing the Lorentz gauge $\partial_\mu A_\mu^a = 0$. Adding this condition to the field equations (24) leads to a simple overdetermined system with the Cartan characters

$$\alpha_{\text{Lor}}^{(4)} = 0\,, \qquad \alpha_{\text{Lor}}^{(3)} = 4d\,, \qquad \alpha_{\text{Lor}}^{(2)} = 12d\,, \qquad \alpha_{\text{Lor}}^{(1)} = 16d\,. \tag{31}$$

A comparison with (30) shows that although the first two characters have the same values as the gauge corrected ones, the lower two are too large. Thus the Lorentz condition does not completely fix the gauge.

7. Einstein's Vacuum Field Equations

As second example, we treat Einstein's vacuum field equations for an n-dimensional spacetime

$$R_{\mu\nu} = 0\,. \tag{32}$$

If we take the components of the metric tensor $g_{\mu\nu}$ as dependent variables, we have because of the symmetries of Ricci tensor and metric, respectively, $n(n+1)/2$ equations for as many dependent variables.

Expanding (32) in a general coordinate system in terms of $g_{\mu\nu}$ leads to very complicated expressions. It is, however, well known, that one can find to any point x_0 of spacetime a coordinate system such that in x_0 the Christoffel symbols vanish and $g_{\mu\nu}(x_0) = \eta_{\mu\nu}$, the constant metric of Minkowski space.

In these locally geodesic coordinates, (32) takes in x_0 the form [14]

$$R_{\mu\nu}(x_0) = \tfrac{1}{2}\eta^{\rho\sigma}\left(\partial_{\mu\rho}g_{\nu\sigma} + \partial_{\nu\sigma}g_{\mu\rho} - \partial_{\mu\nu}g_{\sigma\rho} - \partial_{\sigma\rho}g_{\mu\nu}\right) = 0\,. \tag{33}$$

In this form it is not difficult to show that the Einstein equations are involutive.

Because of the symmetry, we can assume $\mu \geqslant \nu$. We must distinguish four cases in the analysis of the symbol:

(i) $\mu \neq n$ and $\nu \neq n$: the principal derivative is $\partial_{nn}g_{\mu\nu}$.
(ii) $\mu = n$ and $\nu < n-1$: the principal derivative is $\partial_{n,n-1}g_{n-1,\nu}$.
(iii) $\mu = n$ and $\nu = n-1$: the principal derivative is $\partial_{n-1,n-1}g_{11}$.
(iv) $\mu = n$ and $\nu = n$: the principal derivative is $\partial_{n,n-1}g_{11}$.

Besides the last case, this can be seen at once. For (iv) several substitutions of other equations are necessary.

This gives $\beta^{(n)}_{\text{Ein}} = n(n-1)/2$, $\beta^{(n-1)}_{\text{Ein}} = n$. The other $\beta^{(k)}_{\text{Ein}}$ vanish. Thus the Einstein equations do not form a normal system. Their Cartan characters are

$$\alpha^{(n)}_{\text{Ein}} = n\,, \qquad \alpha^{(n-1)}_{\text{Ein}} = n^2\,, \qquad \alpha^{(n-k)}_{\text{Ein}} = \frac{n(n+1)(k+1)}{2}\,. \tag{34}$$

Since x_0 was an arbitrary point, these results remain valid on the entire spacetime.

Equation (32) is invariant under gauge transformations

$$g_{\mu\nu} = \frac{\partial\bar{x}^\rho}{\partial x^\mu}\frac{\partial\bar{x}^\sigma}{\partial x^\nu}\bar{g}_{\rho\sigma}\,, \tag{35}$$

with $\gamma_1 = n$ gauge functions $\bar{x}^\mu$. This yields a gauge corrected strength

$$\bar{Z}^{(1)}_{\text{Ein}} = n(n-1)(n-3)\,. \tag{36}$$

Analogously to our analysis of the Lorentz gauge, we can consider harmonic coordinates, i.e. we pose the gauge condition

$$\Box x^\mu = \frac{1}{\sqrt{-g}}\partial_\nu\left(\sqrt{-g}\,g^{\mu\nu}\right) = 0\,. \tag{37}$$

In locally geodesic coordinates for x_0, this is equivalent to $\partial_\nu g_{\mu\nu}(x_0) = 0$. After prolonging to second order, we get thus n^2 additional equations to (33). One easily sees that always n of them are in the same class. Hence, the Cartan characters of the combined system are

$$\begin{aligned} \alpha_H^{(n)} &= 0, \qquad \alpha_H^{(n-1)} = n(n-1), \\ \alpha_H^{(n-k)} &= \frac{n^2(k+1)+n(k-1)}{2}. \end{aligned} \tag{38}$$

Thus (37) provides no complete gauge fixing, because already $\alpha_H^{(n-1)}$ is too big, as we obtain for the highest gauge corrected Cartan characters

$$\bar{\alpha}_{\text{Ein}}^{(n)} = 0, \qquad \bar{\alpha}_{\text{Ein}}^{(n-1)} = n(n-3). \tag{39}$$

Penney [9] and Mariwalla [5] have already computed the strength of the Maxwell and Einstein equations in arbitrary dimensions using Einstein's approach of counting equations and identities. Penney, however, overlooked the existence of additional identities for the Maxwell equations in higher dimensions and got wrong results. This clearly indicates the advantage of the formal approach. The number of identities is encoded in the Cartan characters. Actually, it is even possible to compute it this way. Pommaret [11] calculated, e.g., with formal methods the number of independent Bianchi identities for a Riemannian manifold of arbitrary dimension.

Acknowledgements

The author wishes to thank J. F. Pommaret for many helpful discussions and P. Winternitz for his hospitality at the CRM in Montréal where most of this work was done. He is supported by Studienstiftung des deutschen Volkes.

References

1. Bluman, G. W. and Kumei, S.: *Symmetries and Differential Equations*, Applied Mathematical Sciences 81, Springer-Verlag, New York, 1989.
2. Einstein, A.: *The Meaning of Relativity*, 5th edn, Princeton University Press, Princeton, 1955.
3. Janet, M.: Sur les systèmes d'équations aux derivées partielles, *J. Math.* **3** (1920), 65–151.
4. John, F.: *Partial Differential Equations*, Applied Mathematical Sciences 1, Springer-Verlag, New York, 1982.
5. Mariwalla, K. H.: Applications of the concept of strength of a system of partial differential equations, *J. Math. Phys.* **15** (1974), 468–473.
6. Meleshko, S. V.: Differential constraints and one-parameter Lie–Bäcklund transformation groups, *Soviet Math. Dokl.* **28** (1983), 37–41.
7. Olver, P. J.: *Applications of Lie Groups to Differential Equations*, Graduate Texts in Mathematics 108, Springer-Verlag, New York, 1985.
8. Olver, P. J. and Rosenau, P.: Group-invariant solutions of differential equations, *SIAM J. Appl. Math.* **47** (1987), 263–278.
9. Penney, R.: On the dimensionality of the real world, *J. Math. Phys.* **6** (1965), 1607–1611.

10. Pommaret, J. F.: *Systems of Partial Differential Equations and Lie Pseudogroups*, Gordon & Breach, London, 1978.
11. Pommaret, J. F.: Explicit calculation of certain differential identities used in mathematical physics, *Proc. Differential Geometry and its Applications, Brno 1986, Math. Appl., East Eur. Ser.* **27** (1987), 271–278.
12. Schü, J., Seiler, W. M., and Calmet, J.: Algorithmic methods for Lie pseudogroups, in N. H. Ibragimov, M. Torrisi and A. Valenti (eds), *Proc. Modern Group Analysis, Advanced Analytical and Computational Methods in Mathematical Physics*, Kluwer, Dordrecht, 1993, pp. 337–344.
13. Seiler, W. M.: On the arbitrariness of the general solution of an involutive partial differential equation, *J. Math. Phys.* **35** (1994), 486–498.
14. Stephani, H.: *Allgemeine Relativitätstheorie*, Deutscher Verlag der Wissenschaften, Berlin, 1980.

Acta Applicandae Mathematicae **41:** 323–339, 1995.

Deformations of Nonassociative Algebras and Integrable Differential Equations

V. V. SOKOLOV and S. I. SVINOLUPOV
Mathematical Institute of Ufa Center of Russian Academy of Sciences, Chernyshevsky str. 112, 450000 Ufa, Russia

(Received: 28 February 1994)

Abstract. A new class of nonassociative algebras related to integrable PDE's and ODE's is introduced. These algebras can be regarded as a noncommutative generalization of Jordan algebras. Their deformations are investigated. Relationships between such algebras and graded Lie algebras are established.

Mathematics Subject Classifications (1991): 16S80, 17A30, 35Q58, 53B05.

Key words: Jordan algebras, left-symmetric algebras, Lie algebras, deformation of nonassociative algebras, generalized chiral field equations.

0. Introduction

A series of investigations devoted to integrable evolution multicomponent PDE's (see [1–3]) and ODE's [4, 5] has shown that there are close relationships between such equations and some special kinds of nonassociative algebras. It turns out that the integrability by the inverse scattering method (**S**-integrability in the terminology of Calogero [6]) gives rise to algebraic structures like Jordan algebras [7], whereas the linearizability (or **C**-integrability) leads to left-symmetric algebras [8].

Recall that a finite-dimensional commutative algebra $\mathfrak{A}$ is said to be Jordan, if the identity

$$W * ((W * W) * V) = (W * W) * (W * V) \tag{0.1}$$

is fulfilled for any $W, V \in \mathfrak{A}$. It is well known that the set of all $m \times m$-matrices becomes a Jordan algebra if we define

$$X * Y = \tfrac{1}{2}(XY + YX). \tag{0.2}$$

An algebra $\mathfrak{A}$ is called left-symmetric if the multiplication $*$ in $\mathfrak{A}$ satisfies the identity

$$\mathrm{As}(X, Y, Z) = \mathrm{As}(Y, X, Z), \tag{0.3}$$

where

$$\mathrm{As}(X, Y, Z) = (X * Y) * Z - X * (Y * Z). \tag{0.4}$$

Note that, in contrast with the definition of the Jordan algebra, it is not assumed that the left-symmetric algebra has to be commutative. Clearly, any associative algebra is the left-symmetric one. It follows from (0.3) that for any left-symmetric algebra the bracket

$$[X,Y] = X * Y - Y * X \tag{0.5}$$

defines the structure of a Lie algebra.

For the important class of examples of nonassociative algebras, the multiplication $*$ is given by the formula

$$X * Y = Q(X,Y)C + L(X)Y + M(Y)X. \tag{0.6}$$

Here X and Y belong to a vector space $\mathfrak{V}$, C is a given vector, Q is a quadratic form, and L and M are linear forms on $\mathfrak{V}$. Thus, $\mathfrak{V}$ equipped with the operation

$$X * Y = -\langle X,Y\rangle C + \langle X,C\rangle Y + \langle Y,C\rangle X, \tag{0.7}$$

where $\langle \cdot,\cdot\rangle$ means the standard scalar product, is a Jordan algebra while the operation

$$X * Y = \langle X,Y\rangle C + \langle X,C\rangle Y \tag{0.8}$$

turns $\mathfrak{V}$ into a left-symmetric algebra.

In the integrability theory, the algebras with the operation (0.6) are related to so-called vector equations (see [9, 10]). For example, the vector KdV equation can be written in terms of (0.7) as

$$\mathbf{u}_t = \mathbf{u}_{xxx} + \mathbf{u} * \mathbf{u}_x,$$

the Burgers vectoral equation has the form

$$\mathbf{u}_t = \mathbf{u}_{xx} + 2\mathbf{u} * \mathbf{u}_x + \mathbf{u} * (\mathbf{u} * \mathbf{u}) - (\mathbf{u} * \mathbf{u}) * \mathbf{u},$$

where $*$ denotes the operation (0.8).

Some classes of multicomponent PDE's can be described in terms of deformations of nonassociative algebras. Consider equations of the form

$$u^i_{xy} = -C^i_{jk}(\mathbf{u})u^j_x u^k_y, \tag{0.9}$$

where $i = 1,\dots,N$ and $\mathbf{u} = (u^1,\dots,u^N)^t$. Here and in the sequel, we assume that summation from 1 to N is carried out over repeated indices.

The class of Equations (0.9) is invariant with respect to an arbitrary point transformation $\mathbf{u} \to \mathbf{F}(\mathbf{u})$. It is easily checked that under such change of coordinates, the functions C^i_{jk} are transformed just as the components of some affine connection Γ. Therefore, an invariant description of integrable cases for (0.9) has to be done in geometrical terms. For instance, the most trivial integrable case is the one which is reduced to a system of independent wave equations by a point-wise transformation. In invariant terms, it is described by the Euclidean affine connection Γ.

In the above example, there exists a preferred coordinate system in which all components of Γ are zeros. It seems that for each 'interesting' system (0.9) one can find a preferred system of coordinates. This coordinate system is singled out by the fact that the functions $C^i_{jk}(\mathbf{u})$ are structural constants of an N-parameter family $\mathfrak{A}(\mathbf{u})$ of nonassociative algebras like Jordan or left-symmetric algebras. Such a family of nonassociative algebras with the structural constants depending on its parameters, we shall consider as a *deformation* of the algebra with structural constants $C^i_{jk}(0)$. It is convenient to represent this family as an algebra $\mathfrak{A}$ in which a multiplication is deformed. Since sometimes we shall deal with different multiplications in the same algebra $\mathfrak{A}$, we shall use a notation $(\mathfrak{A}, *)$ for an algebra $\mathfrak{A}$ with a multiplication $*$.

It is clear that in terms of a multiplication $\circ$ defining by the structural constants $C^i_{jk}(\mathbf{u})$ Equation (0.9) can be written as

$$U_{xy} = -U_x \circ U_y,$$

where $U = U(x, y)$ is a function with the values in $\mathfrak{A}$.

EXAMPLE 1. Let $(\mathfrak{A}, *)$ be a Jordan algebra with the unit e. Consider the equation

$$U_{xy} = (U_x * U^{-1}) * U_y + (U_y * U^{-1}) * U_x - U^{-1} * (U_y * U_x), \tag{0.10}$$

where U^{-1} means the inverse element for U. Recall that the inverse element in a Jordan algebra is a polynomial $p(U)$ such that $U * p(U) = e$. Since all U-polynomials form an associative subalgebra in $\mathfrak{A}$, U^{-1} exists if $\|e - U\|$ is sufficiently small.

According to (0.10), the multiplication $\circ$ is given by the formula

$$-X \circ Y = (X * U^{-1}) * Y + (Y * U^{-1}) * X - U^{-1} * (X * Y). \tag{0.11}$$

It is well known (see, for instance, [11]) that for any Jordan algebra $(\mathfrak{A}, *)$ the algebra $(\mathfrak{A}, \circ)$ is Jordan for any U. It follows from (0.11) that if $U = e$, then the operations $*$ and $\circ$ coincide. Thus, (0.11) defines a N-parameter deformation of the Jordan algebra $\mathfrak{A}$.

If $\mathfrak{A}$ is the set of all matrices with operation (0.2), then (0.11) coincides with the equation of principal chiral field

$$U_{xy} = \tfrac{1}{2}(U_x U^{-1} U_y + U_y U^{-1} U_x). \tag{0.12}$$

For this reason, we will call (0.11) the Jordan chiral field equation.

It is well known that (0.12) possesses a zero-curvature representation of the form

$$\Psi_x = \frac{1}{(1-\lambda)} U_x U^{-1} \Psi, \qquad \Psi_y = \frac{1}{(1+\lambda)} U_y U^{-1} \Psi. \tag{0.13}$$

It is easy to verify that (0.11) admits the following zero-curvature representation

$$\Psi_x = \frac{-2}{(1-\lambda)} L_{U_x} \Psi, \qquad \Psi_y = \frac{-2}{(1+\lambda)} L_{U_y} \Psi. \tag{0.14}$$

Here and below, by L_a we denote the left-multiplication operator mapping x onto $a \circ x$.

Note that (0.14) gives us a zero-curvature representation for (0.12) different from (0.13):

$$\Psi_x = \frac{1}{(1-\lambda)} M\Psi, \qquad \Psi_y = \frac{1}{(1+\lambda)} N\Psi,$$

where Ψ is a matrix and

$$M\Psi = U_x U^{-1} \Psi + \Psi U^{-1} U_x, \qquad N\Psi = U_y U^{-1} \Psi + \Psi U^{-1} U_y.$$

So the deformation (0.11) generates an **S**-integrable Equation (0.9). □

EXAMPLE 2. Let $(\mathfrak{A}, *)$ be a left-symmetric algebra with a left unit e. Let us consider the equation

$$U_{xy} = -R_{e-U}^{-1}(U_x) * U_y. \tag{0.15}$$

Here R_{e-U} is the operator of right multiplication by $e - U$ in $\mathfrak{A}$. The corresponding deformation of multiplication in $\mathfrak{A}$ is given by

$$X \circ Y = R_{e-U}^{-1}(X) * Y. \tag{0.16}$$

It can be checked that the operation (0.16) satisfies (0.3) for any U.

It is not hard to verify that the curvature tensor of the connection Γ associated with (0.15), as well as the curvature tensor of the inverse connection $\overline{\Gamma}$, vanishes. On account of this, Γ-invariant vector fields form a Lie algebra $\mathfrak{G}$, isomorphic to the Lie algebra defined by (0.5). Identifying these vector fields with the left-invariant vector fields on the Lie group G of $\mathfrak{G}$, we can rewrite (0.15) in the form

$$g_{xy} = g_y g^{-1} g_x, \tag{0.17}$$

where $g(x, y)$ is a function with the values in $\mathfrak{G}$. A general solution of (0.17) is of the form $g = g_1(y) g_2(x)$, where g_i are arbitrary $\mathfrak{G}$-valued functions.

Thus, we have a **C**-integrable Equation (0.9) associated with a deformation of left-symmetric algebra. □

In this paper, we present a deformation rule for a wide class of nonassociative algebras. These algebras are defined by the identity

$$[V, X, Y * Z] - [V, X, Y] * Z - Y * [V, X, Z] = 0, \tag{0.18}$$

where

$$[X, Y, Z] = \mathrm{As}(X, Y, Z) - \mathrm{As}(Y, X, Z). \tag{0.19}$$

It interesting to note that all nonassociative algebras naturally arising in connection with integrable systems (Lie algebras, Jordan algebras, left-symmetric algebras, LT-algebras [12]) satisfy our universal identity (0.18).

It has been shown in [5] that the system of ODE's of the form $u^i_t = C^i_{jk} u^j u^k$ with C^i_{jk} being the structural constants of an algebra with identity (0.18) is integrable by a generalization of the factorization method.

The deformations (0.11), (0.16) are special cases of the deformations under consideration. In Section 3, using our deformation rule, we build up integrable equations more general than (0.10). However, the question whether all Equations (0.9) corresponding to considered deformations are integrable, remains open.

1. Covariantly Constant Deformation of Euclidean Connection

In this brief section, a nice geometrical description of the investigated deformations will be given.

Let E be the Euclidean connection on an N-dimensional manifold M, $u = (u^1, \ldots, u^N)$ be the local coordinates on M. Denote by $E^i_{jk}(u)$ the components of E. Let us consider a connection Γ with the components

$$\Gamma^i_{jk} = E^i_{jk} + C^i_{jk}, \tag{1.1}$$

where $C^i_{jk}(u)$ are components of a tensor field C on M.

DEFINITION. The connection Γ is called a *covariantly constant deformation of the Euclidean connection* if the deformation tensor C is covariantly constant with respect to Γ.

It follows from the standard formulas for recalculating the curvature and torsion under the deformation of a connection (see, for example, [13]) that both the curvature tensor of Γ and the torsion can be expressed in terms of the deformation tensor C only. The formulas are of the form

$$T^i_{jk} = C^i_{jk} - C^i_{kj}, \tag{1.2}$$

$$R^i_{mjk} = C^i_{rm} C^r_{jk} - C^i_{rm} C^r_{kj} + C^i_{kr} C^r_{jm} - C^i_{jr} C^r_{km}. \tag{1.3}$$

Since the tensor C is covariantly constant, then M is of a space of covariantly constant curvature and torsion.

Rewriting in terms of the Euclidean connection E, the fact that C is a covariantly constant tensor, we obtain

$$\nabla_m (C^i_{jk}) = C^i_{rk} C^r_{mj} + C^i_{jr} C^r_{mk} - C^i_{mr} C^r_{jk}. \tag{1.4}$$

Here we denote by ∇_m the covariant u^m-derivative in E.

The relations (1.4) will be regarded as an overdetermined system of first-order PDE's with respect to unknown functions $C^i_{jk}(u)$. We intend to investigate the compatibility conditions for (1.4). Of course, they are independent of the choice of coordinate system. For calculation, it is natural to use the coordinate system in which all components of E are identically zero. In this, preferred local coordinates (1.4) takes the form

$$\frac{\partial C^i_{jk}}{\partial u^m} = C^i_{rk}C^r_{mj} + C^i_{jr}C^r_{mk} - C^i_{mr}C^r_{jk}. \tag{1.5}$$

Before we go on to treat (1.5), let us alter our point of view and notations from geometric to algebraic.

2. Compatibility Conditions for Deformation Equation

Let $\mathfrak{V}$ be a vector space with a basis $e_1, \ldots, e_N$. The tensor $C(u)$ gives rise to the N-parameter family of multiplications on $\mathfrak{V}$. Namely, the products of basis vectors are defined by the formula

$$e_j \circ e_k = C^i_{jk}(u)e_i. \tag{2.1}$$

In terms of (2.1), the deformation Equation (1.5) takes the form

$$\partial_X(Y \circ Z) = (X \circ Y) \circ Z + Y \circ (X \circ Z) - X \circ (Y \circ Z). \tag{2.2}$$

Here and below, for any $X = x^i e_i$ we denote by ∂_X a vector field $\sum x^i \partial/\partial u^i$.

Note that (1.2) and (1.3) can be rewritten in the following compact form

$$T(X,Y) = X \circ Y - Y \circ X,$$

$$R(X,Y,Z) = [Y,Z,X].$$

Here a notation of the type $T(X,Y)$ means the value of a tensor T on vectors X and Y. Recall that the operation $[\cdot,\cdot,\cdot]$ is defined by (0.19), (0.4). Of course, we have to change $*$ by $\circ$ everywhere.

THEOREM 1.

(1) *Let* $C^i_{jk}(u^1, \ldots, u^N)$ *be a solution of* (1.5). *Then at any point* $u = (u^1, \ldots, u^N)$, *the multiplication* (2.1) *satisfies the identity* (0.18).

(2) *Let to be an algebra with structural constants* a^i_{jk}, *whose members satisfy the identity* (0.18). *Then a solution of* (1.5) *with initial data*

$$C^i_{jk}(0) = a^i_{jk} \tag{2.3}$$

exists (for sufficiently small u*) and is unique.*

Proof. Let us consider the following linear operators act on $\mathfrak{V}$. For any $X, Y, Z \in \mathfrak{V}$ set

$$L_X(Z) = X \circ Z, \tag{2.4}$$

$$K_{X,Y}(Z) = [X, Y, Z]. \tag{2.5}$$

It follows from (0.19) that

$$K_{X,Y} = L_{X \circ Y - Y \circ X} + [L_Y, L_X]. \tag{2.6}$$

According to (2.1), (2.4), the elements of matrix L_X with $X = x^i e_i$ are of the form $(L_X)^i_j = x^k C^i_{kj}$. Using this formula, one can easily verify that (1.5) is equivalent to the following equation

$$\partial_X(L_Y) = K_{X,Y} + L_{Y \circ X}. \tag{2.7}$$

To prove the first part of the theorem, it suffices to calculate the expression

$$\partial_V(\partial_X(L_Y)) - \partial_X(\partial_V(L_Y)),$$

which must be zero if the system (2.7) is compatible. Of course, the ∂_V and ∂_X derivatives have to be computed by virtue of (2.7). To apply ∂_V to the r.h.s. of (2.7), we need the following formula for differentiating of the operator K:

$$\partial_V(K_{X,Y}) = K_{V \circ X, Y} - K_{V \circ Y, X} + L_{[X,Y,V]}. \tag{2.8}$$

This can be easily deduced from (2.6), (2.7). As the result of computations, we obtain the following relation

$$L_{[V,X,Y]} + [L_Y, K_{V,X}] = 0. \tag{2.9}$$

To convince that (2.9) is equivalent to (0.18), one can apply the left-hand side of (2.9) to Z.

The uniqueness of the solution for the initial problem (1.5), (2.3) is obvious. To prove the existence, we have to verify that an equality obtained by applying ∂_Z to both sides of (2.9) is fulfilled identically with respect to X, Y, Z, V by virtue of (2.7)–(2.9). This can be performed in a straightforward way with the help of the following formula

$$[K_{V,X}, K_{Z,Y}] = K_{X,[Z,Y,V]} - K_{V,[Z,Y,X]}, \tag{2.10}$$

that follows from (2.6), (2.9). □

3. Invariant Subclasses of Nonassociative Algebras

It turns out that the main algebraic properties of the initial algebra $\mathfrak{A}_0$ are preserved under the deformation (1.5), (2.3).

PROPOSITION 1.

(1) *Let I_0 be a left (two-sided) ideal in $\mathfrak{A}_0$, $I(u)$ be the deformation of I_0. Then $I(u)$ is a left (two-sided) ideal in $\mathfrak{A}(u)$;*
(2) *Let e be a left (two-sided) unit of $\mathfrak{A}_0$. Then $e(u) = e - \sum u^i e_i$ is a left (two-sided) unit for $\mathfrak{A}(u)$.*

Proof. Let $e_1, \ldots, e_K$ be a basis of I_0, the elements $e_{K+1}, \ldots, e_N$ complement this basis to a basis of $\mathfrak{A}_0$. Then the structural constants a^i_{jk} of $\mathfrak{A}_0$ satisfy the following conditions: $a^i_{jk} = 0$ if $i > K$, $k \leqslant K$. Put

$$C^i_{jk}(u) = 0 \quad \text{if} \quad i > K, \; k \leqslant K. \tag{3.1}$$

It follows from (1.5) that the other structural constants satisfy the system describing the deformation of I_0. Since the identity (0.18) is valid for I_0, this system is compatible in view of Theorem 1. We have, therefore, constructed the solution of (1.5), (2.3) such that the condition (3.1) is fulfilled. It means that $I(u)$ is a left ideal in $\mathfrak{A}(u)$ for any u. The case of a two-sided ideal can be considered in a similar way.

Proof of item (2). To see that $X \circ e(u) = X$, where $\circ$ is defined by (2.1), it suffices to show that the expression $X \circ e(u)$ does not depend on u. Using (2.2), it is easy to verify that $\partial_Y(X \circ e(u)) = 0$ for any Y. In so doing, it should be kept in mind that the second factor, as distinct from (2.2), depends on u. □

According to Theorem 1, (0.18) must be fulfilled for the initial algebra $\mathfrak{A}_0$. We now present examples of algebras with identity (0.18).

THEOREM 2. (a) *The class of algebras with identity* (0.18) *contains:*

(1) *Associative algebras;*
(2) *Left-symmetric algebras;*
(3) *Lie algebras;*
(4) *Jordan algebras;*
(5) *LT-algebras. Any commutative algebra with identity* (0.18) *is an LT-algebra.*

(b) *The deformation* (1.5) *leaves all these classes of algebras invariant.*

Proof. It follows from (0.19) that $[\cdot, \cdot, \cdot]$ is null operation for any left-symmetric and all the more for any associative algebra. Therefore, (0.18) holds for them.

To prove item (3), we show that

$$[X, Y, Z] = (X * Y) * Z \tag{3.2}$$

for any Lie algebra. According to (0.19), we have

$$[X,Y,Z] = 2(X*Y)*Z + (Y*Z)*X + (Z*X)*Y \tag{3.3}$$

for an algebra with the identity $X*Y = -Y*X$. In view of the Jacobi identity

$$(X*Y)*Z + (Y*Z)*X + (Z*X)*Y = 0,$$

(3.3) coincides with (3.2).

Using (3.2) and the anti-commutativity of $*$, we rewrite (0.18) in the following form

$$(Z*Y)*(V*X) + (Y*(V*X))*Z + ((V*X)*Y)*Z = 0.$$

It is nothing else but the Jacobi identity for Z, Y, $V*X$.

Proof of assertions (4) *and* (5). First of all, we recall that a commutativealgebra is said to be an LT-algebra if the following identity holds

$$\begin{aligned} &2W*(W*(W*V)) + V*(W*(W*W))- \\ &\quad -3W*(V*(W*W)) = 0. \end{aligned} \tag{3.4}$$

It is known that any Jordan algebra is an LT algebra. To prove this, we make the total linearization of (0.1). Namely, substituting

$$W = \lambda X + \mu Y + \nu Z$$

in (0.1) and equating the coefficient of $\lambda\mu\nu$ to zero, we obtain

$$\begin{aligned} &X*(V*(Y*Z)) + Y*(V*(X*Z)) + Z*(V*(X*Y))- \\ &\quad -(X*V)*(Y*Z) - (Y*V)*(X*Z) - (X*Y)*(V*Z) = 0. \end{aligned} \tag{3.5}$$

Putting $X = W$, $Y = W$, $V = W$, $Z = V$, we get (3.4). Thus, (5) implies (4).

In order to prove item (5), we make the total linearization of (3.4). As a result, we obtain

$$\begin{aligned} &Z*(X*(Y*V)) + Z*(Y*(X*V)) + X*(Y*(Z*V))+ \\ &\quad +X*(Z*(Y*V)) + Y*(Z*(X*V)) + Y*(X*(Z*V)) + \\ &\quad +V*(Z*(X*Y)) + V*(X*(Z*Y)) + V*(Y*(X*Z)) - \\ &\quad -3Z*(V*(X*Y)) - 3X*(V*(Z*Y)) - \\ &\quad -3Y*(V*(X*Z)) = 0. \end{aligned} \tag{3.6}$$

Permuting X and V in (3.6) and subtracting the thus-obtained identity from (3.6), we get

$$\begin{aligned} &Z*(X*(Y*V)) + V*(X*(Y*Z)) + Y*(X*(Z*V))- \\ &\quad -Z*(V*(X*Y)) - X*(V*(Z*Y)) - Y*(V*(X*Z)) = 0. \end{aligned} \tag{3.7}$$

It is easy to verify that (3.7) becomes (3.4) if we put $X = V$, $Y = W$, $V = W$, $Z = W$. Thus, (3.7) is equivalent to (3.4).

It follows from (0.19) that, for a commutative algebra, the following equality is valid

$$[X, Y, Z] = \mathrm{As}(X, Z, Y).$$

Using this formula, we rewrite (3.7) in the form

$$Z * [V, X, Y] + Y * [V, X, Z] - [V, X, Z * Y] = 0.$$

It is clear that for a commutative algebra, the last identity coincides with (0.18). This concludes the proof of Part (a).

Now we show that the deformation (0.18) leaves the class of associative algebras invariant. This class can be regarded as a submanifold As of the algebraic manifold of all algebras with identity (0.18). It suffices to make sure that any vector field ∂_X defined by (2.2) is tangent to As at each point of As. To this end, we have to check that $\partial_X(\mathrm{As}(Y, Z, V)) = 0$ for an associative algebra. It follows from the formula

$$\begin{aligned}
&\partial_X(\mathrm{As}(Y, Z, V)) \\
&\quad = \mathrm{As}(V, X \circ Y, Z) + \mathrm{As}(X, V \circ Y, Z) + \mathrm{As}(X, Y, V \circ Z) - \\
&\quad\quad -\mathrm{As}(V, X, Y \circ Z) + \mathrm{As}(V, X, Y) \circ Z,
\end{aligned}$$

which can be derived immediately from (2.2).

The others statements of Part (b) can be proved in much the same way. Note only that the right-hand side of (2.2) vanishes for Lie algebras. Thus, any Lie algebra is a stationary point of the deformation. □

It is not hard to classify the algebras with identity (0.18) of the dimension two. We present the result of classification.

PROPOSITION 2. *Any two-dimensional algebra with identity* (0.18) *can be transformed by a change of basis to one of the following:*

$$e_1 * e_1 = e_1, \quad e_1 * e_2 = (1 - t)e_1 + te_2,$$

$$e_2 * e_1 = te_1 + (1 - t)e_2, \quad e_2 * e_2 = 0; \tag{3.8}$$

$$e_1 * e_1 = e_1, \quad e_1 * e_2 = te_2, \quad e_2 * e_1 = 0, \quad e_2 * e_2 = 0; \tag{3.9}$$

$$e_1 * e_1 = e_1, \quad e_1 * e_2 = te_2, \quad e_2 * e_1 = e_2, \quad e_2 * e_2 = 0; \tag{3.10}$$

$$e_1 * e_1 = 0, \quad e_1 * e_2 = te_1, \quad e_2 * e_1 = e_1, \quad e_2 * e_2 = e_1 + (t + 1)e_2; \tag{3.11}$$

$$e_1 * e_1 = 0, \quad e_1 * e_2 = te_1, \quad e_2 * e_1 = -te_1, \quad e_2 * e_2 = 0, \quad t = 1, 0; \tag{3.12}$$

$$e_1 * e_1 = e_1, \quad e_1 * e_2 = e_1 + 2e_2, \quad e_2 * e_1 = 2e_1 + e_2, \quad e_2 * e_2 = e_2; \tag{3.13}$$

$$e_1 * e_1 = 0, \quad e_1 * e_2 = e_2, \quad e_2 * e_1 = 0, \quad e_2 * e_2 = 0; \tag{3.14}$$

$$e_1 * e_1 = e_1, \quad e_1 * e_2 = 0, \quad e_2 * e_1 = 0, \quad e_2 * e_2 = e_2. \tag{3.15}$$

Commentary. The algebras (3.8)–(3.10) are associative if $t = 1$ or if $t = 0$; (3.8) is a Jordan algebra if $t = 1/2$; (3.11) is a left-symmetric algebra for $t = 0$ and an LT-algebra for $t = 1$. Formula (3.12) describes two-dimensional Lie algebras. The algebras (3.13), (3.14) are (nonassociative) left-symmetric; (3.15) is a commutative associative algebra.

Remark 1. An in-depth study of properties of algebras with identity (0.18) is a subject for a separate paper. To investigate such algebras, one can use the tools of Lie algebras theory (see Section 5). Here we would like only to note the following important fact. The result of the formal joining of a two-sided unit to any algebra with identity (0.18) is again an algebra with identity (0.18). In particular, the joining of the unit to a Lie algebra leads to an algebra with identity (0.18).

Remark 2. It is easy to verify that (0.14) is a zero-curvature representation, not only for (0.10) but also for any system (0.9), where functions $C^i_{jk}(u)$ satisfy (1.5) and $C^i_{jk} = C^i_{kj}$. Thus, according to item (5) of Theorem 2, any LT-algebra gives us the integrable system of a chiral model type.

4. Explicit Formulas for Solutions of Deformation Equation

The deformation Equation (1.5) can be solved explicitly for some special kinds of algebras with identity (0.18).

PROPOSITION 3 (cf. Example 1). *Let* $(\mathfrak{A}, *)$ *be a Jordan algebra with structural constants* a^i_{jk}*, possessing a unit e. Let* $C^i_{jk}(u)$ *be a solution of* (1.5), (2.3). *Then the multiplication* (2.1) *is given by the formula*

$$\begin{aligned} X \circ Y \;=\; & -(e-U)^{-1} * (X * Y) + (X * (e-U)^{-1}) * Y + \\ & +(Y * (e-U)^{-1}) * X. \end{aligned} \tag{4.1}$$

Proof. Let $\phi(U)$ be a solution of the following system

$$\partial_{u_k}(\phi) = -\{\phi,\ e_k,\ \phi\}, \quad k = 1, \ldots, N. \tag{4.2}$$

Here and below, $\{X, Y, Z\}$ means the triple Jordan product

$$\{X, Y, Z\} = X * (Y * Z) + Z * (Y * X) - Y * (X * Z). \tag{4.3}$$

Then the multiplication

$$X \circ Y = -\{X, \phi, Y\} \tag{4.4}$$

satisfies the deformation Equation (2.2). To verify this, we have to use the following well-known (see [14]) identity

$$\{Y, \{W, X, V\}, Z\} - \{\{X, W, Y\}, V, Z\} - \{\{X, W, Z\}, V, Y\}+ \\ +\{X, W, \{Y, V, Z\}\} = 0, \tag{4.5}$$

which is valid for any Jordan algebra. It is easy to check that the result of substituting (4.4) into (2.2) coincides with (4.5), where $\phi(U)$ stands for V and W.

To prove (4.1), it suffices to verify that for any $c \in \mathfrak{A}$ the element $\phi = (U-c)^{-1}$ satisfies (4.2). Without loss of generality, we may assume that $c = 0$.

Recall that

$$U^{-1} = P_U^{-1}(U), \tag{4.6}$$

where for any $Y \in \mathfrak{A}$ the operator $P_Y : \mathfrak{A} \to \mathfrak{A}$ is defined by the following formula

$$P_Y(X) = \{Y, X, Y\}. \tag{4.7}$$

It is well known that

$$P_U P_{U^{-1}} = \mathrm{Id}, \tag{4.8}$$

$$[L_U,\ L_{U^{-1}}] = 0. \tag{4.9}$$

Using (4.6), one can check that U^{-1} satisfies (4.2) iff

$$e_k - 2\{U, U^{-1}, e_k\} + P_U(P_{U^{-1}}(e_k)) = 0.$$

In view (4.8), (4.9), this identity holds. □

EXAMPLE 3. Let us consider the 'vector' Jordan algebra defined by (0.7). If $\langle C, C\rangle \neq 0$, then this Jordan algebra has the unit $e = C/\langle C, C\rangle$. Any element X such that $\langle X, X\rangle \neq 0$ is invertible. The inverse element for X is given by

$$X^{-1} = \frac{2\langle C, X\rangle C - \langle C, C\rangle X}{\langle X, X\rangle\langle C, C\rangle^2}. \tag{4.10}$$

According to (0.7), (4.1), (4.10), the solution of the deformation Equation (1.5) is of the form

$$X \circ Y = -\langle X, Y\rangle a(U) + \langle X, a(U)\rangle Y + \langle Y, a(U)\rangle X,$$

where

$$a(U) = \frac{\langle C, C\rangle (C - \langle C, C\rangle U)}{\langle C - \langle C, C\rangle U,\ C - \langle C, C\rangle U\rangle}. \tag{4.11}$$

Calculating the limit of (4.11) as $\langle C, C\rangle$ tends to zero we get the following formula

$$X \circ Y = \frac{1}{1 - 2\langle U, C\rangle}(-\langle X, Y\rangle C + \langle X, C\rangle Y + \langle Y, C\rangle X),$$

describing the deformed multiplication (2.1) in the case $\langle C, C\rangle = 0$. □

The following statement can be proved straightforwardly.

PROPOSITION 4 (cf. Example 2). *Let* $(\mathfrak{A}, *)$ *be a left-symmetric algebra with a left unit e. Then the multiplication* (2.1) *is given by the formula* (0.16).

EXAMPLE 4. The 'vector' left-symmetric algebra (0.8) does not have a left unit. Therefore, we cannot use formula (0.16). Consider a generic case $\langle C, C\rangle \neq 0$. Using the group of orthogonal transformations, we can bring C to the following form: $C = (1, 0, \ldots, 0)^t$. For such a choice of C, the multiplication (2.1) is given by the following formula:

$$X \circ Y = \frac{1}{z}(\langle X, P\rangle Y + \langle X, (P \otimes P)(Y) + zA(Y)\rangle\ C). \tag{4.12}$$

Here

$$P = (1, u^2, \ldots, u^N)^t, \qquad z = 2 - 2u^1 - \langle P, P\rangle,$$

$$A = \mathrm{diag}(0, 1, 1, \ldots, 1).$$

The multiplication (4.12) generates (see Example 2) a linearizable system (0.9). In the case $N = 2$, the explicit form of (0.9) is given by

$$\begin{aligned} u_{xy} &= \frac{2u_x u_y + v v_y u_x + 2v v_x u_y + (1 - 2u) v_x v_y}{v^2 + 2u - 1}, \\ v_{xy} &= \frac{v_y u_x + v v_x v_y}{v^2 + 2u - 1}. \end{aligned}$$

5. Relationships Between Nonassociative and Lie Algebras

In this section, we show that algebras with identity (0.18) are intimately related to graded Lie algebras.

Let $\mathfrak{A}$ be an algebra with identity (0.18). Denote by $\mathrm{Der}(\mathfrak{A})$ the Lie algebra of all differentiations of $\mathfrak{A}$. Identity (0.18) means that for any $X, Y \in \mathfrak{A}$ operator

$K_{X,Y}$ defined by (2.6) belongs to Der($\mathfrak{A}$). It follows from (2.10) that the vector space generated by operators $K_{X,Y}$ is a subalgebra in Der($\mathfrak{A}$).

Denote by $L(\mathfrak{A})$ the vector space of all left-multiplication operators of $\mathfrak{A}$. Consider a vector space $\mathrm{Strl}(\mathfrak{A}) = L(\mathfrak{A}) \oplus \mathrm{Der}(\mathfrak{A})$. We claim that it becomes a Lie algebra if we define a bracket by

$$[(L_X, a),\ (L_Y, b)] = (L_{X*Y-Y*X+a(Y)-b(X)},\ [a,b] - K_{X,Y}), \tag{5.1}$$

where $a, b \in \mathrm{Der}(\mathfrak{A})$, $X, Y \in \mathfrak{A}$. In fact, the anti-commutativity of (5.1) is clear. The Jacobi identity follows from (2.9). It follows from (5.1) that Strl($\mathfrak{A}$) is a $\mathbb{Z}_2$-graded Lie algebra for any commutative algebra $\mathfrak{A}$.

For Jordan algebras, the algebra Strl($\mathfrak{A}$) thus constructed coincides with the so-called structural Lie algebra (see [15]). It is well known that for any semi-simple Jordan algebra $\mathfrak{A}$, the algebra Strl($\mathfrak{A}$) is a trivial center extension of a semi-simple Lie algebra.

In the case of algebras with identity (0.18), we do not know any generalization of the notion of $\mathbb{Z}_3$-graded superstructural Lie algebra $K(\mathfrak{A}) = \mathfrak{A} \oplus \mathrm{Strl}(\mathfrak{A}) \oplus \mathfrak{A}$ (see [15]) related to any Jordan algebra. But we can build up a $\mathbb{Z}_2$- graded algebra associated with $\mathfrak{A}$ in the following way. It turns out that for any algebra with identity (0.18), a vector space $\mathfrak{G}(\mathfrak{A}) = \mathfrak{A} \oplus \mathrm{Strl}(\mathfrak{A})$ constitutes Lie algebra under the bracket operation

$$\begin{aligned}
&[(x, L_X, a),\ (y, L_Y, b)] \\
&\quad = (Y * x - X * y + a(y) - b(x), \\
&\qquad L_{X*Y-Y*X+a(Y)-b(X)},\ [a,b] - K_{X,Y}),
\end{aligned} \tag{5.2}$$

where $x, y, X, Y \in \mathfrak{A}$, $a, b \in \mathrm{Der}(\mathfrak{A})$.

It is not hard to show that (5.2) implies that:

(a) The vector space $\mathfrak{g}$ generated by all the elements of the form $(0,\ 0,\ a)$ is a Lie subalgebra in $\mathfrak{G}$.
(b) The vector space $N = \{(X,\ -L_X,\ 0)\}$ is a $\mathfrak{g}$-module.
(c) The vector space $\mathfrak{G}_- = \mathfrak{g} \oplus N$ is a Lie subalgebra in $\mathfrak{G}$.

Conversely, the following statement is true (see [5]).

PROPOSITION 5. *Let $\mathfrak{G} = \mathfrak{G}_+ \oplus M$ be a $\mathbb{Z}_2$-graded Lie algebra, such that $[M,\ M] = 0$, and $\mathfrak{G} = \mathfrak{G}_+ + \mathfrak{G}_-$, where $\mathfrak{G}_-$ is a Lie subalgebra in $\mathfrak{G}$. Suppose $\mathfrak{G}_- = \mathfrak{g} \oplus N$, where $\mathfrak{g} = \mathfrak{G}_+ \cap \mathfrak{G}_-$ and N is a $\mathfrak{g}$-module. Let us equip M with a structure of (nonassociative) algebra with the help of the formula $X * Y = [X_+,\ Y]$, where X_+ denotes the projection of X onto $\mathfrak{G}_+$, corresponding the splitting $\mathfrak{G} = \mathfrak{G}_+ \oplus N$. Then the operation $*$ thus defined satisfies the identity* (0.18).

6. Solutions of the Deformation Equation Described by a Deformation of Basis

In this section, a brief discussion is presented of the question of *whether the members of N-parameter family $\mathfrak{A}(u)$ of algebras, defined by* (2.1), *are isomorphic to the initial algebra* $\mathfrak{A}_0$ (see (2.3)) *for sufficiently small u.*

It seems to be true for any algebra with identity (0.18), but we can prove this only for the left-symmetric algebras and for the Jordan algebras with unit.

DEFINITION. The deformation $\mathfrak{A}(u)$ defined by (1.5), (2.3) is called *basic* if there exists a basis $\tilde{e}_i(u)$ for $\mathfrak{A}(u)$ such that all the functions $C^i_{jk}(u)$ do not depend on u.

According to the definition of a basic deformation, one can find an $N \times N$-matrix $J(u)$ such that the multiplication (2.1) is of the form

$$X \circ Y = J^{-1}(J\,X * J\,Y). \tag{6.1}$$

In principle, the operator J is not uniquely determined by (6.1). It is clear that if both J_1 and J_2 satisfy (6.1), then $J_1^{-1}J_2$ is an N-parameter family of automorphisms for $\mathfrak{A}_0$.

THEOREM 3. *Let $\mathfrak{A}$ be an algebra with identity* (0.18). *Assume that there exists a Lie subalgebra $\mathfrak{S}_- \subset$ Strl$(\mathfrak{A})$ such that* Strl$(\mathfrak{A}) = \mathfrak{S}_+ \oplus \mathfrak{S}_-$, *where* $\mathfrak{S}_+ = \{(0,\ a)\}^*$. *Then the deformation* (1.5), (2.3) *of $\mathfrak{A}$ is basic.*

Proof. According to the assumption of Theorem 3, any element of Strl$(\mathfrak{A})$ can be uniquely represented in the form

$$(L_X,\ a) = (0,\ a - \lambda_X) + (L_X, \lambda_X),$$

where $(L_X, \lambda_X) \in \mathfrak{S}_-$. Denote by M_X the operator $L_X - \lambda_X$. It is easy to verify that the fact that $\mathfrak{S}_-$ is a Lie subalgebra is equivalent to the following identity

$$[M_X,\ M_Y] = M_{M_X(Y) - M_Y(X)}. \tag{6.2}$$

Define $J(u)$ as the solution of the initial problem

$$\frac{\partial J}{\partial u^k} = M_{J(e_k)}J, \qquad J(0) = E. \tag{6.3}$$

The compatibility condition for the overdetermined system (6.3) coincides with (6.2). □

COROLLARY. *The deformation of any left-symmetric algebra $\mathfrak{A}_0$ is a basic one.*

---

* Note that the problem of a decomposition of a $\mathbb{Z}_2$-graded Lie algebra into a direct sum of subalgebras is closely related (see [16, 17]) to diverse kinds of the Yang–Baxter equation.

Proof. Let $\mathfrak{S}_-$ be the set $\{(L_X,\ 0)\}$. Since (6.2) with $M_X = L_X$ is an operator form of the definition of left-symmetric algebra (0.3), $\mathfrak{S}_-$ is a Lie algebra. □

Let the conditions of Theorem 3 be satisfied. Then, in accordance with (6.2), the algebra $\mathfrak{L}$ whose operators of left multiplication are M_X is a left-symmetric one. Thus, we have obtained a way to associate a left-symmetric algebra with an algebra satisfying (0.18).

EXAMPLE 5. Let $\mathfrak{A}$ be the vector Jordan algebra from Example 3, e be the unit of $\mathfrak{A}$. Let us denote by Y a vector of unit length orthogonal to e. We can assume, without restriction of generality (see Example 4), that $e = e_1 = (1,0,\ldots,0)^t$ and $Y = e_2 = (0,1,0,\ldots,0)^t$. Put $\lambda_X = [L_Y,\ L_X]$ for any $X \in \mathfrak{A}$. It is easily verified that $\mathfrak{S}_- = \{(L_X, \lambda_X)\}$ is a Lie subalgebra in $\mathrm{Strl}(\mathfrak{A})$. The multiplication table for the associated left-symmetric algebra is of the form

$$\begin{aligned} &e_1 \bullet e_i = e_i \bullet e_1 = e_i; \qquad && e_2 \bullet e_2 = e_1; \qquad && e_2 \bullet e_k = 0; \\ &e_k \bullet e_2 = e_k; && e_k \bullet e_m = 0; && e_k \bullet e_k = e_1 - e_2, \end{aligned}$$

where $k,\ m = 3,4,\ldots,\ k \neq m$.

PROPOSITION 6. *The deformation of any Jordan algebra $\mathfrak{A}$ with unit is a basic one.*

Proof. We shall seek an operator $J(u)$ in the form $J = P_\theta$, where P is defined by (4.7), $\theta(u)$ is an element of $\mathfrak{A}$. Then (6.1) takes the form

$$X \circ Y = P_\theta^{-1}(P_\theta\ X * P_\theta\ Y). \tag{6.4}$$

Using the identity

$$\{Z,Y,Z\} * \{Z,X,Z\} = \{Z,\ \{X,\ Z * Z,\ Y\},\ Z\}$$

(see [11]), we rewrite (6.4) as

$$X \circ Y = \{X,\ \theta * \theta,\ Y\}. \tag{6.5}$$

Compare (6.5) with (4.4). We see that if $\theta^2 = -\phi$, $\theta(0) = e$, where ϕ is defined by (4.2) and the initial data $\phi(0) = -e$, then (6.4) is valid. The existence of such an element $\theta(u)$ is obvious for small u. □

Note that the constructions from Example 5 and Proposition 6 lead to different operators $J(u)$ for the vector Jordan algebra.

Acknowledgements

The authors wish to thank I. Z. Golubchik and I. S. Krasil'shchik for many useful discussions and collaboration. The first author also gratefully acknowledges the financial support of the International Erwin Schrödinger Institute for

Mathematical Physics. The second author was supported in part by an ISF Grant MLY000.

References

1. Svinolupov, S. I.: On the analogues of the Burgers equations, *Phys. Lett. A* **135**(1) (1989), 32–36.
2. Svinolupov, S. I.: Jordan algebras and generalized Korteweg–de Vries equations, *Teoret. Mat. Phys.* **4** (1991), 46–58 (in Russian).
3. Svinolupov, S. I.: Generalized Schrodinger equations and Jordan pairs, *Comm. Math. Phys.* **143** (1992), 559–575.
4. Svinolupov, S. I. and Sokolov, V. V.: Jordan tops and generalization of Lie theorem, *Mat. Zametki* **3** (1993), 115–121 (in Russian).
5. Golubchik, I. Z., Sokolov, V. V., and Svinolupov, S. I.: New class of nonassociative algebras and a generalized factorization method, Preprint ESI, Vienna 53, 1993.
6. Calogero, F.: Why are certain nonlinear PDE's both widely applicable and integrable? in V. E. Zakharov (ed.), *What is Integrability*? Springer-Verlag, Berlin, 1991, pp. 1–62.
7. Jacobson, N.: *Structure and Representations of Jordan Algebras*, Amer. Math. Soc. Colloq. Publ. 30, Amer. Math. Soc., Providence R.I., 1968.
8. Medina, A.: Sur quelques algèbres symetriques a gauche dont l'algèbre de Lie sous-jacente est resoluble, *C. R. Acad. Sci. Paris. Ser. A.* **286**(3) (1978), 173–176.
9. Athorn, C. and Fordy, A. P.: Generalized KdV and MKdV equations associated with symmetric spaces, *J. Phys. A* **20** (1987), 1377–1386.
10. Svinolupov, S. I. and Sokolov, V. V.: Vector generalizations of classical integrable equations, *Teoret. Mat. Fiz.* **100**(2) (1994), 214–218 (in Russian).
11. Zhevlakov, K. A., Slin'ko, A. M., Shestakov, I. P., and Shirshov, A.I.: *Rings Similar to Associative Ones*, Nauka, Moscow, 1974 (in Russian).
12. Osborn, J. M.: Commutative algebras satisfying an identity of degree four, *Proc. Amer. Math. Soc.* **16**(4) (1965), 1114–1120.
13. Helgason, S.: *Differential Geometry, Lie Groups, and Symmetric Spaces*, Academic Press, New York, 1978.
14. Meyberg, K.: Jordan-Tripelsysteme und die Koecher-Konstruktion von Lie-Algebren, *Math. Zeit.* **115**(1) (1970), 58–78.
15. Koecher, M.: Imbedding of Jordan algebras into Lie algebras, 1, 2, *Amer. J. Math.* **89, 90** (1967), 787–816, (1968) 476–510.
16. Drinfel'd, V. G.: Hamiltonian structure on the Lie groups and a geometric sense of classical Yang–Baxter equation, *Dokl. Acad. Nauk SSSR* **268**(2) (1983), 285–287 (in Russian).
17. Semenov-Tian-Shansky, M. A.: What is classical r-matrix? *Funkt. Anal. Prilozh.* **17**(4) (1983), 17–33 (in Russian).

Acta Applicandae Mathematicae **41:** 341–348, 1995.

Constraints of the KP Hierarchy and the Bilinear Method

YI CHENG, YOU-JIN ZHANG
Department of Mathematics, University of Science and Technology of China, Hefei, Anhui 230026, People's Republic of China

and

WALTER STRAMPP
Fachbereich 17–Mathematik/ Informatik, Universität–GH Kassel, Holländische Str. 36, 34109 Kassel, Germany

(Received: 28 February 1994)

Abstract. We consider generalized k-constraints of the KP hierarchy where the Lax operator L is forced to satisfy $L^k_- = q\partial^{-1}r$. We study the effect of those constraints on the bilinear equations.

Mathematics Subject Classifications (1991): 58F07, 35Q58, 15A63.

Key words: KP hierarchy, constraints, bilinear method.

1. Introduction

We are concerned with the Lax equations

$$L_{t_n} = [(L^n)_+, L\,.] \tag{1.1}$$

and the linear problems

$$\psi_{t_n} = (L^n)_+\psi\,, \qquad \psi^*_{t_n} = -(L^{*n})_+\psi^*, \tag{1.2}$$

where the microdifferential operator L takes the form

$$L = \partial + u_2\partial^{-1} + u_3\partial^{-2} + \cdots, \quad (\partial = \partial/\partial x), \tag{1.3}$$

with coefficients u_j, $j \geqslant 2$, depending on the infinite number of variables $t = (t_1, t_2, \ldots)$, $(t_1 = x)$. The linear problems (1.2) possess formal solutions of the form

$$\psi = Pe^{\xi}, \qquad \psi^* = (P^*)^{-1}e^{-\xi}, \qquad \xi(t,\lambda) = \sum_{n=1}^{\infty} t_n\lambda^n, \tag{1.4}$$

with a microdifferential operator

$$P = 1 + w_1\partial^{-1} + w_2\partial^{-2} + \cdots, \tag{1.5}$$

related to L through $L = P\partial P^{-1}$. Note that there is a one-to-one relationship between the functions w_j, $j \geqslant 1$, and the functions u_j, $j \geqslant 2$.

The wave-functions can be expressed in terms of the τ-function as [1, 2]

$$\psi(t,\lambda) = \frac{\tau(t-\varepsilon(\lambda))}{\tau(t)} \mathrm{e}^{\xi(t,\lambda)}, \qquad \psi^*(t,\lambda) = \frac{\tau(t+\varepsilon(\lambda))}{\tau(t)} \mathrm{e}^{-\xi(t,\lambda)}, \tag{1.6}$$

where $\varepsilon(\lambda) = (1/\lambda, 1/2\lambda^2, 1/3\lambda^3, \ldots)$. In what follows, we shall write shortly $\psi(t)$, $\psi^*(t)$. Furthermore, the coefficients u_j can be expressed through the τ-function, for example

$$u_2 = -\left(p_1(-\tilde{\partial})(\log\tau)\right)_x, \tag{1.7}$$

$$u_3 = -\left(p_2(-\tilde{\partial})(\log\tau)\right)_x, \tag{1.8}$$

$$u_4 = -\left(p_3(-\tilde{\partial})(\log\tau)\right)_x - \left((\log\tau)_{xx}\right)^2, \tag{1.9}$$

$$\vdots$$

The wave-functions satisfy the bilinear identity [2]

$$\mathrm{Res}_\lambda\left(\psi(t)\psi^*(t')\right) = 0 \tag{1.10}$$

which gives rise to the bilinear equations for the KP hierarchy.

The procedure of k-reduction imposes the condition

$$(L^k)_- = 0, \tag{1.11}$$

on (1.1) which reduces it to

$$(L^k)_{t_n} = [(L^n)_+, L^k]. \tag{1.12}$$

This leads to Gelfand–Dickey hierarchies of integrable equations in $(1+1)$-dimensions, such as the KdV, Boussinesq and Hirota–Satsuma hierarchies [1].

Recently, it was pointed out that the KP hierarchy admits another type of reductions [3–11] which can be understood as direct generalizations of k-reduction (1.11) through

$$(L^k)_- = q\partial^{-1}r. \tag{1.13}$$

Under (1.13), the KP hierarchy (1.1) is constrained to

$$(L^k)_{t_n} = [(L^n)_+, L^k], \tag{1.14}$$

$$q_{t_n} = (L^n)_+ q, \tag{1.15}$$

$$r_{t_n} = -(L^{*n})_+ r. \tag{1.16}$$

Equations (1.14)–(1.16) lead to commuting $(1+1)$-dimensional hierarchies, such as the AKNS, Yajima–Oikawa and Melnikov hierarchies.

The purpose of this paper is to discuss the bilinear and trilinear structures of the constrained hierarchies (1.14)–(1.16).

2. Bilinear Form of Constraint Equations

It has been shown in [12] that under the condition (1.11) of k-reduction the wave-functions obey the bilinear identity

$$\mathrm{Res}_\lambda\left(\lambda^k\psi(t)\psi^*(t')\right)=0\,. \tag{2.1}$$

In the case of the generalized k-constraint (1.13), this has been generalized in [13] to

$$\mathrm{Res}_\lambda\left(\lambda^k\psi(t)\psi^*(t')\right)=q(t)r(t')\,. \tag{2.2}$$

We obtain (2.2) by using $L^k=P\partial^kP^{-1}=q\partial^{-1}r$ and the following argument

$$\begin{aligned}(-1)^m q(t)\frac{\partial^m r(t)}{\partial x^m} &= \mathrm{Res}_\partial\left(P\partial^kP^{-1}\partial^m\right)\\ &= \mathrm{Res}_\lambda\left((P\partial^k\mathrm{e}^{\xi(t,\lambda)})((-\partial)^mP^{*-1}\mathrm{e}^{-\xi(t,\lambda)})\right)\\ &= (-1)^m\mathrm{Res}_\lambda\left(\lambda^k\psi(t,\lambda)\frac{\partial^m\psi^*(t,\lambda)}{\partial x^m}\right)\,.\end{aligned} \tag{2.3}$$

For obtaining the full bilinear equations in the case of the generalized k-constraint, it is useful to introduce functions ρ and σ through

$$q=\frac{\rho}{\tau},\qquad r=\frac{\sigma}{\tau}\,. \tag{2.4}$$

Then again, after modifying some arguments of the unconstrained case, the bilinear identities become [13]

$$\mathrm{Res}_\lambda\left(\lambda^k\tau(t-\varepsilon(\lambda))\tau(t'+\varepsilon(\lambda))\,\mathrm{e}^{\xi(t-t',\lambda)}\right)=\rho(t)\sigma(t'), \tag{2.5}$$

$$\mathrm{Res}_\lambda\left(\lambda^{-1}\tau(t-\varepsilon(\lambda))\rho(t'+\varepsilon(\lambda))\,\mathrm{e}^{\xi(t-t',\lambda)}\right)=\rho(t)\tau(t'), \tag{2.6}$$

$$\mathrm{Res}_\lambda\left(\lambda^{-1}\sigma(t-\varepsilon(\lambda))\tau(t'+\varepsilon(\lambda))\,\mathrm{e}^{\xi(t-t',\lambda)}\right)=\sigma(t')\tau(t)\,. \tag{2.7}$$

We list a few examples of the bilinear equations:

$$p_{k+1}(\tilde D)D_n\tau\cdot\tau-2p_{n+k+1}(\tilde D)\tau\cdot\tau=D_n\sigma\cdot\rho, \tag{2.8}$$

$$p_{k+1}(\tilde D)\tau\cdot\tau=\rho\sigma, \tag{2.9}$$

$$p_n(\tilde D)\tau\cdot\rho=D_n\tau\cdot\rho, \tag{2.10}$$

$$p_n(\tilde D)\sigma\cdot\tau=D_n\sigma\cdot\tau, \tag{2.11}$$

where $\tilde D=(D_1,\frac12 D_2,\frac13 D_3,\dots)$ and $D_1,D_2,D_3,\dots$ are the usual Hirota bilinear operators.

3. Bilinear Bäcklund Transformations

Consider two microdifferential operators

$$L = P\partial P^{-1}, \qquad \bar{L} = \bar{P}\partial\bar{P}^{-1}, \tag{3.1}$$

satisfying the KP hierarchy and assume the asymptotic behaviour P,

$$\bar{P} \overset{|x|\to\infty}{\longrightarrow} 1.$$

We look for a gauge operator

$$T = \partial + a \overset{|x|\to\infty}{\longrightarrow} \partial - \lambda_0 \quad (\lambda_0 = \text{const}), \tag{3.2}$$

such that

$$\bar{L}T = TL. \tag{3.3}$$

Let

$$\psi = P\mathrm{e}^{\xi}, \qquad \psi^* = (P^*)^{-1}\,\mathrm{e}^{-\xi},$$

$$\bar{\psi} = \bar{P}\mathrm{e}^{\xi}, \qquad \bar{\psi}^* = (\bar{P}^*)^{-1}\,\mathrm{e}^{\xi},$$

and

$$q = \frac{\rho}{\tau}, \qquad r = \frac{\sigma}{\tau}, \qquad \bar{q} = \frac{\bar{\rho}}{\bar{\tau}}, \qquad \bar{r} = \frac{\bar{\sigma}}{\bar{\tau}}.$$

Then the following identities represent the bilinear BT for the k-constrained hierarchy, see [14],

$$\mathrm{Res}_\lambda\,((\lambda - \lambda_0)\bar{\psi}(t)\psi^*(t')) = 0, \tag{3.4}$$
$$(D_1 - \lambda_0)\rho \cdot \bar{\tau} = \bar{\rho}\tau, \tag{3.5}$$
$$-(D_1 + \lambda_0)\bar{\sigma} \cdot \tau = \sigma\bar{\tau}. \tag{3.6}$$

Remark that the BT (3.4)–(3.6) does not depend explicitly on the order k of the constraint. If τ, ρ, σ satisfy the bilinear equations (2.5)–(2.7) for the k-constraint hierarchy and $\bar{\tau}$, $\bar{\rho}$, $\bar{\sigma}$ is determined from (3.4)–(3.6), then we obtain new solutions for the k-constrained hierarchy $\bar{\tau}$, $\bar{\rho}$, $\bar{\sigma}$.

EXAMPLE. The simplest set of equations following from (3.4)–(3.6) is given by

$$(D_2 + D_1^2)\bar{\tau} \cdot \tau = \lambda_0 D_1\bar{\tau} \cdot \tau, \tag{3.7}$$
$$(D_1 - \lambda_0)\rho \cdot \bar{\tau} = \bar{\rho}\tau, \tag{3.8}$$
$$-(D_1 + \lambda_0)\bar{\sigma} \cdot \tau = \sigma\bar{\tau}, \tag{3.9}$$

representing the space part of the BT for the k-constrained hierarchy.

4. τ-Function and Constraints

We shall now study the conditions which the τ-function of the unconstrained KP hierarchy has to satisfy in order to be consistent with the 1-constraint. Let us compare the unconstrained operator (1.3) with the operator of the 1-constrained hierarchy

$$L = \partial + q\partial^{-1}r\,. \tag{4.1}$$

From that comparison, we first obtain

$$u_2 = qr, \tag{4.2}$$

$$u_3 = -qr_x, \tag{4.3}$$

$$u_4 = qr_{xx}\,. \tag{4.4}$$

Expressing u_3, u_4 in terms of qr, r_x/r yields

$$u_3 = -qr\frac{r_x}{r}, \tag{4.5}$$

$$u_4 = qr\left(\left(\frac{r_x}{r}\right)_x + \left(\frac{r_x}{r}\right)^2\right). \tag{4.6}$$

By elimination of r_x/r, we obtain from (4.5) and (4.6)

$$u_2u_4 - u_{2,x}u_3 + u_2u_{3,x} - u_3^2 = 0 \tag{4.7}$$

Now we introduce into Equation (4.7) the τ-function of the unconstrained KP hierarchy through (1.7)–(1.9). The only term in the resulting equation that contains t_3-derivatives looks like $(\tau_{xt_3}\tau - \tau_x\tau_{t_3})(\tau_{xx}\tau - \tau_x^2)$. This allows us to eliminate t_3-derivatives by using the bilinear form of the KP itself leading to the well-known trilinear form for the AKNS system, see [15–17].

$$\begin{vmatrix} p_0(\tilde\partial)p_0(-\tilde\partial)(\tau) & p_0(\tilde\partial)p_1(-\tilde\partial)(\tau) & p_0(\tilde\partial)p_2(-\tilde\partial)(\tau) \\ p_1(\tilde\partial)p_0(-\tilde\partial)(\tau) & p_1(\tilde\partial)p_1(-\tilde\partial)(\tau) & p_1(\tilde\partial)p_2(-\tilde\partial)(\tau) \\ p_2(\tilde\partial)p_0(-\tilde\partial)(\tau) & p_2(\tilde\partial)p_1(-\tilde\partial)(\tau) & p_2(\tilde\partial)p_2(-\tilde\partial)(\tau) \end{vmatrix} = 0\,. \tag{4.8}$$

Here the polynomials $p_j(t)$, $j \geqslant 0$, are defined through

$$e^{\xi(t,\lambda)} = \sum_{j=0}^{\infty} p_j(t)\lambda^n,$$

and $\tilde\partial = (\partial/\partial t_1, \frac{1}{2}\partial/\partial t_2, \frac{1}{3}\partial/\partial t_3, \ldots)$. Note that after solving the trilinear equation, we obtain q and r from

$$qr = -\left(p_1(-\tilde\partial)(\log\tau(t))\right)_x, \tag{4.9}$$

$$-qr_x = -\left(p_2(-\tilde\partial)(\log\tau(t))\right)_x. \tag{4.10}$$

Of course, we can continue the comparison and obtain the full set of trilinear equations of the AKNS system.

Let us now prepare the constraints on the τ-function for the general case of k-constraints. Setting

$$L^k = \sum_{j \leqslant k} v_j(k)\partial^j , \tag{4.11}$$

we obtain from the k-constraint

$$v_{-n}(k) = (-1)^{n-1} q r_{(n-1)x} \,. \tag{4.12}$$

Writing

$$\frac{r_{(n+1)x}}{r} = \left(\frac{r_{nx}}{r}\right)_x + \frac{r_x}{r}\frac{r_{nx}}{r} \tag{4.13}$$

shows that we can recursively express $n \geqslant 1$

$$\frac{r_{(n+1)x}}{r} = \Phi^n \left(\frac{r_x}{r}\right) \tag{4.14}$$

with the operator

$$\Phi = \frac{\partial}{\partial x} + \frac{r_x}{r} \,. \tag{4.15}$$

This gives the set of equations

$$v_{-1}(k) = qr, \tag{4.16}$$

$$v_{-2}(k) = -qr\frac{r_x}{r}, \tag{4.17}$$

$$v_{-3}(k) = qr\left(\left(\frac{r_x}{r}\right)_x + \left(\frac{r_x}{r}\right)^2\right), \tag{4.18}$$

$$\vdots$$

From Equations (4.16) and (4.17) we can eliminate r_x/r, yielding

$$\frac{r_x}{r} = -\frac{v_{-2}(k)}{v_{-1}(k)} \,. \tag{4.19}$$

Inserting (4.19) into Equation (4.18) and so on, gives the following condition ($n \geqslant 3$)

$$v_{-n}(k) = (-1)^{n-1} v_{-1}(k) \Phi \left(-\frac{v_{-2}(k)}{v_{-1}(k)}\right) \,. \tag{4.20}$$

Analogous to the case of the 1-constraint, we now have to insert the expressions for $u_2, u_3, \ldots$ in terms of the τ-function into (4.20).

5. Trilinear Forms

We shall now present a different approach to describe the constraints on the τ-function, namely to consider the bilinear identities (2.5)–(2.7) and to eliminate the functions σ and ρ.

First, we obtain from (2.7)

$$\begin{aligned}&\sigma(t+y+z)\tau(t-y+z)\\&=\mathrm{Res}_\lambda\left(\lambda^{-1}\sigma(t-y+z-\varepsilon(\lambda))\tau(t+y+z+\varepsilon(\lambda))e^{\xi(-2y,\lambda)}\right)\\&=\sum_{j=0}^{\infty}p_j(-2y)p_j(\tilde{\partial}_y)(\sigma(t-y+z)\tau(t+y+z)),\end{aligned} \tag{5.1}$$

with y and z arbitrary. Using (2.5) gives

$$\begin{aligned}&\rho(t-z)\sigma(t-y+z)\\&=\mathrm{Res}_\lambda\left(\lambda^k\tau(t-z-\varepsilon(\lambda))\tau(t-y+z+\varepsilon(\lambda))e^{\xi(y-2z,\lambda)}\right)\\&=\sum_{i=0}^{\infty}p_i(y-2z)p_{i+k+1}(\tilde{\partial}_z)(\tau(t-z)\tau(t-y+z)).\end{aligned} \tag{5.2}$$

Combining (5.1) and (5.2) results in an expression for $\sigma(t+y+z)\tau(t-y+z)\rho(t-z)$ which contains only the τ-function.

Next, (2.5) gives

$$\begin{aligned}&\rho(t-z)\sigma(t+y+z)\\&=\mathrm{Res}_\lambda\left(\lambda^k\tau(t-z-\varepsilon(\lambda))\tau(t+y+z+\varepsilon(\lambda))e^{\xi(-y-2z,\lambda)}\right)\\&=\sum_{j=0}^{\infty}p_j(-y-2z)p_{j+k+1}(\tilde{\partial}_z)\tau(t-z)\tau(t+y+z)\,.\end{aligned} \tag{5.3}$$

Multiplying (5.3) with $\tau(t-y+z)$ yields a second expression for $\sigma(t+y+z)\tau(t-y+z)\rho(t-z)$ which contains only the τ-function. If we compare those expressions, we obtain the trilinear formulation of the k-constraint. For the purpose of making this formulation compact, we introduce the trilinear expression

$$F_k(t,y,z)=\tau(t+y+z)p_k(\tilde{\partial}_z)(\tau(t-z)\tau(t-y+z))\,. \tag{5.4}$$

Then we obtain the following trilinear form

$$\begin{aligned}&\sum_{j=0}^{\infty}\sum_{i=0}^{\infty}p_j(-2y)p_j(\tilde{\partial}_y)\left(p_i(y-2z)F_{k+1+i}(t,y,z)\right)\\&=\sum_{j=0}^{\infty}p_j(-y-2z)F_{k+1+j}(t,-y,z)\,.\end{aligned} \tag{5.5}$$

Note that (5.5) is identically satisfied for $y = 0$.

Of course, the bilinear equations of the KP hierarchy are contained in (5.5). As an example, we consider the case $k = 1$ and the coefficient of y_2 in the Taylor expansion of (5.5) with respect to the variables y and z. This gives

$$\left(\left(\partial_{y_2} - p_2(\tilde{\partial}_y)\right) F_2(t,y,z) - p_1(\tilde{\partial}_y)F_3(t,y,z)\right)_{y=z=0} = 0\,. \tag{5.6}$$

After a few calculations, this becomes

$$\tau(t)\left(p_4(\tilde{D})\tau(t)\cdot\tau(t) - \tfrac{1}{2}D_1D_3\tau(t)\cdot\tau(t)\right) = 0, \tag{5.7}$$

i.e. the KP itself.

References

1. Ohta, Y., Satsuma, J., Takahashi, D., and Tokihiro, T.: *Prog. Theor. Phys. Suppl.* **94** (1988), 219.
2. Date, E., Jimbo, M., Kashiwara, M., and Miwa, T.: in M. Jimbo and T. Miwa (eds), *Nonlinear Integrable Systems – Classical and Quantum Theory*, World Scientific, Singapore, 1983, pp. 39; Jimbo, M. and Miwa, T.: *Publ. RIMS, Kyoto Univ.* **19** (1983), 943.
3. Cheng, Y. and Li, Y. S.: *Phys. Lett. A* **157** (1991), 22; *J. Phys. A* **25** (1992), 419.
4. Konopelchenko, B. G. and Strampp, W.: *Inverse Problems* **7** (1991), L17; *J. Math. Phys.* **33** (1992), 3676.
5. Konopelchenko, B. G., Sidorenko, J., and Strampp, W.: *Phys. Lett. A* **157** (1991), 17.
6. Sidorenko, J. and Strampp, W.: *Inverse Problems* **7** (1991), L37.
7. Xu, B.: *Inverse Problems* **8** (1992), L13; $(1+1)$-dimensional integrable Hamiltonian systems reduced from the KP hierarchy, *Inverse Problems* **9** (1993), 355.
8. Cheng, Y.: *J. Math. Phys.* **33** (1992), 3774.
9. Xu, B. and Li, Y.: *J. Phys. A* **25** (1992), 2957.
10. Sidorenko, J. and Strampp, W.: Multicomponent integrable reductions in the KP hierarchy, Preprint (1991); *J. Math. Phys.* **34** (1993), 1429.
11. Oevel, W. and Strampp, W.: Constrained KP hierarchy and bi-Hamiltonian structures, *Comm. Math. Phys.* **157** (1993), 51.
12. Lebedev, D., Orlov, A., Pakuliak, S., and Zabrodin, A.: *Phys. Lett. A* **160** (1991), 166.
13. Cheng, Y. and Zhang, Y.-J.: Bilinear equations for the constrained KP hierarchy, Preprint, Hefei University, 1992.
14. Cheng, Y., Strampp, W., and Zhang, Y.-J.: Bilinear Bäcklund transformations for the KP and k-constrained KP hierarchy, Preprint, Hefei University, 1993.
15. Matsukidaira, J., Satsuma, J., and Strampp, W.: *Phys. Lett. A* **147** (1990), 467.
16. Satsuma, J., Matsukidaira, J., and Kajiwara, K.: in I. Antoniou and F. J. Lambert (eds), *Solitons and Chaos*, Springer-Verlag, Berlin, 1991, pp. 264–269.
17. Hietarinta, J., Kajiwara, K., Matsukidaira, J., and Satsuma, J.: in M. Boiti, L. Martina and F. Pempinelli (eds), *Nonlinear Evolution Equations and Dynamical Systems*, World Scientific, Singapore, 1992, pp. 30–43.